OXFORD GRADUATE TEXTS IN MATHEMATICS

Operator Theory by Example

STEPHAN RAMON GARCIA

W. M. Keck Distinguished Service Professor and Chair
of the Department of Mathematics and Statistics, Pomona College

JAVAD MASHREGHI

Professor, Laval University

WILLIAM T. ROSS

Richardson Professor of Mathematics, University of Richmond

OXFORD
UNIVERSITY PRESS

Great Clarendon Street, Oxford, OX2 6DP,
United Kingdom

Oxford University Press is a department of the University of Oxford.
It furthers the University's objective of excellence in research, scholarship,
and education by publishing worldwide. Oxford is a registered trade mark of
Oxford University Press in the UK and in certain other countries

Published in the United States of America by Oxford University Press
198 Madison Avenue, New York, NY 10016, United States of America

British Library Cataloguing in Publication Data
Data available

Library of Congress Control Number: 2022946604

ISBN 978–0–19–286386–7(hbk)
ISBN 978–0–19–286387–4(pbk)

DOI: 10.1093/oso/9780192863867.001.0001

Printed and bound by
CPI Group (UK) Ltd, Croydon, CR0 4YY

Links to third party websites are provided by Oxford in good faith and
for information only. Oxford disclaims any responsibility for the materials
contained in any third party website referenced in this work.

To our families:
Gizem, Reyhan, and Altay;
Shahzad, Dorsa, Parisa, and Golsa;
Fiona

CONTENTS

PREFACE

This is the book we wish we had as graduate students. As its name suggests, this book is all about examples. Instead of listing a host of concepts all at once in an abstract setting, we bring ideas along slowly and illustrate each new idea with explicit and instructive examples. As one can see with the chapter titles, the focus of each chapter is on a specific operator and not on a concept. The important topics are covered through concrete operators and settings.

As for style, we take great pains not to talk down to or above our audience. For example, we religiously eschew the dismissive words "obvious" and "trivial," which have caused untold hours of heartache and self-doubt for puzzled graduate students the world over. Our prerequisites are minimal and we take time to highlight arguments and details that are often brushed over in other sources.

In terms of prerequisites, we hope that the reader has had some exposure to Lebesgue's theory of integration. Familiarity with the Lebesgue integral and the three big convergence theorems (Fatou's lemma, the monotone convergence theorem, and the dominated convergence theorem) is sufficient for our purposes. In addition, an undergraduate-level course in complex analysis is needed for some of the chapters. We carefully develop everything else. Moreover, we cover any needed background material as part of the discussion. We do not burden the reader, who is anxious to get to operator theory, with a large volume of preliminary material. Nor do we make them pause their reading to chase down a concept or formula from an appendix.

By "operator theory," we mean the study of bounded operators on Hilbert spaces. We choose to work with Hilbert spaces, not only because of their beauty, ubiquity, and great applicability, but also because they are the stepping stone to more specialized investigations. Interested readers who wish to pursue further studies in some of the topics covered here, but in the Banach-space setting, will be well equipped to do so once Hilbert spaces and their operators are firmly understood. We are primarily concerned with concrete properties of individual operators: norm, spectrum, compactness, invariant subspaces, and so forth. Many of our examples are non-normal operators, and hence lie outside the focus of many standard texts, in which various subclasses of normal operators play a distinguished role. Although algebras of operators occasionally arise in what follows, this is not a book on operator algebras (however, we must admit being influenced by the title of [105]). Nor do we enter into the theory of unbounded operators on Hilbert spaces.

The endnotes for each chapter are filled with historical details which allow the reader to understand the development of each particular topic. We provide copious references in case the reader wishes to consult the original sources or delve deeper into a particular topic they find interesting.

Each chapter comes equipped with dozens of problems. In total, this book contains over 600 problems. Some of them ease the reader into the subject. Others ask the student to supply a proof of some technical detail. More complicated problems, which sometimes explain material not covered in the text, are split into several parts to ensure that the student is not left treading water. We provide hints for many of the problems and it is our intention that the attentive student should be able to work through all of the exercises without outside assistance.

The proofs and examples we present are instructive. We try not to hide behind slick arguments that do not easily generalize. Neither do we hold back on the details. Although everyone may learn something from this book, our primary audience consists of graduate students and entry-level researchers.

Finally, this book is not meant to be a comprehensive treatise on operator theory. That book would comprise many volumes. Our book is a selection of instructive operator-theory vignettes that show a variety of topics that a student may see as they begin to attend conferences or engage in independent research.

So, welcome to operator theory! It is an inspiring subject that has developed over the past 100-plus years and continues to enjoy applications in mathematics, science, and engineering. After reading this book, learning the history of the subject from the endnotes, and working your way through the problems, we hope that you are inspired and excited about the subject as much as we are.

We give special thanks to John B. Conway, Chris Donnay, Elena Kim, Tom Kriete, Artur Nicolau, Ryan O'Dowd, Alan Sola, Dan Timotin, William Verreault, Brett Wick, and Jiahui Yu for giving us useful feedback on the initial draft of this book.

Stephan Ramon Garcia was partially supported by National Science Foundation (US) grants DMS-2054002 and DMS-1800123. Javad Mashreghi was partially supported by an NSERC Discovery Grant.

NOTATION

- $\mathbb{N}$ .... the set of positive integers
- $\mathbb{Z}$ .... the set of integers
- $\mathbb{Q}$ .... the set of rational numbers
- $\mathbb{R}$ .... the set of real numbers
- $\mathbb{C}$ .... the set of complex numbers
- $\mathbb{D}$ .... the open unit disk in $\mathbb{C}$
- $\mathbb{T}$ .... the unit circle in $\mathbb{C}$
- A^- .... the closure of the set A
- $\mathbb{C}[z]$ .... the set of complex polynomials in the variable z
- $\mathbb{C}[z, \overline{z}]$ .... the set of complex polynomials in the variables z and $\overline{z}$
- $\mathscr{P}_n$ .... the set of complex polynomials in z of degree at most n
- $\mathbb{R}^n$ .... real n-dimensional Euclidean space
- $\mathcal{H}, \mathcal{K}$ .... Hilbert spaces
- $\mathcal{U}, \mathcal{V}, \mathcal{W}$ .... vector spaces
- $\mathbf{x}, \mathbf{y}, \mathbf{z}$ .... abstract vectors
- $\mathbb{C}^n$ .... complex n-dimensional Euclidean space (p. 1)
- δ_{jk} .... Kronecker delta function (p. 4)
- ℓ^2 .... the set of square summable infinite sequences (p. 8)
- $\langle \cdot, \cdot \rangle$ .... inner product (p. 8)
- $\langle \cdot, \cdot \rangle_{\mathcal{H}}$ .... inner product on the space $\mathcal{H}$
- $\| \cdot \|_{\mathcal{H}}$ .... norm on the space $\mathcal{H}$
- $\| \cdot \|$ .... norm (p. 8)
- $L^2[0, 1]$ .... Lebesgue space on $[0, 1]$ (p. 10)
- $C[0, 1]$ .... the set of complex-valued continuous functions on $[0, 1]$ (p. 12)
- $\bigvee$ .... closed linear span (p. 19)
- $\operatorname{tr} A$ .... trace of a matrix A (p. 30)
- sgn .... signum function (p. 35)

- $\mathrm{diag}(\lambda_0, \lambda_1, \lambda_2, \ldots)$ .... diagonal matrix (p. 41)
- D_Λ .... diagonal operator with diagonal Λ (p. 42)
- $\|T\|$ .... norm of an operator T (p. 43)
- $\mathcal{B}(\mathcal{H})$ .... the set of bounded operators on a Hilbert space $\mathcal{H}$ (p. 43)
- $\mathcal{B}(\mathcal{H}, \mathcal{K})$ .... the set of bounded operators from $\mathcal{H}$ to $\mathcal{K}$ (p. 46)
- $\mathcal{V}^*$ .... dual of a Banach space $\mathcal{V}$ (p. 47)
- $\ker A$ .... the kernel of a linear transformation A (p. 48)
- $\mathrm{ran}\, A$ .... the range of a linear transformation A (p. 48)
- $\mathcal{M}_n$ .... the set of $n \times n$ complex matrices (p. 48)
- $\sigma(A)$ .... spectrum of an operator A (p. 52)
- $\sigma_p(A)$ .... point spectrum of an operator A (p. 52)
- $\sigma_{ap}(A)$ .... approximate point spectrum of an operator A (p. 52)
- $\mathbf{x} \otimes \mathbf{y}$ .... rank-one tensor (p. 57)
- $\mathcal{Y}^\perp$ .... orthogonal complement of a set $\mathcal{Y}$ (p. 69)
- A^* .... adjoint of an operator A (p. 71)
- M_x .... multiplication by x on $L^2[0,1]$ (p. 93)
- $L^2(\mathbb{T})$ .... Lebesgue space on $\mathbb{T}$ (p. 96)
- $H^2(\mathbb{T})$ .... Hardy space of the unit circle (p. 98)
- M_ξ .... multiplication by ξ on $L^2(\mathbb{T})$ (p. 99)
- H^2 .... Hardy space of the unit disk (p. 113)
- H^∞ .... the set of bounded analytic functions on $\mathbb{D}$ (p. 117)
- $\{A\}'$ .... commutant of an operator A (p. 122)
- $W(A)$ .... numerical range of an operator A (p. 131)
- M_φ .... multiplication by φ on $L^2(\mu)$ (p. 175)
- $L^2(\mu)$ .... Lebesgue space with measure μ (p. 175)
- $L^\infty(\mu)$ .... the set of essentially bounded μ-measurable functions (p. 175)
- $\mathrm{supp}(\mu)$ .... support of a measure μ (p. 175)
- δ_λ .... point mass at λ (p. 176)
- $C(X)$ .... the set of continuous functions on compact set X (p. 176)
- $\mathcal{R}_\varphi$ .... essential range of $\varphi \in L^\infty(\mu)$ (p. 178)
- $r(A)$ .... spectral radius of an operator A (p. 184)
- $\mathcal{D}$ .... Dirichlet space (p. 207)
- A^2 .... Bergman space (p. 225)

- $\mathscr{F}$ .... Fourier transform (p. 246)
- $f * g$ .... convolution of functions f and g (p. 247)
- f_y .... translation $f(x - y)$ (p. 247)
- $\mathscr{Q}$ .... Hilbert transform on the circle (p. 269)
- $\mathscr{H}$ .... Hilbert transform on the real line (p. 269)
- T_α .... Bishop operator (p. 289)
- $\mathcal{H}^{(n)}$ .... direct sum of n copies of a Hilbert space $\mathcal{H}$ (p. 308)
- $\mathcal{H}^{(\infty)}$ .... direct sum of infinitely many copies of $\mathcal{H}$ (p. 308)
- $\mathbb{D}^2$ .... the bidisk (p. 349)
- $H^2(\mathbb{D}^2)$ .... Hardy space of the bidisk (p. 349)
- T_φ .... Toeplitz operator with symbol φ (p. 357)
- P_+ .... Riesz projection (p. 359)
- P_- .... $I - P_+$ (p. 381)
- H_φ .... Hankel operator with symbol φ (p. 381)
- C_φ .... composition operator with symbol φ (p. 404)
- $H^2(\mu)$ .... closure of the polynomials in $L^2(\mu)$ (p. 431)
- N_μ .... multiplication by z on $L^2(\mu)$ (p. 432)
- S_μ .... multiplication by z on $H^2(\mu)$ (p. 432)
- $\mathcal{K}_u$ .... model space $(uH^2)^\perp$ (p. 446)
- S_u .... compressed shift on the model space $\mathcal{K}_u$ (p. 450)

A BRIEF TOUR OF OPERATOR THEORY

Although examples drive this book, we first provide a whirlwind survey of the general concepts of operator theory. We do not expect the student to master these topics now since they are covered in future chapters.

A *Hilbert space* $\mathcal{H}$ is a complex vector space endowed with an inner product $\langle \mathbf{x}, \mathbf{y} \rangle$ that defines a norm $\|\mathbf{x}\| = \sqrt{\langle \mathbf{x}, \mathbf{x} \rangle}$ with respect to which $\mathcal{H}$ is (Cauchy) complete. The inner product on a Hilbert space satisfies the *Cauchy–Schwarz inequality* $|\langle \mathbf{x}, \mathbf{y} \rangle| \leq \|\mathbf{x}\| \|\mathbf{y}\|$ for all $\mathbf{x}, \mathbf{y} \in \mathcal{H}$. Examples of Hilbert spaces include $\mathbb{C}^n$ (complex Euclidean space), ℓ^2 (the space of square-summable complex sequences), and $L^2[0, 1]$ (the Lebesgue space of square-integrable, complex-valued functions on $[0, 1]$).

Vectors $\mathbf{x}, \mathbf{y}$ in a Hilbert space $\mathcal{H}$ are *orthogonal* if $\langle \mathbf{x}, \mathbf{y} \rangle = 0$. The *dimension* of a Hilbert space $\mathcal{H}$ is the cardinality of a maximal set of nonzero orthogonal vectors. This book is almost exclusively concerned with Hilbert spaces of countable dimension. Every such Hilbert space has an *orthonormal basis* $(\mathbf{u}_n)_{n=1}^{\infty}$, a (possibly finite) maximal orthogonal set such that $\langle \mathbf{u}_m, \mathbf{u}_n \rangle = \delta_{mn}$ for all $m, n \geq 1$. With respect to an orthonormal basis $(\mathbf{u}_n)_{n=1}^{\infty}$, each $\mathbf{x} \in \mathcal{H}$ enjoys a generalized Fourier expansion $\mathbf{x} = \sum_{n=1}^{\infty} \langle \mathbf{x}, \mathbf{u}_n \rangle \mathbf{u}_n$ that satisfies Parseval's identity $\|\mathbf{x}\|^2 = \sum_{n=1}^{\infty} |\langle \mathbf{x}, \mathbf{u}_n \rangle|^2$.

A *subspace* (a norm-closed linear submanifold) of $\mathcal{H}$ is itself a Hilbert space with the operations inherited from $\mathcal{H}$. If $(\mathbf{w}_n)_{n=1}^{\infty}$ is a (possibly finite) orthonormal basis for a subspace $\mathcal{M}$ of $\mathcal{H}$ and $\mathbf{x} \in \mathcal{H}$, then $P_{\mathcal{M}} \mathbf{x} = \sum_{n=1}^{\infty} \langle \mathbf{x}, \mathbf{w}_n \rangle \mathbf{w}_n$ belongs to $\mathcal{M}$ and satisfies $\|\mathbf{x} - P_{\mathcal{M}} \mathbf{x}\| \leq \|\mathbf{x} - \mathbf{y}\|$ for every $\mathbf{y} \in \mathcal{M}$. In short, $P_{\mathcal{M}} \mathbf{x}$ is the unique closest vector to $\mathbf{x}$ in $\mathcal{M}$. Furthermore, $P_{\mathcal{M}}$ defines a linear transformation on $\mathcal{H}$ whose range is $\mathcal{M}$. It is called the *orthogonal projection* of $\mathcal{H}$ onto $\mathcal{M}$ and it satisfies $P_{\mathcal{M}}^2 = P_{\mathcal{M}}$ and $\langle P_{\mathcal{M}} \mathbf{x}, \mathbf{y} \rangle = \langle \mathbf{x}, P_{\mathcal{M}} \mathbf{y} \rangle$ for all $\mathbf{x}, \mathbf{y} \in \mathcal{H}$.

Let $\mathcal{H}$ and $\mathcal{K}$ be Hilbert spaces. A linear transformation $A : \mathcal{H} \to \mathcal{K}$ is *bounded* if $\|A\| = \sup\{\|A\mathbf{x}\|_{\mathcal{K}} : \|\mathbf{x}\|_{\mathcal{H}} = 1\}$ is finite. Let $\mathcal{B}(\mathcal{H}, \mathcal{K})$ denote the set of bounded linear operators from $\mathcal{H}$ to $\mathcal{K}$. We write $\mathcal{B}(\mathcal{H})$ for $\mathcal{B}(\mathcal{H}, \mathcal{H})$. The quantity $\|A\|$ is the *norm* of A. Since $\mathcal{B}(\mathcal{H})$ is closed under addition and scalar multiplication, it is a vector space. Furthermore, since $\|A + B\| \leq \|A\| + \|B\|$ and $\|cA\| = |c| \|A\|$ for all $A, B \in \mathcal{B}(\mathcal{H})$ and $c \in \mathbb{C}$, it follows that $\mathcal{B}(\mathcal{H})$ is a normed vector space. Endowed with this norm, $\mathcal{B}(\mathcal{H})$ is complete and thus forms a Banach space. Moreover, the composition AB belongs to $\mathcal{B}(\mathcal{H})$ and $\|AB\| \leq \|A\| \|B\|$ for all $A, B \in \mathcal{B}(\mathcal{H})$. Therefore, $\mathcal{B}(\mathcal{H})$ is a Banach algebra.

For each $A \in \mathcal{B}(\mathcal{H}, \mathcal{K})$, there is a unique $A^* \in \mathcal{B}(\mathcal{K}, \mathcal{H})$ such that $\langle A\mathbf{x}, \mathbf{y} \rangle_{\mathcal{K}} = \langle \mathbf{x}, A^*\mathbf{y} \rangle_{\mathcal{H}}$ for all $\mathbf{x} \in \mathcal{H}$ and $\mathbf{y} \in \mathcal{K}$. The operator A^* is the *adjoint* of A; it is the analogue of the conjugate transpose of a matrix. One can show that $A \mapsto A^*$ is conjugate linear, that $A^{**} = A$, $\|A\| = \|A^*\|$, and $\|A^*A\| = \|A\|^2$. This additional structure upgrades $\mathcal{B}(\mathcal{H})$ from a Banach algebra to a C^*-algebra. One can exploit adjoints to obtain information about the kernel and range of an operator.

For most of the operators $A \in \mathcal{B}(\mathcal{H})$ covered in this book, we give the *matrix representation* $[A] = [\langle A\mathbf{u}_j, \mathbf{u}_i \rangle]_{i,j=1}^{\infty}$ with respect to an orthonormal basis $(\mathbf{u}_n)_{n=1}^{\infty}$ for $\mathcal{H}$. This matrix representation $[A]$ defines a bounded operator $\mathbf{x} \mapsto [A]\mathbf{x}$ on the Hilbert space ℓ^2 of square summable sequences that is structurally identical to A. Schur's theorem helps us determine which infinite matrices define bounded operators on ℓ^2. Many of the operators covered in this book, such as the Cesàro operator, the Volterra operator, weighted shifts, Toeplitz operators, and Hankel operators, have fascinating structured-matrix representations.

An important class of operators is the *compact operators*. These are the $A \in \mathcal{B}(\mathcal{H})$ such that $(A\mathbf{x}_n)_{n=1}^{\infty}$ has a convergent subsequence whenever $(\mathbf{x}_n)_{n=1}^{\infty}$ is a bounded sequence in $\mathcal{H}$. Equivalently, an operator is compact if it takes each bounded set to one whose closure is compact. Each finite-rank operator is compact and every compact operator is the norm limit of finite-rank operators. The compact operators form a norm-closed, $*$-closed ideal within $\mathcal{B}(\mathcal{H})$.

Some operators have a particularly close relationship with their adjoint. For example, the operator M on $L^2[0,1]$ defined by $(Mf)(x) = xf(x)$ satisfies $M^* = M$. Such operators are *selfadjoint*. If μ is a positive finite compactly supported Borel measure on $\mathbb{C}$, then the operator N on $L^2(\mu)$ defined by $(Nf)(z) = zf(z)$ satisfies $(N^*f)(z) = \bar{z}f(z)$, and thus $N^*N = NN^*$. Such operators are *normal*. The operator $(Uf)(e^{i\theta}) = e^{i\theta}f(e^{i\theta})$ on $L^2(\mathbb{T})$ satisfies $U^*U = UU^* = I$. Such operators are *unitary*.

Unitary operators preserve the ambient structure of Hilbert spaces and can serve as a vehicle to relate $A \in \mathcal{B}(\mathcal{H})$ with $B \in \mathcal{B}(\mathcal{K})$. We say that A is *unitarily equivalent* to B if there is a unitary $U \in \mathcal{B}(\mathcal{H}, \mathcal{K})$ such that $UAU^* = B$. Unitary equivalence is often used to identify seemingly complicated operators with relatively simple ones.

An operator $A \in \mathcal{B}(\mathcal{H})$ is *invertible* if there is a $B \in \mathcal{B}(\mathcal{H})$ such that $AB = BA = I$, where I is the identity operator on $\mathcal{H}$. If $\mathcal{H}$ is finite dimensional, then the conditions "A is invertible", "A is surjective", and "A is injective" are equivalent. If $\mathcal{H}$ is infinite dimensional, invertibility is a more delicate matter. The *spectrum* of A, denoted by $\sigma(A)$, is the set of $\lambda \in \mathbb{C}$ such that $A - \lambda I$ is not invertible in $\mathcal{B}(\mathcal{H})$. If $\mathcal{H}$ is finite dimensional, then $\sigma(A)$ is the set of eigenvalues of A. If $\mathcal{H}$ is infinite dimensional, it is possible for an operator to have no eigenvalues. Nevertheless, $\sigma(A)$ is always a nonempty compact subset of $\mathbb{C}$. Unitarily equivalent operators have the same spectrum.

The spectrum plays an important role in the functional calculus of an operator. For $A \in \mathcal{B}(\mathcal{H})$ and a polynomial $p(z) = c_0 + c_1 z + c_2 z^2 + \cdots + c_n z^n$, one can define the operator $p(A) = c_0 I + c_1 A + c_2 A^2 + \cdots + c_n A^n$. The Riesz functional calculus says that if f is analytic on an open neighborhood of $\sigma(A)$, one can define $f(A) \in \mathcal{B}(\mathcal{H})$. If A is a normal operator, one can define $f(A)$ for all Borel-measurable functions on $\sigma(A)$.

One of the great gems of operator theory is the *spectral theorem* for normal operators. It says that any normal operator N is unitarily equivalent to a multiplication operator $M_\varphi f = \varphi f$ on some $L^2(X, \mu)$ space. Under certain circumstances, X can be taken to be $\sigma(N)$ and μ has support on $\sigma(N)$. There is also the spectral multiplicity theorem which determines when two normal operators are unitarily equivalent.

A subspace $\mathcal{M}$ of $\mathcal{H}$ is *invariant* for $A \in \mathcal{B}(\mathcal{H})$ if $A\mathcal{M} \subseteq \mathcal{M}$. For example, $\{0\}$ and $\mathcal{H}$ are invariant subspaces for any $A \in \mathcal{B}(\mathcal{H})$. The most famous open problem in operator theory, the *invariant subspace problem*, asks whether every $A \in \mathcal{B}(\mathcal{H})$, where dim $\mathcal{H} \geqslant 2$, possesses an invariant subspace besides the two listed above. Most of the operators in this book have an abundance of invariant subspaces that permit a concrete description.

There are several natural topologies on $\mathcal{B}(\mathcal{H})$. Most of the time, we can discuss these concepts in terms of sequences and convergence instead of getting into bases and subbases for the respective topologies. First and foremost, there is the *norm topology*, where $A_n \to A$ if $\|A_n - A\| \to 0$. Next comes the *strong operator topology* (SOT), where $A_n \to A$ (SOT) if $\|A_n\mathbf{x} - A\mathbf{x}\| \to 0$ for each $\mathbf{x} \in \mathcal{H}$. There is also the *weak operator topology* (WOT), where $A_n \to A$ (WOT) if $\langle (A_n - A)\mathbf{x}, \mathbf{y} \rangle \to 0$ for each $\mathbf{x}, \mathbf{y} \in \mathcal{H}$. Norm convergence implies SOT convergence and SOT convergence implies WOT convergence. The converses do not hold. These topologies appear when determining the commutant of an operator. For $A \in \mathcal{B}(\mathcal{H})$, the *commutant* $\{A\}'$ is the set of all bounded operators that commute with A. One can see that $p(A)$ belongs to $\{A\}'$ (where $p \in \mathbb{C}[z]$ is a polynomial) as does either the strong or weak closure of $\{p(A) : p \in \mathbb{C}[z]\}$. For some operators, neither of these closures comprise the entire commutant.

There is certainly much more to be said, many examples to work through, and numerous connections to complex analysis that we have not yet touched upon (although these form an important component of the book). However, we hope that the preceding brief summary of the basic definitions has shed some light on the path forward. These definitions will be introduced and discussed, in due course and in great depth, as we work our way through twenty chapters full of instructive examples.

1

. . ● . .

Hilbert Spaces

Key Concepts: Inner product, norm, inner product space, $\mathbb{C}^n$, ℓ^2, $L^2[0,1]$, Hilbert space, Cauchy–Schwarz inequality, triangle inequality, orthogonal projection, orthonormal basis, Banach space.

Outline: This chapter explores the basics of Hilbert spaces by using $\mathbb{C}^n$ (n-dimensional Euclidean space), ℓ^2 (the space of square-summable complex sequences), and $L^2[0,1]$ (the space of square-integrable, complex-valued Lebesgue-measurable functions on $[0,1]$) as examples. In addition, this chapter covers the Cauchy–Schwarz and triangle inequalities, orthonormal bases, and orthogonal projections. Since Banach spaces play a role in the subsequent chapters, this chapter also covers a few Banach-space basics. Our approach is pedagogical and not aimed at optimal efficiency. Some results are covered multiple times, in increasing levels of generality, in order to illustrate alternate proofs or different perspectives.

1.1 Euclidean Space

Let $\mathbb{C}^n$, *n-dimensional Euclidean space*, denote the set of vectors $\mathbf{a} = (a_1, a_2, ..., a_n)$, where each $a_i \in \mathbb{C}$. With the operations of addition $\mathbf{a} + \mathbf{b} = (a_1 + b_1, a_2 + b_2, ..., a_n + b_n)$ and scalar multiplication $\lambda \mathbf{a} = (\lambda a_1, \lambda a_2, ..., \lambda a_n)$, along with the zero element $\mathbf{0} = (0, 0, ..., 0)$, $\mathbb{C}^n$ is a vector space. It also comes equipped with an inner product and corresponding norm

$$\langle \mathbf{a}, \mathbf{b} \rangle = \sum_{i=1}^{n} a_i \overline{b_i} \qquad \text{and} \qquad \|\mathbf{a}\| = \Big(\sum_{i=1}^{n} |a_i|^2 \Big)^{\frac{1}{2}},$$

respectively, where $\overline{z}$ denotes the complex conjugate of $z \in \mathbb{C}$. In particular, $\langle \mathbf{a}, \mathbf{a} \rangle = \|\mathbf{a}\|^2$. The inner product satisfies the following for $\mathbf{a}, \mathbf{b}, \mathbf{c} \in \mathbb{C}^n$ and $\lambda \in \mathbb{C}$.

(a) $\langle \mathbf{a}, \mathbf{a} \rangle \geq 0$.

(b) $\langle \mathbf{a}, \mathbf{a} \rangle = 0$ if and only if $\mathbf{a} = \mathbf{0}$.

(c) $\langle \mathbf{a}, \mathbf{b} \rangle = \overline{\langle \mathbf{b}, \mathbf{a} \rangle}$.

(d) $\langle \mathbf{a} + \mathbf{b}, \mathbf{c} \rangle = \langle \mathbf{a}, \mathbf{c} \rangle + \langle \mathbf{b}, \mathbf{c} \rangle$.

(e) $\langle \lambda \mathbf{a}, \mathbf{b} \rangle = \lambda \langle \mathbf{a}, \mathbf{b} \rangle$.

The properties above ensure that the inner product is linear in the first slot:

$$\langle \lambda \mathbf{a} + \mu \mathbf{b}, \mathbf{c} \rangle = \lambda \langle \mathbf{a}, \mathbf{c} \rangle + \mu \langle \mathbf{b}, \mathbf{c} \rangle,$$

and conjugate linear in the second slot:

$$\langle \mathbf{a}, \lambda \mathbf{b} + \mu \mathbf{c} \rangle = \overline{\lambda} \langle \mathbf{a}, \mathbf{b} \rangle + \overline{\mu} \langle \mathbf{a}, \mathbf{c} \rangle.$$

The inner product on $\mathbb{C}^n$ also satisfies the following fundamental inequality. Variants and generalizations of this inequality in other settings, and with different proofs, appear throughout this chapter.

Proposition 1.1.1 (Cauchy–Schwarz inequality). *If* $\mathbf{a}, \mathbf{b} \in \mathbb{C}^n$, *then*

$$|\langle \mathbf{a}, \mathbf{b} \rangle| \leqslant \|\mathbf{a}\| \|\mathbf{b}\|. \tag{1.1.2}$$

Equality holds if and only if $\mathbf{a}$ *and* $\mathbf{b}$ *are linearly dependent.*

Proof If $x_i, y_i \in \mathbb{R}$ for $1 \leqslant i \leqslant n$, then

$$\sum_{i=1}^{n} \sum_{j=1}^{n} (x_i y_j - x_j y_i)^2 = \sum_{i=1}^{n} \sum_{j=1}^{n} (x_i^2 y_j^2 - 2x_i x_j y_i y_j + x_j^2 y_i^2)$$

$$= \sum_{i=1}^{n} x_i^2 \sum_{j=1}^{n} y_j^2 + \sum_{i=1}^{n} y_i^2 \sum_{j=1}^{n} x_j^2 - 2 \sum_{i=1}^{n} x_i y_i \sum_{j=1}^{n} x_j y_j$$

$$= 2 \left(\sum_{i=1}^{n} x_i^2 \right) \left(\sum_{i=1}^{n} y_i^2 \right) - 2 \left(\sum_{i=1}^{n} x_i y_i \right)^2.$$

Since the left side is nonnegative, it follows that

$$\left(\sum_{i=1}^{n} x_i y_i \right)^2 \leqslant \left(\sum_{i=1}^{n} x_i^2 \right) \left(\sum_{i=1}^{n} y_i^2 \right). \tag{1.1.3}$$

To obtain (1.1.2), apply (1.1.3) to $x_i = |a_i|$ and $y_i = |b_i|$ for $1 \leqslant i \leqslant n$. Exercise 1.10.3 requests a proof of the second part of the proposition. ∎

An important consequence of the Cauchy–Schwarz inequality is the following inequality, so named because of the image in Figure 1.1.1.

Proposition 1.1.4 (Triangle inequality). *If* $\mathbf{a}, \mathbf{b} \in \mathbb{C}^n$, *then* $\|\mathbf{a} + \mathbf{b}\| \leqslant \|\mathbf{a}\| + \|\mathbf{b}\|$. *Equality holds if and only if* $\mathbf{a}$ *or* $\mathbf{b}$ *is a nonnegative multiple of the other.*

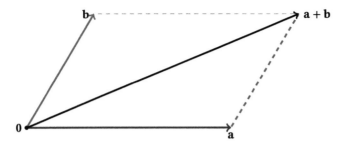

Figure 1.1.1 The triangle inequality.

Proof The Cauchy–Schwarz inequality yields

$$\sum_{i=1}^{n} \text{Re}(a_i \overline{b_i}) = \text{Re}\left(\sum_{i=1}^{n} a_i \overline{b_i}\right) = \text{Re}\langle \mathbf{a}, \mathbf{b} \rangle \leqslant |\langle \mathbf{a}, \mathbf{b} \rangle| \leqslant \|\mathbf{a}\|\|\mathbf{b}\|.$$

Therefore,

$$\|\mathbf{a} + \mathbf{b}\|^2 = \sum_{i=1}^{n} |a_i + b_i|^2$$

$$= \sum_{i=1}^{n} \left(|a_i|^2 + 2\,\text{Re}(a_i \overline{b_i}) + |b_i|^2\right)$$

$$\leqslant \|\mathbf{a}\|^2 + \|\mathbf{b}\|^2 + 2\|\mathbf{a}\|\|\mathbf{b}\|$$

$$= (\|\mathbf{a}\| + \|\mathbf{b}\|)^2.$$

Take square roots above and deduce the triangle inequality. Exercise 1.10.4 requests a proof of the second part of the proposition. ∎

The norm on $\mathbb{C}^n$ defines a metric $d(\mathbf{a}, \mathbf{b}) = \|\mathbf{a} - \mathbf{b}\|$ with respect to which $\mathbb{C}^n$ is *Cauchy complete*. That is, every Cauchy sequence in $\mathbb{C}^n$ converges (Exercise 1.10.7). The metric notation $d(\mathbf{a}, \mathbf{b})$ is usually suppressed in favor of $\|\mathbf{a} - \mathbf{b}\|$, which more clearly suggests its translation invariance:

$$d(\mathbf{a}, \mathbf{b}) = \|\mathbf{a} - \mathbf{b}\| = \|(\mathbf{a} - \mathbf{c}) - (\mathbf{b} - \mathbf{c})\| = d(\mathbf{a} - \mathbf{c}, \mathbf{b} - \mathbf{c}).$$

The inner product on $\mathbb{C}^n$ is the complex version of the dot product on $\mathbb{R}^n$. Recall that two vectors in $\mathbb{R}^n$ are orthogonal if and only if their dot product is zero. This inspires the following definition.

Definition 1.1.5. Vectors $\mathbf{a}, \mathbf{b} \in \mathbb{C}^n$ are *orthogonal*, written $\mathbf{a} \perp \mathbf{b}$, if $\langle \mathbf{a}, \mathbf{b} \rangle = 0$.

The structure imparted upon $\mathbb{C}^n$ by the inner product yields analogues of some familiar results from Euclidean geometry (Figure 1.1.2).

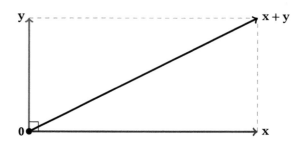

Figure 1.1.2 The Pythagorean theorem: $\|\mathbf{x}\|^2 + \|\mathbf{y}\|^2 = \|\mathbf{x} + \mathbf{y}\|^2$ if $\mathbf{x} \perp \mathbf{y}$.

Proposition 1.1.6 (Pythagorean theorem). *If* $\mathbf{a}, \mathbf{b} \in \mathbb{C}^n$ *and* $\mathbf{a} \perp \mathbf{b}$, *then*

$$\|\mathbf{a} + \mathbf{b}\|^2 = \|\mathbf{a}\|^2 + \|\mathbf{b}\|^2.$$

Proof By the properties of the inner product discussed earlier, observe that

$$\begin{aligned}
\|\mathbf{a} + \mathbf{b}\|^2 &= \langle \mathbf{a} + \mathbf{b}, \mathbf{a} + \mathbf{b} \rangle \\
&= \langle \mathbf{a}, \mathbf{a} \rangle + \langle \mathbf{a}, \mathbf{b} \rangle + \langle \mathbf{b}, \mathbf{a} \rangle + \langle \mathbf{b}, \mathbf{b} \rangle \\
&= \|\mathbf{a}\|^2 + 0 + 0 + \|\mathbf{b}\|^2 \\
&= \|\mathbf{a}\|^2 + \|\mathbf{b}\|^2,
\end{aligned}$$

which completes the proof. ∎

Suppose that $(\mathbf{a}_i)_{i=1}^n \in \mathbb{C}^n$ is a basis, in the sense of linear algebra, that is, $\mathbf{a}_1, \mathbf{a}_2, ..., \mathbf{a}_n$ are linearly independent and span$\{\mathbf{a}_1, \mathbf{a}_2, ..., \mathbf{a}_n\} = \mathbb{C}^n$. The Gram–Schmidt process (see Theorem 1.5.1 below) produces a basis $(\mathbf{u}_i)_{i=1}^n$ such that $\langle \mathbf{u}_j, \mathbf{u}_k \rangle = \delta_{jk}$, where

$$\delta_{jk} = \begin{cases} 1 & \text{if } j = k, \\ 0 & \text{if } j \neq k, \end{cases}$$

is the *Kronecker delta function*. In other words, the $\mathbf{u}_i$ are pairwise orthogonal and have unit length. Such a basis is an *orthonormal basis*.

Proposition 1.1.7. *Let* $(\mathbf{u}_i)_{i=1}^n$ *be an orthonormal basis for* $\mathbb{C}^n$. *Then the following hold for each* $\mathbf{x} \in \mathbb{C}^n$.

(a) $\mathbf{x} = \displaystyle\sum_{i=1}^n \langle \mathbf{x}, \mathbf{u}_i \rangle \mathbf{u}_i.$

(b) $\|\mathbf{x}\|^2 = \displaystyle\sum_{i=1}^n |\langle \mathbf{x}, \mathbf{u}_i \rangle|^2.$

Proof (a) Since span$\{\mathbf{u}_1, \mathbf{u}_2, ..., \mathbf{u}_n\} = \mathbb{C}^n$, for each $\mathbf{x} \in \mathbb{C}^n$ there are scalars a_i such that

$$\mathbf{x} = \sum_{i=1}^{n} a_i \mathbf{u}_i.$$

For each fixed k, the orthonormality of $(\mathbf{u}_i)_{i=1}^{n}$ ensures that

$$\langle \mathbf{x}, \mathbf{u}_k \rangle = \left\langle \sum_{i=1}^{n} a_i \mathbf{u}_i, \mathbf{u}_k \right\rangle$$

$$= \sum_{i=1}^{n} a_i \langle \mathbf{u}_i, \mathbf{u}_k \rangle$$

$$= \sum_{i=1}^{n} a_i \delta_{ik}$$

$$= a_k.$$

(b) From part (a),

$$\langle \mathbf{x}, \mathbf{x} \rangle = \left\langle \sum_{i=1}^{n} \langle \mathbf{x}, \mathbf{u}_i \rangle \mathbf{u}_i, \sum_{j=1}^{n} \langle \mathbf{x}, \mathbf{u}_j \rangle \mathbf{u}_j \right\rangle$$

$$= \sum_{i,j=1}^{n} \langle \mathbf{x}, \mathbf{u}_i \rangle \overline{\langle \mathbf{x}, \mathbf{u}_j \rangle} \langle \mathbf{u}_i, \mathbf{u}_j \rangle$$

$$= \sum_{i,j=1}^{n} \langle \mathbf{x}, \mathbf{u}_i \rangle \overline{\langle \mathbf{x}, \mathbf{u}_j \rangle} \delta_{ij}$$

$$= \sum_{i=1}^{n} |\langle \mathbf{x}, \mathbf{u}_i \rangle|^2,$$

which completes the proof. ∎

A *subspace* of $\mathbb{C}^n$ is a nonempty subset of $\mathbb{C}^n$ that is closed under vector addition and scalar multiplication. In $\mathbb{C}^n$, such a set is also topologically closed and hence this does not conflict with Definition 1.4.7 below of a subspace in the Hilbert-space setting. If $\mathcal{M} \subseteq \mathbb{C}^n$ is a subspace of dimension k, then the Gram–Schmidt process (see Theorem 1.5.1 below) provides an orthonormal basis $(\mathbf{v}_i)_{i=1}^{k}$ for $\mathcal{M}$.

Proposition 1.1.8. *Let $(\mathbf{v}_i)_{i=1}^{k}$ be an orthonormal basis for a subspace $\mathcal{M}$ of $\mathbb{C}^n$. For each $\mathbf{x} \in \mathbb{C}^n$, define*

$$P_{\mathcal{M}}\mathbf{x} = \sum_{i=1}^{k} \langle \mathbf{x}, \mathbf{v}_i \rangle \mathbf{v}_i. \tag{1.1.9}$$

Then the following hold.

(a) $\|\mathbf{x} - P_{\mathcal{M}}\mathbf{x}\| \leqslant \|\mathbf{x} - \mathbf{v}\|$ *for every* $\mathbf{v} \in \mathcal{M}$.

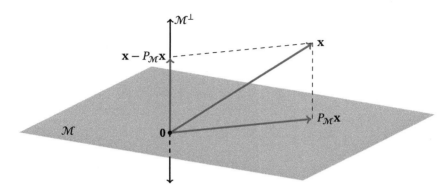

Figure 1.1.3 The orthogonal projection onto the subspace $\mathcal{M}$.

(b) $(\mathbf{x} - P_{\mathcal{M}}\mathbf{x}) \perp \mathbf{v}$ *for every* $\mathbf{v} \in \mathcal{M}$.

(c) $\sum_{i=1}^{k} |\langle \mathbf{x}, \mathbf{v}_i \rangle|^2 \leqslant \|\mathbf{x}\|^2$.

(d) $P_{\mathcal{M}}\mathbf{x} = \mathbf{x}$ *if and only if* $\mathbf{x} \in \mathcal{M}$.

Proof (a) For any $c_1, c_2, \ldots, c_k \in \mathbb{C}$, Exercise 1.10.10 yields

$$\left\| \mathbf{x} - \sum_{i=1}^{k} c_i \mathbf{v}_i \right\|^2 = \|\mathbf{x}\|^2 - \sum_{i=1}^{k} |\langle \mathbf{x}, \mathbf{v}_i \rangle|^2 + \sum_{i=1}^{k} |\langle \mathbf{x}, \mathbf{v}_i \rangle - c_i|^2. \qquad (1.1.10)$$

This expression is minimized precisely when $c_i = \langle \mathbf{x}, \mathbf{v}_i \rangle$ for all $1 \leqslant i \leqslant k$.
(b) For any $1 \leqslant i \leqslant k$,

$$\langle \mathbf{x} - P_{\mathcal{M}}\mathbf{x}, \mathbf{v}_i \rangle = \left\langle \mathbf{x} - \sum_{j=1}^{k} \langle \mathbf{x}, \mathbf{v}_j \rangle \mathbf{v}_j, \mathbf{v}_i \right\rangle \qquad (1.1.11)$$

$$= \langle \mathbf{x}, \mathbf{v}_i \rangle - \left\langle \sum_{j=1}^{k} \langle \mathbf{x}, \mathbf{v}_j \rangle \mathbf{v}_j, \mathbf{v}_i \right\rangle$$

$$= \langle \mathbf{x}, \mathbf{v}_i \rangle - \sum_{j=1}^{k} \langle \mathbf{x}, \mathbf{v}_j \rangle \langle \mathbf{v}_j, \mathbf{v}_i \rangle$$

$$= \langle \mathbf{x}, \mathbf{v}_i \rangle - \sum_{j=1}^{k} \langle \mathbf{x}, \mathbf{v}_j \rangle \delta_{ij}$$

$$= \langle \mathbf{x}, \mathbf{v}_i \rangle - \langle \mathbf{x}, \mathbf{v}_i \rangle = 0. \qquad (1.1.12)$$

Thus, $(\mathbf{x} - P_{\mathcal{M}}\mathbf{x}) \perp \mathbf{v}_i$ for all $1 \leqslant i \leqslant k$. Since $(\mathbf{v}_i)_{i=1}^{k}$ is a basis for $\mathcal{M}$, the conjugate linearity of the inner product in the second slot ensures that $(\mathbf{x} - P_{\mathcal{M}}\mathbf{x}) \perp \mathbf{v}$ for every $\mathbf{v} \in \mathcal{M}$.

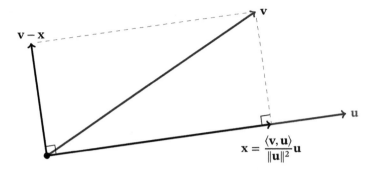

Figure 1.1.4 The orthogonal projection of one vector onto another.

(c) For each $1 \leqslant i \leqslant k$, set $c_i = \langle \mathbf{x}, \mathbf{v}_i \rangle$ in (1.1.10) and obtain

$$0 \leqslant \|\mathbf{x} - P_{\mathcal{M}}\mathbf{x}\|^2 = \|\mathbf{x}\|^2 - \sum_{i=1}^{k} |\langle \mathbf{x}, \mathbf{v}_i \rangle|^2.$$

The desired result follows.

(d) If $P_{\mathcal{M}}\mathbf{x} = \mathbf{x}$, then $\mathbf{x} \in \mathcal{M}$ because $P_{\mathcal{M}}\mathbf{x}$ is a linear combination of $\mathbf{v}_1, \mathbf{v}_2, \ldots, \mathbf{v}_k$, each of which belongs to $\mathcal{M}$. If $\mathbf{x} \in \mathcal{M}$, then (a) ensures that $\|\mathbf{x} - P_{\mathcal{M}}\mathbf{x}\| \leqslant \|\mathbf{x} - \mathbf{x}\| = 0$. Therefore, $P_{\mathcal{M}}\mathbf{x} = \mathbf{x}$. ∎

The vector $P_{\mathcal{M}}\mathbf{x}$ from (1.1.9) is the *orthogonal projection* of $\mathbf{x}$ onto $\mathcal{M}$ (Figure 1.1.3) and is discussed again later in a more general setting. If $\mathcal{M} = \text{span}\{\mathbf{u}\}$, then the orthogonal projection of $\mathbf{v}$ onto $\mathbf{u}$ is

$$\mathbf{x} = \frac{\langle \mathbf{v}, \mathbf{u} \rangle}{\|\mathbf{u}\|^2} \mathbf{u}.$$

This important relation is depicted in Figure 1.1.4.

The alert reader might notice that the definition of $P_{\mathcal{M}}\mathbf{x}$ from Proposition 1.1.8 appears to depend on the choice of orthonormal basis $(\mathbf{v}_i)_{i=1}^{k}$ for $\mathcal{M}$. It does not.

Corollary 1.1.13. *Suppose that $(\mathbf{u}_i)_{i=1}^{k}$ and $(\mathbf{v}_i)_{i=1}^{k}$ are orthonormal bases for a subspace $\mathcal{M}$ of $\mathbb{C}^n$. For $\mathbf{x} \in \mathbb{C}^n$ define*

$$\mathbf{p} = \sum_{i=1}^{k} \langle \mathbf{x}, \mathbf{u}_i \rangle \mathbf{u}_i \quad \text{and} \quad \mathbf{q} = \sum_{i=1}^{k} \langle \mathbf{x}, \mathbf{v}_i \rangle \mathbf{v}_i.$$

Then $\mathbf{p} = \mathbf{q}$.

Proof Proposition 1.1.8b implies that $\mathbf{x} - \mathbf{p}$ and $\mathbf{x} - \mathbf{q}$ are orthogonal to every vector in $\mathcal{M}$. The linearity of the inner product in the first slot ensures that $\mathbf{p} - \mathbf{q} = (\mathbf{x} - \mathbf{q}) - (\mathbf{x} - \mathbf{p})$ is orthogonal to every vector in $\mathcal{M}$. Since $\mathcal{M}$ is a subspace of $\mathbb{C}^n$, it follows that $\mathbf{p} - \mathbf{q} \in \mathcal{M}$. Therefore, $0 = \langle \mathbf{p} - \mathbf{q}, \mathbf{p} - \mathbf{q} \rangle = \|\mathbf{p} - \mathbf{q}\|^2$, and hence $\mathbf{p} = \mathbf{q}$. ∎

The following result is useful for determining if a subspace $\mathcal{M}$ is a proper subset of $\mathbb{C}^n$.

Corollary 1.1.14. *Let $\mathcal{M}$ be a subspace of $\mathbb{C}^n$. The following are equivalent.*

(a) $\mathcal{M} \neq \mathbb{C}^n$.

(b) *There is a $\mathbf{y} \in \mathbb{C}^n \backslash \{\mathbf{0}\}$ such that $\mathbf{y} \perp \mathbf{v}$ for all $\mathbf{v} \in \mathcal{M}$.*

Proof (a) $\Rightarrow$ (b) Let $\mathbf{x} \notin \mathcal{M}$ and set $\mathbf{y} = \mathbf{x} - P_{\mathcal{M}}\mathbf{x}$. Then $\mathbf{y} \neq \mathbf{0}$ and $\mathbf{y} \perp \mathbf{v}$ for all $\mathbf{v} \in \mathcal{M}$ (Proposition 1.1.8).
(b) $\Rightarrow$ (a) Suppose $\mathbf{y} \neq \mathbf{0}$ and is orthogonal to each $\mathbf{v} \in \mathcal{M}$. Then $\mathbf{y} \notin \mathcal{M}$ since otherwise $\|\mathbf{y}\|^2 = \langle \mathbf{y}, \mathbf{y} \rangle = 0$. Thus, $\mathcal{M} \neq \mathbb{C}^n$. ∎

1.2 The Sequence Space ℓ^2

This next space, the sequence space ℓ^2, is infinite dimensional. It is of great importance since it is a natural generalization of $\mathbb{C}^n$ and because most of the other Hilbert spaces considered in this book are, in a certain sense, heavily disguised versions of ℓ^2.

Definition 1.2.1. Let ℓ^2 be the set of all sequences $\mathbf{a} = (a_n)_{n=0}^{\infty}$ of complex numbers such that $\sum_{n=0}^{\infty} |a_n|^2$ is finite.

The inner product and the corresponding norm on ℓ^2 are

$$\langle \mathbf{a}, \mathbf{b} \rangle = \sum_{n=0}^{\infty} a_n \overline{b_n} \quad \text{and} \quad \|\mathbf{a}\| = \Big(\sum_{n=0}^{\infty} |a_n|^2 \Big)^{\frac{1}{2}}, \tag{1.2.2}$$

respectively, where $\mathbf{a} = (a_n)_{n=0}^{\infty}$ and $\mathbf{b} = (b_n)_{n=0}^{\infty}$ belong to ℓ^2. Note that $\langle \mathbf{a}, \mathbf{a} \rangle = \|\mathbf{a}\|^2$. To show that ℓ^2 is a vector space, and that the proposed norm and inner product are well defined, requires some work. The definition of ℓ^2 ensures that the infinite series that defines $\|\mathbf{a}\|$ converges. However, we must justify why the infinite series that defines $\langle \mathbf{a}, \mathbf{b} \rangle$ in (1.2.2) converges. For each $N \geq 1$, the Cauchy–Schwarz inequality (Proposition 1.1.1) says that

$$\sum_{n=0}^{N} |a_n b_n| \leq \Big(\sum_{n=0}^{N} |a_n|^2 \Big)^{\frac{1}{2}} \Big(\sum_{n=0}^{N} |b_n|^2 \Big)^{\frac{1}{2}}.$$

Let $N \to \infty$ and conclude the first series in (1.2.2) converges absolutely, and hence converges. Thus, $\langle \mathbf{a}, \mathbf{b} \rangle$ is well defined for all $\mathbf{a}, \mathbf{b} \in \ell^2$. This also establishes the following version of Proposition 1.1.1 for ℓ^2.

Proposition 1.2.3 (Cauchy–Schwarz inequality). *If $\mathbf{a}, \mathbf{b} \in \ell^2$, then $|\langle \mathbf{a}, \mathbf{b} \rangle| \leq \|\mathbf{a}\| \|\mathbf{b}\|$. Equality holds if and only if $\mathbf{a}$ and $\mathbf{b}$ are linearly dependent.*

To show that ℓ^2 is a vector space with the addition and scalar multiplication operations defined by

$$\mathbf{a} + \mathbf{b} = (a_0 + b_0, a_1 + b_1, \ldots) \quad \text{and} \quad c\mathbf{a} = (ca_0, ca_1, \ldots),$$

one must verify that $\mathbf{a}+\mathbf{b}$ belongs to ℓ^2. The proof of Proposition 1.1.4, applied to ℓ^2, yields the desired result.

Proposition 1.2.4 (Triangle inequality). *If $\mathbf{a}, \mathbf{b} \in \ell^2$, then $\mathbf{a} + \mathbf{b} \in \ell^2$ and $\|\mathbf{a} + \mathbf{b}\| \leqslant \|\mathbf{a}\| + \|\mathbf{b}\|$. Equality holds if and only if $\mathbf{a}$ or $\mathbf{b}$ is a nonnegative multiple of the other.*

We are now in a position to prove that ℓ^2 is (Cauchy) complete with respect to the norm induced by the inner product.

Proposition 1.2.5. *ℓ^2 is complete.*

Proof Suppose that $\mathbf{a}^{(1)}, \mathbf{a}^{(2)}, \mathbf{a}^{(3)}, \ldots$ is a Cauchy sequence in ℓ^2. Each $\mathbf{a}^{(n)}$ is itself a sequence of complex numbers:

$$\mathbf{a}^{(n)} = (a_0^{(n)}, a_1^{(n)}, a_2^{(n)}, a_3^{(n)}, \ldots).$$

For fixed k,

$$|a_k^{(m)} - a_k^{(n)}| \leqslant \|\mathbf{a}^{(m)} - \mathbf{a}^{(n)}\|$$

and hence $a_k^{(1)}, a_k^{(2)}, a_k^{(3)}, \ldots$ is a Cauchy sequence in $\mathbb{C}$. For each k, let

$$a_k = \lim_{n \to \infty} a_k^{(n)}$$

and define $\mathbf{a} = (a_0, a_1, a_2, \ldots)$, the prospective limit of the sequence $(\mathbf{a}^{(n)})_{n=1}^{\infty}$.
To prove this, let $\varepsilon > 0$. Since $(\mathbf{a}^{(n)})_{n=1}^{\infty}$ is a Cauchy sequence in ℓ^2 there is an N such that

$$\left(\sum_{i=0}^{k} |a_i^{(m)} - a_i^{(n)}|^2 \right)^{\frac{1}{2}} \leqslant \|\mathbf{a}^{(m)} - \mathbf{a}^{(n)}\| < \varepsilon \quad \text{for } m, n \geqslant N \text{ and } k \geqslant 0.$$

Let $n \to \infty$ and obtain

$$\left(\sum_{i=0}^{k} |a_i^{(m)} - a_i|^2 \right)^{\frac{1}{2}} \leqslant \varepsilon. \tag{1.2.6}$$

For all k, m, it follows from the triangle inequality on $\mathbb{C}^k$ that

$$\left(\sum_{i=0}^{k} |a_i|^2 \right)^{\frac{1}{2}} \leqslant \left(\sum_{i=0}^{k} |a_i^{(m)} - a_i|^2 \right)^{\frac{1}{2}} + \left(\sum_{i=0}^{k} |a_i^{(m)}|^2 \right)^{\frac{1}{2}} \leqslant \varepsilon + \|\mathbf{a}^{(m)}\|.$$

Thus, letting $k \to \infty$ yields $\|\mathbf{a}\| \leqslant \varepsilon + \|\mathbf{a}^{(m)}\|$ for all $m \geqslant N$. In particular, this shows that $\mathbf{a} \in \ell^2$. By (1.2.6),

$$\|\mathbf{a}^{(m)} - \mathbf{a}\| = \left(\sum_{i=0}^{\infty} |a_i^{(m)} - a_i|^2 \right)^{\frac{1}{2}} \leqslant \varepsilon \quad \text{for all } m \geqslant N, \tag{1.2.7}$$

and hence $\mathbf{a}^{(n)}$ converges to $\mathbf{a}$ in the norm of ℓ^2. ■

As with $\mathbb{C}^n$ in Definition 1.1.5, $\mathbf{a}, \mathbf{b} \in \ell^2$ are *orthogonal*, written $\mathbf{a} \perp \mathbf{b}$, if $\langle \mathbf{a}, \mathbf{b} \rangle = 0$. The *standard basis vectors*

$$\mathbf{e}_0 = (1, 0, 0, \ldots), \quad \mathbf{e}_1 = (0, 1, 0, \ldots), \quad \mathbf{e}_2 = (0, 0, 1, 0, \ldots), \ldots \tag{1.2.8}$$

are orthonormal in ℓ^2; that is, $\langle \mathbf{e}_m, \mathbf{e}_n \rangle = \delta_{mn}$. Moreover,

$$\mathbf{a} = \sum_{n=0}^{\infty} \langle \mathbf{a}, \mathbf{e}_n \rangle \mathbf{e}_n \quad \text{and} \quad \|\mathbf{a}\|^2 = \sum_{n=0}^{\infty} |\langle \mathbf{a}, \mathbf{e}_n \rangle|^2$$

for every $\mathbf{a} \in \ell^2$. These are the ℓ^2 analogues of the formulas from Proposition 1.1.7.

Due to convergence issues, concepts such as orthonormal bases, subspaces, and orthogonal projections are more subtle than in the Euclidean setting and are discussed in a more unified context in Section 1.7.

1.3 The Lebesgue Space $L^2[0, 1]$

The Lebesgue space $L^2[0, 1]$ is the set of Lebesgue-measurable, complex-valued functions f on $[0, 1]$ such that

$$\int_0^1 |f(x)|^2 dx < \infty.$$

As is traditional in the subject, we identify functions that are equal almost everywhere. A student who needs a review of Lebesgue measure and integration should consult [317] or [319].

Define the inner product and corresponding norm on $L^2[0, 1]$ by

$$\langle f, g \rangle = \int_0^1 f(x)\overline{g(x)}\,dx \quad \text{and} \quad \|f\| = \left(\int_0^1 |f(x)|^2\,dx \right)^{\frac{1}{2}}.$$

Note that $\langle f, f \rangle = \|f\|^2$. As with ℓ^2, there are convergence issues to address in the definition of the inner product. These are resolved with the following integral version of the Cauchy–Schwarz inequality.

Proposition 1.3.1 (Cauchy–Schwarz inequality). *If $f, g \in L^2[0, 1]$, then $|\langle f, g \rangle| \leq \|f\| \|g\|$. Equality holds if and only if f and g are linearly dependent.*

Proof Assume that neither f nor g is the zero function, since the inequality holds otherwise. For $a, b \geq 0$, note that

$$ab \leq \frac{1}{2}(a^2 + b^2) \tag{1.3.2}$$

since $(a - b)^2 \geq 0$. Apply this to

$$a = \frac{|f(x)|}{\|f\|} \quad \text{and} \quad b = \frac{|g(x)|}{\|g\|},$$

and get

$$\frac{|f(x)g(x)|}{\|f\|\|g\|} \leqslant \frac{1}{2}\left(\frac{|f(x)|^2}{\|f\|^2} + \frac{|g(x)|^2}{\|g\|^2}\right).$$

Integrating both sides yields

$$\frac{1}{\|f\|\|g\|}\int_0^1 |f(x)g(x)|dx \leqslant \frac{1}{2}\left(\frac{\|f\|^2}{\|f\|^2} + \frac{\|g\|^2}{\|g\|^2}\right) = 1,$$

which shows that fg is integrable, and thus the inner product is well defined. Moreover, the Cauchy–Schwarz inequality follows. Exercise 1.10.5 requests a proof of the second part of the proposition. ∎

Next we prove the triangle inequality for $L^2[0,1]$, which shows that $L^2[0,1]$ is closed under the operation of function addition. Since $L^2[0,1]$ is also closed under scalar multiplication, it is a vector space.

Proposition 1.3.3 (Triangle inequality). *If $f,g \in L^2[0,1]$, then $f+g \in L^2[0,1]$ and*

$$\|f+g\| \leqslant \|f\| + \|g\|.$$

Equality holds if and only if f or g is a nonnegative multiple of the other.

Proof For $f,g \in L^2[0,1]$ and $x \in [0,1]$,

$$\begin{aligned}
|f(x)+g(x)|^2 &= |f(x)|^2 + 2\,\mathrm{Re}\left(f(x)\overline{g(x)}\right) + |g(x)|^2 \\
&\leqslant |f(x)|^2 + 2|f(x)||g(x)| + |g(x)|^2 \\
&\leqslant 2(|f(x)|^2 + |g(x)|^2),
\end{aligned}$$

by (1.3.2). Integrating the inequality above reveals that $f+g \in L^2[0,1]$. It is left to prove the triangle inequality. For this, observe that

$$\begin{aligned}
\|f+g\|^2 &= \langle f+g, f+g \rangle \\
&= \|f\|^2 + 2\,\mathrm{Re}\langle f,g \rangle + \|g\|^2 \\
&\leqslant \|f\|^2 + 2\|f\|\|g\| + \|g\|^2 \qquad \text{(Cauchy–Schwarz)} \\
&= (\|f\| + \|g\|)^2.
\end{aligned}$$

Take square roots of both sides of the inequality above and obtain the triangle inequality. Exercise 1.10.6 requests a proof of the second statement of the proposition. ∎

As with $\mathbb{C}^n$ and ℓ^2, Proposition 1.3.3 shows that $L^2[0,1]$ is a vector space with an inner product. The next result asserts that $L^2[0,1]$ is complete.

Proposition 1.3.4 (Riesz–Fischer). *$L^2[0,1]$ is complete.*

Proof We follow a proof from [319, Ch. 3]. Let $(f_n)_{n=1}^\infty$ be a Cauchy sequence in $L^2[0,1]$. Choose a subsequence $(f_{n_i})_{i=1}^\infty$ such that

$$\|f_{n_{i+1}} - f_{n_i}\| \leqslant \frac{1}{2^i} \quad \text{for } i \geqslant 1.$$

For each $k \geqslant 1$ define

$$g_k(x) = \sum_{i=1}^{k} |f_{n_{i+1}}(x) - f_{n_i}(x)| \quad \text{and} \quad g(x) = \sum_{i=1}^{\infty} |f_{n_{i+1}}(x) - f_{n_i}(x)|.$$

Observe that $g_k(x) \to g(x)$ for each $x \in [0, 1]$. The triangle inequality (Proposition 1.3.3) applied $k-1$ times shows that

$$\|g_k\| \leqslant \sum_{i=1}^{k} \|f_{n_{i+1}} - f_{n_i}\| \leqslant \sum_{i=1}^{k} \frac{1}{2^i} < 1.$$

Since $g_k \to g$ pointwise, Fatou's lemma yields

$$\|g\| \leqslant \liminf_{k \to \infty} \|g_k\| \leqslant 1.$$

In particular, g is finite almost everywhere and

$$f_{n_1} + \sum_{j=1}^{\infty} (f_{n_{j+1}} - f_{n_j})$$

converges absolutely almost everywhere and defines a measurable function f. Therefore,

$$\lim_{k \to \infty} f_{n_k} = \lim_{k \to \infty} \left(f_{n_1} + \sum_{j=1}^{k-1} (f_{n_{j+1}} - f_{n_j}) \right) = f$$

almost everywhere.

To complete the proof, it suffices to show that $f \in L^2[0, 1]$ and $\|f_n - f\| \to 0$. Let $\varepsilon > 0$. Since $(f_n)_{n=1}^{\infty}$ is a Cauchy sequence, there is an $N \geqslant 1$ such that $\|f_n - f_m\| \leqslant \varepsilon$ for all $m, n \geqslant N$. Fatou's lemma implies that

$$\|f - f_m\| \leqslant \liminf_{i \to \infty} \|f_{n_i} - f_m\| \leqslant \varepsilon$$

for all $m \geqslant N$. Therefore, $f \in L^2[0, 1]$ and $\|f_m - f\| \to 0$. ∎

For further work, it is important to know an explicit and convenient dense subset of $L^2[0, 1]$. Let $C[0, 1]$ denote the set of continuous, complex-valued functions on $[0, 1]$. The extreme value theorem ensures that

$$\|f\|_{\infty} = \sup_{0 \leqslant x \leqslant 1} |f(x)|$$

is finite for each $f \in C[0, 1]$. In fact, this defines a norm on $C[0, 1]$ with respect to which $C[0, 1]$ is complete. A sequence $(f_n)_{n=1}^{\infty}$ converges in $C[0, 1]$ with respect to this norm if and only if it converges uniformly on $[0, 1]$. Also important is the inequality

$$\|f\| \leqslant \|f\|_{\infty} \quad \text{for all } f \in C[0, 1]. \tag{1.3.5}$$

Proposition 1.3.6. $C[0,1]$ *is dense in* $L^2[0,1]$.

Proof Let $f \in L^2[0,1]$ and $\varepsilon > 0$. For each $N \geqslant 0$, let

$$f_N(x) = \begin{cases} f(x) & \text{if } |f(x)| \leqslant N, \\ 0 & \text{if } |f(x)| > N. \end{cases}$$

Observe that

$$\int_0^1 |f - f_N|^2 dx = \int_{|f|>N} |f|^2 dx.$$

Choose N large enough such that

$$\|f - f_N\| \leqslant \frac{\varepsilon}{2}. \tag{1.3.7}$$

For this fixed N, Lusin's theorem [319, Ch. 2] produces a closed set $E \subseteq [0,1]$ such that $f_N|_E$ is continuous and $|[0,1]\backslash E| \leqslant \varepsilon^2/(16N^2)$ (here $|A|$ denotes the Lebesgue measure of $A \subseteq [0,1]$). The Tietze extension theorem [319, Ch. 20] produces a $g \in C[0,1]$ such that $|g| \leqslant N$ on $[0,1]$ and $g = f_N$ on E. Then,

$$\begin{aligned} \|f_N - g\|^2 &= \int_0^1 |f_N - g|^2 \, dx \\ &= \int_E |f_N - g|^2 \, dx + \int_{[0,1]\backslash E} |f_N - g|^2 \, dx \\ &= \int_{[0,1]\backslash E} |f_N - g|^2 \, dx \\ &\leqslant \|f_N - g\|_\infty^2 \int_{[0,1]\backslash E} 1 \, dx \qquad \text{(by (1.3.5))} \\ &\leqslant (\|g\|_\infty + \|f_N\|_\infty)^2 \int_{[0,1]\backslash E} 1 \, dx \\ &\leqslant 4N^2 |[0,1]\backslash E| \\ &\leqslant 4N^2 \frac{\varepsilon^2}{16N^2} \\ &= \frac{\varepsilon^2}{4}. \end{aligned}$$

Thus,

$$\|f_N - g\| \leqslant \frac{\varepsilon}{2}. \tag{1.3.8}$$

Finally,

$$\begin{aligned} \|f - g\| &\leqslant \|f - f_N\| + \|f_N - g\| \\ &\leqslant \frac{\varepsilon}{2} + \|f_N - g\| \qquad \text{(by (1.3.7))} \end{aligned}$$

$$\leqslant \frac{\varepsilon}{2} + \frac{\varepsilon}{2} = \varepsilon. \hspace{3cm} \text{(by (1.3.8))}$$

Hence, $C[0, 1]$ is dense in $L^2[0, 1]$. ∎

Finding an orthonormal basis (to be formally defined in a moment) for $L^2[0, 1]$ is harder than with $\mathbb{C}^n$ and ℓ^2. In particular, $L^2[0, 1]$ does not come prelabeled with a distinguished orthonormal basis. We rectify this with the following proposition.

Theorem 1.3.9. *For each $n \in \mathbb{Z}$, let $f_n(x) = e^{2\pi inx}$. The following hold for every $f \in L^2[0, 1]$.*

(a) $(f_n)_{n=-\infty}^{\infty}$ *is an orthonormal sequence in $L^2[0, 1]$.*

(b) $\left\| f - \sum_{n=-N}^{N} \langle f, f_n \rangle f_n \right\| \to 0$ *as $N \to \infty$.*

(c) $\|f\|^2 = \sum_{n=-\infty}^{\infty} |\langle f, f_n \rangle|^2.$

Proof (a) For $m \neq n$,

$$\begin{aligned}
\langle f_m, f_n \rangle &= \int_0^1 e^{2\pi imx} \overline{e^{2\pi inx}} \, dx \\
&= \int_0^1 e^{2\pi imx} e^{-2\pi inx} \, dx \\
&= \int_0^1 e^{2\pi i(m-n)x} \, dx \\
&= \frac{1}{2\pi i(m-n)} e^{2\pi i(m-n)x} \Big|_{x=0}^{x=1} \\
&= 0.
\end{aligned}$$

Moreover,

$$\langle f_n, f_n \rangle = \int_0^1 e^{2\pi i(n-n)} \, dx = \int_0^1 1 \, dx = 1.$$

(b) Let $f \in L^2[0, 1]$ and $\varepsilon > 0$. Proposition 1.3.6 provides a $g \in C[0, 1]$ such that $\|f - g\| < \varepsilon/2$. The Stone–Weierstrass theorem [320, Ch. 5] asserts that the span of $\{f_n : n \in \mathbb{Z}\}$ is dense in $C[0, 1]$ and hence there is an

$$h = \sum_{n=-N}^{N} c_n f_n$$

such that $\|g - h\|_\infty < \varepsilon/2$. The analogue of (1.4.12) below implies that

$$\left\| f - \sum_{n=-N}^{N} \langle f, f_n \rangle f_n \right\| \leqslant \left\| f - \sum_{n=-N}^{N} c_n f_n \right\|.$$

Thus,

$$\left\| f - \sum_{n=-N}^{N} \langle f, f_n \rangle f_n \right\| \leq \| f - h \|$$

$$\leq \| f - g \| + \| g - h \|$$

$$\leq \frac{\varepsilon}{2} + \| g - h \|_\infty \qquad \text{(by (1.3.5))}$$

$$\leq \frac{\varepsilon}{2} + \frac{\varepsilon}{2} = \varepsilon. \qquad (1.3.10)$$

For any $M \geq N$, apply (1.4.12) again to see that

$$\left\| f - \sum_{n=-M}^{M} \langle f, f_n \rangle f_n \right\| \leq \left\| f - \sum_{n=-M}^{M} c_n f_n \right\|,$$

where $c_n = \langle f, f_n \rangle$ when $-N \leq n \leq N$ and $c_n = 0$ when $N < |n| \leq M$. From (1.3.10) it follows that this last quantity is at most ε.

(c) Observe that (b) yields

$$f = \sum_{n=-\infty}^{\infty} \langle f, f_n \rangle f_n,$$

where the convergence is in $L^2[0,1]$. If $h_n \to h$ and $k_n \to k$ in $L^2[0,1]$, an exercise with the Cauchy–Schwarz inequality (see Exercise 1.10.25) implies that

$$\langle h_n, k_n \rangle \to \langle h, k \rangle. \qquad (1.3.11)$$

Use the orthonormality of $(f_n)_{n=-\infty}^{\infty}$, the linearity of the inner product in the first slot, the conjugate linearity in the second slot, and (1.3.11), to compute

$$\| f \|^2 = \langle f, f \rangle$$

$$= \left\langle \sum_{n=-\infty}^{\infty} \langle f, f_n \rangle f_n, \sum_{m=-\infty}^{\infty} \langle f, f_m \rangle f_m \right\rangle$$

$$= \sum_{m,n=-\infty}^{\infty} \langle f, f_n \rangle \overline{\langle f, f_m \rangle} \langle f_n, f_m \rangle$$

$$= \sum_{m,n=-\infty}^{\infty} \langle f, f_n \rangle \overline{\langle f, f_m \rangle} \delta_{mn}$$

$$= \sum_{n=-\infty}^{\infty} |\langle f, f_n \rangle|^2,$$

which completes the proof. ∎

Remark 1.3.12. The upper and lower limits in (b) may tend to ∞ independently. We use N and $-N$ for convenience. In (c), absolute convergence ensures that any interpretation of the sum yields the same result.

1.4 Abstract Hilbert Spaces

The norms and inner products on $\mathbb{C}^n$, ℓ^2, and $L^2[0,1]$ generalize to other settings. The general framework of Hilbert spaces is one of the most fruitful concepts in modern mathematics. It begins with the following notion.

Definition 1.4.1. Let $\mathcal{H}$ be a complex vector space. Then $\Phi : \mathcal{H} \times \mathcal{H} \to \mathbb{C}$ is an *inner product* if, for all $\mathbf{x}, \mathbf{y}, \mathbf{z} \in \mathcal{H}$ and $c \in \mathbb{C}$, the following hold.

(a) $\Phi(\mathbf{x}, \mathbf{x}) \geqslant 0$.

(b) $\Phi(\mathbf{x}, \mathbf{x}) = 0$ if and only if $\mathbf{x} = \mathbf{0}$.

(c) $\Phi(\mathbf{x} + \mathbf{y}, \mathbf{z}) = \Phi(\mathbf{x}, \mathbf{z}) + \Phi(\mathbf{y}, \mathbf{z})$.

(d) $\Phi(c\mathbf{x}, \mathbf{y}) = c\Phi(\mathbf{x}, \mathbf{y})$.

(e) $\Phi(\mathbf{x}, \mathbf{y}) = \overline{\Phi(\mathbf{y}, \mathbf{x})}$.

A vector space endowed with an inner product is an *inner-product space*.

If Φ is an inner product, it is customary to use the notation

$$\langle \mathbf{x}, \mathbf{y} \rangle = \Phi(\mathbf{x}, \mathbf{y}).$$

This inner product determines a norm

$$\|\mathbf{x}\| = \langle \mathbf{x}, \mathbf{x} \rangle^{\frac{1}{2}}.$$

Here are several properties of inner product spaces that generalize those enjoyed by $\mathbb{C}^n$, ℓ^2, and $L^2[0,1]$.

Proposition 1.4.2. *Let* $\mathbf{x}, \mathbf{y}$ *belong to an inner product space.*

(a) Cauchy–Schwarz inequality: $|\langle \mathbf{x}, \mathbf{y} \rangle| \leqslant \|\mathbf{x}\|\|\mathbf{y}\|$. *Equality holds if and only if* $\mathbf{x}$ *and* $\mathbf{y}$ *are linearly dependent.*

(b) Triangle inequality: $\|\mathbf{x} + \mathbf{y}\| \leqslant \|\mathbf{x}\| + \|\mathbf{y}\|$. *Equality holds if and only if* $\mathbf{x}$ *or* $\mathbf{y}$ *is a nonnegative multiple of the other.*

Proof (a) The inequality holds if $\mathbf{x} = \mathbf{0}$ or $\mathbf{y} = \mathbf{0}$. Without loss of generality, assume $\mathbf{y} \neq \mathbf{0}$. Then

$$0 \leqslant \left\langle \mathbf{x} - \langle \mathbf{x}, \mathbf{y} \rangle \frac{\mathbf{y}}{\|\mathbf{y}\|^2}, \mathbf{x} - \langle \mathbf{x}, \mathbf{y} \rangle \frac{\mathbf{y}}{\|\mathbf{y}\|^2} \right\rangle$$

$$= \langle \mathbf{x}, \mathbf{x} \rangle - \frac{\langle \mathbf{x}, \mathbf{y} \rangle \langle \mathbf{y}, \mathbf{x} \rangle}{\|\mathbf{y}\|^2} - \frac{\overline{\langle \mathbf{x}, \mathbf{y} \rangle} \langle \mathbf{x}, \mathbf{y} \rangle}{\|\mathbf{y}\|^2} + \frac{\langle \mathbf{x}, \mathbf{y} \rangle \overline{\langle \mathbf{x}, \mathbf{y} \rangle} \langle \mathbf{y}, \mathbf{y} \rangle}{\|\mathbf{y}\|^4}$$

$$= \|\mathbf{x}\|^2 - \frac{|\langle \mathbf{x}, \mathbf{y} \rangle|^2}{\|\mathbf{y}\|^2} - \frac{|\langle \mathbf{x}, \mathbf{y} \rangle|^2}{\|\mathbf{y}\|^2} + \frac{|\langle \mathbf{x}, \mathbf{y} \rangle|^2 \|\mathbf{y}\|^2}{\|\mathbf{y}\|^4}$$

$$= \|\mathbf{x}\|^2 - \frac{|\langle \mathbf{x}, \mathbf{y} \rangle|^2}{\|\mathbf{y}\|^2}, \tag{1.4.3}$$

which proves that $|\langle \mathbf{x}, \mathbf{y} \rangle| \leqslant \|\mathbf{x}\| \|\mathbf{y}\|$. From (1.4.3) it follows that $|\langle \mathbf{x}, \mathbf{y} \rangle| = \|\mathbf{x}\| \|\mathbf{y}\|$ if and only if

$$\mathbf{x} - \langle \mathbf{x}, \mathbf{y} \rangle \frac{\mathbf{y}}{\|\mathbf{y}\|^2} = \mathbf{0},$$

in other words, if and only if $\mathbf{x}$ is a multiple of $\mathbf{y}$.

(b) Use (a) and obtain

$$\begin{aligned}
\|\mathbf{x} + \mathbf{y}\|^2 &= \langle \mathbf{x} + \mathbf{y}, \mathbf{x} + \mathbf{y} \rangle \\
&= \|\mathbf{x}\|^2 + 2\operatorname{Re}\langle \mathbf{x}, \mathbf{y} \rangle + \|\mathbf{y}\|^2 \\
&\leqslant \|\mathbf{x}\|^2 + 2\|\mathbf{x}\| \|\mathbf{y}\| + \|\mathbf{y}\|^2 \\
&= (\|\mathbf{x}\|^2 + \|\mathbf{y}\|^2)^2.
\end{aligned}$$

Take square roots of both sides and obtain the triangle inequality. Equality holds if and only if $\operatorname{Re}\langle \mathbf{x}, \mathbf{y} \rangle = \|\mathbf{x}\| \|\mathbf{y}\|$ and this is equivalent to

$$\langle \mathbf{x}, \mathbf{y} \rangle = \|\mathbf{x}\| \|\mathbf{y}\|. \tag{1.4.4}$$

If one of $\mathbf{x}$ or $\mathbf{y}$ is a nonnegative multiple of the other, then (1.4.4) holds. Conversely, if (1.4.4) holds, then the condition for equality in the Cauchy–Schwarz inequality says that $\mathbf{x}$ or $\mathbf{y}$ is a multiple of each other. Condition (1.4.4) ensures that the constant involved is nonnegative. ∎

The notion of orthogonality in $\mathbb{C}^n$, ℓ^2, and $L^2[0, 1]$ generalizes to inner product spaces.

Definition 1.4.5. Vectors $\mathbf{x}, \mathbf{y}$ in an inner product space are *orthogonal*, written $\mathbf{x} \perp \mathbf{y}$, if $\langle \mathbf{x}, \mathbf{y} \rangle = 0$.

Proposition 1.4.6 (Pythagorean theorem). *If $\mathbf{x}$ and $\mathbf{y}$ are orthogonal vectors in an inner product space, then $\|\mathbf{x} + \mathbf{y}\|^2 = \|\mathbf{x}\|^2 + \|\mathbf{y}\|^2$.*

Proof The proof is identical to that of Proposition 1.1.6. ∎

The spaces $\mathbb{C}^n$, ℓ^2, and $L^2[0, 1]$ are complete. For a general inner product space, one needs to impose this condition as an axiom since not every inner product space is complete (Exercise 1.10.13).

Definition 1.4.7. A *Hilbert space* is a complete inner product space. More specifically, a Hilbert space is an inner product space that is (Cauchy) complete with respect to the norm $\|\mathbf{x}\| = \sqrt{\langle \mathbf{x}, \mathbf{x} \rangle}$ induced by the inner product.

Examples of Hilbert spaces include $\mathbb{C}^n$, ℓ^2, and $L^2[0, 1]$. With ℓ^2, the sequence $(\mathbf{e}_n)_{n=0}^{\infty}$ is an orthonormal set and every $\mathbf{a} \in \ell^2$ can be written uniquely as

$$\mathbf{a} = \sum_{n=0}^{\infty} \langle \mathbf{a}, \mathbf{e}_n \rangle \mathbf{e}_n,$$

in which

$$\|\mathbf{a}\|^2 = \sum_{n=0}^{\infty} |a_n|^2.$$

In Theorem 1.3.9, we saw something similar for $L^2[0,1]$: every $f \in L^2[0,1]$ can be written

$$f = \sum_{n=-\infty}^{\infty} \langle f, f_n \rangle f_n \quad \text{where} \quad f_n(x) = e^{2\pi i n x},$$

the series converges in norm (see Theorem 1.3.9), and

$$\|f\|^2 = \sum_{n=-\infty}^{\infty} |\langle f, f_n \rangle|^2.$$

Thus, the following definition is a natural one.

Definition 1.4.8. An orthonormal sequence $(\mathbf{x}_n)_{n=1}^{\infty}$ in a Hilbert space $\mathcal{H}$ is an *orthonormal basis* if every $\mathbf{x} \in \mathcal{H}$ can be written as

$$\mathbf{x} = \sum_{n=1}^{\infty} \langle \mathbf{x}, \mathbf{x}_n \rangle \mathbf{x}_n.$$

The sum above converges in the norm of $\mathcal{H}$, in the sense that

$$\lim_{N \to \infty} \left\| \mathbf{x} - \sum_{n=1}^{N} \langle \mathbf{x}, \mathbf{x}_n \rangle \mathbf{x}_n \right\| = 0.$$

Some of our orthonormal bases are indexed by $\mathbb{Z}$ (Theorem 1.3.9) while others are indexed by $\mathbb{N}$ or $\mathbb{N} \cup \{0\}$ (see (1.2.8)). We state most of our results under the assumption that the orthonormal sets considered are infinite. The reader is advised that most of these results apply equally well in the finite-dimensional setting.

For an orthonormal sequence in $\mathcal{H}$ (not necessarily an orthonormal basis) there is the following result.

Theorem 1.4.9. *Let* $(\mathbf{x}_n)_{n=1}^{\infty}$ *be an orthonormal sequence in a Hilbert space* $\mathcal{H}$.

(a) Bessel's inequality: *For every* $\mathbf{x} \in \mathcal{H}$,

$$\sum_{n=1}^{\infty} |\langle \mathbf{x}, \mathbf{x}_n \rangle|^2 \leqslant \|\mathbf{x}\|^2. \tag{1.4.10}$$

(b) Parseval's theorem: *If* $(\mathbf{x}_n)_{n=1}^{\infty}$ *is an orthonormal basis for* $\mathcal{H}$, *then*

$$\|\mathbf{x}\|^2 = \sum_{n=1}^{\infty} |\langle \mathbf{x}, \mathbf{x}_n \rangle|^2 \tag{1.4.11}$$

for every $\mathbf{x} \in \mathcal{H}$.

Proof (a) For any $c_1, c_2, ..., c_k \in \mathbb{C}$, Exercise 1.10.10 yields

$$\left\| \mathbf{x} - \sum_{i=1}^{k} c_i \mathbf{x}_i \right\|^2 = \|\mathbf{x}\|^2 - \sum_{i=1}^{k} |\langle \mathbf{x}, \mathbf{x}_i \rangle|^2 + \sum_{i=1}^{k} |\langle \mathbf{x}, \mathbf{x}_i \rangle - c_i|^2. \qquad (1.4.12)$$

Let $c_i = \langle \mathbf{x}, \mathbf{x}_i \rangle$ for all $1 \leqslant i \leqslant N$ and obtain

$$\left\| \mathbf{x} - \sum_{i=1}^{k} \langle \mathbf{x}, \mathbf{x}_i \rangle \mathbf{x}_i \right\|^2 = \|\mathbf{x}\|^2 - \sum_{i=1}^{k} |\langle \mathbf{x}, \mathbf{x}_i \rangle|^2. \qquad (1.4.13)$$

Thus, $\sum_{i=1}^{k} |\langle \mathbf{x}, \mathbf{x}_i \rangle|^2 \leqslant \|\mathbf{x}\|^2$ for all $k \geqslant 1$. Let $k \to \infty$ to obtain Bessel's inequality.
(b) By the definition of an orthonormal basis, the left side of (1.4.13) goes to zero as $k \to \infty$. This yields Parseval's theorem. ∎

To develop a useful criterion for determining whether an orthonormal sequence is an orthonormal basis, we need the next definition.

Definition 1.4.14. If $\mathcal{S}$ is a subset of a Hilbert space $\mathcal{H}$, its *closed span* $\bigvee \mathcal{S}$ is the closure of the set of all finite linear combinations of elements of $\mathcal{S}$.

Proposition 1.4.15. *For an orthonormal sequence $(\mathbf{x}_n)_{n=1}^{\infty}$ in a Hilbert space $\mathcal{H}$, the following are equivalent.*

(a) *$(\mathbf{x}_n)_{n=1}^{\infty}$ is an orthonormal basis.*

(b) *$\bigvee \{\mathbf{x}_n : n \geqslant 1\} = \mathcal{H}$.*

Proof (a) $\Rightarrow$ (b) This follows from Definition 1.4.8.
(b) $\Rightarrow$ (a) Given $\varepsilon > 0$ and $\mathbf{x} \in \mathcal{H}$, the hypothesis provides a linear combination $\sum_{n=1}^{N} c_n \mathbf{x}_n$ such that

$$\left\| \mathbf{x} - \sum_{n=1}^{N} c_n \mathbf{x}_n \right\| < \varepsilon.$$

For any $m \geqslant N$, use (1.4.12) to see that

$$\left\| \mathbf{x} - \sum_{n=1}^{m} \langle \mathbf{x}, \mathbf{x}_n \rangle \mathbf{x}_n \right\| \leqslant \left\| \mathbf{x} - \sum_{n=1}^{m} c_n \mathbf{x}_n \right\|,$$

where $c_{N+1} = \cdots = c_m = 0$. Therefore,

$$\left\| \mathbf{x} - \sum_{n=1}^{m} \langle \mathbf{x}, \mathbf{x}_n \rangle \mathbf{x}_n \right\| < \varepsilon,$$

and hence

$$\mathbf{x} = \sum_{n=1}^{\infty} \langle \mathbf{x}, \mathbf{x}_n \rangle \mathbf{x}_n.$$

We conclude that $(\mathbf{x}_n)_{n=1}^{\infty}$ is an orthonormal basis for $\mathcal{H}$. ∎

Does every Hilbert space have an orthonormal basis? Is an orthonormal basis for a Hilbert space necessarily countable? We address these questions in the next two sections and in Exercises 1.10.37, 1.10.38, and 1.10.39.

1.5 The Gram–Schmidt Process

The Gram–Schmidt process takes a linearly independent list of vectors in a Hilbert space and returns an orthonormal list with the same span. In fact, something more is true: for each $k \geq 1$, the span of the first k vectors in each list is the same. We state the result below for infinite lists, although the proof works just as well for finite lists.

Theorem 1.5.1 (Gram–Schmidt). *Let $(\mathbf{v}_i)_{i=1}^{\infty}$ be a linearly independent sequence of vectors in a Hilbert space $\mathcal{H}$. Then there is an orthonormal sequence of vectors $(\mathbf{u}_i)_{i=1}^{\infty}$ such that*

$$\mathrm{span}\{\mathbf{v}_1, \mathbf{v}_2,..., \mathbf{v}_k\} = \mathrm{span}\{\mathbf{u}_1, \mathbf{u}_2,..., \mathbf{u}_k\} \quad \text{for all } k \geq 1. \tag{1.5.2}$$

In particular, $\bigvee\{\mathbf{v}_n : n \geq 1\} = \bigvee\{\mathbf{u}_n : n \geq 1\}$.

Proof Proceed by induction on k. For $k = 1$, define

$$\mathbf{u}_1 = \frac{\mathbf{v}_1}{\|\mathbf{v}_1\|},$$

a unit vector with $\mathrm{span}\{\mathbf{v}_1\} = \mathrm{span}\{\mathbf{u}_1\}$. For the induction hypothesis, suppose that given $k - 1$ linearly independent vectors $\mathbf{v}_1, \mathbf{v}_2,..., \mathbf{v}_{k-1}$, there exists orthonormal vectors $\mathbf{u}_1, \mathbf{u}_2,..., \mathbf{u}_{k-1}$ such that

$$\mathrm{span}\{\mathbf{v}_1, \mathbf{v}_2,..., \mathbf{v}_j\} = \mathrm{span}\{\mathbf{u}_1, \mathbf{u}_2,..., \mathbf{u}_j\} \quad \text{for all } 1 \leq j \leq k - 1. \tag{1.5.3}$$

Since $\mathbf{v}_1, \mathbf{v}_2,..., \mathbf{v}_k$ are linearly independent,

$$\mathbf{v}_k \notin \mathrm{span}\{\mathbf{v}_1, \mathbf{v}_2,..., \mathbf{v}_{k-1}\} = \mathrm{span}\{\mathbf{u}_1, \mathbf{u}_2,..., \mathbf{u}_{k-1}\},$$

and hence

$$\mathbf{x}_k = \mathbf{v}_k - \sum_{i=1}^{k-1} \langle \mathbf{v}_k, \mathbf{u}_i \rangle \mathbf{u}_i \neq \mathbf{0}.$$

Define $\mathbf{u}_k = \mathbf{x}_k / \|\mathbf{x}_k\|$ and observe that for each $1 \leq j \leq k - 1$,

$$\langle \mathbf{u}_j, \mathbf{x}_k \rangle = \left\langle \mathbf{u}_j, \mathbf{v}_k - \sum_{i=1}^{k-1} \langle \mathbf{v}_k, \mathbf{u}_i \rangle \mathbf{u}_i \right\rangle$$

$$= \langle \mathbf{u}_j, \mathbf{v}_k \rangle - \left\langle \mathbf{u}_j, \sum_{i=1}^{k-1} \langle \mathbf{v}_k, \mathbf{u}_i \rangle \mathbf{u}_i \right\rangle$$

$$= \langle \mathbf{u}_j, \mathbf{v}_k \rangle - \sum_{i=1}^{k-1} \overline{\langle \mathbf{v}_k, \mathbf{u}_i \rangle} \langle \mathbf{u}_j, \mathbf{u}_i \rangle$$

$$= \langle \mathbf{u}_j, \mathbf{v}_k \rangle - \sum_{i=1}^{k-1} \langle \mathbf{u}_i, \mathbf{v}_k \rangle \delta_{ij}$$

$$= \langle \mathbf{u}_j, \mathbf{v}_k \rangle - \langle \mathbf{u}_j, \mathbf{v}_k \rangle$$

$$= 0.$$

Because $\mathbf{u}_k$ is a linear combination of $\mathbf{v}_1, \mathbf{v}_2, ..., \mathbf{v}_k$, the induction hypothesis (1.5.3) ensures that

$$\text{span}\{\mathbf{u}_1, \mathbf{u}_2, ..., \mathbf{u}_k\} \subseteq \text{span}\{\mathbf{v}_1, \mathbf{v}_2, ..., \mathbf{v}_k\}. \tag{1.5.4}$$

Since $\mathbf{u}_1, \mathbf{u}_2, ..., \mathbf{u}_k$ are orthonormal, they are linearly independent (Exercise 1.10.21) and hence (1.5.4) is an equality since the subspaces involved are both k-dimensional.
∎

The proof of Theorem 1.5.1 suggests the following algorithm, which works well for theoretical purposes, even though it is subject to numerical instability in some real-world applications. First set

$$\mathbf{u}_1 = \frac{\mathbf{v}_1}{\|\mathbf{v}_1\|}.$$

Then for $k \geq 2$, let

$$\mathbf{u}_k = \frac{\mathbf{v}_k - \langle \mathbf{v}_k, \mathbf{u}_1 \rangle \mathbf{u}_1 - \cdots - \langle \mathbf{v}_k, \mathbf{u}_{k-1} \rangle \mathbf{u}_{k-1}}{\|\mathbf{v}_k - \langle \mathbf{v}_k, \mathbf{u}_1 \rangle \mathbf{u}_1 - \cdots - \langle \mathbf{v}_k, \mathbf{u}_{k-1} \rangle \mathbf{u}_{k-1}\|}.$$

Observe that if $(\mathbf{v}_n)_{n=1}^{\infty}$ is already orthonormal, then the Gram–Schmidt process returns the original list (Exercise 1.10.34).

Example 1.5.5. The monomials $(x^n)_{n=0}^{\infty}$ are a linearly independent sequence in $L^2[0,1]$. Indeed, a finite linear combination of these monomials that equals the zero function is a polynomial with infinitely many roots and hence must be the zero polynomial. However, $(x^n)_{n=0}^{\infty}$ is not orthonormal since, for example, $\|x\| = 1/\sqrt{3}$ and $\langle x, x^2 \rangle = 1/4$. The Gram–Schmidt process, when applied to these monomials, returns the orthogonal polynomials

$$u_0(x) = 1, \quad u_1(x) = \sqrt{3}(2x-1), \quad u_2(x) = \sqrt{5}(6x^2 - 6x + 1), \ ....$$

Moreover, (1.5.2) ensures that the degree of each $u_n(x)$ is n.

1.6 Orthonormal Bases and Total Orthonormal Sets

Each of the Hilbert spaces $\mathbb{C}^n$, ℓ^2, and $L^2[0,1]$ has an orthonormal basis. Does every Hilbert space have an orthonormal basis? We begin with a standard definition that applies equally well to any metric space.

Definition 1.6.1. A Hilbert space is *separable* if it contains a countable dense set.

Theorem 1.6.2. *For a Hilbert space $\mathcal{H}$, the following are equivalent.*

(a) $\mathcal{H}$ *is separable.*

(b) $\mathcal{H}$ has a countable orthonormal basis.

Proof (a) $\Rightarrow$ (b) Since $\mathcal{H}$ is separable it has a countable dense set $(\mathbf{x}_n)_{n=1}^{\infty}$. Refine this sequence as follows. If $\mathbf{x}_k \in \text{span}\{\mathbf{x}_1, \mathbf{x}_2,..., \mathbf{x}_{k-1}\}$, omit $\mathbf{x}_k$ from the sequence and relabel it $(\mathbf{x}_n)_{n=1}^{\infty}$. Proceeding in this manner results in a countable linearly independent sequence $(\mathbf{x}_n)_{n=1}^{\infty}$ (possibly finite). Now apply the Gram–Schmidt process (Theorem 1.5.1) to the resulting list to obtain a (countable) orthonormal sequence $(\mathbf{u}_n)_{n=1}^{\infty}$ such that

$$\bigvee\{\mathbf{u}_n : n \geqslant 1\} = \bigvee\{\mathbf{x}_n : n \geqslant 1\}.$$

Since $(\mathbf{x}_n)_{n=1}^{\infty}$ is dense in $\mathcal{H}$, it follows that $\bigvee\{\mathbf{x}_n : n \geqslant 1\} = \mathcal{H}$ and hence, $\bigvee\{\mathbf{u}_n : n \geqslant 1\} = \mathcal{H}$. By Proposition 1.4.15, $\mathcal{H}$ has a countable orthonormal basis.

(b) $\Rightarrow$ (a) If $\mathcal{H}$ has an orthonormal basis $(\mathbf{u}_n)_{n=1}^{\infty}$, then

$$\left\{ \sum_{n=1}^{N} (a_n + ib_n)\mathbf{u}_n : a_n, b_n \in \mathbb{Q}, N \geqslant 0 \right\}$$

is countable and dense in $\mathcal{H}$ (here $\mathbb{Q}$ denotes the rational numbers). ∎

The Hilbert spaces $\mathbb{C}^n$, ℓ^2, and $L^2[0,1]$ are separable since each has a countable orthonormal basis. For the most part, the Hilbert spaces considered in this book are separable. For an example of a nonseparable Hilbert space, see Exercise 1.10.37. Some authors define a separable Hilbert space as one with a countable orthonormal basis, as opposed to a countable dense subset. The previous theorem proves that these two definitions are equivalent.

For general Hilbert spaces, including the nonseparable ones, there is the following definition.

Definition 1.6.3. A set of vectors $\{\mathbf{x}_\alpha : \alpha \in \Gamma\}$ in a Hilbert space $\mathcal{H}$ is a *total orthonormal set* if it satisfies the following conditions.

(a) $\langle \mathbf{x}_\alpha, \mathbf{x}_\beta \rangle = \delta_{\alpha\beta}$ for all $\alpha, \beta \in \Gamma$.

(b) $\bigvee\{\mathbf{x}_\alpha : \alpha \in \Gamma\} = \mathcal{H}$.

It is important to note here that the index set Γ can be of any cardinality. The following is a version of Theorem 1.6.2 for general (possibly nonseparable) Hilbert spaces. The reader is invited to work through the proof in Exercise 1.10.38.

Proposition 1.6.4. *Every Hilbert space has a total orthonormal set.*

1.7 Orthogonal Projections

Proposition 1.1.8 covered orthogonal projections onto subspaces of $\mathbb{C}^n$. For Hilbert spaces, the same sort of results hold, but the definitions and the reasoning are more delicate.

Definition 1.7.1. A subset $\mathcal{M}$ of a Hilbert space $\mathcal{H}$ is a *subspace* if the following hold.

(a) $\mathcal{M} \neq \varnothing$.

(b) $\mathcal{M}$ is closed under addition and scalar multiplication.

(c) $\mathcal{M}$ is norm closed in $\mathcal{H}$.

A "subspace" in this context is closed under the vector-space operations and is topologically closed with respect to the norm of $\mathcal{H}$. A subspace of a Hilbert space is itself a Hilbert space when endowed with the inherited inner product from the larger space. Exercise 1.10.36 proves the following.

Proposition 1.7.2. *A subspace of a separable Hilbert space is also separable and thus has a countable orthonormal basis.*

Below is a generalization of Proposition 1.1.8. The reader is reminded that unless stated otherwise, our Hilbert spaces are separable and infinite dimensional. The adjustments needed in the finite-dimensional case are minor, and the non-separable case rarely concerns us.

Proposition 1.7.3. *If $\mathcal{M}$ is a subspace of separable Hilbert space $\mathcal{H}$ and $\mathbf{x} \in \mathcal{H}$, then there is a unique vector $P_{\mathcal{M}}\mathbf{x} \in \mathcal{H}$ such that the following hold.*

(a) $P_{\mathcal{M}}\mathbf{x} \in \mathcal{M}$.

(b) $\|\mathbf{x} - P_{\mathcal{M}}\mathbf{x}\| \leqslant \|\mathbf{x} - \mathbf{v}\|$ *for all* $\mathbf{v} \in \mathcal{M}$.

(c) $(\mathbf{x} - P_{\mathcal{M}}\mathbf{x}) \perp \mathbf{v}$ *for all* $\mathbf{v} \in \mathcal{M}$.

(d) $P_{\mathcal{M}}\mathbf{x} = \mathbf{x}$ *if and only if* $\mathbf{x} \in \mathcal{M}$.

(e) *If* $(\mathbf{v}_n)_{n=1}^{\infty}$ *is any orthonormal basis for* $\mathcal{M}$, *then*

$$P_{\mathcal{M}}\mathbf{x} = \sum_{i=1}^{\infty} \langle \mathbf{x}, \mathbf{v}_i \rangle \mathbf{v}_i.$$

In particular, $P_{\mathcal{M}}\mathbf{x}$ is independent of the choice of orthonormal basis $(\mathbf{v}_n)_{n=1}^{\infty}$.

Proof (a) The Gram–Schmidt process provides a countable orthonormal basis $(\mathbf{v}_n)_{n=1}^{\infty}$ for $\mathcal{M}$. Bessel's inequality (1.4.10) says that

$$P_{\mathcal{M}}\mathbf{x} = \sum_{i=1}^{\infty} \langle \mathbf{x}, \mathbf{v}_i \rangle \mathbf{v}_i$$

is well defined. It belongs to $\mathcal{M}$ since it belongs to $\bigvee\{\mathbf{v}_i : i \geqslant 1\}$ and $\mathcal{M}$ is closed.
(b) Since $(\mathbf{v}_n)_{n=1}^{\infty}$ is an orthonormal basis for $\mathcal{M}$, each $\mathbf{v} \in \mathcal{M}$ is of the form

$$\mathbf{v} = \sum_{i=1}^{\infty} \langle \mathbf{v}, \mathbf{v}_i \rangle \mathbf{v}_i.$$

For each $N \geqslant 1$, Exercise 1.10.10 yields

$$\left\| \mathbf{x} - \sum_{i=1}^{N} \langle \mathbf{x}, \mathbf{v}_i \rangle \mathbf{v}_i \right\| \leqslant \left\| \mathbf{x} - \sum_{i=1}^{N} \langle \mathbf{v}, \mathbf{v}_i \rangle \mathbf{v}_i \right\|.$$

Let $N \to \infty$ and obtain $\|\mathbf{x} - P_{\mathcal{M}}\mathbf{x}\| \leqslant \|\mathbf{x} - \mathbf{v}\|$.

(c) This proof is essentially identical to the proof of Proposition 1.1.8b.

(d) If $\mathbf{x} = P_{\mathcal{M}}\mathbf{x}$, then (a) ensures that $\mathbf{x} \in \mathcal{M}$. If $\mathbf{x} \in \mathcal{M}$, then (a) implies that $P_{\mathcal{M}}\mathbf{x} \in \mathcal{M}$, and hence $\mathbf{x} - P_{\mathcal{M}}\mathbf{x} \in \mathcal{M}$. Then (c) yields $\|\mathbf{x} - P_{\mathcal{M}}\mathbf{x}\|^2 = \langle \mathbf{x} - P_{\mathcal{M}}\mathbf{x}, \mathbf{x} - P_{\mathcal{M}}\mathbf{x}\rangle = 0$. Therefore, $\mathbf{x} = P_{\mathcal{M}}\mathbf{x}$.

(e) This proof is essentially identical to the proof of Corollary 1.1.13. ∎

The proof of Corollary 1.1.14 carries over directly to prove the following.

Corollary 1.7.4. *For a subspace $\mathcal{M}$ of a Hilbert space $\mathcal{H}$, the following are equivalent.*

(a) $\mathcal{M} \neq \mathcal{H}$.

(b) *There is a $\mathbf{y} \in \mathcal{H}\backslash\{\mathbf{0}\}$ such that $\mathbf{y} \perp \mathbf{v}$ for all $\mathbf{v} \in \mathcal{M}$.*

Below are some fundamental properties of the mapping $\mathbf{x} \mapsto P_{\mathcal{M}}\mathbf{x}$. The proof below is for separable Hilbert spaces (the only type that we regularly consider in this book). The modifications necessary to treat the non-separable case are mostly typographical.

Proposition 1.7.5. *Let $\mathcal{M}$ be a subspace of a Hilbert space $\mathcal{H}$.*

(a) *The mapping $\mathbf{x} \mapsto P_{\mathcal{M}}\mathbf{x}$ is linear on $\mathcal{H}$.*

(b) $\langle P_{\mathcal{M}}\mathbf{x}, \mathbf{y}\rangle = \langle \mathbf{x}, P_{\mathcal{M}}\mathbf{y}\rangle$ *for all $\mathbf{x}, \mathbf{y} \in \mathcal{H}$.*

(c) $P_{\mathcal{M}}(P_{\mathcal{M}}\mathbf{x}) = P_{\mathcal{M}}\mathbf{x}$ *for all $\mathbf{x} \in \mathcal{H}$.*

Proof In what follows, we frequently pass an infinite series through an inner product. This is justified by Exercise 1.10.25:

$$\mathbf{x}_n \to \mathbf{x} \quad \text{and} \quad \mathbf{y}_n \to \mathbf{y} \quad \Longrightarrow \quad \langle \mathbf{x}_n, \mathbf{y}_n\rangle \to \langle \mathbf{x}, \mathbf{y}\rangle. \tag{1.7.6}$$

(a) Let $(\mathbf{v}_n)_{n=1}^{\infty}$ be an orthonormal basis for $\mathcal{M}$. By Proposition 1.7.3,

$$P_{\mathcal{M}}\mathbf{x} = \sum_{i=1}^{\infty} \langle \mathbf{x}, \mathbf{v}_i\rangle \mathbf{v}_i \quad \text{for all } \mathbf{x} \in \mathcal{H}.$$

For any $\mathbf{x}, \mathbf{y} \in \mathcal{M}$ and $c \in \mathbb{C}$,

$$\begin{aligned}
P_{\mathcal{M}}(\mathbf{x} + c\mathbf{y}) &= \sum_{i=1}^{\infty} \langle \mathbf{x} + c\mathbf{y}, \mathbf{v}_i\rangle \mathbf{v}_i \\
&= \sum_{i=1}^{\infty} (\langle \mathbf{x}, \mathbf{v}_i\rangle + c\langle \mathbf{y}, \mathbf{v}_i\rangle)\mathbf{v}_i \\
&= \sum_{i=1}^{\infty} \langle \mathbf{x}, \mathbf{v}_i\rangle \mathbf{v}_i + c\sum_{i=1}^{\infty} \langle \mathbf{y}, \mathbf{v}_i\rangle \mathbf{v}_i \\
&= P_{\mathcal{M}}\mathbf{x} + cP_{\mathcal{M}}\mathbf{y}.
\end{aligned}$$

(b) Use the properties of the inner product from Definition 1.4.1, along with (1.7.6), and obtain

$$\langle P_{\mathcal{M}}\mathbf{x}, \mathbf{y}\rangle = \Big\langle \sum_{i=1}^{\infty}\langle \mathbf{x}, \mathbf{v}_i\rangle \mathbf{v}_i, \mathbf{y}\Big\rangle$$

$$= \sum_{i=1}^{\infty}\langle \mathbf{x}, \mathbf{v}_i\rangle\langle \mathbf{v}_i, \mathbf{y}\rangle$$

$$= \sum_{i=1}^{\infty}\langle \mathbf{x}, \mathbf{v}_i\rangle\overline{\langle \mathbf{y}, \mathbf{v}_i\rangle}$$

$$= \sum_{i=1}^{\infty}\langle \mathbf{x}, \langle \mathbf{y}, \mathbf{v}_i\rangle \mathbf{v}_i\rangle$$

$$= \Big\langle \mathbf{x}, \sum_{i=1}^{\infty}\langle \mathbf{y}, \mathbf{v}_i\rangle \mathbf{v}_i\Big\rangle$$

$$= \langle \mathbf{x}, P_{\mathcal{M}}\mathbf{y}\rangle.$$

(c) From (b) observe that

$$P_{\mathcal{M}}(P_{\mathcal{M}}\mathbf{x}) = \sum_{i=1}^{\infty}\langle P_{\mathcal{M}}\mathbf{x}, \mathbf{v}_i\rangle \mathbf{v}_i$$

$$= \sum_{i=1}^{\infty}\langle \mathbf{x}, P_{\mathcal{M}}\mathbf{v}_i\rangle \mathbf{v}_i$$

$$= \sum_{i=1}^{\infty}\langle \mathbf{x}, \mathbf{v}_i\rangle \mathbf{v}_i \qquad (\text{since } P_{\mathcal{M}}\mathbf{v}_i = \mathbf{v}_i)$$

$$= P_{\mathcal{M}}\mathbf{x},$$

which completes the proof. ∎

1.8 Banach Spaces

Closely related to Hilbert spaces are Banach spaces: complete normed vector spaces. Although general Banach spaces are not the main focus of this book, they do play a critical role. As with Hilbert spaces, we start off with some representative examples of Banach spaces before defining them formally.

Example 1.8.1. For $1 < p < \infty$, let

$$\ell^p = \Big\{\mathbf{a} = (a_n)_{n=0}^{\infty} : \sum_{n=0}^{\infty}|a_n|^p < \infty\Big\} \quad \text{and} \quad \|\mathbf{a}\|_p = \Big(\sum_{n=0}^{\infty}|a_n|^p\Big)^{\frac{1}{p}}.$$

A generalization of the Cauchy–Schwarz inequality, known as *Hölder's inequality* [319], says that if q satisfies

$$\frac{1}{p} + \frac{1}{q} = 1,$$

(q is the *conjugate exponent* to p), then

$$\sum_{n=0}^{\infty} |z_n w_n| \leqslant \|\mathbf{z}\|_p \|\mathbf{w}\|_q \quad \text{for all } \mathbf{z} \in \ell^p \text{ and } \mathbf{w} \in \ell^q.$$

From here, one can show that ℓ^p is a vector space (with similar addition and scalar multiplication operations as with ℓ^2) such that $\|\mathbf{a} + \mathbf{b}\|_p \leqslant \|\mathbf{a}\|_p + \|\mathbf{b}\|_p$ for all $\mathbf{a}, \mathbf{b} \in \ell^p$ and $\|c\mathbf{a}\|_p = |c| \|\mathbf{a}\|_p$ for all $c \in \mathbb{C}$. Furthermore, ℓ^p is complete.

Example 1.8.2. Let ℓ^∞ denote the set of all bounded sequences $\mathbf{a} = (a_n)_{n=0}^{\infty}$ endowed with the norm

$$\|\mathbf{a}\|_\infty = \sup_{n \geqslant 0} |a_n|.$$

One can show that ℓ^∞ is a complete normed vector space.

Example 1.8.3. For $1 < p < \infty$, let $L^p[0, 1]$ be the set of Lebesgue-measurable, complex-valued functions f on $[0, 1]$ such that

$$\|f\|_p = \left(\int_0^1 |f(x)|^p \, dx \right)^{\frac{1}{p}} < \infty.$$

The corresponding version of Hölder's inequality,

$$\int_0^1 |h(x)k(x)| \, dx \leqslant \|h\|_p \|k\|_q \quad \text{for all } h \in L^p[0, 1] \text{ and } k \in L^q[0, 1], \tag{1.8.4}$$

implies that $L^p[0, 1]$ is a vector space (with similar operations of function addition and scalar multiplication as with $L^2[0, 1]$) such that $\|f + g\|_p \leqslant \|f\|_p + \|g\|_p$ and $\|cf\|_p = |c| \|f\|_p$ for all $c \in \mathbb{C}$. One can show that the $L^p[0, 1]$ spaces are complete normed vector spaces.

Example 1.8.5. Let $L^\infty[0, 1]$ denote the space of Lebesgue-measurable, complex-valued functions f on $[0, 1]$ with bounded *essential supremum*

$$\|f\|_\infty = \sup\{a \geqslant 0 : |\{x \in [0, 1] : |f(x)| > a\}| > 0\}. \tag{1.8.6}$$

In the above, $|A|$ denotes the Lebesgue measure of a set $A \subseteq [0, 1]$. Note that $L^\infty[0, 1] \subseteq L^p[0, 1]$ for all $p \geqslant 1$ and (see [319, Ch. 3])

$$\lim_{p \to \infty} \|f\|_p = \|f\|_\infty \quad \text{for all } f \in L^\infty[0, 1].$$

Example 1.8.7. From Section 1.3 recall the space $C[0, 1]$ of complex-valued continuous functions f on $[0, 1]$ with norm

$$\|f\|_\infty = \max_{0 \leqslant x \leqslant 1} |f(x)|.$$

One can show that $C[0, 1]$ is a complete normed vector space (see Definition 1.8.8). Note that $C[0, 1] \subseteq L^\infty[0, 1]$ and the essential supremum norm from (1.8.6) equals the supremum norm above.

All of the vector spaces defined above are examples of Banach spaces.

Definition 1.8.8. Let V be a complex vector space. Suppose that the function $\Phi : V \to [0, \infty)$ satisfies the following for all $\mathbf{u}, \mathbf{v} \in V$ and $c \in \mathbb{C}$.

(a) $\Phi(\mathbf{v}) = 0$ if and only if $\mathbf{v} = \mathbf{0}$.

(b) $\Phi(\mathbf{u} + \mathbf{v}) \leqslant \Phi(\mathbf{u}) + \Phi(\mathbf{v})$.

(c) $\Phi(c\mathbf{u}) = |c|\Phi(\mathbf{u})$.

Then V is a *normed vector space* with *norm* Φ. If V is also (Cauchy) complete with respect to the norm Φ, then V is a *Banach space*.

One traditionally writes the norm as $\|\mathbf{v}\| = \Phi(\mathbf{v})$ for $\mathbf{v} \in V$. Every Hilbert space is a Banach space, but the converse is false.

Theorem 1.8.9 (Jordan–von Neumann [206]). *Let V be a Banach space with norm $\| \cdot \|$. The following are equivalent.*

(a) *V is a Hilbert space, meaning there exists an inner product $\langle \cdot, \cdot \rangle$ on V such that $\langle \mathbf{v}, \mathbf{v} \rangle = \|\mathbf{v}\|^2$ for all $\mathbf{v} \in V$.*

(b) *$\|\mathbf{u} + \mathbf{v}\|^2 + \|\mathbf{u} - \mathbf{v}\|^2 = 2(\|\mathbf{u}\|^2 + \|\mathbf{v}\|^2)$ for all $\mathbf{u}, \mathbf{v} \in V$.*

The identity in (b) is the *parallelogram identity*. There is an alternate characterization of Hilbert spaces due to Fréchet [133]. Exercises 1.10.26, 1.10.27, 1.10.28, and 1.10.29 yield the following.

Corollary 1.8.10. *The Banach spaces ℓ^p and $L^p[0, 1]$ for $p \neq 2$, as well as $C[0, 1]$, are not Hilbert spaces.*

Below is a useful criterion for determining if a normed vector space is a Banach space.

Theorem 1.8.11. *Let V be a normed vector space with norm $\| \cdot \|$. The following are equivalent.*

(a) *V is a Banach space.*

(b) *Every absolutely convergent series in V converges, that is, for every $(\mathbf{x}_n)_{n=1}^{\infty}$ in V such that $\sum_{n=1}^{\infty} \|\mathbf{x}_n\| < \infty$, the series $\sum_{n=1}^{\infty} \mathbf{x}_n$ converges in V.*

Proof (a) $\Rightarrow$ (b) Let $(\mathbf{x}_n)_{n=1}^{\infty}$ be such that $\sum_{n=1}^{\infty} \|\mathbf{x}_n\|$ converges. For each $m, n \geqslant N$,

$$\left\| \sum_{k=1}^{m} \mathbf{x}_k - \sum_{k=1}^{n} \mathbf{x}_k \right\| \leqslant \sum_{k=N}^{\infty} \|\mathbf{x}_k\|.$$

Since $\lim_{N \to \infty} \sum_{k=N}^{\infty} \|\mathbf{x}_k\| = 0$, the sequence $(\sum_{k=1}^{n} \mathbf{x}_k)_{n=1}^{\infty}$ of partial sums is Cauchy. Since V is complete, $\sum_{n=1}^{\infty} \mathbf{x}_n$ converges.

(b) $\Rightarrow$ (a) Let $(\mathbf{x}_n)_{n=1}^{\infty}$ be a Cauchy sequence. It suffices to show that a subsequence of the Cauchy sequence $(\mathbf{x}_n)_{n=1}^{\infty}$ converges since this would imply the convergence of $(\mathbf{x}_n)_{n=1}^{\infty}$ to the same limit.

Use the fact that $(\mathbf{x}_n)_{n=1}^{\infty}$ is Cauchy to pick a subsequence $(\mathbf{x}_{n_k})_{k=1}^{\infty}$ such that

$$\|\mathbf{x}_{n_{k+1}} - \mathbf{x}_{n_k}\| \leqslant \frac{1}{2^k} \quad \text{for all } k \geqslant 1.$$

If $\mathbf{y}_k = \mathbf{x}_{n_{k+1}} - \mathbf{x}_{n_k}$, then

$$\sum_{k=1}^{\infty} \|\mathbf{y}_k\| = \sum_{k=1}^{\infty} \|\mathbf{x}_{n_{k+1}} - \mathbf{x}_{n_k}\| \leqslant \sum_{k=1}^{\infty} \frac{1}{2^k} < \infty.$$

By hypothesis, $\sum_{k=1}^{\infty} \mathbf{y}_k$ converges. However,

$$\sum_{k=1}^{\ell} \mathbf{y}_k = \sum_{k=1}^{\ell} (\mathbf{x}_{n_{k+1}} - \mathbf{x}_{n_k}) = \mathbf{x}_{n_{\ell+1}} - \mathbf{x}_{n_1} \quad \text{for all } \ell \geqslant 1.$$

Therefore, the subsequence $(\mathbf{x}_{n_k})_{k=1}^{\infty}$ converges. ∎

1.9 Notes

In 1906, Hilbert examined the unit ball $\{\mathbf{a} \in \ell^2 : \|\mathbf{a}\| \leqslant 1\}$ of ℓ^2 as part of his investigation of quadratic forms in infinitely many variables $\xi_1, \xi_2, \xi_3, \ldots$ that satisfy $\xi_1^2 + \xi_2^2 + \xi_3^2 + \cdots \leqslant 1$ [196] (see also [197]). It was F. Riesz [303] who referred to the space of square-summable sequences as *l'espace hilbertien*, while the symbol ℓ_2 (some authors prefer subscripts and others superscripts to describe these spaces) first appeared in Banach's famous book [33].

In 1907, Riesz [298] also established the following important relationship between ℓ^2 and $L^2[0, 1]$. Let $(\varphi_n)_{n=1}^{\infty}$ be an orthonormal sequence in $L^2[0, 1]$. If $(a_n)_{n=1}^{\infty}$ is a sequence of real numbers, then there is an $f \in L^2[0, 1]$ such that

$$a_n = \int_0^1 f(x)\varphi_n(x)\,dx \quad \text{for all } n \geqslant 1$$

if and only if $\sum_{n=1}^{\infty} |a_n|^2$ is finite. Also in 1907, Fischer [130] proved the completeness of $L^2[0, 1]$ (Proposition 1.3.4). In 1908, Schmidt [332] explored the geometric properties of ℓ^2, including orthogonality. It is important to emphasize that although this chapter presented ℓ^2 first (for pedagogical purposes), $L^2[0, 1]$ was actually studied first. The axiomatic version of what is now known as a Hilbert space came from von Neumann in 1930 [369].

The Cauchy–Schwarz inequality plays an important role in Hilbert-space theory and was first developed for individual cases. For Euclidean space $\mathbb{C}^n$, the result goes back to Cauchy in 1821, while the L^2 version was discovered by Buniakowsky in 1859 [73] and Schwarz in 1885 [335]. The ℓ^2 version is due to Schmidt in 1908 [332]. Finally, von Neumann [369] proved the Cauchy–Schwarz inequality for general Hilbert spaces in 1930.

Schmidt [332] contributed to the concepts of orthogonal projections and orthonormalization. He also extended Bessel's inequality from an earlier trigonometric version that appeared in an 1828 paper of Bessel [49].

Pietsch's book [270] is a thorough source for the history of functional analysis.

1.10 Exercises

Exercise 1.10.1. If $\mathbf{x}, \mathbf{y}$ belong to an inner product space $\mathcal{V}$, prove that

$$\|\mathbf{x} + \mathbf{y}\|^2 = \|\mathbf{x}\|^2 + 2\operatorname{Re}\langle \mathbf{x}, \mathbf{y} \rangle + \|\mathbf{y}\|^2.$$

Exercise 1.10.2. If $\mathbf{x}, \mathbf{y}$ belong to an inner product space $\mathcal{V}$, prove that $\mathbf{x} \perp \mathbf{y}$ if and only if $\|\mathbf{x} + w\mathbf{y}\| \geqslant \|\mathbf{x}\|$ for all $w \in \mathbb{C}$.
Remark: This criterion is used to develop a notion of "orthogonality" (*Birkhoff–James orthogonality*) in a Banach space [82].

Exercise 1.10.3. For $\mathbf{a}, \mathbf{b} \in \mathbb{C}^n$, prove that $|\langle \mathbf{a}, \mathbf{b} \rangle| = \|\mathbf{a}\| \|\mathbf{b}\|$ if and only if $\mathbf{a}$ and $\mathbf{b}$ are linearly dependent. Only use the proof of Proposition 1.1.1.

Exercise 1.10.4. For $\mathbf{a}, \mathbf{b} \in \mathbb{C}^n$, prove that $\|\mathbf{a} + \mathbf{b}\| = \|\mathbf{a}\| + \|\mathbf{b}\|$ if and only if $\mathbf{a}$ or $\mathbf{b}$ is a nonnegative multiple of the other. Only use the proof of Proposition 1.1.4.

Exercise 1.10.5. For $f, g \in L^2[0, 1]$, prove that

$$\left| \int_0^1 f(x)g(x)\,dx \right| = \left(\int_0^1 |f(x)|^2\,dx \right)^{\frac{1}{2}} \left(\int_0^1 |g(x)|^2\,dx \right)^{\frac{1}{2}}$$

if and only if f and g are linearly dependent. Only use the proof of Proposition 1.3.1.

Exercise 1.10.6. For $f, g \in L^2[0, 1]$, prove that $\|f + g\| = \|f\| + \|g\|$ if and only if f or g is a nonnegative multiple of the other. Only use the proof of Proposition 1.3.3.

Exercise 1.10.7. Prove that $\mathbb{C}^n$ is complete.

Exercise 1.10.8. If $a_i, b_i > 0$ for all $1 \leqslant i \leqslant n$, prove that

$$\frac{a_1^2}{b_1} + \frac{a_2^2}{b_2} + \cdots + \frac{a_n^2}{b_n} \geqslant \frac{(a_1 + a_2 + \cdots + a_n)^2}{b_1 + b_2 + \cdots + b_n}.$$

Exercise 1.10.9. For $f \in C[0, 1]$, prove that

$$\left| \int_0^1 t f(t)\,dt \right| \leqslant \frac{1}{\sqrt{2}} \left(\int_0^1 |f(t)|^2\,dt \right)^{\frac{1}{2}}.$$

For what f does equality hold?

Exercise 1.10.10. Let $\mathcal{M} \subseteq \mathbb{C}^n$ be a k-dimensional subspace with orthonormal basis $(\mathbf{v}_i)_{i=1}^k$.

(a) For any $c_1, c_2, \dots, c_k \in \mathbb{C}$ and $\mathbf{x} \in \mathbb{C}^n$, prove that

$$\left\| \mathbf{x} - \sum_{i=1}^k c_i \mathbf{v}_i \right\|^2 = \|\mathbf{x}\|^2 - \sum_{i=1}^k |\langle \mathbf{x}, \mathbf{v}_i \rangle|^2 + \sum_{i=1}^k |\langle \mathbf{x}, \mathbf{v}_i \rangle - c_i|^2.$$

(b) Prove that

$$\left\| \mathbf{x} - \sum_{i=1}^{k} \langle \mathbf{x}, \mathbf{v}_i \rangle \, \mathbf{v}_i \right\|^2 \leqslant \left\| \mathbf{x} - \sum_{i=1}^{k} c_i \mathbf{v}_i \right\|^2 \quad \text{for all } c_1, c_2, ..., c_k \in \mathbb{C}.$$

(c) Prove that $\sum_{i=1}^{k} \langle \mathbf{x}, \mathbf{v}_i \rangle \mathbf{v}_i$ is the closest point in $\mathcal{M}$ to $\mathbf{x}$.

Exercise 1.10.11. This problem concerns the difference in the Pythagorean theorem for the settings $\mathbb{R}^n$ and $\mathbb{C}^n$.

(a) For $\mathbf{a}, \mathbf{b} \in \mathbb{R}^n$, prove that $\|\mathbf{a} + \mathbf{b}\|^2 = \|\mathbf{a}\|^2 + \|\mathbf{b}\|^2$ if and only if $\mathbf{a} \perp \mathbf{b}$.

(b) Proposition 1.1.6 shows that if $\mathbf{a}, \mathbf{b} \in \mathbb{C}^n$ and $\mathbf{a} \perp \mathbf{b}$, then $\|\mathbf{a} + \mathbf{b}\|^2 = \|\mathbf{a}\|^2 + \|\mathbf{b}\|^2$. In $\mathbb{C}^n$, prove that the converse is not true.

Exercise 1.10.12. Let

$$\mathcal{M} = \left\{ \mathbf{x} \in \ell^2(\mathbb{N}) : \sum_{n=1}^{\infty} \frac{x_n}{n} = 0 \right\} \quad \text{and} \quad \mathcal{N} = \left\{ \mathbf{x} \in \ell^2(\mathbb{N}) : \sum_{n=1}^{\infty} \frac{x_n}{\sqrt{n}} = 0 \right\}.$$

(a) Which of $\mathcal{M}$ and $\mathcal{N}$ is closed in ℓ^2?

(b) Which of $\mathcal{M}$ and $\mathcal{N}$ is dense in ℓ^2?

Exercise 1.10.13.

(a) Prove that $C[0,1]$ is an inner product space when endowed with the inner product

$$\langle f, g \rangle = \int_0^1 f(x)\overline{g(x)} \, dx.$$

(b) Prove that $C[0,1]$ is not complete with respect to the norm $\|f\| = \sqrt{\langle f, f \rangle}$ and is therefore not a Hilbert space.

Exercise 1.10.14. Let $\mathcal{M}_n$ denote the set of $n \times n$ complex matrices. The *trace* of $A = [a_{ij}] \in \mathcal{M}_n$ is $\text{tr}(A) := a_{11} + a_{22} + \cdots + a_{nn}$. Prove that $\langle A, B \rangle = \text{tr}(B^*A)$ defines an inner product on $\mathcal{M}_n$.

Exercise 1.10.15. Let W be the set of absolutely continuous functions f on $[0,1]$ such that $f' \in L^2[0,1]$. Define an inner product on W by

$$\langle f, g \rangle = \int_0^1 \left(f(x)\overline{g(x)} + f'(x)\overline{g'(x)} \right) dx.$$

Prove that W is a Hilbert space.
Remark: A reader needing a review of the notion of absolute continuity and the Lebesgue differentiation theorem should consult [317, Ch. 5]. The space W is a special example of a class of spaces called *Sobolev spaces* which are important in differential equations. This problem continues in Exercise 10.7.34.

Exercise 1.10.16. Let V be an inner product space.

(a) Prove that

$$\|x + y\|^2 + \|x - y\|^2 = 2(\|x\|^2 + \|y\|^2) \quad \text{for all } x, y \in V.$$

This is the *parallelogram identity*.

(b) Interpret this result geometrically.

Exercise 1.10.17. For an inner product space V prove that

$$\langle x, y \rangle = \frac{1}{4}(\|x + y\|^2 - \|x - y\|^2 - i\|x - iy\|^2 + i\|x + iy\|^2) \quad \text{for all } x, y \in V.$$

This is the *polarization identity*.

Exercise 1.10.18. Let V be an inner product space and, for $n \geqslant 3$, let $\omega = e^{2\pi ik/n}$, where $\gcd(k, n) = 1$. Prove the following generalized polarization identity:

$$\langle x, y \rangle = \frac{1}{n} \sum_{j=1}^{n} \omega^j \|x + \omega^j y\|^2 \quad \text{for all } x, y \in V.$$

Exercise 1.10.19. Let V be an inner product space. Prove the following integral version of the polarization identity:

$$\langle x, y \rangle = \int_0^{2\pi} e^{i\theta} \|x + e^{i\theta} y\|^2 \frac{d\theta}{2\pi} \quad \text{for all } x, y \in V.$$

Exercise 1.10.20. Let x, y, z be vectors in an inner product space V.

(a) Prove the *Apollonius identity*:

$$\|x - z\|^2 + \|y - z\|^2 = \frac{1}{2}\|x - y\|^2 + 2\left\|z - \frac{x + y}{2}\right\|^2.$$

(b) Interpret this result geometrically.

Exercise 1.10.21. Let $\mathcal{H}$ be a separable Hilbert space.

(a) Prove that every orthonormal set in $\mathcal{H}$ is linearly independent.

(b) Prove that every orthonormal basis for $\mathcal{H}$ has the same cardinality.

Exercise 1.10.22. Let $(u_n)_{n=1}^{\infty}$ be an orthonormal sequence in a separable Hilbert space $\mathcal{H}$. For $x \in \mathcal{H}$, prove that the following are equivalent.

(a) $\|x\|^2 = \sum_{n=1}^{\infty} |\langle x, u_n \rangle|^2.$

(b) $x = \sum_{n=1}^{\infty} \langle x, u_n \rangle u_n.$

Exercise 1.10.23. Let $(\mathbf{u}_n)_{n=1}^\infty$ be an orthonormal sequence in a separable Hilbert space $\mathcal{H}$. Prove that the following are equivalent.

(a) $(\mathbf{u}_n)_{n=1}^\infty$ is an orthonormal basis.

(b) For all $\mathbf{x} \in \mathcal{H}$, $\|\mathbf{x}\|^2 = \displaystyle\sum_{n=1}^\infty |\langle \mathbf{x}, \mathbf{u}_n \rangle|^2$.

(c) For all $\mathbf{x} \in \mathcal{E}$, where $\mathcal{E}$ is a set whose span is dense in $\mathcal{H}$, $\|\mathbf{x}\|^2 = \displaystyle\sum_{n=1}^\infty |\langle \mathbf{x}, \mathbf{u}_n \rangle|^2$.

Exercise 1.10.24. Let $(\mathbf{u}_n)_{n=1}^\infty$ be an orthonormal sequence in a Hilbert space $\mathcal{H}$.

(a) Prove that $\displaystyle\sum_{n=1}^\infty |\langle \mathbf{x}, \mathbf{u}_n \rangle \langle \mathbf{y}, \mathbf{u}_n \rangle| \leqslant \|\mathbf{x}\|\|\mathbf{y}\|$ for all $\mathbf{x}, \mathbf{y} \in \mathcal{H}$.

(b) Prove that $\displaystyle\sum_{n=1}^\infty \langle \mathbf{x}, \mathbf{u}_n \rangle \langle \mathbf{u}_n, \mathbf{y} \rangle = \langle \mathbf{x}, \mathbf{y} \rangle$ for all $\mathbf{x}, \mathbf{y} \in \mathcal{H}$. if and only if $(\mathbf{u}_n)_{n=1}^\infty$ is an orthonormal basis.

Exercise 1.10.25. Let $(\mathbf{x}_n)_{n=1}^\infty$ and $(\mathbf{y}_n)_{n=1}^\infty$ be sequences in a Hilbert space $\mathcal{H}$. If $\mathbf{x}_n \to \mathbf{x}$ and $\mathbf{y}_n \to \mathbf{y}$, prove that $\langle \mathbf{x}_n, \mathbf{y}_n \rangle \to \langle \mathbf{x}, \mathbf{y} \rangle$.

Exercise 1.10.26. Consider the vector space $\mathbb{C}^n$ endowed with the norm

$$\|\mathbf{x}\|_\infty = \max_{1 \leqslant i \leqslant n} |x_i| \quad \text{for } \mathbf{x} = (x_1, x_2, ..., x_n) \in \mathbb{C}^n.$$

Is this norm derived from an inner product? That is, does there exist an inner product $\langle \cdot, \cdot \rangle$ on $\mathbb{C}^n$ such that $\|\mathbf{x}\|_\infty^2 = \langle \mathbf{x}, \mathbf{x} \rangle$ for all $\mathbf{x} \in \mathbb{C}^n$?

Exercise 1.10.27. Answer Exercise 1.10.26 when $\mathbb{C}^n$ is endowed with the norm $\|\mathbf{x}\|_1 = \sum_{i=1}^n |x_i|$.

Exercise 1.10.28. Is $C[0, 1]$ with the norm $\|f\|_\infty = \max_{x \in [0,1]} |f(x)|$ an inner product space? That is, does there exist an inner product $\langle \cdot, \cdot \rangle$ on $C[0, 1]$ such that $\|f\|_\infty^2 = \langle f, f \rangle$ for all $f \in C[0, 1]$?

Exercise 1.10.29. Answer Exercise 1.10.28 when $C[0, 1]$ is endowed with the norm

$$\|f\|_1 = \int_0^1 |f(x)| dx.$$

Exercise 1.10.30. A subset $\mathcal{E}$ of a Hilbert space $\mathcal{H}$ is *convex* if for any $\mathbf{x}, \mathbf{y} \in \mathcal{E}$, the line segment $\{t\mathbf{x} + (1 - t)\mathbf{y} : 0 \leqslant t \leqslant 1\}$ is contained in $\mathcal{E}$.

(a) Prove that a subspace of $\mathcal{H}$ is convex.

(b) Prove that the closed unit ball $\{\mathbf{x} \in \mathcal{H} : \|\mathbf{x}\| \leqslant 1\}$ is convex.

(c) Prove that the intersection of two convex sets is convex.

Exercise 1.10.31. If $\mathcal{E}$ is a nonempty closed convex subset of a Hilbert space $\mathcal{H}$, use the following steps to prove that $\mathcal{E}$ has a unique element of smallest norm.

(a) Let $\delta = \inf_{\mathbf{x} \in \mathcal{E}} \|\mathbf{x}\|$. For $\mathbf{u}, \mathbf{v} \in \mathcal{E}$, use the parallelogram identity to prove that

$$\|\mathbf{u} - \mathbf{v}\|^2 \leqslant 2(\|\mathbf{u}\|^2 + \|\mathbf{v}\|^2) - 4\delta^2.$$

(b) Suppose that $(\mathbf{v}_n)_{n=1}^{\infty}$ is a sequence in $\mathcal{E}$ such that $\|\mathbf{v}_n\| \to \delta$. Prove that $(\mathbf{v}_n)_{n=1}^{\infty}$ is a Cauchy sequence and thus $\mathbf{v}_n$ tends to a limit $\mathbf{v}$.

(c) Prove that $\mathbf{v}$ is the unique element of $\mathcal{E}$ of smallest norm.

Exercise 1.10.32. If $\mathcal{E}$ is a nonempty closed convex subset of a Hilbert space $\mathcal{H}$, prove that for any $\mathbf{x} \in \mathcal{H}$, there is a unique $\mathbf{y} \in \mathcal{E}$ such that

$$\text{dist}(\mathbf{x}, \mathcal{E}) := \inf_{\mathbf{z} \in \mathcal{E}} \|\mathbf{x} - \mathbf{z}\| = \|\mathbf{x} - \mathbf{y}\|.$$

Exercise 1.10.33. Let $\mathcal{E} \subseteq \ell^2$ consist of finitely supported sequences (that is, only a finite number of terms in the sequence are nonzero) $\mathbf{x} = (x_n)_{n=0}^{\infty} \in \ell^2$ such that $\sum_{n=0}^{\infty} x_n = 1$.

(a) Prove that $\mathcal{E}$ is nonempty and convex.

(b) Prove that $\mathbf{0} \in \mathcal{E}^-$ (the closure of $\mathcal{E}$), but $\mathbf{0} \notin \mathcal{E}$. Consequently, the hypothesis that $\mathcal{E}$ is closed is important in Exercise 1.10.32.

Exercise 1.10.34. If $(\mathbf{v}_n)_{n=1}^{\infty}$ is an orthonormal sequence in a Hilbert space $\mathcal{H}$, prove that the Gram–Schmidt process returns $(\mathbf{v}_n)_{n=1}^{\infty}$.

Exercise 1.10.35. Compute $\displaystyle \inf_{a,b,c \in \mathbb{C}} \int_0^1 |t^3 - a - bt - ct^2|^2 \, dt$.

Exercise 1.10.36. Show that a subspace $\mathcal{M}$ of a separable Hilbert space $\mathcal{H}$ is also separable. Use the following steps.

(a) Let $\{\mathbf{x}_1, \mathbf{x}_2, \dots\}$ be a countable dense set in $\mathcal{H}$ and let $(r_n)_{n=1}^{\infty}$ be an enumeration of the positive rational numbers. If $B(\mathbf{x}, r) = \{\mathbf{y} \in \mathcal{H} : \|\mathbf{y} - \mathbf{x}\| < r\}$, prove that $W = \{(m, n) : B(\mathbf{x}_m, r_n) \cap \mathcal{M} \neq \emptyset\} \neq \emptyset$.

(b) For each $(m, n) \in W$, let $\mathbf{y}_{m,n} \in B(\mathbf{x}_m, r_n) \cap \mathcal{M}$ and let $A = \{\mathbf{y}_{m,n} : (m, n) \in W\}$. Prove that A is a countable and dense in $\mathcal{M}$.

Exercise 1.10.37. The Hilbert spaces routinely considered in this book are separable. However, there are non-separable Hilbert spaces that arise in the study of almost periodic functions [48, 56]. For complex-valued, Lebesgue-measurable functions f and g on $\mathbb{R}$, define

$$\langle f, g \rangle = \lim_{R \to \infty} \frac{1}{2R} \int_{-R}^{R} f(x)\overline{g(x)} \, dx$$

whenever the limit exists. Similarly, define

$$\|f\|^2 = \lim_{R \to \infty} \frac{1}{2R} \int_{-R}^{R} |f(x)|^2 dx$$

whenever the limit exists, and let $\Omega = \{f : \|f\| < \infty\}$.

(a) Prove that each $\sum_{n=1}^{N} c_n e^{i\lambda_n x}$, where $c_n \in \mathbb{C}$ and $\lambda_n \in \mathbb{R}$, belongs to Ω.

(b) Prove that $\|f\| = 0$ is possible without f being zero almost everywhere.

(c) Let $\mathcal{N} = \{f \in \Omega : \|f\| = 0\}$ and define an inner product on the cosets $f + \mathcal{N}$, $g + \mathcal{N}$ of the quotient space $\Omega/\mathcal{N}$ by $\langle f + \mathcal{N}, g + \mathcal{N} \rangle = \langle f, g \rangle$. Prove that $\Omega/\mathcal{N}$ is a Hilbert space.

(d) Compute the inner product $\langle \sin(sx) + \mathcal{N}, \sin(tx) + \mathcal{N} \rangle$ for $s, t \in \mathbb{R}$ and deduce that $\Omega/\mathcal{N}$ is a non-separable Hilbert space.

Exercise 1.10.38. Show that every Hilbert space $\mathcal{H}$ has a total orthonormal set (Definition 1.6.3) using the following steps. The reader should review the terminology and statement of Zorn's lemma from set theory.

(a) Let Ω denote the set of all orthonormal subsets of $\mathcal{H}$ and order the elements of Ω by set inclusion. Prove that $\Omega \neq \emptyset$ and that every chain in Ω has an upper bound.

(b) Zorn's lemma says that Ω has a maximal element $\omega \in \Omega$. Prove that the closed span of the elements of ω is equal to $\mathcal{H}$.

Exercise 1.10.39. This exercise proves a version of Parseval's theorem for total orthonormal sets $\{x_\alpha : \alpha \in \Gamma\}$ in potentially non-separable Hilbert spaces.

(a) For any $x \in \mathcal{H}$, show that $\langle x, x_\alpha \rangle \neq 0$ for at most countably many $\alpha \in \Gamma$.

(b) For each $x \in \mathcal{H}$, let $\Gamma_x = \{\alpha \in \Gamma : \langle x, x_\alpha \rangle \neq 0\}$. Prove that $\|x\|^2 = \sum_{\alpha \in \Gamma_x} |\langle x, x_\alpha \rangle|^2$.

Exercise 1.10.40. Consider the function

$$\psi(t) = \begin{cases} 1 & \text{if } 0 \leqslant t < \frac{1}{2}, \\ -1 & \text{if } \frac{1}{2} \leqslant t < 1, \\ 0 & \text{otherwise.} \end{cases}$$

Prove that

$$\psi_{n,k}(t) = 2^{n/2} \psi(2^n t - k) \quad \text{for } t \in [0, 1], n \geqslant 0, \text{ and } 0 \leqslant k \leqslant 2^n - 1,$$

along with the constant function 1, form an orthonormal basis for $L^2[0, 1]$ (Figure 1.10.1).

Remark: These functions form the *Haar basis* [165] for $L^2[0, 1]$, which is an example of a *wavelet system* [84].

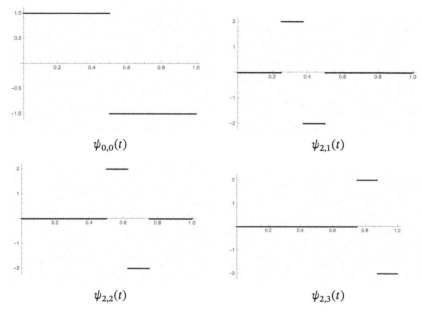

$\psi_{0,0}(t)$

$\psi_{2,1}(t)$

$\psi_{2,2}(t)$

$\psi_{2,3}(t)$

Figure 1.10.1 Graphs of a few of the Haar basis elements.

Exercise 1.10.41. The *Rademacher functions* are defined as follows. Let $r_0 \equiv 1$ and $r_n(t) = \text{sgn}(\sin(2^n \pi t))$ for $n \geqslant 1$, where

$$\text{sgn}(x) = \begin{cases} 1 & \text{if } x > 0, \\ 0 & \text{if } x = 0, \\ -1 & \text{if } x < 0, \end{cases}$$

denotes the *signum function* (see Figure 1.10.2).

(a) Prove that $\displaystyle\int_0^1 |r_n(t)|^2 \, dt = 1$ for all $n \geqslant 0$.

(b) Prove the following fascinating orthogonality property: for any distinct nonnegative integers $n_1, n_2, \ldots, n_k$, where $k \geqslant 2$,

$$\int_0^1 r_{n_1}(t) r_{n_2}(t) \cdots r_{n_k}(t) \, dt = 0.$$

Remark: See [26] and [279] for more on this.

Exercise 1.10.42. Prove that the Rademacher family $(r_n)_{n=0}^\infty$ from Exercise 1.10.41 is not complete in $L^2[0, 1]$, meaning their span is not dense in $L^2[0, 1]$.

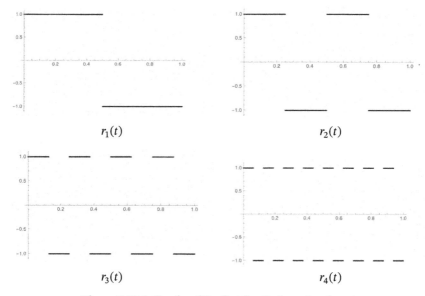

Figure 1.10.2 Graphs of the first few Rademacher functions.

Exercise 1.10.43. Let $\rho(x) = \sqrt{1-x^2}$ and let $L^2(\rho)$ be the set of Lebesgue-measurable functions on $[-1,1]$ such that

$$\|f\|^2 = \int_{-1}^{1} |f(x)|^2 \rho(x)\,dx < \infty.$$

With the norm $\|\cdot\|$ and corresponding inner product, $L^2(\rho)$ is a Hilbert space.

(a) Consider the Taylor series expansion (in the variable t) of

$$\frac{1}{1-2xt+t^2} = \sum_{n=0}^{\infty} u_n(x)t^n$$

and prove that each $u_n(x)$ is a polynomial of degree n. These polynomials are the *Chebyshev polynomials of the second kind.*

(b) Compute u_0, u_1, u_2, u_3, u_4.

(c) Prove that $u_n(\cos\theta) = \dfrac{\sin((n+1)\theta)}{\sin\theta}$ for all $n \geqslant 1$.

(d) Use a trigonometric substitution to show that $\displaystyle\int_{-1}^{1} u_m(x)u_n(x)\rho(x)\,dx = \frac{\pi}{2}\delta_{mn}$.

(e) Prove that $\bigvee\{u_n : n \geqslant 0\}$ contains every polynomial.

(f) Conclude that $(\sqrt{2/\pi}\,u_n)_{n=0}^{\infty}$ is an orthonormal basis for $L^2(\rho)$.

Remark: These polynomials play an important role in Chapter 15.

Exercise 1.10.44. Let $w(x) = \exp(-x^2/2)$ and let $L^2(w)$ be the set of Lebesgue-measurable functions on $\mathbb{R}$ such that

$$\|f\|^2 = \int_{-\infty}^{\infty} |f(x)|^2 w(x)\, dx < \infty.$$

With this norm $\|f\|$ and corresponding inner product, $L^2(w)$ is a Hilbert space.

(a) Consider the Taylor series expansion (in the variable t) of

$$e^{xt - \frac{t^2}{2}} = \sum_{n=0}^{\infty} H_n(x) \frac{t^n}{n!}$$

and prove that each $H_n(x)$ is a polynomial of degree n. These are the *Hermite polynomials*.

(b) Compute H_0, H_1, H_2, H_3, H_4.

(c) Prove that $H_n(x) = (-1)^n e^{\frac{x^2}{2}} \dfrac{d^n}{dx^n} e^{-\frac{x^2}{2}}$.

(d) Use integration by parts to prove that $\displaystyle\int_{-\infty}^{\infty} H_m(x) H_n(x) w(x)\, dx = \sqrt{2\pi} n! \delta_{mn}$.

(e) Prove that $\bigvee\{H_n : n \geqslant 0\}$ contains every polynomial.

(f) Conclude that $(c_n H_n)_{n=0}^{\infty}$, where

$$c_n = \frac{1}{\sqrt[4]{2\pi} \sqrt{n!}},$$

is an orthonormal basis for $L^2(w)$.

Remark: These polynomials play an important role in Chapter 11.

Exercise 1.10.45. Let $(f_n)_{n=1}^{\infty}$ be an orthonormal sequence in $L^2[a, b]$. Prove that $(f_n)_{n=1}^{\infty}$ is an orthonormal basis if and only if

$$\sum_{n=1}^{\infty} \left| \int_a^x f_n(t)\, dt \right|^2 = x - a \quad \text{for all } x \in [a, b].$$

Remark: This is a result of Vitali [366].

Exercise 1.10.46. Let $(f_n)_{n=1}^{\infty}$ be an orthonormal sequence in $L^2[a, b]$. Prove that $(f_n)_{n=1}^{\infty}$ is an orthonormal basis if and only if

$$\sum_{n=1}^{\infty} \int_a^b \left| \int_a^x f_n(t)\, dt \right|^2 dx = \frac{(b-a)^2}{2}.$$

Remark: This is a result from [104].

Exercise 1.10.47. Prove that $\bigvee\{e^{inx} : n \in \mathbb{Z}\} \neq L^2[-a, a]$ if $a > \pi$.

Exercise 1.10.48. Let $\mathcal{H}$ be a Hilbert space. The *Gram determinant* of $x_1, x_2, ..., x_n \in \mathcal{H}$ is

$$G(x_1, x_2, ..., x_n) := \det \begin{bmatrix} \langle x_1, x_1 \rangle & \langle x_1, x_2 \rangle & \cdots & \langle x_1, x_n \rangle \\ \langle x_2, x_1 \rangle & \langle x_2, x_2 \rangle & \cdots & \langle x_2, x_n \rangle \\ \vdots & \vdots & \ddots & \vdots \\ \langle x_n, x_1 \rangle & \langle x_n, x_2 \rangle & \cdots & \langle x_n, x_n \rangle \end{bmatrix}.$$

(a) Prove that $G(x_1, x_2, ..., x_n) \geq 0$.

(b) Prove that $x_1, x_2, ..., x_n$ are linearly dependent if and only if $G(x_1, x_2, ..., x_n) = 0$.

(c) Prove that the Cauchy–Schwarz inequality is equivalent to $G(x_1, x_2) \geq 0$. Therefore,

$$G(x_1, x_2, ..., x_n) \geq 0 \quad \text{for } x_1, x_2, ..., x_n \in \mathcal{H}$$

is a generalization of the Cauchy–Schwarz inequality.

(d) Let $\mathcal{M} = \text{span}\{x_1, x_2, ..., x_n\}$ and let $x \in \mathcal{H}$. Prove that

$$\text{dist}(x, \mathcal{M}) = \inf_{y \in \mathcal{M}} \|x - y\| = \left(\frac{G(x, x_1, x_2, ..., x_n)}{G(x_1, x_2, ..., x_n)} \right)^{\frac{1}{2}}.$$

Exercise 1.10.49. Recall Exercise 1.10.48 for the definition and properties of the Gram determinant. Let $\mathcal{H}$ be a Hilbert space, let $x_1, x_2, ..., x_n \in \mathcal{H}$ be linearly independent, and let $c_1, c_2, ..., c_n \in \mathbb{C}$.

(a) Prove that the *moment problem*: find an $x \in \mathcal{H}$ for with $\langle x, x_i \rangle = c_i$, for all $1 \leq i \leq n$, has a unique solution with minimal norm.

(b) Prove that the solution to (a) is

$$x = -\frac{1}{G(x_1, x_2, ..., x_n)} \det \begin{bmatrix} 0 & c_1 & c_2 & \cdots & c_n \\ x_1 & \langle x_1, x_1 \rangle & \langle x_1, x_2 \rangle & \cdots & \langle x_1, x_n \rangle \\ x_2 & \langle x_2, x_1 \rangle & \langle x_2, x_2 \rangle & \cdots & \langle x_2, x_n \rangle \\ \vdots & \vdots & \vdots & \ddots & \vdots \\ x_n & \langle x_n, x_1 \rangle & \langle x_n, x_2 \rangle & \cdots & \langle x_n, x_n \rangle \end{bmatrix}.$$

Exercise 1.10.50. Let $f \in L^2[0, \frac{\pi}{2}]$. Extend f to $[-\pi, \pi]$ as follows. First extend it to $[0, \pi]$ such that f is even with respect to the line $x = \pi/2$, that is, $f(x) = f(\pi - x)$ for $x \in [0, \pi]$. Then extend it to $[-\pi, \pi]$ such that f is an odd function, that is, $f(x) = -f(-x)$ for $x \in [-\pi, \pi]$.

(a) Expand f with respect to the orthogonal basis $\{1, \cos nx, \sin nx : n \geq 1\}$ on $[-\pi, \pi]$.

(b) Use part (a) to prove that $\{\sin((2n - 1)x) : n \geq 1\}$ is an orthogonal basis in $L^2[0, \pi/2]$.

(c) Prove that $\left\{\sin((n - \frac{1}{2})x) : n \geqslant 1\right\}$ is an orthogonal basis in $L^2[0, \pi]$.

(d) Deduce that $\left\{1, \cos(nx), \sin((n - \frac{1}{2})x) : n \geqslant 1\right\}$ is an orthogonal basis in $L^2[-\pi, \pi]$.

Remark: The system (d) plays a major role in the proof of Kadec's $\frac{1}{4}$-Theorem [208]: if $(\lambda_n)_{n \in \mathbb{Z}} \subseteq \mathbb{R}$ and $\sup_{n \in \mathbb{Z}} |\lambda_n - n| < \frac{1}{4}$, then $(e^{i\lambda_n x})_{n=-\infty}^{\infty}$ is a Riesz basis for $L^2[-\pi, \pi]$.

Exercise 1.10.51. Prove that the *Hilbert cube*

$$\mathcal{E} = \left\{ \mathbf{z} \in \ell^2(\mathbb{N}) : |z_n| \leqslant \frac{1}{n}, n \geqslant 1 \right\}$$

is a compact subset of $\ell^2(\mathbb{N})$.

Exercise 1.10.52. There are several modes of convergence that frequently arise in Hilbert-space theory. We say that $\mathbf{x}_n \to \mathbf{0}$ in *norm* if $\|\mathbf{x}_n\| \to 0$ and *weakly* if $\langle \mathbf{x}_n, \mathbf{y} \rangle \to 0$ for every $\mathbf{y} \in \mathcal{H}$.

(a) Prove that if $\mathbf{x}_n \to \mathbf{0}$ in norm, then $\mathbf{x}_n \to \mathbf{0}$ weakly.

(b) If $(\mathbf{x}_n)_{n=1}^{\infty}$ is an orthonormal sequence in $\mathcal{H}$, prove that $\mathbf{x}_n \to \mathbf{0}$ weakly but not in norm.

(c) Prove that a weakly convergent sequence is bounded.

(d) If $\mathbf{x}_n \to \mathbf{x}$ weakly and $\|\mathbf{x}_n\| \to \|\mathbf{x}\|$, prove that $\|\mathbf{x}_n - \mathbf{x}\| \to 0$.

(e) If $\|\mathbf{x}_n\| \leqslant 1$, $\|\mathbf{y}_n\| \leqslant 1$ for all $n \geqslant 1$, and $\langle \mathbf{x}_n, \mathbf{y}_n \rangle \to 1$, prove that $\|\mathbf{x}_n - \mathbf{y}_n\| \to 0$.

Remark: One can endow $\mathcal{H}$ with a topology that makes it a topological vector space such that a sequence converges with respect to this topology precisely when it converges weakly. See [94] for the details.

Exercise 1.10.53. Let $(\mathbf{x}_n)_{n=1}^{\infty}$ be a sequence in a Hilbert space $\mathcal{H}$. Prove that $(\mathbf{x}_n)_{n=1}^{\infty}$ is weakly convergent if and only if $\sup_{n \geqslant 1} \|\mathbf{x}_n\| < \infty$ and $\lim_{n \to \infty} \langle \mathbf{x}_m, \mathbf{x}_n \rangle$ exists for each $m \geqslant 1$.

1.11 Hints for the Exercises

Hint for Ex. 1.10.7: Simplify the proof of Proposition 1.2.5.

Hint for Ex. 1.10.11: Use Exercise 1.10.1.

Hint for Ex. 1.10.12: Consult Corollary 1.7.4.

Hint for Ex. 1.10.14: Prove that $\operatorname{tr}(B^*A) = \sum_{i,j=1}^{n} a_{ij}\overline{b_{ij}}$.

Hint for Ex. 1.10.15: Recall that if f is absolutely continuous, then f' exists almost everywhere and $f(x) = f(0) + \int_0^x f'(t)dt$.

Hint for Ex. 1.10.19: Expand $\|\mathbf{x} + e^{i\theta}\mathbf{y}\|^2$.

Hint for Ex. 1.10.22: Let $\mathbf{y} = \mathbf{x} - \sum_{n=1}^{\infty} \langle \mathbf{x}, \mathbf{u}_n \rangle \mathbf{u}_n$ and use orthogonality.

Hint for Ex. 1.10.23: Use Exercise 1.10.22.

Hint for Ex. 1.10.24: Use the Cauchy–Schwarz and Bessel inequalities, along with Exercise 1.10.22.

Hint for Ex. 1.10.26: Use Theorem 1.8.9.

Hint for Ex. 1.10.31: Show that $(\mathbf{v}_n)_{n=1}^{\infty}$ is a Cauchy sequence.

Hint for Ex. 1.10.32: Use a translation.

Hint for Ex. 1.10.33: Consider $\mathbf{x}_n = (\frac{1}{n}, \frac{1}{n}, ..., \frac{1}{n}, 0, 0, 0, ...)$.

Hint for Ex. 1.10.35: Orthonormalize $1, t, t^2$ in $L^2[0, 1]$.

Hint for Ex. 1.10.38: If $\mathcal{M}$ is the closed span of ω and $\mathcal{M} \neq \mathcal{H}$, then $\mathcal{M}^{\perp} \neq \{\mathbf{0}\}$. If $\mathbf{x}$ is a nonzero element of $\mathcal{M}^{\perp}$, use $\mathbf{x}$ to contradict the maximality of ω.

Hint for Ex. 1.10.39: For each $n \geqslant 1$, consider the $\alpha \in \Gamma$ such that $|\langle \mathbf{x}, \mathbf{x}_\alpha \rangle| > \frac{1}{n}$. Now use Bessel's inequality.

Hint for Ex. 1.10.42: Consider $f(x) = \cos(2\pi x)$ and Corollary 1.7.4.

Hint for Ex. 1.10.45: Use Exercise 1.10.23 and note that

$$\int_a^x \overline{f_n(t)}\, dt = \int_a^b \chi_{[a,x]}(t)\overline{f_n(t)}\, dt = \langle \chi_{[a,x]}, f_n \rangle.$$

Hint for Ex. 1.10.46: Use Exercise 1.10.45 and Bessel's inequality.

Hint for Ex. 1.10.49: Show that if $\mathbf{x}$ is a solution of minimal norm, then it is unique and belongs to $\text{span}\{\mathbf{x}_1, \mathbf{x}_2, ..., \mathbf{x}_n\}$. Show directly that the vector defined in the exercise satisfies the required conditions.

Hint for Ex. 1.10.51: Suppose $(\mathbf{z}_n)_{n=1}^{\infty}$ is a bounded sequence in the Hilbert cube. Use a diagonalization argument to show $(\mathbf{z}_n)_{n=1}^{\infty}$ has a convergent subsequence.

Hint for Ex. 1.10.52: The proof of (c) requires the principle of uniform boundedness (Theorem 2.2.3).

2

$\cdot$ $\cdot$ $\bullet$ $\cdot$ $\cdot$

Diagonal Operators

Key Concepts: Bounded linear transformation, operator norm, $\mathcal{B}(\mathcal{H})$, closed graph theorem, Hahn–Banach theorem, uniform boundedness principle, kernel and range of an operator, invertible operator, spectrum, point spectrum, approximate point spectrum, compact operator, compact selfadjoint operator, spectral theorem.

Outline: For a bounded sequence of complex numbers $\Lambda = (\lambda_n)_{n=0}^{\infty}$, the linear transformation $D_\Lambda : \ell^2 \to \ell^2$ is defined formally by

$$D_\Lambda\left(\sum_{n=0}^{\infty} a_n \mathbf{e}_n \right) = \sum_{n=0}^{\infty} \lambda_n a_n \mathbf{e}_n,$$

where $(\mathbf{e}_n)_{n=0}^{\infty}$ is the standard orthonormal basis for ℓ^2. The matrix representation of D_Λ with respect to this basis is the infinite diagonal matrix $\mathrm{diag}(\lambda_0, \lambda_1, \lambda_2, ...)$. Consequently, D_Λ is called a *diagonal operator*. This chapter explores the properties of D_Λ (norm, eigenvalues, spectrum, compactness) and extends these concepts to general Hilbert-space operators.

2.1 Diagonal Operators

Diagonal matrices play an important representational role in linear algebra as part of the spectral theorem for normal matrices. When generalizing diagonal matrices to ℓ^2, the matrix $\mathrm{diag}(\lambda_0, \lambda_1, \lambda_2, ...)$ acts formally on $(a_n)_{n=0}^{\infty} \in \ell^2$ by

$$\begin{bmatrix} \lambda_0 & 0 & 0 & 0 & \cdots \\ 0 & \lambda_1 & 0 & 0 & \cdots \\ 0 & 0 & \lambda_2 & 0 & \cdots \\ 0 & 0 & 0 & \lambda_3 & \cdots \\ \vdots & \vdots & \vdots & \vdots & \ddots \end{bmatrix} \begin{bmatrix} a_0 \\ a_1 \\ a_2 \\ a_3 \\ \vdots \end{bmatrix} = \begin{bmatrix} \lambda_0 a_0 \\ \lambda_1 a_1 \\ \lambda_2 a_2 \\ \lambda_3 a_3 \\ \vdots \end{bmatrix}.$$

However, it is not immediately clear that the right side belongs to ℓ^2. For a sequence $\Lambda = (\lambda_n)_{n=0}^{\infty}$ of complex numbers and an $\mathbf{a} = (a_n)_{n=0}^{\infty} \in \ell^2$, define the sequence $D_\Lambda \mathbf{a}$ by

$$D_\Lambda \mathbf{a} = (\lambda_n a_n)_{n=0}^{\infty}.$$

This first proposition characterizes when D_Λ maps ℓ^2 to itself. Recall from the previous chapter that the ℓ^2 norm of a sequence $\mathbf{a} = (a_n)_{n=0}^{\infty}$ is

$$\|\mathbf{a}\| = \Big(\sum_{n=0}^{\infty} |a_n|^2 \Big)^{\frac{1}{2}}.$$

Proposition 2.1.1. $D_\Lambda \mathbf{a} \in \ell^2$ *for all* $\mathbf{a} \in \ell^2$ *if and only if* Λ *is bounded. Moreover,*

$$\sup_{\|\mathbf{a}\|=1} \|D_\Lambda \mathbf{a}\| = \sup_{n \geqslant 0} |\lambda_n|. \tag{2.1.2}$$

Proof ($\implies$): We prove the contrapositive. If Λ is not bounded, then there is a subsequence $(\lambda_{n_k})_{k=1}^{\infty}$ such that $|\lambda_{n_k}| > k$ for all $k \geqslant 1$. The sequence $\mathbf{a} = (a_m)_{m=0}^{\infty}$ defined by

$$a_m = \begin{cases} \dfrac{1}{\lambda_m} & \text{if } m = n_k, \\ 0 & \text{otherwise,} \end{cases}$$

satisfies

$$\sum_{m=0}^{\infty} |a_m|^2 = \sum_{k=1}^{\infty} |a_{n_k}|^2 = \sum_{k=1}^{\infty} \frac{1}{\lambda_{n_k}^2} \leqslant \sum_{k=1}^{\infty} \frac{1}{k^2} < \infty$$

and hence belongs to ℓ^2. However, $(D_\Lambda \mathbf{a})_m$, the mth element of the sequence $D_\Lambda \mathbf{a}$, is

$$(D_\Lambda \mathbf{a})_m = \begin{cases} 1 & \text{if } m = n_k, \\ 0 & \text{otherwise.} \end{cases}$$

Since there are an infinite number of 1s in this sequence, $\|D_\Lambda \mathbf{a}\| = \infty$. Therefore, $D_\Lambda \mathbf{a} \notin \ell^2$.

($\impliedby$): If Λ is bounded, then $D_\Lambda \mathbf{a} \in \ell^2$ for all $\mathbf{a} \in \ell^2$ since

$$\|D_\Lambda \mathbf{a}\|^2 = \sum_{n=0}^{\infty} |\lambda_n a_n|^2 \leqslant \Big(\sup_{n \geqslant 0} |\lambda_n|^2 \Big) \sum_{n=0}^{\infty} |a_n|^2 = \Big(\sup_{n \geqslant 0} |\lambda_n| \Big)^2 \|\mathbf{a}\|^2.$$

It follows that

$$\sup_{\|\mathbf{a}\|=1} \|D_\Lambda \mathbf{a}\| \leqslant \sup_{n \geqslant 0} |\lambda_n|. \tag{2.1.3}$$

It remains to prove (2.1.2). Since $|\lambda_n| = \|D_\Lambda \mathbf{e}_n\|$, it follows that

$$\sup_{n \geqslant 0} |\lambda_n| = \sup_{n \geqslant 0} \|D_\Lambda \mathbf{e}_n\| \leqslant \sup_{\|\mathbf{a}\|=1} \|D_\Lambda \mathbf{a}\|,$$

which verifies one direction of (2.1.2). The reverse inequality is (2.1.3). ∎

If Λ is bounded, the proposition above ensures that

$$D_\Lambda : \ell^2 \to \ell^2, \quad \text{where } D_\Lambda \mathbf{a} = (\lambda_n a_n)_{n=0}^\infty$$

is well defined and satisfies

$$\|D_\Lambda \mathbf{a}\| \leqslant \left(\sup_{n \geqslant 0} |\lambda_n| \right) \|\mathbf{a}\| \quad \text{for all } \mathbf{a} \in \ell^2.$$

Moreover, D_Λ is a linear transformation because

$$D_\Lambda(\mathbf{a} + c\mathbf{b}) = D_\Lambda \mathbf{a} + c D_\Lambda \mathbf{b} \quad \text{for all } \mathbf{a}, \mathbf{b} \in \ell^2 \text{ and } c \in \mathbb{C}.$$

Since D_Λ satisfies

$$\|D_\Lambda \mathbf{a} - D_\Lambda \mathbf{b}\| = \|D_\Lambda(\mathbf{a} - \mathbf{b})\| \leqslant \left(\sup_{n \geqslant 0} |\lambda_n| \right) \|\mathbf{a} - \mathbf{b}\| \quad \text{for all } \mathbf{a}, \mathbf{b} \in \ell^2, \tag{2.1.4}$$

it is a continuous linear transformation from ℓ^2 to itself.

The operator D_Λ is called a *diagonal operator* since the matrix representation of D_Λ with respect to the standard orthonormal basis $(\mathbf{e}_n)_{n=0}^\infty$ for ℓ^2 is the diagonal matrix

$$\text{diag}(\lambda_0, \lambda_1, \lambda_2, \ldots).$$

The previous discussion suggests an important definition.

Definition 2.1.5. Let T be a linear transformation on a Hilbert space $\mathcal{H}$.

(a) T is *bounded* if there is a constant $c > 0$ such that

$$\|T\mathbf{x}\| \leqslant c\|\mathbf{x}\| \quad \text{for all } \mathbf{x} \in \mathcal{H}. \tag{2.1.6}$$

(b) The set of all bounded linear transformations on $\mathcal{H}$ is denoted by $\mathcal{B}(\mathcal{H})$.

(c) The *operator norm* of T is

$$\|T\| := \sup_{\|\mathbf{x}\|=1} \|T\mathbf{x}\|.$$

Example 2.1.7. If $\Lambda = (\lambda_n)_{n=0}^\infty$ is bounded, (2.1.2) says that

$$\|D_\Lambda\| = \sup_{n \geqslant 0} |\lambda_n|.$$

The following lemma shows that $\|T\|$ is the smallest admissible constant c in (2.1.6). Furthermore, $\|T\mathbf{x}\| \leqslant \|T\|\|\mathbf{x}\|$ for all $\mathbf{x} \in \mathcal{H}$.

Lemma 2.1.8. *If $T \in \mathcal{B}(\mathcal{H})$, then*

$$\|T\| = \inf\{c > 0 : \|T\mathbf{x}\| \leqslant c\|\mathbf{x}\| \text{ for all } \mathbf{x} \in \mathcal{H}\}. \tag{2.1.9}$$

Proof Let $\tilde{c}$ denote the infimum in (2.1.9). If $\|T\mathbf{x}\| \leqslant c\|\mathbf{x}\|$ for all $\mathbf{x} \in \mathcal{H}$, then

$$\|T\| = \sup_{\|\mathbf{x}\|=1} \|T\mathbf{x}\| \leqslant c \sup_{\|\mathbf{x}\|=1} \|\mathbf{x}\| = c.$$

Thus, $\|T\| \leqslant \tilde{c}$. Given any $\varepsilon > 0$, there is a unit vector $\mathbf{x}_\varepsilon \in \mathcal{H}$ such that

$$\tilde{c} - \varepsilon < \|T\mathbf{x}_\varepsilon\| \leqslant \sup_{\|\mathbf{x}\|=1} \|T\mathbf{x}\| = \|T\|.$$

Since this holds for every $\varepsilon > 0$, it follows that $\tilde{c} \leqslant \|T\|$. ∎

The inequality in (2.1.4) shows that for the diagonal operator D_Λ, boundedness implies continuity. The converse is true and moreover, this phenomenon occurs with all bounded operators.

Lemma 2.1.10. *For a linear transformation T on a Hilbert space $\mathcal{H}$, the following are equivalent.*

(a) *T is bounded.*

(b) *T is continuous on $\mathcal{H}$.*

(c) *T is continuous at $\mathbf{0}$.*

Proof (a) $\Rightarrow$ (b) If $\mathbf{x}_n \to \mathbf{x}$, then the linearity of T and (2.1.6) ensure that

$$\|T\mathbf{x}_n - T\mathbf{x}\| = \|T(\mathbf{x}_n - \mathbf{x})\| \leqslant \|T\|\|\mathbf{x}_n - \mathbf{x}\| \to 0.$$

Thus, T is continuous.

(b) $\Rightarrow$ (c) If T is continuous on $\mathcal{H}$, then it is continuous at $\mathbf{0}$.

(c) $\Rightarrow$ (a) Suppose that T is continuous at $\mathbf{0}$. Since a linear transformation maps $\mathbf{0}$ to $\mathbf{0}$, there is a $\delta > 0$ such that $\|T\mathbf{x}\| = \|T\mathbf{x} - T\mathbf{0}\| \leqslant 1$ for all $\|\mathbf{x}\| = \|\mathbf{x} - \mathbf{0}\| \leqslant \delta$. For $\mathbf{x} \neq \mathbf{0}$,

$$\|T\mathbf{x}\| = \left\| \frac{\|\mathbf{x}\|}{\delta} T\left(\frac{\delta \mathbf{x}}{\|\mathbf{x}\|}\right) \right\| = \frac{\|\mathbf{x}\|}{\delta} \left\| T\left(\frac{\delta \mathbf{x}}{\|\mathbf{x}\|}\right) \right\| \leqslant \frac{1}{\delta} \|\mathbf{x}\|.$$

Thus, T is bounded. ∎

For $A, B \in \mathcal{B}(\mathcal{H})$ and $c \in \mathbb{C}$, the operations of addition, scalar multiplication, and composition in $\mathcal{B}(\mathcal{H})$ are defined by

$$(A + B)\mathbf{x} = A\mathbf{x} + B\mathbf{x}, \quad (cA)\mathbf{x} = cA\mathbf{x}, \quad \text{and} \quad (AB)\mathbf{x} = A(B\mathbf{x}) \quad \text{for } \mathbf{x} \in \mathcal{H},$$

respectively. There *zero operator* 0 and the *identity operator* I on $\mathcal{H}$ are defined by

$$0\mathbf{x} = \mathbf{0} \quad \text{and} \quad I\mathbf{x} = \mathbf{x} \quad \text{for } \mathbf{x} \in \mathcal{H}, \tag{2.1.11}$$

respectively. Exercise 2.8.12 verifies the following facts which prove that $\mathcal{B}(\mathcal{H})$ is a normed algebra.

Proposition 2.1.12. $\mathcal{B}(\mathcal{H})$ *satisfies the following.*

(a) $\mathcal{B}(\mathcal{H})$ *is a complex vector space with the operations of addition $A + B$ and scalar multiplication cA defined above. Moreover, $\|A + B\| \leq \|A\| + \|B\|$ and $\|cA\| = |c|\|A\|$.*

(b) *If $A, B \in \mathcal{B}(\mathcal{H})$, then $AB \in \mathcal{B}(\mathcal{H})$ and $\|AB\| \leq \|A\|\|B\|$.*

(c) $\|A\| = 0$ *if and only if $A = 0$.*

For an integer $n \geq 0$, define

$$A^0 = I \quad \text{and} \quad A^n = AAA \cdots A \quad (n \text{ times}).$$

The definition of the norm implies that $\|I\| = 1$ and the previous proposition ensures that

$$\|A^n\| \leq \|A\|^n \quad \text{for all } n \geq 0. \tag{2.1.13}$$

The previous proposition also confirms that $\mathcal{B}(\mathcal{H})$ is a normed vector space. The next proposition shows that it is a Banach space (see Definition 1.8.8).

Proposition 2.1.14. $\mathcal{B}(\mathcal{H})$ *is a Banach space.*

Proof Proposition 2.1.12 shows that $\mathcal{B}(\mathcal{H})$ is a vector space with norm $\|\cdot\|$. It suffices to show that $\mathcal{B}(\mathcal{H})$ is complete. Let $(A_n)_{n=1}^{\infty}$ be a Cauchy sequence in $\mathcal{B}(\mathcal{H})$. Then for each $\mathbf{x} \in \mathcal{H}$,

$$\|A_n\mathbf{x} - A_m\mathbf{x}\| \leq \|A_n - A_m\|\|\mathbf{x}\| \to 0.$$

Therefore, $(A_n\mathbf{x})_{n=1}^{\infty}$ is a Cauchy sequence in $\mathcal{H}$. Since $\mathcal{H}$ is complete, $A_n\mathbf{x}$ converges to some $A\mathbf{x} \in \mathcal{H}$. Then for $\mathbf{x}, \mathbf{y} \in \mathcal{H}$ and $c \in \mathbb{C}$,

$$
\begin{aligned}
A(\mathbf{x} + c\mathbf{y}) &= \lim_{n \to \infty} A_n(\mathbf{x} + c\mathbf{y}) \\
&= \lim_{n \to \infty} (A_n\mathbf{x} + cA_n\mathbf{y}) \\
&= \lim_{n \to \infty} A_n\mathbf{x} + c \lim_{n \to \infty} A_n\mathbf{y} \\
&= A\mathbf{x} + cA\mathbf{y}.
\end{aligned}
$$

Thus, $\mathbf{x} \mapsto A\mathbf{x}$ is a linear transformation on $\mathcal{H}$, which we denote by A.

It remains to show that A is bounded and that $\|A_n - A\| \to 0$. Since $(A_n)_{n=1}^{\infty}$ is Cauchy, it is a bounded sequence in $\mathcal{B}(\mathcal{H})$. Therefore,

$$M = \sup_{n \geq 1} \|A_n\| < \infty.$$

Thus, for any $n \geq 1$ and $\mathbf{x} \in \mathcal{H}$,

$$
\begin{aligned}
\|A\mathbf{x}\| &= \|A\mathbf{x} - A_n\mathbf{x} + A_n\mathbf{x}\| \\
&\leq \|A\mathbf{x} - A_n\mathbf{x}\| + \|A_n\mathbf{x}\|
\end{aligned}
$$

$$\leqslant \|A\mathbf{x} - A_n\mathbf{x}\| + \|A_n\|\|\mathbf{x}\|$$
$$\leqslant \|A\mathbf{x} - A_n\mathbf{x}\| + M\|\mathbf{x}\|.$$

Since $\|A\mathbf{x} - A_n\mathbf{x}\| \to 0$, it follows that $\|A\mathbf{x}\| \leqslant M\|\mathbf{x}\|$ for all $\mathbf{x} \in \mathcal{H}$. Therefore, A is bounded and $\|A\| \leqslant M$.

Given $\varepsilon > 0$, choose N such that $\|A_m - A_n\| \leqslant \varepsilon/2$ for all $m, n \geqslant N$. Since $\|A_n\mathbf{x} - A\mathbf{x}\| \to 0$ for every $\mathbf{x} \in \mathcal{H}$, there is an $m(\mathbf{x}) \geqslant N$ such that $\|A_{m(\mathbf{x})}\mathbf{x} - A\mathbf{x}\| \leqslant \frac{\varepsilon}{2}\|\mathbf{x}\|$. Then for any $n \geqslant N$ and $\mathbf{x} \in \mathcal{H}$,

$$\|A_n\mathbf{x} - A\mathbf{x}\| = \|A_n\mathbf{x} - A_{m(\mathbf{x})}\mathbf{x} + A_{m(\mathbf{x})}\mathbf{x} - A\mathbf{x}\|$$
$$\leqslant \|A_n\mathbf{x} - A_{m(\mathbf{x})}\mathbf{x}\| + \|A_{m(\mathbf{x})}\mathbf{x} - A\mathbf{x}\|$$
$$\leqslant \|A_n - A_{m(\mathbf{x})}\|\|\mathbf{x}\| + \frac{\varepsilon}{2}\|\mathbf{x}\|$$
$$\leqslant \frac{\varepsilon}{2}\|\mathbf{x}\| + \frac{\varepsilon}{2}\|\mathbf{x}\| = \varepsilon\|\mathbf{x}\|.$$

Thus,

$$\|A_n - A\| = \sup_{\|\mathbf{x}\|=1} \|A_n\mathbf{x} - A\mathbf{x}\| \leqslant \varepsilon \quad \text{for all } n \geqslant N$$

and hence $A_n \to A$ in the operator norm. ∎

Another important detail is the continuity of multiplication.

Proposition 2.1.15. *If $A_n \to A \in \mathcal{B}(\mathcal{H})$ and $B_n \to B \in \mathcal{B}(\mathcal{H})$, then $A_nB_n \to AB$.*

Proof Since $B_n \to B$, there is an $M > 0$ such that $\|B_n\| \leqslant M$ for all n. Then,

$$\|A_nB_n - AB\| \leqslant \|A_nB_n - AB_n + AB_n - AB\|$$
$$= \|(A_n - A)B_n + A(B_n - B)\|$$
$$\leqslant \|A_n - A\|\|B_n\| + \|A\|\|B_n - B\| \quad \text{(by Prop. 2.1.12)}$$
$$\leqslant M\|A_n - A\| + \|A\|\|B_n - B\| \to 0.$$

Thus, $A_nB_n \to AB$. ∎

One can consider operators between different Hilbert spaces in the same manner. Let $\mathcal{H}$ and $\mathcal{K}$ be Hilbert spaces. A linear transformation $T : \mathcal{H} \to \mathcal{K}$ is *bounded* if

$$\|T\| = \sup_{\|\mathbf{x}\|_{\mathcal{H}}=1} \|T\mathbf{x}\|_{\mathcal{K}} \tag{2.1.16}$$

is finite. The set of bounded linear operators from $\mathcal{H}$ to $\mathcal{K}$ is denoted by $\mathcal{B}(\mathcal{H}, \mathcal{K})$. If $\mathcal{H} = \mathcal{K}$, this is just $\mathcal{B}(\mathcal{H})$. Although $\mathcal{B}(\mathcal{H}, \mathcal{K})$ is not an algebra (unless $\mathcal{H} = \mathcal{K}$), it is a Banach space with respect to the *operator norm* defined in (2.1.16).

2.2 Banach-Space Interlude

This section covers several important Banach-space theorems that appear throughout this book. A good reference for these is [94]. The notion of boundedness of a linear transformation makes sense in any normed vector space. Indeed, the definitions of the previous section make no use of the inner-product structure on a Hilbert space, only the norm. Consequently, we may speak of bounded linear operators between Banach spaces (throughout this book, a "bounded operator" is understood to be linear). The set of bounded operators on a Banach space V is denoted by $\mathcal{B}(V)$. The set of bounded operators from a Banach space V to a Banach space W is denoted by $\mathcal{B}(V, W)$.

Theorem 2.2.1 (Open mapping theorem). *Let V be a Banach space and $T \in \mathcal{B}(V)$ be surjective. Then $T(\mathcal{U})$ is open whenever $\mathcal{U} \subseteq V$ is open.*

It is often cumbersome to show that a linear transformation is bounded via Definition 2.1.5. The next result provides another method.

Theorem 2.2.2 (Closed graph theorem). *Let T be a linear transformation from a Banach space V to a Banach space W. Then the following are equivalent.*

(a) *$T \in \mathcal{B}(V, W)$.*

(b) *If $\mathbf{x}_n \to \mathbf{x}$ in V and $T\mathbf{x}_n \to \mathbf{y}$ in W, then $\mathbf{y} = T\mathbf{x}$.*

The theorem above is called the closed graph theorem since it says that T is bounded if and only if its graph $\{(\mathbf{x}, T\mathbf{x}) : \mathbf{x} \in V\}$ is closed in the Banach space $V \oplus W = \{(\mathbf{v}, \mathbf{w}) : \mathbf{v} \in V, \mathbf{w} \in W\}$ with norm $\|(\mathbf{u}, \mathbf{v})\|_{V \oplus W} = \|\mathbf{u}\|_V + \|\mathbf{v}\|_W$.

Theorem 2.2.3 (Principle of uniform boundedness). *Let V, W be Banach spaces and $\mathscr{F} \subseteq \mathcal{B}(V, W)$. If $\sup\{\|T\mathbf{x}\|_W : T \in \mathscr{F}\} < \infty$ for each $\mathbf{x} \in V$, then $\sup\{\|T\| : T \in \mathscr{F}\} < \infty$.*

The previous theorem says something remarkable: a pointwise bounded set of linear operators is uniformly bounded.

Definition 2.2.4. The *dual space* of a Banach space V is $V^* := \mathcal{B}(V, \mathbb{C})$. An element $\varphi \in V^*$ is a *bounded linear functional* and

$$\|\varphi\| = \sup_{\|\mathbf{x}\|=1} |\varphi(\mathbf{x})|.$$

In the course of our discussion of the adjoint of a Hilbert-space operator in Chapter 3, Theorem 3.1.3 characterizes the dual space of a Hilbert space.

Two versions of the Hahn–Banach theorem play important roles in functional analysis. The first says that a linear functional on a subspace of a Banach space can be extended to the whole space without increasing its norm. As with Hilbert spaces in Definition 1.7.1, a *subspace* of a Banach space V is a nonempty, norm-closed, vector subspace of V.

Theorem 2.2.5 (Hahn–Banach extension theorem). *Let $\mathcal{M}$ be a subspace of a Banach space V and let $\varphi \in \mathcal{M}^*$. Then there is a $\psi \in V^*$ such that $\psi|_{\mathcal{M}} = \varphi$ and $\|\psi\| = \|\varphi\|$.*

The second version of the Hahn–Banach theorem is often used to determine if a subset of a Banach space has a dense linear span.

Theorem 2.2.6 (Hahn–Banach separation theorem). *Let $\mathcal{M}$ be a subspace of a Banach space V. If $\mathbf{x} \in V \backslash \mathcal{M}$, then there exists a $\varphi \in V^*$ such that $\varphi|_{\mathcal{M}} = 0$ and $\varphi(\mathbf{x}) = 1$.*

2.3 Inverse of an Operator

This section covers the invertibility properties of Hilbert-space operators.

Definition 2.3.1. For $A \in \mathcal{B}(\mathcal{H})$, the *kernel* of A is $\ker A := \{\mathbf{x} \in \mathcal{H} : A\mathbf{x} = \mathbf{0}\}$ and the *range* of A is $\operatorname{ran} A := \{A\mathbf{x} : \mathbf{x} \in \mathcal{H}\}$.

Observe that $\ker A$ is a subspace of $\mathcal{H}$, in particular, it is (topologically) closed (Exercise 2.8.2). Although $\operatorname{ran} A$ is a vector space, it may not be closed (Exercise 2.8.3).

Recall that $\mathcal{M}_n$ denotes the set of $n \times n$ matrices with complex entries. The symbol I denotes the $n \times n$ identity matrix. See [141] for a review of linear algebra.

Proposition 2.3.2. *For $A \in \mathcal{M}_n$, the following are equivalent.*

(a) *There exists a $B \in \mathcal{M}_n$ such that $BA = I$.*

(b) *There exists a $B \in \mathcal{M}_n$ such that $AB = I$.*

(c) $\ker A = \{\mathbf{0}\}$.

(d) $\operatorname{ran} A = \mathbb{C}^n$.

A matrix A that satisfies any of the equivalent conditions above is *invertible* and its *inverse*, denoted by A^{-1}, is the unique $n \times n$ matrix such that $AA^{-1} = A^{-1}A = I$. For bounded operators on infinite-dimensional Hilbert spaces, invertibility is more subtle.

Example 2.3.3. Consider the forward and backward shift operators S, T on ℓ^2 defined by

$$S(a_0, a_1, a_2, a_3, \ldots) = (0, a_0, a_1, a_2, \ldots) \quad \text{and} \quad T(a_0, a_1, a_2, a_3, \ldots) = (a_1, a_2, a_3, a_4, \ldots),$$

respectively. These operators are discussed in great detail in Chapter 5 where it is shown that

$$TS = I \quad \text{but} \quad ST \neq I.$$

Furthermore,

$$\ker S = \{\mathbf{0}\} \quad \text{but} \quad \operatorname{ran} S \neq \ell^2,$$

while

$$\ker T = \operatorname{span}\{\mathbf{e}_0\} \neq \{\mathbf{0}\} \quad \text{but} \quad \operatorname{ran} T = \ell^2.$$

Example 2.3.3 shows that left invertibility and right invertibility are not equivalent. Therefore, the definition of invertibility for Hilbert-space operators insists upon both conditions. Below, we let I denote the identity operator on a Hilbert space $\mathcal{H}$ from (2.1.11).

Definition 2.3.4. $A \in \mathcal{B}(\mathcal{H})$ is *invertible* if there is a $B \in \mathcal{B}(\mathcal{H})$ such that $AB = BA = I$. The operator B is the *inverse* of A and is denoted by A^{-1}.

Exercise 2.8.4 shows that if an inverse of $A \in \mathcal{B}(\mathcal{H})$ exists, it is unique. Therefore, one speaks of "the" inverse of an operator. The next result provides a condition for invertibility that is often easier to check than the definition itself, where one is required to explicitly produce the inverse.

Lemma 2.3.5. *For $A \in \mathcal{B}(\mathcal{H})$, the following are equivalent.*

(a) *A is invertible.*

(b) $\ker A = \{\mathbf{0}\}$ *and* $\operatorname{ran} A = \mathcal{H}$.

(c) $\operatorname{ran} A$ *is dense in $\mathcal{H}$ and* $\inf\limits_{\|\mathbf{x}\|=1} \|A\mathbf{x}\| > 0$.

Proof (a) $\Rightarrow$ (b) Suppose that $A \in \mathcal{B}(\mathcal{H})$ is invertible. Then $\operatorname{ran} A = \mathcal{H}$ since

$$A(A^{-1}\mathbf{y}) = (AA^{-1})\mathbf{y} = \mathbf{y} \quad \text{for all } \mathbf{y} \in \mathcal{H}.$$

If $\mathbf{x} \in \ker A$, then $A\mathbf{x} = \mathbf{0}$ from which it follows that

$$\mathbf{0} = A^{-1}\mathbf{0} = A^{-1}(A\mathbf{x}) = (A^{-1}A)\mathbf{x} = I\mathbf{x} = \mathbf{x}.$$

Therefore, $\ker A = \{\mathbf{0}\}$.

(b) $\Rightarrow$ (a) Since $\ker A = \{\mathbf{0}\}$ and $\operatorname{ran} A = \mathcal{H}$, there is a linear transformation A^{-1} on $\mathcal{H}$ such that $AA^{-1} = A^{-1}A = I$. The next step is to show that $A^{-1} \in \mathcal{B}(\mathcal{H})$. Since A is surjective, the open mapping theorem (Theorem 2.2.1) implies that $A(\mathcal{U})$ is open whenever $\mathcal{U} \subseteq \mathcal{H}$ is open. Thus, the topological characterization of continuity ensures that A^{-1} is a continuous linear transformation. By Lemma 2.1.10, continuity and boundedness are equivalent for linear transformations on Hilbert spaces. Therefore, $A^{-1} \in \mathcal{B}(\mathcal{H})$.

(a) $\Rightarrow$ (c) Suppose that $A \in \mathcal{B}(\mathcal{H})$ is invertible. Then $\operatorname{ran} A = \mathcal{H}$, which is dense in $\mathcal{H}$. If

$$\inf_{\|\mathbf{x}\|=1} \|A\mathbf{x}\| = 0,$$

then there is a sequence $(\mathbf{x}_n)_{n=1}^{\infty}$ of unit vectors such that $\|A\mathbf{x}_n\| \to 0$. Then

$$1 = \|\mathbf{x}_n\| = \|A^{-1}A\mathbf{x}_n\| \leqslant \|A^{-1}\|\|A\mathbf{x}_n\| \to 0,$$

which is a contradiction.

(c) $\Rightarrow$ (b) The condition

$$\delta = \inf_{\|\mathbf{x}\|=1} \|A\mathbf{x}\| > 0$$

implies that $\ker A = \{\mathbf{0}\}$. If $(A\mathbf{x}_n)_{n=1}^{\infty}$ is a Cauchy sequence, then

$$\|A\mathbf{x}_m - A\mathbf{x}_n\| = \|A(\mathbf{x}_m - \mathbf{x}_n)\| \geq \delta\|\mathbf{x}_m - \mathbf{x}_n\| \quad \text{for all } m, n \geq 1.$$

Thus, $(\mathbf{x}_n)_{n=1}^{\infty}$ is a Cauchy sequence in $\mathcal{H}$ and hence converges to some $\mathbf{x} \in \mathcal{H}$. Since A is bounded, it is continuous (Lemma 2.1.10) and hence $A\mathbf{x}_n \to A\mathbf{x}$. Therefore, $\operatorname{ran} A$ is closed. By hypothesis, $\operatorname{ran} A$ is dense in $\mathcal{H}$ and thus $\operatorname{ran} A = \mathcal{H}$. ∎

A proof of the following corollary is requested in Exercise 2.8.5.

Corollary 2.3.6. *Let $A, B \in \mathcal{B}(\mathcal{H})$.*

(a) *If A and B are invertible, then so is AB and $(AB)^{-1} = B^{-1}A^{-1}$.*

(b) *If A is invertible, then so is A^{-1} and $(A^{-1})^{-1} = A$.*

An $A \in \mathcal{B}(\mathcal{H})$ such that

$$\inf_{\|\mathbf{x}\|=1} \|A\mathbf{x}\| > 0 \tag{2.3.7}$$

is *bounded below*. Thus, Lemma 2.3.5 says that $A \in \mathcal{B}(\mathcal{H})$ is invertible if and only if it is bounded below and has dense range. See Exercises 2.8.37 and 2.8.38 for characterizations of left and right invertibility.

Corollary 2.3.8. *If $A \in \mathcal{B}(\mathcal{H})$ is invertible, then $\|A^{-1}\| \geq \|A\|^{-1}$.*

Proof Apply Proposition 2.1.12 to $AA^{-1} = I$ and deduce that

$$1 = \|I\| = \|AA^{-1}\| \leq \|A\|\|A^{-1}\|,$$

which completes the proof. ∎

Proposition 2.3.9. *If $A \in \mathcal{B}(\mathcal{H})$ and $\|I - A\| < 1$, then A is invertible and*

$$\|A^{-1}\| \leq \frac{1}{1 - \|I - A\|}. \tag{2.3.10}$$

Proof Since $\|I - A\| < 1$ and $\|(I - A)^n\| \leq \|I - A\|^n$, by (2.1.13), the series

$$\sum_{n=0}^{\infty} \|(I - A)^n\| \leq \sum_{n=0}^{\infty} \|I - A\|^n$$

converges. Proposition 2.1.14 and Theorem 1.8.11 ensure that $\sum_{n=0}^{\infty}(I - A)^n$ converges in $\mathcal{B}(\mathcal{H})$ to some B. The continuity of multiplication (Proposition 2.1.15) implies that

$$AB = \left(I - (I - A)\right)\left(\lim_{N \to \infty} \sum_{n=0}^{N}(I - A)^n\right)$$

$$= \lim_{N \to \infty}\left[\sum_{n=0}^{N}(I - A)^n - \sum_{n=0}^{N}(I - A)^{n+1}\right]$$

$$= \lim_{N \to \infty} \left[I - (I - A)^{N+1} \right] = I,$$

since $\|(I - A)^{N+1}\| \to 0$ as $N \to \infty$. Similarly, $BA = I$, so A is invertible and $A^{-1} = B$. Moreover,

$$\|A^{-1}\| = \|B\| \leqslant \sum_{n=0}^{\infty} \|I - A\|^n = \frac{1}{1 - \|I - A\|},$$

which completes the proof. ∎

Exercise 2.8.5 shows that the invertible elements of $\mathcal{B}(\mathcal{H})$ form a group.

Proposition 2.3.11. *Let $\mathcal{G}$ denote the group of invertible operators in $\mathcal{B}(\mathcal{H})$. Then $\mathcal{G}$ is an open set in $\mathcal{B}(\mathcal{H})$ and inversion is continuous on $\mathcal{G}$.*

Proof If $A \in \mathcal{G}$ and $\varepsilon > 0$, let

$$\delta = \min \left\{ \frac{\varepsilon}{2 \|A^{-1}\|^2}, \frac{1}{2 \|A^{-1}\|} \right\}.$$

If $B \in \mathcal{B}(\mathcal{H})$ and $\|A - B\| < \delta$, then

$$\|I - A^{-1}B\| = \|A^{-1}(A - B)\| \leqslant \|A^{-1}\| \, \|A - B\| < \|A^{-1}\| \delta \leqslant \frac{1}{2}.$$

Proposition 2.3.9 ensures that $A^{-1}B$ is invertible. Thus, by Corollary 2.3.6, $B = A(A^{-1}B)$ is invertible as well and hence $\mathcal{G}$ is open. Furthermore, (2.3.10) implies that

$$\|(A^{-1}B)^{-1}\| \leqslant \frac{1}{1 - \|I - A^{-1}B\|} < 2,$$

which gives

$$\begin{aligned}
\|B^{-1}\| &= \|B^{-1}AA^{-1}\| \\
&\leqslant \|B^{-1}A\| \|A^{-1}\| \\
&\leqslant \|(A^{-1}B)^{-1}\| \|A^{-1}\| \qquad \text{(by Corollary 2.3.6)} \\
&< 2\|A^{-1}\|.
\end{aligned}$$

Therefore,

$$\begin{aligned}
\|A^{-1} - B^{-1}\| &= \|A^{-1}(A - B)B^{-1}\| \\
&\leqslant \|A^{-1}\| \|A - B\| \|B^{-1}\| \\
&\leqslant 2\|A^{-1}\|^2 \|A - B\| \\
&< 2\|A^{-1}\|^2 \delta \\
&\leqslant \varepsilon.
\end{aligned}$$

Thus, inversion is continuous on $\mathcal{G}$. ∎

2.4 Spectrum of an Operator

A complex number λ is an *eigenvalue* of $A \in \mathcal{M}_n$ if there is a nonzero $\mathbf{x} \in \mathbb{C}^n$ such that $A\mathbf{x} = \lambda\mathbf{x}$, that is, if $\ker(A - \lambda I) \neq \{\mathbf{0}\}$. Proposition 2.3.2 implies the following.

Corollary 2.4.1. *For $A \in \mathcal{M}_n$ and $\lambda \in \mathbb{C}$, the following are equivalent.*

(a) λ *is an eigenvalue of A.*

(b) $A - \lambda I$ *is not invertible.*

Every $n \times n$ matrix has an eigenvalue (Exercise 2.8.6). In fact, it has at most a finite number of them. The situation is different in the infinite-dimensional setting. First let us formally state the definition of an eigenvalue for Hilbert-space operators.

Definition 2.4.2. A complex number λ is an *eigenvalue* of $A \in \mathcal{B}(\mathcal{H})$ if $\ker(A - \lambda I) \neq \{\mathbf{0}\}$. The *multiplicity* of λ as an eigenvalue of A is $\dim \ker(A - \lambda I)$.

Example 2.4.3. For a diagonal operator D_Λ, note that $D_\Lambda \mathbf{e}_n = \lambda_n \mathbf{e}_n$ for all $n \geqslant 0$, and hence every element of Λ is an eigenvalue of D_Λ. In particular, D_Λ may have infinitely many distinct eigenvalues, or eigenvalues of infinite multiplicity.

Example 2.4.4. Recall the forward and backward shifts S and T from Example 2.3.3. It turns out that S has no eigenvalues (Proposition 5.1.4), whereas each point in the open unit disk $\mathbb{D} = \{z \in \mathbb{C} : |z| < 1\}$ is an eigenvalue of T (Proposition 5.2.4).

For $A \in \mathcal{M}_n$, the matrix $A - \lambda I$ is not invertible if and only if λ is an eigenvalue of A. In the infinite-dimensional setting, $A - \lambda I$ may fail to be invertible for several reasons.

Definition 2.4.5. Let $A \in \mathcal{B}(\mathcal{H})$.

(a) The *spectrum* of A, denoted by $\sigma(A)$, is the set of $\lambda \in \mathbb{C}$ such that $A - \lambda I$ does not have an inverse in $\mathcal{B}(\mathcal{H})$.

(b) The *point spectrum* of A, denoted by $\sigma_p(A)$, is the set of $\lambda \in \mathbb{C}$ such that $A - \lambda I$ is not injective.

(c) The *approximate point spectrum* of A, denoted by $\sigma_{ap}(A)$, is the set of $\lambda \in \mathbb{C}$ such that

$$\inf_{\|\mathbf{x}\|=1} \|(A - \lambda I)\mathbf{x}\| = 0.$$

In particular, observe that $\sigma_p(A)$ is the set of eigenvalues of A. The next proposition gives a series of containments for the sets introduced in the previous definition.

Proposition 2.4.6. *For $A \in \mathcal{B}(\mathcal{H})$, we have $\sigma_p(A) \subseteq \sigma_{ap}(A) \subseteq \sigma(A)$.*

Proof If $\lambda \in \sigma_p(A)$, there is a unit vector $\mathbf{x} \in \mathcal{H}$ such that $(A - \lambda I)\mathbf{x} = \mathbf{0}$ and hence $\lambda \in \sigma_{ap}(A)$. Thus, $\sigma_p(A) \subseteq \sigma_{ap}(A)$. Lemma 2.3.5c implies that if $\lambda \in \sigma_{ap}(A)$, then $A - \lambda I$ is not invertible. Thus, $\sigma_{ap}(A) \subseteq \sigma(A)$. ∎

For a diagonal operator D_Λ, the next theorem describes $\sigma_p(D_\Lambda)$, $\sigma_{ap}(D_\Lambda)$, and $\sigma(D_\Lambda)$.

Theorem 2.4.7. *For a bounded sequence $\Lambda = (\lambda_n)_{n=0}^\infty$ in $\mathbb{C}$, the following hold.*

(a) $\sigma_p(D_\Lambda) = \Lambda$.

(b) $\sigma_{ap}(D_\Lambda) = \sigma(D_\Lambda) = \Lambda^-$, *where Λ^- denotes the closure of Λ.*

Proof (a) Since $D_\Lambda \mathbf{e}_n = \lambda_n \mathbf{e}_n$ for all $n \geq 0$, it follows that $\Lambda \subseteq \sigma_p(D_\Lambda)$. On the other hand, if $\lambda \notin \Lambda$ and

$$\mathbf{x} = \sum_{n=0}^\infty a_n \mathbf{e}_n \in \ell^2,$$

then

$$(D_\Lambda - \lambda I)\mathbf{x} = \sum_{n=0}^\infty (\lambda_n - \lambda) a_n \mathbf{e}_n$$

and hence

$$\|(D_\Lambda - \lambda I)\mathbf{x}\|^2 = \sum_{n=0}^\infty |\lambda_n - \lambda|^2 |a_n|^2.$$

This implies that $\|(D_\Lambda - \lambda I)\mathbf{x}\| = 0$ if and only if $a_n = 0$ for all $n \geq 0$. In other words, $\mathbf{x} = \mathbf{0}$. Therefore, $\sigma_p(D_\lambda) = \Lambda$.

(b) If $\lambda \notin \Lambda^-$, there is some $\delta > 0$ such that $|\lambda - \lambda_n| \geq \delta$ for all $n \geq 0$. Thus, the sequence

$$\tilde{\Lambda} = \left(\frac{1}{\lambda_n - \lambda} \right)_{n=0}^\infty$$

is bounded and $(D_\Lambda - \lambda I)^{-1} = D_{\tilde{\Lambda}}$. Therefore, $\sigma_{ap}(D_\Lambda) \subseteq \sigma(D_\Lambda) \subseteq \Lambda^-$ by Proposition 2.4.6. On the other hand, if $\lambda \in \Lambda^-$, (a) provides a subsequence $\lambda_{n_k} \in \Lambda = \sigma_p(D_\Lambda)$ such that $\lambda_{n_k} \to \lambda$. Then

$$\|(D_\Lambda - \lambda I)\mathbf{e}_{n_k}\| = \|(D_\Lambda - \lambda_{n_k} I)\mathbf{e}_{n_k} + (\lambda_{n_k} - \lambda)\mathbf{e}_{n_k}\|$$
$$= \|\mathbf{0} + (\lambda_{n_k} - \lambda)\mathbf{e}_{n_k}\|$$
$$= \|(\lambda_{n_k} - \lambda)\mathbf{e}_{n_k}\|$$
$$= |\lambda_{n_k} - \lambda| \to 0$$

and hence $\lambda \in \sigma_{ap}(D_\Lambda) \subseteq \sigma(D_\Lambda)$. ∎

Example 2.4.8. Computations from Chapter 5 confirm that the operators S and T from Example 2.3.3 satisfy $\sigma(S) = \mathbb{D}^-$, $\sigma_p(S) = \emptyset$, and $\sigma_{ap}(S) = \mathbb{T}$; while $\sigma(T) = \mathbb{D}^-$, $\sigma_p(T) = \mathbb{D}$, and $\sigma_{ap}(T) = \mathbb{D}^-$.

Here are a few important facts about the spectrum of an operator.

Theorem 2.4.9. *Let $A \in \mathcal{B}(\mathcal{H})$.*

(a) $\sigma(A) \subseteq \{z : |z| \leq \|A\|\}$.

(b) $\sigma(A)$ *is compact.*

(c) $\sigma(A) \neq \emptyset$.

Proof (a) If $|z| > \|A\|$, then $zI - A = z(I - z^{-1}A)$. Since $\|z^{-1}A\| \leqslant |z|^{-1}\|A\| < 1$, it follows that $\|I - (I - z^{-1}A)\| = \|z^{-1}A\| < 1$. Proposition 2.3.9 ensures that $I - z^{-1}A$, and hence $A - zI$, is invertible. Thus, $z \notin \sigma(A)$.

(b) The function $f : \mathbb{C} \to \mathcal{B}(\mathcal{H})$ defined by $f(z) = A - zI$ is continuous since

$$\|f(z) - f(w)\| = \|(A - zI) - (A - wI)\| = |z - w| \quad \text{for all } z, w \in \mathbb{C}.$$

Since $\mathcal{G}$, the group of invertible elements of $\mathcal{B}(\mathcal{H})$, is open (Proposition 2.3.11), it follows that $\mathcal{G}^c = \mathcal{B}(\mathcal{H})\backslash\mathcal{G}$ is closed. Thus, the inverse image $f^{-1}(\mathcal{G}^c) = \sigma(A)$ is closed, and hence compact by (a).

(c) Suppose toward a contradiction that $\sigma(A) = \emptyset$. Let $\varphi \in \mathcal{B}(\mathcal{H})^*$ and consider the function $g : \mathbb{C} \to \mathbb{C}$ defined by

$$g(z) = \varphi((A - zI)^{-1}).$$

Fix $z_0 \in \mathbb{C}$ and note that

$$
\begin{aligned}
(A - zI)^{-1} - (A - z_0I)^{-1} &= (A - zI)^{-1}[I - (A - zI)(A - z_0I)^{-1}] \\
&= (A - zI)^{-1}[(A - z_0I) - (A - zI)](A - z_0I)^{-1} \\
&= (A - zI)^{-1}(zI - z_0I)(A - z_0I)^{-1} \\
&= (z - z_0)(A - zI)^{-1}(A - z_0I)^{-1}. \quad (2.4.10)
\end{aligned}
$$

Inversion is continuous on $\mathcal{G}$ (Proposition 2.3.11) and φ is linear and continuous. Thus,

$$
\begin{aligned}
\lim_{z \to z_0} \frac{g(z) - g(z_0)}{z - z_0} &= \lim_{z \to z_0} \frac{\varphi((A - zI)^{-1}) - \varphi((A - z_0I)^{-1})}{z - z_0} \\
&= \varphi\left(\lim_{z \to z_0} \frac{(A - zI)^{-1} - (A - z_0I)^{-1}}{z - z_0} \right) \\
&= \varphi\left(\lim_{z \to z_0} (A - zI)^{-1}(A - z_0I)^{-1} \right) \quad \text{(by (2.4.10))} \\
&= \varphi((A - z_0I)^{-2}),
\end{aligned}
$$

so g is differentiable at z_0. Since $z_0 \in \mathbb{C}$ is arbitrary, g is an entire function. For $|z| > \|A\|$,

$$
\begin{aligned}
|g(z)| &= |\varphi((A - zI)^{-1})| \\
&= |\varphi(z^{-1}(I - z^{-1}A)^{-1})| \\
&= |z|^{-1}|\varphi((I - z^{-1}A)^{-1})| \\
&\leqslant |z|^{-1}\|\varphi\|\|(I - z^{-1}A)^{-1}\|
\end{aligned}
$$

$$\leqslant |z|^{-1}\|\varphi\|\left\|\sum_{n=0}^{\infty} z^{-n}A^n\right\|$$

$$\leqslant |z|^{-1}\|\varphi\|\sum_{n=0}^{\infty} \|z^{-n}A^n\|$$

$$\leqslant |z|^{-1}\|\varphi\|\sum_{n=0}^{\infty} |z|^{-n}\|A\|^n$$

$$\leqslant |z|^{-1}\frac{\|\varphi\|}{1 - \|z^{-1}A\|}$$

$$= \frac{\|\varphi\|}{|z| - \|A\|},$$

which tends to zero as $z \to \infty$. In particular, g is bounded on the region $\{|z| > 2\|A\|\}$. Since g is continuous on $|z| \leqslant 2\|A\|$, it is bounded there as well. Thus, g is a bounded entire function. Liouville's Theorem [92, p. 77] ensures that g is constant. Moreover, this constant must be zero by the limiting argument above.

Putting this all together implies that $\varphi(A^{-1}) = g(0) = 0$ for all $\varphi \in \mathcal{B}(\mathcal{H})^*$. The Hahn–Banach separation theorem (Theorem 2.2.6) yields $A^{-1} = 0$, a contradiction. Therefore, $\sigma(A) \neq \emptyset$. ∎

Definition 2.4.11. The *resolvent set* of $A \in \mathcal{B}(\mathcal{H})$ is $\mathbb{C}\backslash\sigma(A)$.

Theorem 2.4.9 ensures that for each $A \in \mathcal{B}(\mathcal{H})$, the resolvent set is nonempty and that the operator-valued function $z \mapsto (zI - A)^{-1}$ is analytic on $\mathbb{C}\backslash\sigma(A)$. This function is called the *resolvent* of A.

2.5 Compact Diagonal Operators

The most tractable Hilbert-space operators are the compact operators. The following result, concerning the diagonal operators D_Λ, motivates the definition of a compact operator (Definition 2.5.3).

Theorem 2.5.1. *Let $\Lambda = (\lambda_n)_{n=0}^{\infty}$ be bounded. Then the following are equivalent.*

(a) *For every bounded sequence $(\mathbf{a}_n)_{n=1}^{\infty}$ in ℓ^2, $(D_\Lambda \mathbf{a}_n)_{n=1}^{\infty}$ has a convergent subsequence.*

(b) $\lambda_n \to 0$.

Proof (a) $\Rightarrow$ (b) The proof proceeds by contraposition. If λ_n does not approach zero, there is a subsequence $(\lambda_{n_k})_{k=1}^{\infty}$ and a $\delta > 0$ such that $|\lambda_{n_k}| \geqslant \delta$ for all $k \geqslant 1$. The corresponding subsequence of standard basis vectors $(\mathbf{e}_{n_k})_{k=1}^{\infty}$ is bounded but

$$\|D\mathbf{e}_{n_k} - D\mathbf{e}_{n_\ell}\| = \|\lambda_{n_k}\mathbf{e}_{n_k} - \lambda_{n_\ell}\mathbf{e}_{n_\ell}\| = \sqrt{|\lambda_{n_k}|^2 + |\lambda_{n_\ell}|^2} \geqslant \sqrt{2}\delta > 0.$$

Thus, $(D\mathbf{e}_{n_k})_{k=1}^{\infty}$ has no convergent subsequence.

(b) $\Rightarrow$ (a) Let $D = D_\Lambda$ and $D_N = D_{\Lambda_N}$, where

$$\Lambda_N = (\lambda_0, \lambda_1, \lambda_2, ..., \lambda_N, 0, 0, 0, ...).$$

Suppose that $\lambda_n \to 0$. Then (2.1.2) implies that

$$\|D - D_N\| = \sup_{n \geqslant N+1} |\lambda_n| \to 0 \quad \text{as } N \to \infty.$$

Fix N and a bounded sequence $(\mathbf{a}_n)_{n=1}^\infty$ in ℓ^2. Observe that $D_N \mathbf{a}_n$ is contained in the $(N+1)$-dimensional subspace span$\{\mathbf{e}_j : 0 \leqslant j \leqslant N\}$ of ℓ^2. For $N = 0$, the Heine–Borel theorem says that $(D_0 \mathbf{a}_n)_{n=1}^\infty$ has a convergent subsequence

$$D_0 \mathbf{a}_{01}, \ D_0 \mathbf{a}_{02}, \ D_0 \mathbf{a}_{03}, ...,$$

where $\mathbf{a}_{0k} = \mathbf{a}_m$ for some m that depends on D_0 and k. Since $(D_1 \mathbf{a}_{0k})_{k=1}^\infty$ is bounded, there is a subsequence

$$\mathbf{a}_{11}, \ \mathbf{a}_{12}, \ \mathbf{a}_{13}, ...$$

of the sequence $\mathbf{a}_{0k}$ such that $D_1 \mathbf{a}_{1k}$ converges. Since $(D_2 \mathbf{a}_{1k})_{k=1}^\infty$ is bounded, there is a subsequence

$$\mathbf{a}_{21}, \ \mathbf{a}_{22}, \ \mathbf{a}_{23}, ...$$

of the sequence $\mathbf{a}_{1k}$ such that $D_2 \mathbf{a}_{2k}$ converges. Continue in this manner to $n = 3, 4, 5, ...$ to create $(\mathbf{a}_{nk})_{k=1}^\infty$ such that $(D_n \mathbf{a}_{nk})_{k=1}^\infty$ is a convergent sequence. It suffices to show that $(D\mathbf{a}_{mm})_{m=1}^\infty$ is a Cauchy sequence and hence is convergent. For any N, observe that

$$\|D\mathbf{a}_{mm} - D\mathbf{a}_{kk}\| \leqslant \|D\mathbf{a}_{mm} - D_N \mathbf{a}_{mm}\| + \|D_N \mathbf{a}_{mm} - D_N \mathbf{a}_{kk}\|$$
$$+ \|D_N \mathbf{a}_{kk} - D\mathbf{a}_{kk}\|$$
$$\leqslant \|D - D_N\| \|\mathbf{a}_{mm}\| + \|D_N \mathbf{a}_{mm} - D_N \mathbf{a}_{kk}\|$$
$$+ \|D - D_N\| \|\mathbf{a}_{kk}\|. \tag{2.5.2}$$

Since $(\mathbf{a}_n)_{n=1}^\infty$ is bounded, so is the subsequence $(\mathbf{a}_{mm})_{m=1}^\infty$ and thus there is some $c > 0$ such that $\|\mathbf{a}_{mm}\| \leqslant c$ for all m. Given $\varepsilon > 0$, choose N large enough so that $\|D - D_N\| \leqslant \varepsilon/(3c)$. Since $(D_N \mathbf{a}_{mm})_{m=1}^\infty$ is a Cauchy sequence, there is an M such that $\|D_N \mathbf{a}_{mm} - D_N \mathbf{a}_{kk}\| \leqslant \varepsilon/3$ for all $m, k \geqslant M$. Apply these estimates to (2.5.2) and deduce that

$$\|D\mathbf{a}_{mm} - D\mathbf{a}_{kk}\| \leqslant \frac{\varepsilon}{3c} c + \frac{\varepsilon}{3} + \frac{\varepsilon}{3c} c = \varepsilon.$$

Thus, $(D\mathbf{a}_{mm})_{m=1}^\infty$ is a Cauchy sequence in ℓ^2 and hence converges. ∎

This inspires the following definition for Hilbert-space operators.

Definition 2.5.3. $T \in \mathcal{B}(\mathcal{H})$ is *compact* if, for every bounded sequence $(\mathbf{x}_n)_{n=1}^\infty$ in $\mathcal{H}$, $(T\mathbf{x}_n)_{n=1}^\infty$ has a convergent subsequence.

In the previous definition, the requirement that T is bounded is superfluous (Exercise 2.8.18). The ideas in the proof of Theorem 2.5.1 can be used to prove the following (see Exercises 2.8.19, 2.8.22, and 2.8.23).

Proposition 2.5.4. *Any finite-rank operator is compact.*

Proposition 2.5.5. *The product of a compact operator and a bounded operator is compact.*

Proposition 2.5.6. *If $(A_n)_{n=1}^{\infty}$ is a sequence of compact operators and $A_n \to A$ in norm, then A is compact.*

Other properties of compact operators are mentioned throughout this book. In fact, certain compact operators (the selfadjoint ones – see the next section) can be described with diagonal operators. Chapter 8 covers cyclic vectors and invariant subspaces for diagonal operators.

2.6 Compact Selfadjoint Operators

The spectrum and norm of a diagonal operator are readily computable. Moreover, there is a practical criterion for compactness. It is for these reasons diagonal operators are often used as models for various types of operators. For example, it is sometimes the case that $A \in \mathcal{B}(\mathcal{H})$ has an orthonormal basis $(\mathbf{x}_n)_{n=1}^{\infty}$ of eigenvectors corresponding to the sequence of eigenvalues $\Lambda = (\lambda_n)_{n=1}^{\infty}$ of A. Since A is bounded, so is Λ. Furthermore, since $(\mathbf{x}_n)_{n=1}^{\infty}$ is an orthonormal basis for $\mathcal{H}$, every $\mathbf{x} \in \mathcal{H}$ is of the form

$$\mathbf{x} = \sum_{n=1}^{\infty} \langle \mathbf{x}, \mathbf{x}_n \rangle \mathbf{x}_n,$$

and hence

$$A\mathbf{x} = \sum_{n=1}^{\infty} \lambda_n \langle \mathbf{x}, \mathbf{x}_n \rangle \mathbf{x}_n. \tag{2.6.1}$$

The (i, j) entry of the matrix representation of A with respect to this basis is

$$\langle A\mathbf{x}_j, \mathbf{x}_i \rangle = \begin{cases} 0 & \text{if } i \neq j, \\ \lambda_i & \text{if } i = j, \end{cases}$$

and thus the matrix representation of A is $\text{diag}(\lambda_1, \lambda_2, \lambda_3, \dots)$. Furthermore, A is compact if and only if $\lambda_n \to 0$ (Theorem 2.5.1).

It is common to use the notation $\mathbf{u} \otimes \mathbf{v}$ for the rank-one operator defined on $\mathcal{H}$ by

$$(\mathbf{u} \otimes \mathbf{v})(\mathbf{x}) = \langle \mathbf{x}, \mathbf{v} \rangle \mathbf{u} \quad \text{for } \mathbf{x} \in \mathcal{H}. \tag{2.6.2}$$

With this notation, one writes (2.6.1) as

$$A = \sum_{n=1}^{\infty} \lambda_n (\mathbf{x}_n \otimes \mathbf{x}_n).$$

Only certain operators are "diagonalizable" in the sense above. Compact selfadjoint operators enjoy this property. This is explored below and in greater detail in Chapter 8.

Definition 2.6.3. $A \in \mathcal{B}(\mathcal{H})$ is *selfadjoint* if $\langle A\mathbf{x}, \mathbf{y} \rangle = \langle \mathbf{x}, A\mathbf{y} \rangle$ for all $\mathbf{x}, \mathbf{y} \in \mathcal{H}$.

The adjoint of an operator is discussed in greater detail in Chapter 3. A diagonal operator D_Λ with $\Lambda = (\lambda_n)_{n=0}^\infty$ is selfadjoint if and only if $\lambda_n \in \mathbb{R}$ for all $n \geqslant 0$ (Exercise 2.8.26). Moreover, $\sigma_p(D_\Lambda) = \Lambda$ (Theorem 2.4.7). This is true for compact selfadjoint operators.

Lemma 2.6.4. *Let $A \in \mathcal{B}(\mathcal{H})$ be selfadjoint.*

(a) $\sigma_p(A) \subseteq \mathbb{R}$.

(b) *Eigenvectors corresponding to distinct eigenvalues are orthogonal.*

(c) *If A is compact and $\lambda \in \sigma_p(A) \backslash \{0\}$, then $\dim \ker(A - \lambda I) < \infty$, that is, λ has finite multiplicity.*

Proof (a) Let $\lambda \in \sigma_p(A)$ and $\mathbf{x}$ be a unit eigenvector for λ. Then

$$\lambda = \langle \lambda \mathbf{x}, \mathbf{x} \rangle = \langle A\mathbf{x}, \mathbf{x} \rangle = \langle \mathbf{x}, A\mathbf{x} \rangle = \langle \mathbf{x}, \lambda \mathbf{x} \rangle = \overline{\lambda}.$$

Thus, $\sigma_p(A) \subseteq \mathbb{R}$.

(b) Suppose $\lambda, \mu \in \sigma_p(A)$ and $\lambda \neq \mu$. If $A\mathbf{x} = \lambda \mathbf{x}$ and $A\mathbf{y} = \mu \mathbf{y}$, where $\mathbf{x}, \mathbf{y} \neq \mathbf{0}$, then

$$\begin{aligned}
0 &= \langle A\mathbf{x}, \mathbf{y} \rangle - \langle A\mathbf{x}, \mathbf{y} \rangle \\
&= \langle A\mathbf{x}, \mathbf{y} \rangle - \langle \mathbf{x}, A\mathbf{y} \rangle \\
&= \langle \lambda \mathbf{x}, \mathbf{y} \rangle - \langle \mathbf{x}, \mu \mathbf{y} \rangle \\
&= \lambda \langle \mathbf{x}, \mathbf{y} \rangle - \overline{\mu} \langle \mathbf{x}, \mathbf{y} \rangle \\
&= (\lambda - \mu) \langle \mathbf{x}, \mathbf{y} \rangle. \qquad \qquad \text{(by (a))}
\end{aligned}$$

Since $\lambda - \mu \neq 0$, it follows that $\langle \mathbf{x}, \mathbf{y} \rangle = 0$.

(c) See Exercise 2.8.25. ■

Lemma 2.6.5. *Let $A \in \mathcal{B}(\mathcal{H})$.*

(a) *If A is selfadjoint, then $\|A\|$ or $-\|A\|$ belong to $\sigma_{ap}(A)$.*

(b) *If A is selfadjoint and compact, then $\|A\|$ or $-\|A\|$ belong to $\sigma_p(A)$.*

Proof Exercise 2.8.29 shows that

$$\|A\| = \sup_{\|\mathbf{x}\|=1} |\langle A\mathbf{x}, \mathbf{x} \rangle|. \qquad (2.6.6)$$

Let $(\mathbf{x}_n)_{n=1}^\infty$ be a sequence of unit vectors such that $|\langle A\mathbf{x}_n, \mathbf{x}_n \rangle| \to \|A\|$. Since A is selfadjoint, it follows that $\langle A\mathbf{x}_n, \mathbf{x}_n \rangle = \langle \mathbf{x}_n, A\mathbf{x}_n \rangle = \overline{\langle A\mathbf{x}_n, \mathbf{x}_n \rangle}$ and thus is real. Passing to a subsequence if necessary, we may assume that $\langle A\mathbf{x}_n, \mathbf{x}_n \rangle \to \|A\|$ or $\langle A\mathbf{x}_n, \mathbf{x}_n \rangle \to -\|A\|$. Without loss of generality, assume that $\langle A\mathbf{x}_n, \mathbf{x}_n \rangle \to \|A\|$. Then

$$
\begin{aligned}
0 \leqslant \|(A - \|A\|I)\mathbf{x}_n\|^2 \\
= \|A\mathbf{x}_n - \|A\|\mathbf{x}_n\|^2 \\
= \langle A\mathbf{x}_n, A\mathbf{x}_n \rangle - \|A\|\langle A\mathbf{x}_n, \mathbf{x}_n \rangle - \|A\|\langle \mathbf{x}_n, A\mathbf{x}_n \rangle + \|A\|^2 \langle \mathbf{x}_n, \mathbf{x}_n \rangle \\
= \|A\mathbf{x}_n\|^2 - 2\|A\|\langle A\mathbf{x}_n, \mathbf{x}_n \rangle + \|A\|^2 \|\mathbf{x}_n\|^2 \\
\leqslant \|A\|^2 \|\mathbf{x}_n\|^2 - 2\|A\|\langle A\mathbf{x}_n, \mathbf{x}_n \rangle + \|A\|^2 \|\mathbf{x}_n\|^2 \\
= 2\|A\|^2 - 2\|A\|\langle A\mathbf{x}_n, \mathbf{x}_n \rangle,
\end{aligned}
$$

which tends to zero as $n \to \infty$. Therefore, $\|A\| \in \sigma_{ap}(A)$. This proves (a). Exercise 2.8.27 ensures that when A is also compact, then $\|A\| \in \sigma_p(A)$, which proves (b). ∎

Theorem 2.6.7 (Spectral theorem for compact selfadjoint operators). *Let $A \in \mathcal{B}(\mathcal{H})$ be a compact selfadjoint operator on a separable, infinite-dimensional Hilbert space. There exists a real sequence $(\lambda_n)_{n=1}^{\infty}$ tending to zero and an orthonormal basis $(\mathbf{x}_n)_{n=1}^{\infty}$ for $\mathcal{H}$ such that*

$$
A\mathbf{x} = \sum_{n=1}^{\infty} \lambda_n \langle \mathbf{x}, \mathbf{x}_n \rangle \mathbf{x}_n \quad \text{for all } \mathbf{x} \in \mathcal{H}. \tag{2.6.8}
$$

With respect to the orthonormal basis $(\mathbf{x}_n)_{n=1}^{\infty}$, the operator A has the matrix representation D_Λ on $\ell^2(\mathbb{N})$, where $\Lambda = (\lambda_n)_{n=1}^{\infty}$.

The proof of this theorem requires the following concept. For a subset $\mathcal{Y}$ of $\mathcal{H}$, define

$$
\mathcal{Y}^{\perp} = \{\mathbf{x} \in \mathcal{H} : \langle \mathbf{x}, \mathbf{y} \rangle = 0 \text{ for all } \mathbf{y} \in \mathcal{Y}\}.
$$

This set is the *orthogonal complement* of $\mathcal{Y}$ and is formally defined and discussed in Definition 3.1.1.

Proof We assume that A is injective; see Exercise 3.6.41 for the general case. Suppose that $A \in \mathcal{B}(\mathcal{H})$ is compact and selfadjoint with $\ker A = \{0\}$. This proof constructs the $\mathbf{x}_n$ inductively. Lemma 2.6.5 says that A has an nonzero eigenvalue $\lambda_1 = \|A\|$ or $\lambda_1 = -\|A\|$. Let $\mathbf{x}_1$ be a corresponding unit eigenvector to λ_1. Since span$\{\mathbf{x}_1\}$ is A-invariant and A is selfadjoint, $\mathcal{H}_2 = (\text{span}\{\mathbf{x}_1\})^{\perp}$ is also A-invariant. Indeed, if $\mathbf{x} \in \mathcal{H}_2$, that is, $\langle \mathbf{x}, \mathbf{x}_1 \rangle = 0$, then

$$
\langle A\mathbf{x}, \mathbf{x}_1 \rangle = \langle \mathbf{x}, A\mathbf{x}_1 \rangle = \langle \mathbf{x}, \lambda_1 \mathbf{x}_1 \rangle = \overline{\lambda_1} \langle \mathbf{x}, \mathbf{x}_1 \rangle = 0
$$

and hence $A\mathbf{x} \in \mathcal{H}_2$.

Let $A_2 = A|_{\mathcal{H}_2}$ denote the restriction of A to $\mathcal{H}_2$. Then A_2 is compact and selfadjoint (and injective). Lemma 2.6.5 yields a nonzero eigenvalue $\lambda_2 = \|A_2\|$ or $\lambda_2 = -\|A_2\|$ and corresponding unit eigenvector $\mathbf{x}_2 \in \mathcal{H}_2$. Continue this process to obtain a sequence $(\lambda_n)_{n=1}^{\infty}$ of nonzero real numbers and corresponding orthonormal eigenvectors $(\mathbf{x}_n)_{n=1}^{\infty}$ such that the restriction $A_n = A|_{\mathcal{H}_n}$ of A to $\mathcal{H}_n = (\text{span}\{\mathbf{x}_1, \mathbf{x}_2, ..., \mathbf{x}_{n-1}\})^{\perp}$ satisfies $\|A_n\| = |\lambda_n|$.

We claim that that $\lambda_n \to 0$. If λ_n does not approach zero, there is a subsequence λ_{n_k} of distinct terms (note the use of Lemma 2.6.4c) and a $\delta > 0$ such that $|\lambda_{n_k}| \geqslant \delta$ for all k. The corresponding subsequence of unit eigenvectors $\mathbf{x}_{n_k}$, which are orthogonal by Lemma 2.6.4b, is bounded, but

$$\|A\mathbf{x}_{n_k} - A\mathbf{x}_{n_\ell}\| = \|\lambda_{n_k}\mathbf{x}_{n_k} - \lambda_{n_\ell}\mathbf{x}_{n_\ell}\| = \sqrt{|\lambda_{n_k}|^2 + |\lambda_{n_\ell}|^2} \geqslant \sqrt{2}\delta.$$

Thus, $(A\mathbf{x}_{n_k})_{k=1}^\infty$ has no convergent subsequence. This contradicts the assumption that A is compact.

Next we verify (2.6.8) and that $(\mathbf{x}_n)_{n=1}^\infty$ is an orthonormal basis for $\mathcal{H}$. For each $\mathbf{x} \in \mathcal{H}$, apply the Pythagorean theorem (Proposition 1.4.6) to

$$\mathbf{x} = \left(\mathbf{x} - \sum_{i=1}^{n-1} \langle \mathbf{x}, \mathbf{x}_i \rangle \mathbf{x}_i\right) + \sum_{i=1}^{n-1} \langle \mathbf{x}, \mathbf{x}_i \rangle \mathbf{x}_i,$$

where $\mathbf{y}_n$, the first summand above, belongs to $\mathcal{H}_n$ and the second summand belongs to $\mathcal{H}_n^\perp$, and obtain

$$\|\mathbf{x}\|^2 = \|\mathbf{y}_n\|^2 + \sum_{i=1}^{n-1} |\langle \mathbf{x}, \mathbf{x}_i \rangle|^2.$$

Consequently, $\|\mathbf{y}_n\| \leqslant \|\mathbf{x}\|$ for all $n \geqslant 1$. For each $n \geqslant 1$,

$$\left\|A\mathbf{x} - \sum_{i=1}^{n-1} \lambda_i \langle \mathbf{x}, \mathbf{x}_i \rangle \mathbf{x}_i\right\| = \|A\mathbf{y}_n\| = \|A_n\mathbf{y}_n\| \leqslant \|A_n\| \|\mathbf{y}_n\| \leqslant |\lambda_n| \|\mathbf{x}\|.$$

Since this tends to zero as $n \to \infty$, (2.6.8) follows.

Finally notice that $\bigvee\{\mathbf{x}_n : n \geqslant 1\} = \mathcal{H}$ since if $\mathbf{x} \perp \mathbf{x}_n$ for all n, then (2.6.8) shows that $\mathbf{x} \in \ker A = \{\mathbf{0}\}$. ∎

See [339] for the proof of the following important theorem of Riesz which says that the nonzero elements of the spectrum of a compact operator are eigenvalues. Note that Exercise 2.8.28 ensures that 0 belongs to the spectrum of any compact operator on an infinite-dimensional Hilbert space. It need not be an eigenvalue, however, although it does belong to the approximate point spectrum. See Chapter 7 for an example of a nonzero compact operator with no eigenvalues and whose spectrum is $\{0\}$.

Theorem 2.6.9 (Riesz). *Suppose $A \in \mathcal{B}(\mathcal{H})$ is compact. If $\lambda \in \sigma(A) \backslash \{0\}$, then $\lambda \in \sigma_p(A)$.*

A wide variety of compact operators are discussed in this book. This includes compact Hankel operators, compact composition operators, and the Volterra operator.

2.7 Notes

The integral operators

$$f(x) \mapsto \int_a^b K(x, y) f(y) \, dy \tag{2.7.1}$$

and the matrix operators $\mathbf{x} \mapsto A\mathbf{x}$ were some of the first operators to be studied.

Schmidt, Toeplitz, and Hellinger [191] examined linear transformations as matrices (finite or infinite). In fact, according to Friedrichs, around 1920 Schmidt advised von Neumann *"Nein! Nein! Sagen Sie nicht Operator, sagen Sie Matrix!"* ("No! No! Don't say operator, say matrix!") [270]. By 1913, Riesz [303] stressed the importance of *substitutions linéaires* (linear substitutions) on ℓ^2. He defined the *borne de la substitution* (bound of the substitution) for the matrix operator $A\mathbf{x} = \mathbf{x}'$ as the smallest constant M_A such that

$$\sum_{n=1}^{\infty} |x_k'|^2 \leqslant M_A^2 \sum_{n=1}^{\infty} |x_k|^2,$$

along with the facts (not formally stated but used implicitly) that

$$M_{AB} \leqslant M_A M_B \quad \text{and} \quad M_{A+B} \leqslant M_A + M_B$$

(Proposition 2.1.12). Hildebrandt in 1931 [198] and Stone in 1932 [353] used the terms "linear limited" to denote linear transformations on Hilbert and Banach spaces and the term "modulus" to describe $\|A\|$.

The series

$$I + A + A^2 + A^2 + \cdots = (I - A)^{-1} \tag{2.7.2}$$

for $\|A\| < 1$, known as a *Neumann series*, was investigated in 1877 by C. Neumann [248]. In the study of potential theory, he looked at operators A whose special properties imply that $\mathbf{x} + A\mathbf{x} + A^2\mathbf{x} + A^3\mathbf{x} + \cdots$ converges for all $\mathbf{x}$. Riesz [303] observed the convergence of the Neumann series and the identity (2.7.2). In 1918, Riesz [304] showed that an operator divides the complex plane into two parts: its spectrum and its resolvent set (the complement of the spectrum). The terms "spectrum" and "resolvent" go back to Hilbert [196]. Riesz also observed that the mapping $z \mapsto A(I - zA)^{-1}$ *montrent le caractéres d'une fonction holomorphe en z* ("shows the behavior of a holomorphic function in z") which, as seen in the proof of Theorem 2.4.9, plays an important role in showing that the spectrum is a nonempty compact subset of $\mathbb{C}$. He also showed that the nonzero elements of the spectrum of a compact operator consist only of eigenvalues (Theorem 2.6.9). This important observation is used in Chapter 18.

In the past, finite-rank operators were sometimes called "degenerate" operators [200]. They are all of the form

$$T = \sum_{i=1}^{n} \mathbf{u}_i \otimes \mathbf{v}_i.$$

This tensor notation appeared in work of Schauder [330]. Finite-rank integral operators of the form (2.7.1) with kernels

$$K(x, y) = \sum_{i=1}^{n} g_i(x) h_i(y)$$

were studied by Goursat [159] and Schmidt [331].

For a Hilbert-space operator A, the following properties are equivalent.

(a) A is "completely continuous" in the following sense: if $(\mathbf{x}_n)_{n=1}^{\infty}$ is a sequence such that $\langle \mathbf{x}_n, \mathbf{y} \rangle \to \langle \mathbf{x}, \mathbf{y} \rangle$ for all $\mathbf{y}$, then $A\mathbf{x}_n \to A\mathbf{x}$ in norm.

(b) A is the norm limit of finite-rank operators.

(c) A is compact (Definition 2.5.3).

These equivalences were studied by Hilbert [196] and Riesz [303] in the ℓ^2 setting, Schmidt [331] for integral operators, and Hildebrandt [198] in the abstract setting.

Diagonal operators also appear in a result of Weyl, von Neumann, and Berg [105] which says that any normal operator N (Chapter 8) on a separable Hilbert space can be decomposed as $N = D + K$, where D is a diagonal operator and K is compact.

2.8 Exercises

Exercise 2.8.1. Let $\Lambda = \left(\frac{n}{n+1}\right)_{n=0}^{\infty}$. Prove there is no $\mathbf{x} \in \ell^2 \setminus \{\mathbf{0}\}$ such that $\|D_\Lambda \mathbf{x}\| = \|D_\Lambda\| \|\mathbf{x}\|$.

Exercise 2.8.2. Let $A \in \mathcal{B}(\mathcal{H})$. Prove that $\ker A$ is a (closed) subspace of $\mathcal{H}$.

Exercise 2.8.3. If $\mathcal{H}$ is an infinite-dimensional Hilbert space, find an $A \in \mathcal{B}(\mathcal{H})$ whose range is not closed.

Exercise 2.8.4. Prove that an inverse of $A \in \mathcal{B}(\mathcal{H})$, if it exists, is unique.

Exercise 2.8.5. Show that the set $\mathcal{G}$ of invertible elements of $\mathcal{B}(\mathcal{H})$ forms a group. That is, $\mathcal{G}$ is closed under multiplication and inversion, contains the identity, and its multiplication is associative.

Exercise 2.8.6. Use the steps below to prove that every $A \in \mathcal{M}_n$ has an eigenvalue.

(a) Prove that $I, A, A^2, \ldots, A^{n^2}$ are linearly dependent.

(b) Prove there is a nonzero polynomial p such that $p(A) = 0$.

(c) Factor p into linear factors and use this to show that $A - \lambda I$ is not invertible for some $\lambda \in \mathbb{C}$.

Exercise 2.8.7. What are necessary and sufficient conditions for a diagonal operator to be invertible in $\mathcal{B}(\ell^2)$?

Exercise 2.8.8. What are necessary and sufficient conditions for a diagonal operator D_Λ on ℓ^2 to be an *isometry*, that is, $\|D_\Lambda \mathbf{x}\| = \|\mathbf{x}\|$ for all $\mathbf{x} \in \ell^2$?

Exercise 2.8.9. If $T \in \mathcal{B}(\mathcal{H})$ is an isometry, prove that $\operatorname{ran} T$ is closed.

Exercise 2.8.10.

(a) If D_Λ is a diagonal operator on ℓ^2 with distinct eigenvalues, describe the operators $A \in \mathcal{B}(\ell^2)$ such that $A D_\Lambda = D_\Lambda A$.

(b) What happens in (a) if the eigenvalues of D_Λ are not distinct?

Exercise 2.8.11. Show that every diagonal operator D_Λ on ℓ^2 has a square root, meaning there exists an $A \in \mathcal{B}(\ell^2)$ such that $A^2 = D_\Lambda$. Does a square root of a diagonal operator need to be another diagonal operator?

Exercise 2.8.12. Let $A, B \in \mathcal{B}(\mathcal{H})$ and $c \in \mathbb{C}$.

(a) Prove that $A + B \in \mathcal{B}(\mathcal{H})$ and $\|A + B\| \leqslant \|A\| + \|B\|$.

(b) Prove that $cA \in \mathcal{B}(\mathcal{H})$ and $\|cA\| = |c|\|A\|$.

(c) Prove that $AB \in \mathcal{B}(\mathcal{H})$ and $\|AB\| \leqslant \|A\|\|B\|$.

Remark: This shows that $\mathcal{B}(\mathcal{H})$ is a normed algebra.

Exercise 2.8.13. *von Neumann's inequality* says that if $p(z) = a_0 + a_1 z + a_2 z^2 + \cdots + a_n z^n$ and $T \in \mathcal{B}(\mathcal{H})$ with $\|T\| \leqslant 1$, then $p(T) = a_0 I + a_1 T + a_2 T^2 + \cdots + a_n T^n$ satisfies

$$\|p(T)\| \leqslant \sup_{|z| \leqslant 1} |p(z)|.$$

Prove von Neumann's inequality if T is a diagonal operator.
Remark: See [144, p. 213] for a proof of von Neumann's inequality for a general bounded operator.

Exercise 2.8.14. Prove that in Proposition 2.3.9, the assumption $\|I - A\| < 1$ can be replaced with $\limsup_{n \to \infty} \|(I - A)^n\|^{\frac{1}{n}} < 1$.

Exercise 2.8.15. Let $A \in \mathcal{B}(\mathcal{H})$ be invertible. Prove that any $B \in \mathcal{B}(\mathcal{H})$ such that $\|A - B\| < \|A\|$ is also invertible.

Exercise 2.8.16. Let $A \in \mathcal{B}(\mathcal{H})$ be a compact operator. Assume that $\mathbf{x}_n \to \mathbf{x}$ and $\mathbf{y}_n \to \mathbf{y}$ weakly in $\mathcal{H}$. Prove that $\langle A\mathbf{x}_n, \mathbf{y}_m \rangle \to \langle A\mathbf{x}, \mathbf{y} \rangle$ as $m, n \to \infty$.

Exercise 2.8.17. Let T be a linear transformation on a Hilbert space whose range is finite dimensional. Prove that T is bounded.

Exercise 2.8.18. Let T be a linear transformation on a Hilbert space $\mathcal{H}$ such that for every bounded sequence $(\mathbf{x}_n)_{n=1}^\infty$ in $\mathcal{H}$, $(T\mathbf{x}_n)_{n=1}^\infty$ has a convergent subsequence. Prove that T is bounded.

Exercise 2.8.19. Prove that any finite-rank operator is compact.

Exercise 2.8.20. Prove that an orthogonal projection is compact if and only if its range is finite dimensional.

Exercise 2.8.21. Let $\mathcal{H}$ be an infinite-dimensional Hilbert space and let K be a compact subset of $\mathbb{C}$. Find a $T \in \mathcal{B}(\mathcal{H})$ such that $\sigma(T) = K$.

Exercise 2.8.22. Prove that if $(A_n)_{n=0}^\infty$ is a sequence of compact operators in $\mathcal{B}(\mathcal{H})$ and $\|A_n - A\| \to 0$, then A is compact.

Exercise 2.8.23. Prove that the product of a compact operator and a bounded operator is compact.

Exercise 2.8.24. Prove that every compact operator $A \in \mathcal{B}(\mathcal{H})$ is the limit of finite-rank operators using the following steps.

(a) Choose an orthonormal basis $(\mathbf{u}_n)_{n=1}^{\infty}$ of $\mathcal{H}$ and let P_N be the orthogonal projection of $\mathcal{H}$ onto span$\{\mathbf{u}_1, \mathbf{u}_2, ..., \mathbf{u}_N\}$. Prove that $P_N A$ is compact.

(b) Prove that $\|P_N A - A\|$ is a decreasing function of N.

(c) Prove that if $\|P_N A - A\|$ does not tend to zero, then there are unit vectors $\mathbf{x}_N$ and a constant $c > 0$ such that $\|P_N A \mathbf{x}_N - A\mathbf{x}_N\| \geqslant c$ for all N.

(d) Use the compactness of A to derive a contradiction.

Exercise 2.8.25. Prove that if $A \in \mathcal{B}(\mathcal{H})$ is compact, then $\ker(A - \lambda I)$ is finite dimensional for each $\lambda \in \mathbb{C}\backslash\{0\}$.

Exercise 2.8.26. Prove that D_Λ satisfies $\langle D_\Lambda \mathbf{x}, \mathbf{y} \rangle = \langle \mathbf{x}, D_\Lambda \mathbf{y} \rangle$ for all $\mathbf{x}, \mathbf{y} \in \ell^2$ if and only if $\Lambda \subseteq \mathbb{R}$.

Exercise 2.8.27. If $A \in \mathcal{B}(\mathcal{H})$ is compact and $\lambda \in \sigma_{ap}(A)\backslash\{0\}$, prove that $\lambda \in \sigma_p(A)$.

Exercise 2.8.28. Prove that if $\mathcal{H}$ is an infinite-dimensional Hilbert space and $A \in \mathcal{B}(\mathcal{H})$ is compact, then $0 \in \sigma(A)$.

Exercise 2.8.29. If $A \in \mathcal{B}(\mathcal{H})$ is selfadjoint, prove that $\|A\| = \sup_{\|\mathbf{x}\|=1} |\langle A\mathbf{x}, \mathbf{x} \rangle|$.

Exercise 2.8.30. Mimic the proof of Lemma 2.6.5 to show that if $A \in \mathcal{B}(\mathcal{H})$ is selfadjoint and compact with $\{\langle A\mathbf{x}, \mathbf{x} \rangle : \|\mathbf{x}\| = 1\} = [a, b]$, then a and b are eigenvalues of A.

Exercise 2.8.31. Let $A \in \mathcal{B}(\mathcal{H})$ and $\|A\| < 1$.

(a) If $f(z) = \sum_{n=0}^{\infty} a_n z^n$ has radius of convergence 1 or greater, prove that $f(A) = a_0 I + a_1 A + a_2 A^2 + a_3 A^3 + \cdots$ converges in operator norm.

(b) Prove that $\cos^2 A + \sin^2 A = I$.

Exercise 2.8.32. The Axiom of Choice ensures that every vector space V has a *Hamel basis*, a linearly independent set $\{\mathbf{x}_\alpha\}_{\alpha \in I}$ such that each $\mathbf{v} \in V$ can be written uniquely as a finite linear combination of Hamel-basis vectors. Assuming this result, prove that every infinite-dimensional Hilbert space $\mathcal{H}$ admits an unbounded linear functional.

Exercise 2.8.33. Use the Axiom of Choice to prove that not every $f : \mathbb{R} \to \mathbb{R}$ that satisfies $f(x + y) = f(x) + f(y)$ for all $x, y \in \mathbb{R}$ is of the form $f(x) = cx$ for some $c \in \mathbb{R}$.

Exercise 2.8.34. A sequence $(\mathbf{x}_n)_{n=1}^{\infty}$ in a Hilbert space $\mathcal{H}$ is *complete* if $\bigvee\{\mathbf{x}_n : n \geqslant 1\} = \mathcal{H}$. Let $(\mathbf{u}_n)_{n=1}^{\infty}$ be an orthonormal basis for $\mathcal{H}$ and $(\mathbf{x}_n)_{n=1}^{\infty}$ be a sequence in $\mathcal{H}$ such that

$$\sum_{n=1}^{\infty} \|\mathbf{u}_n - \mathbf{x}_n\|^2 < 1.$$

Prove that $(\mathbf{x}_n)_{n=1}^{\infty}$ is complete in $\mathcal{H}$.

Exercise 2.8.35. Let $(\mathbf{x}_n)_{n=1}^{\infty}$ be an orthonormal basis for a Hilbert space $\mathcal{H}$ and let $(\mathbf{y}_n)_{n=1}^{\infty}$ be an orthonormal sequence in $\mathcal{H}$. If

$$\sum_{n=1}^{\infty} \|\mathbf{x}_n - \mathbf{y}_n\|^2 < \infty,$$

prove that $(\mathbf{y}_n)_{n=1}^{\infty}$ is an orthonormal basis.

Exercise 2.8.36. This exercise is a continuation of Exercise 2.8.34.

(a) Prove that $(t^{2n})_{n=0}^{\infty}$ is complete in $L^2[0, 1]$.

(b) Is $(t^{2n})_{n=0}^{\infty}$ complete in $L^2[-1, 1]$?

Exercise 2.8.37. Let $A \in \mathcal{B}(\mathcal{H})$. Prove that the following are equivalent.

(a) There is a $B \in \mathcal{B}(\mathcal{H})$ such that $BA = I$.

(b) A is injective and ran A is closed.

(c) $\inf_{\|\mathbf{x}\|=1} \|A\mathbf{x}\| > 0$.

Remark: A is left invertible if it satisfies any of the conditions above.

Exercise 2.8.38. Let $A \in \mathcal{B}(\mathcal{H})$. Prove that the following are equivalent.

(a) There is a $B \in \mathcal{B}(\mathcal{H})$ such that $AB = I$.

(b) A is surjective.

Remark: A is right invertible if it satisfies any of the conditions above.

Exercise 2.8.39. Recall that $(e_n)_{n=-\infty}^{\infty}$, where $e_n(x) = e^{2\pi i n x}$, is an orthonormal basis for $L^2[0, 1]$ (Theorem 1.3.9). Let $\Lambda = (\lambda_n)_{n=-\infty}^{\infty}$ be a bounded sequence of complex numbers and define $T_{\Lambda}(e_n) = \lambda_n e_n$ for all $n \in \mathbb{Z}$.

(a) Prove that T_{Λ} extends by linearity to a bounded operator on $L^2[0, 1]$ with norm $\|T_{\Lambda}\| = \sup_{n \in \mathbb{Z}} |\lambda_n|$.

(b) For $a \in \mathbb{R}$, define $(\tau_a e_n)(x) = e_n(x - a)$. Show that τ_a extends by linearity to a bounded operator on $L^2[0, 1]$ that commutes with T_{Λ}.

(c) Prove that if $T \in \mathcal{B}(L^2[0, 1])$ satisfies $T\tau_a = \tau_a T$ for all $a \in \mathbb{R}$, then $T = T_{\Lambda}$ for some bounded sequence Λ.

Remark: The operator T_{Λ} above is a *Fourier multiplier*.

Exercise 2.8.40. For $1 \leqslant p \leqslant \infty$, let X_p denote the vector space $C[0, 1]$ equipped with the norm

$$\|f\|_p = \begin{cases} \left(\int_0^1 |f(x)|^p \, dx \right)^{\frac{1}{p}} & \text{for } 1 \leqslant p < \infty, \\ \max_{x \in [0,1]} |f(x)| & \text{for } p = \infty. \end{cases}$$

Define the inclusion map $i_{pq} : X_p \to X_q$ by $i_{pq}(f) = f$. For what pairs (p, q) is i_{pq} bounded?

Exercise 2.8.41. For $A \in \mathcal{B}(\mathcal{H})$, the *numerical range* of A is $W(A) = \{\langle A\mathbf{x}, \mathbf{x} \rangle : \|\mathbf{x}\| = 1\}$. The Toeplitz–Hausdorff theorem [144, p. 222] says that $W(A)$ is convex. If $(\lambda_n)_{n=0}^{\infty}$ is a sequence of positive real numbers such that λ_n decreases to zero, prove that $W(D_\Lambda) = (0, \lambda_0]$ as follows.

(a) If $(\mathbf{e}_n)_{n=0}^{\infty}$ is the standard basis for ℓ^2, prove that $|\langle \mathbf{x}, \mathbf{e}_n \rangle| \leqslant 1$ for all unit vectors $\mathbf{x} \in \ell^2$ and $n \geqslant 0$.

(b) Prove that $W(D_\Lambda) = \left\{ \sum_{n=0}^{\infty} \lambda_n a_n : 0 \leqslant a_n \leqslant 1, \sum_{n=0}^{\infty} a_n = 1 \right\}$.

Exercise 2.8.42. Let A be a compact selfadjoint operator. Describe $W(A)$.

Exercise 2.8.43. For a Hilbert space $\mathcal{H}$ and $\mathbf{y} \in \mathcal{H}$ consider the linear transformation $\ell_{\mathbf{y}} : \mathcal{H} \to \mathbb{C}$ defined by $\ell_{\mathbf{y}}(\mathbf{x}) = \langle \mathbf{x}, \mathbf{y} \rangle$. Prove that $\ell_{\mathbf{y}} \in \mathcal{B}(\mathcal{H}, \mathbb{C})$ and $\|\ell_{\mathbf{y}}\| = \|\mathbf{y}\|$.

Exercise 2.8.44. Recall the Banach space ℓ^{∞} from Example 1.8.2 and let $c \subseteq \ell^{\infty}$ denote the vector space of convergent sequences in ℓ^{∞}, that is the $\mathbf{a} = (a_n)_{n=0}^{\infty} \in \ell^{\infty}$ such that $\lim_{n \to \infty} a_n$ exists.

(a) Prove that c is a subspace of ℓ^{∞}. In particular, prove that c is topologically closed.

(b) Prove that the linear transformation $T : c \to \mathbb{C}$ defined by $T(\mathbf{a}) = \lim_{n \to \infty} a_n$ belongs to $\mathcal{B}(c, \mathbb{C})$.

(c) Find $\|T\|$.

(d) Find $\ker T$.

2.9 Hints for the Exercises

Hint for Ex. 2.8.3: Let $(\mathbf{u}_n)_{n=1}^{\infty}$ be an orthonormal basis for $\mathcal{H}$ and consider the operator

$$T\left(\sum_{n=1}^{\infty} a_n \mathbf{u}_n \right) = \sum_{n=1}^{\infty} \frac{a_n}{n} \mathbf{u}_n, \quad \text{where } (a_n)_{n=1}^{\infty} \in \ell^2.$$

Hint for Ex. 2.8.10: Any bounded diagonal operator commutes with D_Λ. To show the converse, if $AD_\Lambda = D_\Lambda A$, then $Ap(D_\Lambda) = p(D_\Lambda)A$ for any polynomial p. Evaluate both sides of the equation above at $\mathbf{e}_N$ and choose an appropriate interpolating polynomial p depending on N.

Hint for Ex. 2.8.11: For $-1 \leqslant x \leqslant 1$ consider the square of the matrix

$$\begin{bmatrix} x & \sqrt{1-x^2} \\ \sqrt{1-x^2} & -x \end{bmatrix}.$$

Hint for Ex. 2.8.15: Consider $I - A^{-1}B$.

Hint for Ex. 2.8.16: Use the polarization identity from Exercise 1.10.17.

Hint for Ex. 2.8.21: Choose an appropriate diagonal operator.

Hint for Ex. 2.8.22: Use a diagonalization argument similar to the one in the proof Theorem 2.5.1.

Hint for Ex. 2.8.27: Suppose that $\|\mathbf{x}_n\| = 1$ and $\|(A - \lambda I)\mathbf{x}_n\| \to 0$. Show that there is a $\mathbf{y} \neq \mathbf{0}$ and a subsequence $\mathbf{x}_{n_k} \to \lambda^{-1}\mathbf{y}$.

Hint for Ex. 2.8.28: Proceed by contradiction and assume that A is invertible.

Hint for Ex. 2.8.29: Start with $\langle A\mathbf{x}, \mathbf{y} \rangle = \frac{1}{4}\langle A(\mathbf{x} + \mathbf{y}), \mathbf{x} + \mathbf{y} \rangle - \frac{1}{4}\langle A(\mathbf{x} - \mathbf{y}), \mathbf{x} - \mathbf{y} \rangle$ and use the polarization identity from Exercise 1.10.17.

Hint for Ex. 2.8.32: Let $(\mathbf{x}_\alpha)_{\alpha \in I}$ be a Hamel basis for $\mathcal{H}$ and define the functional's action on Hamel-basis vectors.

Hint for Ex. 2.8.34: Consider $I - T$, where

$$T : \mathcal{H} \to \mathcal{H}, \quad T\Big(\sum_{n=1}^{\infty} a_n \mathbf{u}_n \Big) = \sum_{n=1}^{\infty} a_n (\mathbf{u}_n - \mathbf{x}_n).$$

Hint for Ex. 2.8.35: Construct a bounded invertible operator $T \in \mathcal{B}(\mathcal{H})$ such that $T\mathbf{x}_n = \mathbf{y}_n$ for all $n \geqslant 1$.

Hint for Ex. 2.8.39: Start with $(Te_n)(x) = \sum_{m=-\infty}^{\infty} c_m(n)e_m(x)$ and use the fact that $\tau_a Te_n = T\tau_a e_n$ for all a to prove that $c_m(n) = 0$ for all $m \neq n$.

Hint for Ex. 2.8.40: Use Hölder's inequality.

3

$$\cdot \quad \cdot \quad \bullet \quad \cdot \quad \cdot$$

Infinite Matrices

Key Concepts: Riesz representation theorem, adjoint of an operator, operators defined by infinite matrices, Hilbert matrix, Cesàro matrix, Schur's test, matrices defining compact operators, matrices defining contractions.

Outline: This chapter concerns operators on ℓ^2 defined by infinite matrices (not necessarily diagonal). When are such operators well defined? Bounded? Compact? Schur's test yields a tangible solution to some of these questions.

3.1 Adjoint of an Operator

Let $A = [a_{ij}]_{i,j=1}^n \in \mathcal{M}_n$, the set of $n \times n$ matrices with complex entries. The adjoint A^* is its conjugate transpose $A^* = [\overline{a_{ji}}]_{i,j=1}^n$ and it satisfies $\langle A\mathbf{x}, \mathbf{y}\rangle = \langle \mathbf{x}, A^*\mathbf{y}\rangle$ for all $\mathbf{x}, \mathbf{y} \in \mathbb{C}^n$. If $\mathcal{H}$ is a Hilbert space and $A \in \mathcal{B}(\mathcal{H})$, we want to find an $A^* \in \mathcal{B}(\mathcal{H})$ such that

$$\langle A\mathbf{x}, \mathbf{y}\rangle = \langle \mathbf{x}, A^*\mathbf{y}\rangle \quad \text{for all } \mathbf{x}, \mathbf{y} \in \mathcal{H}.$$

The existence of such an A^* is not immediately clear.

We begin with the following observation. For $A \in \mathcal{B}(\mathcal{H})$, the Cauchy–Schwarz inequality and the definition of the operator norm say that

$$|\langle A\mathbf{x}, \mathbf{y}\rangle| \leqslant \|A\mathbf{x}\|\|\mathbf{y}\| \leqslant \|A\|\|\mathbf{x}\|\|\mathbf{y}\| \quad \text{for all } \mathbf{x}, \mathbf{y} \in \mathcal{H}.$$

Thus, for each fixed $\mathbf{y} \in \mathcal{H}$, the linear functional $\mathbf{x} \mapsto \langle A\mathbf{x}, \mathbf{y}\rangle$ is bounded, meaning it defines an element of $\mathcal{B}(\mathcal{H}, \mathbb{C})$. Theorem 3.1.3 below describes the bounded linear functionals on $\mathcal{H}$. To develop this, we need to further explore the following concept first seen in Chapter 2.

Definition 3.1.1. For a Hilbert space $\mathcal{H}$ and a nonempty subset $\mathcal{Y} \subseteq \mathcal{H}$, let

$$\mathcal{Y}^\perp = \{\mathbf{x} \in \mathcal{H} : \langle \mathbf{x}, \mathbf{y}\rangle = 0 \text{ for all } \mathbf{y} \in \mathcal{Y}\}$$

denote the *orthogonal complement* of $\mathcal{Y}$ (see Figure 3.1.1).

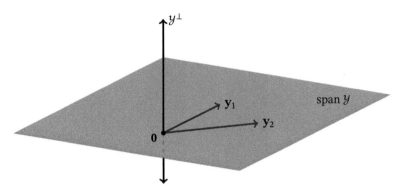

Figure 3.1.1 The orthogonal complement of $\mathcal{Y} = \{\mathbf{y}_1, \mathbf{y}_2\}$ is the subspace $\mathcal{Y}^\perp$.

Here are a few facts about orthogonal complements (Exercise 3.6.6).

Proposition 3.1.2. *Let $\mathcal{Y}$ be a nonempty subset of a Hilbert space $\mathcal{H}$.*

(a) $\{\mathbf{0}\}^\perp = \mathcal{H}$.

(b) $\mathcal{H}^\perp = \{\mathbf{0}\}$.

(c) $\mathcal{Y}^\perp$ *is a subspace of $\mathcal{H}$.*

(d) $(\mathcal{Y}^\perp)^\perp = \bigvee \mathcal{Y}$.

(e) $\mathcal{Y}$ *has dense linear span in $\mathcal{H}$ if and only if $\mathcal{Y}^\perp = \{\mathbf{0}\}$.*

(f) *If $\mathcal{M}$ is a subspace of $\mathcal{H}$, then $\mathcal{M} \neq \mathcal{H}$ if and only if $\mathcal{M}^\perp \neq \{\mathbf{0}\}$.*

(g) *If $P_\mathcal{M}$ is the orthogonal projection onto $\mathcal{M}$, then $I - P_\mathcal{M}$ is the orthogonal projection onto $\mathcal{M}^\perp$.*

The next theorem identifies the dual space (recall Definition 2.2.4) $\mathcal{H}^* = \mathcal{B}(\mathcal{H}, \mathbb{C})$ of a Hilbert space $\mathcal{H}$. It can be identified, in a conjugate-linear fashion, with $\mathcal{H}$ itself.

Theorem 3.1.3 (Riesz representation theorem). *For a Hilbert space $\mathcal{H}$, the following hold.*

(a) *Every bounded linear functional $\varphi : \mathcal{H} \to \mathbb{C}$ is of the form $\varphi_\mathbf{y}(\mathbf{x}) = \langle \mathbf{x}, \mathbf{y} \rangle$ for a unique $\mathbf{y} \in \mathcal{H}$.*

(b) *The map $\mathbf{y} \mapsto \varphi_\mathbf{y}$ is a conjugate-linear isometry from $\mathcal{H}$ onto $\mathcal{H}^*$.*

Proof (a) Since φ is bounded, $\ker \varphi$ is a subspace of $\mathcal{H}$. Without loss of generality we may assume that $\ker \varphi \neq \mathcal{H}$, since otherwise φ is the zero functional. In this case, $\varphi(\mathbf{x}) = \langle \mathbf{x}, \mathbf{0} \rangle$ for all $\mathbf{x} \in \mathcal{H}$ and $\mathbf{0}$ is the only vector with this property (Proposition 3.1.2b).

Since $\ker \varphi \neq \mathcal{H}$, Proposition 3.1.2f implies that $(\ker \varphi)^\perp \neq \{\mathbf{0}\}$ is a subspace of $\mathcal{H}$. Let $\mathbf{y}_0 \in (\ker \varphi)^\perp$ satisfy $\varphi(\mathbf{y}_0) = 1$. In particular, $\mathbf{y}_0 \neq \mathbf{0}$. Then for all $\mathbf{x} \in \mathcal{H}$,

$$\varphi(\mathbf{x} - \varphi(\mathbf{x})\mathbf{y}_0) = \varphi(\mathbf{x}) - \varphi(\mathbf{x})\varphi(\mathbf{y}_0) = 0.$$

Thus, $\mathbf{x} - \varphi(\mathbf{x})\mathbf{y}_0 \in \ker \varphi$ and

$$0 = \langle \mathbf{x} - \varphi(\mathbf{x})\mathbf{y}_0, \mathbf{y}_0 \rangle = \langle \mathbf{x}, \mathbf{y}_0 \rangle - \varphi(\mathbf{x}) \|\mathbf{y}_0\|^2.$$

Rearrange the preceding to obtain

$$\varphi(\mathbf{x}) = \|\mathbf{y}_0\|^{-2} \langle \mathbf{x}, \mathbf{y}_0 \rangle = \left\langle \mathbf{x}, \|\mathbf{y}_0\|^{-2} \mathbf{y}_0 \right\rangle = \langle \mathbf{x}, \mathbf{y} \rangle,$$

where $\mathbf{y} = \|\mathbf{y}_0\|^{-2} \mathbf{y}_0$. This proves existence. For uniqueness, suppose that $\varphi(\mathbf{x}) = \langle \mathbf{x}, \mathbf{y}' \rangle$ for all $\mathbf{x} \in \mathcal{H}$. Then,

$$0 = \langle \mathbf{x}, \mathbf{y} \rangle - \langle \mathbf{x}, \mathbf{y}' \rangle = \langle \mathbf{x}, \mathbf{y} - \mathbf{y}' \rangle \quad \text{for all } \mathbf{x} \in \mathcal{H}.$$

Let $\mathbf{x} = \mathbf{y} - \mathbf{y}'$ and conclude that $\|\mathbf{y} - \mathbf{y}'\|^2 = 0$ (Proposition 3.1.2b). Thus, $\mathbf{y} = \mathbf{y}'$.
(b) For fixed $\mathbf{y} \in \mathcal{H}$, the Cauchy–Schwarz inequality implies that

$$|\varphi_{\mathbf{y}}(\mathbf{x})| = |\langle \mathbf{x}, \mathbf{y} \rangle| \leqslant \|\mathbf{x}\|\|\mathbf{y}\| \quad \text{for all } \mathbf{x} \in \mathcal{H}$$

and hence,

$$\|\varphi_{\mathbf{y}}\| = \sup_{\|\mathbf{x}\|=1} |\varphi_{\mathbf{y}}(\mathbf{x})| \leqslant \|\mathbf{y}\|.$$

Furthermore, assuming $\mathbf{y} \neq \mathbf{0}$,

$$\varphi_{\mathbf{y}}\left(\frac{\mathbf{y}}{\|\mathbf{y}\|}\right) = \left\langle \frac{\mathbf{y}}{\|\mathbf{y}\|}, \mathbf{y} \right\rangle = \|\mathbf{y}\|$$

and thus $\|\varphi_{\mathbf{y}}\| = \|\mathbf{y}\|$. Since the inner product is conjugate linear in the second slot, the map $\mathbf{y} \mapsto \varphi_{\mathbf{y}}$ is a conjugate linear isometry. Since every $\varphi \in \mathcal{H}$ is equal to $\varphi_{\mathbf{y}}$ for some unique $\mathbf{y} \in \mathcal{H}$, this conjugate-linear isometry is surjective. ∎

Let $A \in \mathcal{B}(\mathcal{H})$. For each $\mathbf{y} \in \mathcal{H}$, the Riesz representation theorem provides a unique vector, denoted $A^*\mathbf{y}$, such that

$$\langle A\mathbf{x}, \mathbf{y} \rangle = \langle \mathbf{x}, A^*\mathbf{y} \rangle \quad \text{for all } \mathbf{x} \in \mathcal{H}. \tag{3.1.4}$$

The next proposition shows that $\mathbf{x} \mapsto A^*\mathbf{x}$ is a bounded linear transformation on $\mathcal{H}$. We use A^* to denote this map, which is uniquely determined by (3.1.4).

Proposition 3.1.5. *Let $A \in \mathcal{B}(\mathcal{H})$.*

(a) $\mathbf{x} \mapsto A^*\mathbf{x}$ *defines a linear transformation on $\mathcal{H}$.*

(b) $A^* \in \mathcal{B}(\mathcal{H})$.

(c) $A^{**} = A$.

(d) $\|A^*\| = \|A\|$.

(e) $\|A^*A\| = \|A\|^2$.

Proof (a) If $\mathbf{x}, \mathbf{y}, \mathbf{z} \in \mathcal{H}$ and $c \in \mathbb{C}$, one sees from (3.1.4) that

$$\langle A^*(\mathbf{x} + c\mathbf{y}), \mathbf{z} \rangle = \langle \mathbf{x} + c\mathbf{y}, A\mathbf{z} \rangle$$
$$= \langle \mathbf{x}, A\mathbf{z} \rangle + c \langle \mathbf{y}, A\mathbf{z} \rangle$$
$$= \langle A^*\mathbf{x}, \mathbf{z} \rangle + c \langle A^*\mathbf{y}, \mathbf{z} \rangle$$
$$= \langle A^*\mathbf{x} + cA^*\mathbf{y}, \mathbf{z} \rangle.$$

Thus, $A^*(\mathbf{x} + c\mathbf{y}) = A^*\mathbf{x} + cA^*\mathbf{y}$ and hence A^* is a linear transformation on $\mathcal{H}$. Note the use of the fact that if $\langle \mathbf{u}, \mathbf{z} \rangle = \langle \mathbf{v}, \mathbf{z} \rangle$ for every $\mathbf{z} \in \mathcal{H}$, then $\mathbf{u} = \mathbf{v}$.

(b) For all $\mathbf{y} \in \mathcal{H}$, the Cauchy–Schwarz inequality and the definition of the operator norm ensure that

$$\|A^*\mathbf{y}\| = \left\langle \frac{A^*\mathbf{y}}{\|A^*\mathbf{y}\|}, A^*\mathbf{y} \right\rangle$$
$$\leqslant \sup_{\|\mathbf{x}\|=1} |\langle \mathbf{x}, A^*\mathbf{y} \rangle|$$
$$= \sup_{\|\mathbf{x}\|=1} |\langle A\mathbf{x}, \mathbf{y} \rangle|$$
$$\leqslant \|\mathbf{y}\| \sup_{\|\mathbf{x}\|=1} \|A\mathbf{x}\|$$
$$= \|\mathbf{y}\| \|A\|.$$

This shows that $\|A^*\| \leqslant \|A\|$ and hence A^* is bounded.

(c) From (3.1.4) and the previous parts of this proposition, it follows that $A^{**} = (A^*)^*$ is the unique bounded linear operator on $\mathcal{H}$ such that $\langle A^{**}\mathbf{x}, \mathbf{y} \rangle = \langle \mathbf{x}, A^*\mathbf{y} \rangle$ for all $\mathbf{x}, \mathbf{y}$. However, A also satisfies this property, so $A^{**} = A$.

(d) The proof of (b) shows that $\|A^*\| \leqslant \|A\|$. Apply this to A^* in place of A and use (c) to conclude that $\|A\| = \|A^{**}\| = \|(A^*)^*\| \leqslant \|A^*\|$. Therefore, $\|A\| = \|A^*\|$.

(e) Note that

$$\|A\|^2 = \sup_{\|\mathbf{x}\|=1} \|A\mathbf{x}\|^2$$
$$= \sup_{\|\mathbf{x}\|=1} \langle A\mathbf{x}, A\mathbf{x} \rangle$$
$$= \sup_{\|\mathbf{x}\|=1} \langle A^*A\mathbf{x}, \mathbf{x} \rangle$$
$$\leqslant \sup_{\|\mathbf{x}\|=1} \|A^*A\mathbf{x}\| \qquad \text{(by Cauchy–Schwarz)}$$
$$= \|A^*A\|$$
$$\leqslant \|A^*\| \|A\| \qquad \text{(by Proposition 2.1.12)}$$
$$= \|A\|^2 \qquad \text{(from (d))}.$$

Thus, the previous inequalities must be equalities, and hence $\|A^*A\| = \|A\|^2$. ∎

Exercise 3.6.5 establishes the following properties of the adjoint operation. In particular, the adjoint operation is conjugate linear and reverses the order of operator multiplication.

Proposition 3.1.6. *Let $A, B \in \mathcal{B}(\mathcal{H})$ and $\lambda \in \mathbb{C}$.*

(a) $\sigma(A^*) = \{\bar{\lambda} : \lambda \in \sigma(A)\}$.

(b) $(A + \lambda B)^* = A^* + \bar{\lambda} B^*$.

(c) $(AB)^* = B^* A^*$.

Another set of useful facts is how the kernel and range of an operator behave with respect to orthogonal complements and adjoints.

Proposition 3.1.7. *Let $A \in \mathcal{B}(\mathcal{H})$.*

(a) $\ker A = (\operatorname{ran} A^*)^\perp$.

(b) $\ker A^* = (\operatorname{ran} A)^\perp$.

(c) $(\ker A)^\perp = (\operatorname{ran} A^*)^-$.

(d) $(\ker A^*)^\perp = (\operatorname{ran} A)^-$.

Proof (a) Suppose that $\mathbf{x} \in \ker A$. Then $A\mathbf{x} = \mathbf{0}$, and so

$$0 = \langle \mathbf{0}, \mathbf{y} \rangle = \langle A\mathbf{x}, \mathbf{y} \rangle = \langle \mathbf{x}, A^*\mathbf{y} \rangle \quad \text{for all } \mathbf{y} \in \mathcal{H}.$$

Thus, $\mathbf{x} \in (\operatorname{ran} A^*)^\perp$. Conversely, if $\mathbf{x} \in (\operatorname{ran} A^*)^\perp$, then

$$0 = \langle \mathbf{x}, A^*\mathbf{y} \rangle = \langle A\mathbf{x}, \mathbf{y} \rangle \quad \text{for all } \mathbf{y} \in \mathcal{H}.$$

Thus, $A\mathbf{x} = \mathbf{0}$ and hence $\mathbf{x} \in \ker A$.

(b) Apply (a) to A^* and recall that $A^{**} = A$.

(c) Since $\operatorname{ran} A^*$ is a vector space (but not necessarily a subspace since it may not be topologically closed), Proposition 3.1.2 and (a) imply that

$$(\ker A)^\perp = ((\operatorname{ran} A^*)^\perp)^\perp = (\operatorname{ran} A^*)^-.$$

(d) Apply (a) to A^* and recall that $A^{**} = A$. ∎

The previous material can be generalized to linear transformations between two Hilbert spaces $\mathcal{H}$ and $\mathcal{K}$. If $A \in \mathcal{B}(\mathcal{H}, \mathcal{K})$, then there is a unique $A^* \in \mathcal{B}(\mathcal{K}, \mathcal{H})$ such that

$$\langle A\mathbf{x}, \mathbf{y} \rangle_{\mathcal{K}} = \langle \mathbf{x}, A^*\mathbf{y} \rangle_{\mathcal{H}} \quad \text{for all } \mathbf{x} \in \mathcal{H} \text{ and } \mathbf{y} \in \mathcal{K}.$$

We mostly use the adjoint of an $A \in \mathcal{B}(\mathcal{H})$ but, from time to time, we need the adjoint of $A \in \mathcal{B}(\mathcal{H}, \mathcal{K})$.

3.2 Special Case of Schur's Test

To discuss when an infinite matrix $A = [a_{ij}]_{i,j=0}^{\infty}$ induces a bounded operator $\mathbf{x} \mapsto A\mathbf{x}$ on ℓ^2, we need the following indirect way to determine if a sequence $\mathbf{x} = (x_n)_{n=0}^{\infty}$ belongs to ℓ^2. Since ℓ^2 is indexed starting with 0, we index the infinite matrices below starting with $(0, 0)$. These choices reflect the fact that later in the book we frequently identify a sequence $\mathbf{a} = (a_n)_{n=0}^{\infty}$ with a power series $\sum_{n=0}^{\infty} a_n z^n$.

Proposition 3.2.1. *For a sequence* $\mathbf{x} = (x_n)_{n=0}^{\infty}$, *the following are equivalent.*

(a) $\mathbf{x} \in \ell^2$.

(b) $\displaystyle\sum_{n=0}^{\infty} x_n y_n$ *converges for every* $\mathbf{y} \in \ell^2$.

Furthermore, if there is an $M > 0$ *such that for each* $\mathbf{y} = (y_n)_{n=0}^{\infty} \in \ell^2$, *the series* $\sum_{n=0}^{\infty} x_n y_n$ *converges and*

$$\left| \sum_{n=0}^{\infty} x_n y_n \right| \leqslant M \|\mathbf{y}\|, \tag{3.2.2}$$

then $\mathbf{x} \in \ell^2$ *with* $\|\mathbf{x}\| \leqslant M$.

Proof (a) $\Rightarrow$ (b) The Cauchy–Schwarz inequality ensures that the series $\sum_{n=0}^{\infty} x_n y_n$ converges absolutely:

$$\sum_{n=0}^{\infty} |x_n| |y_n| \leqslant \left(\sum_{n=0}^{\infty} |x_n|^2 \right)^{\frac{1}{2}} \left(\sum_{n=0}^{\infty} |y_n|^2 \right)^{\frac{1}{2}}.$$

Therefore, $\sum_{n=0}^{\infty} x_n y_n$ converges.

(b) $\Rightarrow$ (a) For fixed $\mathbf{x} = (x_n)_{n=0}^{\infty}$ and $N \geqslant 1$, define the linear functional $T_N : \ell^2 \to \mathbb{C}$ by

$$T_N(\mathbf{y}) = \sum_{n=0}^{N} x_n y_n \quad \text{for } \mathbf{y} = (y_n)_{n=0}^{\infty} \in \ell^2.$$

Theorem 3.1.3 shows that $T_N \in \mathcal{B}(\ell^2, \mathbb{C})$ and

$$\|T_N\| = \left(\sum_{n=0}^{N} |x_n|^2 \right)^{\frac{1}{2}}. \tag{3.2.3}$$

The assumption in (b) says that $\lim_{N\to\infty} T_N(\mathbf{y})$ exists for each $\mathbf{y} \in \ell^2$. Consequently, the principle of uniform boundedness (Theorem 2.2.3) implies that

$$L = \sup_{N \geqslant 0} \|T_N\| < \infty.$$

Then (3.2.3) yields

$$\sum_{n=0}^{N} |x_n|^2 \leqslant L^2 \quad \text{for all } N \geqslant 0.$$

Let $N \to \infty$ and conclude that (3.2.2) holds and $\mathbf{x} \in \ell^2$.

For the second part of the proposition, observe that (3.2.2) implies that $\|T_N\| \leqslant M$ for all $N \geqslant 0$. Therefore,

$$\|\mathbf{x}\| = \lim_{N \to \infty} \Big(\sum_{n=0}^{N} |x_n|^2 \Big)^{\frac{1}{2}} = \lim_{N \to \infty} \|T_N\| \leqslant M,$$

which completes the proof. ∎

The main result of this section concerns the case where the infinite matrix $A = [a_{ij}]_{i,j=0}^{\infty}$ is selfadjoint, that is,

$$a_{ij} = \overline{a_{ji}} \quad \text{for all } i, j \geqslant 0.$$

Proposition 3.2.4. *Let $A = [a_{ij}]_{i,j=0}^{\infty}$ be selfadjoint. If there is an $M > 0$ such that*

$$\sum_{j=0}^{\infty} |a_{ij}| \leqslant M \quad \text{for all } i \geqslant 0, \tag{3.2.5}$$

then A defines a bounded operator $\mathbf{x} \mapsto A\mathbf{x}$ on ℓ^2 and $\|A\| \leqslant M$.

Proof Let $\mathbf{x} \in \ell^2$. For each $i \geqslant 0$, (3.2.5) ensures that the number

$$z_i = \sum_{j=0}^{\infty} a_{ij} x_j$$

is well defined. For each $\mathbf{w} = (w_n)_{n=0}^{\infty} \in \ell^2$,

$$\sum_{i=0}^{\infty} |z_i w_i| = \sum_{i=0}^{\infty} \Big| \sum_{j=0}^{\infty} a_{ij} x_j w_i \Big|$$

$$\leqslant \sum_{i,j=0}^{\infty} |a_{ij} x_j w_i|$$

$$= \sum_{i,j=0}^{\infty} \Big(|a_{ij}|^{\frac{1}{2}} |x_j| \Big) \Big(|a_{ij}|^{\frac{1}{2}} |w_i| \Big)$$

$$\leqslant \Big(\sum_{i,j=0}^{\infty} |a_{ij}| |x_j|^2 \Big)^{\frac{1}{2}} \Big(\sum_{i,j=0}^{\infty} |a_{ij}| |w_i|^2 \Big)^{\frac{1}{2}}. \tag{3.2.6}$$

The selfadjointness of A and (3.2.5) imply

$$\sum_{i=0}^{\infty} \sum_{j=0}^{\infty} |a_{ij}| |w_i|^2 = \sum_{i=0}^{\infty} \Big(\sum_{j=0}^{\infty} |a_{ij}| \Big) |w_i|^2 \leqslant M \sum_{i=0}^{\infty} |w_i|^2$$

and

$$\sum_{i=0}^{\infty} \sum_{j=0}^{\infty} |a_{ij}| |x_j|^2 = \sum_{j=0}^{\infty} \Big(\sum_{i=0}^{\infty} |a_{ij}| \Big) |x_j|^2 \leqslant M \sum_{j=0}^{\infty} |x_j|^2.$$

Inserting the two estimates above into (3.2.6) yields

$$\sum_{i=0}^{\infty} |z_i w_i| \leqslant \left(M \sum_{j=0}^{\infty} |x_j|^2\right)^{\frac{1}{2}} \left(M \sum_{i=0}^{\infty} |w_i|^2\right)^{\frac{1}{2}}$$

$$= M \|\mathbf{x}\| \|\mathbf{w}\|.$$

Proposition 3.2.1 implies that $\mathbf{z} = A\mathbf{x} \in \ell^2$ and $\|A\mathbf{x}\| \leqslant M\|\mathbf{x}\|$ for all $\mathbf{x} \in \ell^2$. Thus, A is a bounded operator on ℓ^2 and $\|A\| \leqslant M$. ∎

Example 3.2.7. Consider

$$T = \begin{bmatrix} 0 & 1 & 0 & 0 & 0 & \cdots \\ 1 & 0 & 1 & 0 & 0 & \cdots \\ 0 & 1 & 0 & 1 & 0 & \cdots \\ 0 & 0 & 1 & 0 & 1 & \cdots \\ 0 & 0 & 0 & 1 & 0 & \cdots \\ \vdots & \vdots & \vdots & \vdots & \vdots & \ddots \end{bmatrix},$$

which is an example of a Toeplitz matrix (Chapter 16). Apply Proposition 3.2.4 with $M = 2$ to conclude that T defines a bounded operator on ℓ^2 and $\|T\| \leqslant 2$. One can see that $\|T\| = 2$ as follows. Let

$$\mathbf{x}_n = \frac{1}{\sqrt{n}}(1, 1, \ldots, 1, 0, 0, 0, \ldots),$$

where there are n ones in the vector above. Then $\mathbf{x}_n$ is a unit vector and

$$T\mathbf{x}_n = \frac{1}{\sqrt{n}}(1, 2, 2, 2, \ldots, 2, 1, 1, 0, 0, 0, \ldots),$$

where there are $n - 3$ 2s in the vector above. Moreover,

$$\lim_{n \to \infty} \|T\mathbf{x}_n\| = \lim_{n \to \infty} \frac{1}{\sqrt{n}}\sqrt{3 + 4(n-3)} = 2$$

and hence $\|T\| = 2$.

Example 3.2.8. In Proposition 3.2.4, the hypothesis that A is selfadjoint is necessary. For example,

$$A = \begin{bmatrix} 1 & 0 & 0 & 0 & \cdots \\ 1 & 0 & 0 & 0 & \cdots \\ 1 & 0 & 0 & 0 & \cdots \\ 1 & 0 & 0 & 0 & \cdots \\ \vdots & \vdots & \vdots & \vdots & \ddots \end{bmatrix}$$

has entries that are summable along each row, but it does not define a bounded operator on ℓ^2 since $A\mathbf{e}_0 = (1, 1, 1, \ldots)$ does not belong to ℓ^2.

3.3 Schur's Test

Proposition 3.2.4 applies only to selfadjoint matrices. The next theorem of Schur is for arbitrary infinite matrices.

Theorem 3.3.1 (Schur). *Let $A = [a_{ij}]_{i,j=0}^{\infty}$. Suppose there are positive numbers α and β and sequences $(p_i)_{i=0}^{\infty}$ and $(q_i)_{i=0}^{\infty}$ of positive numbers such that*

$$\sum_{i=0}^{\infty} |a_{ij}| p_i \leqslant \alpha q_j \ \text{ for } j \geqslant 0 \quad \text{and} \quad \sum_{j=0}^{\infty} |a_{ij}| q_j \leqslant \beta p_i \ \text{ for } i \geqslant 0. \qquad (3.3.2)$$

Then A defines a bounded operator on ℓ^2 and $\|A\| \leqslant \sqrt{\alpha\beta}$.

Proof Let $\mathbf{x} = (x_i)_{i=0}^{\infty}$ and $\mathbf{w} = (w_i)_{i=0}^{\infty}$ belong to ℓ^2 and write

$$\sum_{i,j=0}^{\infty} |a_{ij} x_j w_i| = \sum_{i,j=0}^{\infty} \left(\frac{\sqrt{p_i}}{\sqrt{q_j}}|a_{ij}|^{\frac{1}{2}}|x_j|\right)\left(\frac{\sqrt{q_j}}{\sqrt{p_i}}|a_{ij}|^{\frac{1}{2}}|w_i|\right).$$

The Cauchy–Schwarz inequality implies

$$\sum_{i,j=0}^{\infty} |a_{ij} x_j w_i| \leqslant \left(\sum_{i,j=0}^{\infty} \frac{p_i}{q_j}|a_{ij}| |x_j|^2\right)^{\frac{1}{2}}\left(\sum_{i,j=0}^{\infty} \frac{q_j}{p_i}|a_{ij}| |w_i|^2\right)^{\frac{1}{2}}.$$

The assumptions in (3.3.2) say that

$$\sum_{i,j=0}^{\infty} \frac{q_j}{p_i}|a_{ij}| |w_i|^2 = \sum_{i=0}^{\infty}\left(\sum_{j=0}^{\infty} q_j|a_{ij}|\right)\frac{|w_i|^2}{p_i} \leqslant \beta \sum_{i=0}^{\infty} |w_i|^2$$

and

$$\sum_{i,j=0}^{\infty} \frac{p_i}{q_j}|a_{ij}| |x_j|^2 = \sum_{j=0}^{\infty}\left(\sum_{i=0}^{\infty} p_i|a_{ij}|\right)\frac{|x_j|^2}{q_j} \leqslant \alpha \sum_{j=0}^{\infty} |x_j|^2.$$

Thus,

$$\sum_{i,j=0}^{\infty} |a_{ij} x_j w_i| \leqslant \sqrt{\alpha\beta}\|\mathbf{x}\|\|\mathbf{w}\|.$$

This shows that

$$y_i = \sum_{j=0}^{\infty} a_{ij} x_j$$

is a well-defined convergent series and thus the product $\mathbf{y} = A\mathbf{x}$ is well defined for each $\mathbf{x} \in \ell^2$. Proposition 3.2.1 ensures that $\mathbf{y} \in \ell^2$ and $\|A\mathbf{x}\| \leqslant \sqrt{\alpha\beta}\|\mathbf{x}\|$. In other words, A is a bounded operator on ℓ^2 and $\|A\| \leqslant \sqrt{\alpha\beta}$. ∎

Example 3.3.3. Consider

$$C = [c_{ij}]_{i,j=0}^{\infty} = \begin{bmatrix} 1 & 0 & 0 & 0 & 0 & \cdots \\ \frac{1}{2} & \frac{1}{2} & 0 & 0 & 0 & \cdots \\ \frac{1}{3} & \frac{1}{3} & \frac{1}{3} & 0 & 0 & \cdots \\ \frac{1}{4} & \frac{1}{4} & \frac{1}{4} & \frac{1}{4} & 0 & \cdots \\ \vdots & \vdots & \vdots & \vdots & \vdots & \ddots \end{bmatrix}.$$

This is the matrix representation of the Cesàro operator (Chapter 6). To apply Theorem 3.3.1, we follow an argument from [69] and let

$$p_n = q_n = \frac{1}{\sqrt{n+1}} \quad \text{for } n \geqslant 0.$$

Then for $j > 0$,

$$\sum_{i=0}^{\infty} c_{ij} p_i = \sum_{i=j}^{\infty} c_{ij} p_i$$

$$= \sum_{i=j}^{\infty} \frac{1}{i+1} \frac{1}{\sqrt{i+1}}$$

$$\leqslant \int_{j-1}^{\infty} \frac{dx}{(x+1)^{3/2}}$$

$$= \frac{2}{\sqrt{j}}$$

$$= \frac{\sqrt{j+1}}{\sqrt{j}} \frac{2}{\sqrt{j+1}}$$

$$\leqslant 2\sqrt{2} \, q_j.$$

For $j = 0$,

$$\sum_{i=0}^{\infty} c_{i0} p_i = 1 + \sum_{i=1}^{\infty} c_{i0} p_i \leqslant 1 + 2 = 3q_0 \quad (\text{since } q_0 = 1).$$

Furthermore,

$$\sum_{j=0}^{\infty} c_{ij} q_j = \sum_{j=0}^{i} \frac{1}{i+1} \frac{1}{\sqrt{j+1}}$$

$$\leqslant \frac{1}{i+1} \int_0^i \frac{dx}{\sqrt{x}}$$

$$= \frac{1}{i+1} 2\sqrt{i}$$

$$\leqslant \frac{1}{i+1} 2\sqrt{i+1}$$

$$= 2p_i.$$

Apply Schur's test with $\alpha = 3$ and $\beta = 2$ to deduce that C defines a bounded operator on ℓ^2 and $\|C\| \leqslant \sqrt{\alpha\beta} = \sqrt{3 \cdot 2} = \sqrt{6} \approx 2.49$. Proposition 6.2.9 asserts that $\|C\| = 2$.

Example 3.3.4. The infinite *Hilbert matrix*

$$
H = \begin{bmatrix}
1 & \frac{1}{2} & \frac{1}{3} & \frac{1}{4} & \cdots \\
\frac{1}{2} & \frac{1}{3} & \frac{1}{4} & \frac{1}{5} & \cdots \\
\frac{1}{3} & \frac{1}{4} & \frac{1}{5} & \frac{1}{6} & \cdots \\
\frac{1}{4} & \frac{1}{5} & \frac{1}{6} & \frac{1}{7} & \cdots \\
\vdots & \vdots & \vdots & \vdots & \ddots
\end{bmatrix}
$$

is an example of a Hankel matrix (Chapter 17). Here $H = [h_{ij}]_{i,j=0}^{\infty}$, where

$$
h_{ij} = \frac{1}{i + j + 1}.
$$

Apply Schur's test with $p_i = q_i = \dfrac{1}{\sqrt{i+1}}$ and $\alpha = \beta = 5$. To see how this works, let

$$
f(x) = \frac{1}{(x + j + 1)\sqrt{x + 1}}.
$$

The integral test says that

$$
\sum_{i=0}^{\infty} f(i) = f(0) + \sum_{i=1}^{\infty} f(i)
$$

$$
\leqslant f(0) + \int_0^{\infty} f(x)\,dx
$$

$$
= \frac{1}{j+1} + 2\frac{\tan^{-1}(\sqrt{j})}{\sqrt{j}}
$$

$$
\leqslant \frac{1}{\sqrt{j+1}} + 2\frac{\tan^{-1}(\sqrt{j})}{\sqrt{j}}
$$

$$
\leqslant \frac{1}{\sqrt{j+1}} + \frac{4}{\sqrt{j+1}}
$$

$$
= \frac{5}{\sqrt{j+1}}.
$$

In a similar manner,

$$
\sum_{j=1}^{\infty} \frac{1}{i + j + 1} \frac{1}{\sqrt{j+1}} \leqslant \frac{5}{\sqrt{i+1}}.
$$

Deduce from Theorem 3.3.1 that $\|H\| \leqslant 5$. In fact, $\|H\| = \pi$ (Chapter 17).

The following is a particular version of Schur's test that is somewhat easier to apply.

Corollary 3.3.5. Let $A = [a_{ij}]_{i,j=0}^{\infty}$. Suppose there are positive constants α and β such that

$$\sum_{i=0}^{\infty} |a_{ij}| \leqslant \alpha \ \text{ for } \ j \geqslant 0 \quad \text{and} \quad \sum_{j=0}^{\infty} |a_{ij}| \leqslant \beta \ \text{ for } \ i \geqslant 0.$$

Then A is a bounded operator on ℓ^2 and $\|A\| \leqslant \sqrt{\alpha\beta}$.

Proof Apply Theorem 3.3.1 with $q_j = 1$ and $p_i = 1$ for all i, j. ■

Remark 3.3.6. A slight abuse of notation is standard. We identify a matrix $A = [a_{ij}]_{i,j=0}^{\infty}$ with the linear operator it induces on ℓ^2 with respect to the standard basis. In this context, the adjoint notation is suggestive: the conjugate transpose $[\overline{a_{ji}}]_{i,j=0}^{\infty}$ of the matrix A is the matrix representation (with respect to the standard basis for ℓ^2) of the adjoint operator A^*. Thus, A^* may refer to the adjoint of the operator A or to the conjugate transpose of the matrix A without confusion.

3.4 Compactness and Contractions

A diagonal operator D_Λ is compact if and only if its eigenvalues tend to zero (Theorem 2.5.1). Here is a useful compactness result for certain types of infinite matrices. This is used in Proposition 18.1.3 to study compact composition operators.

Theorem 3.4.1. Let $A = [a_{ij}]_{i,j=0}^{\infty}$. If

$$M = \sum_{i,j=0}^{\infty} |a_{ij}| < \infty,$$

then A is a compact operator on ℓ^2.

Proof Corollary 3.3.5 ensures that A is bounded. Let $\varepsilon > 0$ and pick N such that

$$\sum_{i,j=0}^{N} |a_{ij}| > M - \varepsilon.$$

Now define

$$A_N = \begin{bmatrix} a_{00} & a_{01} & a_{02} & a_{03} & \cdots & a_{0N} & 0 & 0 & 0 & \cdots \\ a_{10} & a_{11} & a_{12} & a_{13} & \cdots & a_{1N} & 0 & 0 & 0 & \cdots \\ a_{20} & a_{21} & a_{22} & a_{23} & \cdots & a_{2N} & 0 & 0 & 0 & \cdots \\ a_{30} & a_{31} & a_{32} & a_{33} & \cdots & a_{3N} & 0 & 0 & 0 & \cdots \\ \vdots & \vdots & \vdots & \vdots & \ddots & \vdots & 0 & 0 & 0 & \cdots \\ a_{N0} & a_{N1} & a_{N2} & a_{N3} & \cdots & a_{NN} & 0 & 0 & 0 & \cdots \\ 0 & 0 & 0 & 0 & \cdots & 0 & 0 & 0 & 0 & \cdots \\ 0 & 0 & 0 & 0 & \cdots & 0 & 0 & 0 & 0 & \cdots \\ 0 & 0 & 0 & 0 & \cdots & 0 & 0 & 0 & 0 & \cdots \\ \vdots & \vdots & \vdots & \vdots & \cdots & \vdots & \vdots & \vdots & \vdots & \ddots \end{bmatrix}.$$

Since $\operatorname{ran} A_N \subseteq \operatorname{span}\{\mathbf{e}_0, \mathbf{e}_1, \dots, \mathbf{e}_N\}$, it follows that A_N has finite rank and is therefore compact (Proposition 2.5.4). Corollary 3.3.5 says that $A - A_N$ is a bounded operator on ℓ^2. Moreover,

$$\|A - A_N\| \leqslant \sum_{i,j=0}^{\infty} |a_{ij}| - \sum_{i,j=0}^{N} |a_{ij}| < M - (M - \varepsilon) = \varepsilon.$$

Thus, A is the norm limit of the finite-rank operators A_N and hence is compact (Proposition 2.5.6). ∎

Definition 3.4.2. *$T \in \mathcal{B}(\mathcal{H})$ is a* contraction *if $\|T\mathbf{x}\| \leqslant \|\mathbf{x}\|$ for all $\mathbf{x} \in \mathcal{H}$.*

If T is nonzero, then cT is a contraction for $|c| \leqslant \|T\|^{-1}$. We have considered operators on ℓ^2 given by infinite matrices $A = [a_{ij}]_{i,j=0}^{\infty}$, where $A\mathbf{x} = \mathbf{y} = (y_i)_{i=0}^{\infty}$ satisfies

$$y_i = \sum_{j=0}^{\infty} a_{ij} x_j \quad \text{for } i \geqslant 0.$$

From time to time, starting with the next lemma, we consider operators on $\ell^2(\mathbb{Z})$ given by infinite matrices $A = [a_{ij}]_{i,j=-\infty}^{\infty}$, where $A\mathbf{x} = \mathbf{y} = (y_i)_{i=-\infty}^{\infty}$ satisfies

$$y_i = \sum_{j=-\infty}^{\infty} a_{ij} x_j \quad \text{for } i \in \mathbb{Z}.$$

Lemma 3.4.3. *Let $A = [a_{ij}]_{i,j=-\infty}^{\infty}$ be such that for each fixed $i_0, j_0 \in \mathbb{Z}$, the submatrix $A(i_0, j_0) = [a_{ij}]_{i \geqslant i_0, j \geqslant j_0}$ is a contraction on ℓ^2. Then $[a_{ij}]_{i,j=-\infty}^{\infty}$ is a contraction on $\ell^2(\mathbb{Z})$.*

Proof The matrix $A(i_0, j_0)$ is a contraction and hence $A(i_0, j_0)\mathbf{e}_j$, the jth column of $A(i_0, j_0)$, has norm at most 1. Observe that

$$A(i_0, j_0)^* = [\overline{a_{ji}}]_{i \geqslant i_0, j \geqslant j_0}$$

(Remark 3.3.6) is also a contraction (Proposition 3.1.5) and thus $\|A(i_0, j_0)^* \mathbf{e}_j\| \leqslant 1$. In other words,

$$\sum_{j=j_0}^{\infty} |a_{ij}|^2 \leqslant 1 \quad \text{for all } i, j_0 \in \mathbb{Z}.$$

Let $j_0 \to -\infty$ and deduce that

$$\sum_{j=-\infty}^{\infty} |a_{ij}|^2 \leqslant 1 \quad \text{for } i \in \mathbb{Z}. \tag{3.4.4}$$

Given $\mathbf{x} = (x_j)_{j=-\infty}^{\infty} \in \ell^2(\mathbb{Z})$, (3.4.4) ensures that for each fixed $i \in \mathbb{Z}$,

$$\sum_{j=-\infty}^{\infty} |a_{ij} x_j| \leqslant \Big(\sum_{j=-\infty}^{\infty} |a_{ij}|^2 \Big)^{\frac{1}{2}} \Big(\sum_{j=-\infty}^{\infty} |x_j|^2 \Big)^{\frac{1}{2}} \leqslant \|\mathbf{x}\| < \infty.$$

Hence,

$$y_i = \sum_{j=-\infty}^{\infty} a_{ij} x_j$$

is well defined and finite. If $\mathbf{y} = (y_n)_{n=-\infty}^{\infty}$, then formally $\mathbf{y} = A\mathbf{x}$. To finish the proof, it suffices to show that $\mathbf{y} \in \ell^2(\mathbb{Z})$ and $\|\mathbf{y}\| \leqslant \|\mathbf{x}\|$.

For each $i_0, j_0 \in \mathbb{Z}$, the assumption that $A(i_0, j_0)$ is a contraction implies that

$$\sum_{i=i_0}^{\infty} \left| \sum_{j=j_0}^{\infty} a_{ij} x_j \right|^2 = \|A(i_0, j_0)\mathbf{x}\|^2 \leqslant \|\mathbf{x}\|^2.$$

The inequality above holds uniformly with respect to i_0 and j_0. Note that

$$\sum_{i=i_0}^{\infty} |y_i|^2 = \sum_{i=i_0}^{\infty} \lim_{j_0 \to -\infty} \left| \sum_{j=j_0}^{\infty} a_{ij} x_j \right|^2$$

$$\leqslant \liminf_{j_0 \to -\infty} \sum_{i=i_0}^{\infty} \left| \sum_{j=j_0}^{\infty} a_{ij} x_j \right|^2 \qquad \text{(Fatou's lemma)}$$

$$\leqslant \|\mathbf{x}\|^2.$$

Now let $i_0 \to -\infty$ to get

$$\|\mathbf{y}\|_{\ell^2(\mathbb{Z})}^2 = \sum_{i=-\infty}^{\infty} |y_i|^2 \leqslant \|\mathbf{x}\|^2$$

which completes the proof. ∎

3.5 Notes

The idea of the adjoint of an operator goes back to F. Riesz [300] who used the term *Transponierte* for what we now call the adjoint of a linear transformation on an L^p space. He also showed that an operator and its adjoint have the same norm. Adjoints of operators on abstract Banach spaces were explored by Banach [32], Schauder [330], and Hildebrandt [198].

In 1910, Hellinger and Toeplitz [191] explored the boundedness of matrix operators on ℓ^2. Schur's test for the boundedness of matrix operators appeared in his 1911 paper [333]. See [94, 169] for modern treatments of Schur's test. Other aspects of compactness for operators on ℓ^2 were studied by Hilbert [196] and Riesz [303].

With regards to the Hilbert matrix H from Example 3.3.4 (which will be explored further in Chapter 17), Hilbert showed that $\|H\| \leqslant 2\pi$ and Schur proved that $\|H\| = \pi$. There is a generalization H_λ of the Hilbert matrix whose (i, j) entry is

$$\frac{1}{i+j+\lambda} \quad \text{for } \lambda \in \mathbb{R} \backslash \{\ldots, -3, -2, -1, 0\}.$$

Results from [199, 204, 237, 333] show that

$$\|H_\lambda\| = \begin{cases} \pi \csc \pi\lambda & \text{if } \lambda < \frac{1}{2}, \\ \pi & \text{if } \lambda \geq \frac{1}{2}. \end{cases}$$

3.6 Exercises

Exercise 3.6.1. If $A, B \in \mathcal{B}(\mathcal{H})$ and $\langle Ax, y \rangle = \langle Bx, y \rangle$ for all $x, y \in \mathcal{H}$, prove that $A = B$.

Exercise 3.6.2. Prove that if $A, B \in \mathcal{B}(\mathcal{H})$ and the matrix representations of A and B with respect to an orthonormal basis $(u_n)_{n=1}^\infty$ are the same, then $A = B$.

Exercise 3.6.3. Let x, y belong to a Hilbert space $\mathcal{H}$.

(a) Prove that $\|x \otimes y\| = \|x\|\|y\|$.

(b) Prove that $(x \otimes y)^* = y \otimes x$.

Exercise 3.6.4. Let $(u_n)_{n=1}^\infty$ be an orthonormal basis for a Hilbert space $\mathcal{H}$. For each $x, y \in \mathcal{H}$, write down the matrix representation of $x \otimes y$ with respect to $(u_n)_{n=1}^\infty$ in terms of $\langle x, u_n \rangle$ and $\langle y, u_n \rangle$.

Exercise 3.6.5. For $A, B \in \mathcal{B}(\mathcal{H})$ and $\lambda \in \mathbb{C}$, prove the following.

(a) $\sigma(A^*) = \{\bar{\lambda} : \lambda \in \sigma(A)\}$.

(b) $(A + \lambda B)^* = A^* + \bar{\lambda} B^*$.

(c) $(AB)^* = B^* A^*$.

Exercise 3.6.6. Let $\mathcal{Y}$ be a nonempty subset of a Hilbert space $\mathcal{H}$. Prove the following.

(a) $\{0\}^\perp = \mathcal{H}$.

(b) $\mathcal{H}^\perp = \{0\}$.

(c) $\mathcal{Y}^\perp$ is a subspace of $\mathcal{H}$.

(d) $(\mathcal{Y}^\perp)^\perp = \bigvee \mathcal{Y}$.

(e) $\mathcal{Y}$ has dense linear span in $\mathcal{H}$ if and only if $\mathcal{Y}^\perp = \{0\}$.

Exercise 3.6.7. If $\mathcal{Y}$ and $\mathcal{Z}$ are nonempty subsets of a Hilbert space $\mathcal{H}$, prove that if $\mathcal{Y} \subseteq \mathcal{Z}$, then $\mathcal{Z}^\perp \subseteq \mathcal{Y}^\perp$.

Exercise 3.6.8. Let $T \in \mathcal{B}(\mathcal{H})$ and suppose there exist $m, M > 0$ such that $m\|x\|^2 \leq \langle T^*Tx, x \rangle \leq M\|x\|^2$ and $m\|x\|^2 \leq \langle TT^*x, x \rangle \leq M\|x\|^2$ for all $x \in \mathcal{H}$. Prove that T is invertible.

Exercise 3.6.9. The next two problems give an alternate criterion for compactness of an operator. Use the following steps to prove that if $A \in \mathcal{B}(\mathcal{H})$ is compact, then A has the following property: if $\mathbf{x}_n \to \mathbf{0}$ weakly, then $A\mathbf{x}_n \to \mathbf{0}$ in norm. This latter property is called *complete continuity*.

(a) If A is compact and $\mathbf{x}_n \to \mathbf{0}$ weakly, prove that there is a subsequence $(\mathbf{x}_{n_k})_{k=1}^{\infty}$ and a $\mathbf{y} \in \mathcal{H}$ such that $\|A\mathbf{x}_{n_k} - \mathbf{y}\| \to 0$.

(b) Prove that $A\mathbf{x}_n \to \mathbf{0}$ weakly and hence $\mathbf{y} = \mathbf{0}$.

(c) Prove that $\|A\mathbf{x}_n\| \to 0$.

Remark: See Exercise 1.10.52 for a review of weak convergence in a Hilbert space.

Exercise 3.6.10. This is a continuation of Exercise 3.6.9. Use the following steps to prove that if $A \in \mathcal{B}(\mathcal{H})$ is completely continuous, then A is compact.

(a) Prove the following theorem of Banach [94]: If $(\mathbf{x}_n)_{n=1}^{\infty}$ is a bounded sequence in a Hilbert space $\mathcal{H}$ then there is an $\mathbf{x} \in \mathcal{H}$ and a subsequence $(\mathbf{x}_{n_k})_{k=1}^{\infty}$ such that $\mathbf{x}_{n_k} \to \mathbf{x}$ weakly.

(b) Prove that A is compact.

Remark: Banach's theorem mentioned in (a) says that the closed unit ball $\{\mathbf{x} \in \mathcal{H} : \|\mathbf{x}\| \leqslant 1\}$ is weakly sequentially compact.

Exercise 3.6.11. Follow these steps to show that if $A \in \mathcal{B}(\mathcal{H})$ is compact, then so is A^*.

(a) If $(\mathbf{x}_n)_{n=1}^{\infty}$ is a bounded sequence, prove that $(AA^*\mathbf{x}_n)_{n=1}^{\infty}$ has a convergence subsequence $(AA^*\mathbf{x}_{n_k})_{k=1}^{\infty}$.

(b) Prove that $\|A^*\mathbf{x}_{n_k} - A^*\mathbf{x}_{n_\ell}\|^2 \leqslant \|AA^*\mathbf{x}_{n_k} - AA^*\mathbf{x}_{n_\ell}\| \cdot \sup_{k,\ell} \|\mathbf{x}_{n_k} - \mathbf{x}_{n_\ell}\|$.

(c) Prove that A^* is compact.

Exercise 3.6.12. For $A \in \mathcal{B}(\mathcal{H})$, prove that A is compact if and only if A^*A is compact.

Exercise 3.6.13. Let $A \in \mathcal{B}(\mathcal{H})$.

(a) Prove that A is an isometry if and only if $A^*A = I$.

(b) Is the adjoint of an isometry necessarily an isometry?

Exercise 3.6.14. For a sequence $(b_n)_{n=0}^{\infty}$ of complex numbers, consider the *terraced matrix*

$$T = \begin{bmatrix} b_0 & 0 & 0 & 0 & 0 & \cdots \\ b_1 & b_1 & 0 & 0 & 0 & \cdots \\ b_2 & b_2 & b_2 & 0 & 0 & \cdots \\ b_3 & b_3 & b_3 & b_3 & 0 & \cdots \\ \vdots & \vdots & \vdots & \vdots & \vdots & \ddots \end{bmatrix}.$$

If $b_n = (n+1)^{-3}$, show that T is a compact operator on ℓ^2.
Remark: See [289] for more on this.

Exercise 3.6.15. The following matrix is an example of a Toeplitz matrix (see Chapter 16).

(a) If $|\alpha| < 1$ and $|\beta| < 1$, prove that

$$
\begin{bmatrix}
1 & \alpha & \alpha^2 & \alpha^3 & \alpha^4 & \cdots \\
\beta & 1 & \alpha & \alpha^2 & \alpha^3 & \cdots \\
\beta^2 & \beta & 1 & \alpha & \alpha^2 & \cdots \\
\beta^3 & \beta^2 & \beta & 1 & \alpha & \cdots \\
\beta^4 & \beta^3 & \beta^2 & \beta & 1 & \cdots \\
\vdots & \vdots & \vdots & \vdots & \vdots & \ddots
\end{bmatrix}
$$

defines a bounded operator on ℓ^2.

(b) Obtain an upper bound for its norm.

Exercise 3.6.16. Let $T = [t_{ij}]_{i,j=0}^{\infty} \in \mathcal{B}(\ell^2)$.

(a) If T is an upper triangular matrix, prove that every diagonal entry t_{ii} is an eigenvalue of T.

(b) Is this still true if T is lower triangular?

Exercise 3.6.17. Let H be the Hilbert matrix from Example 3.3.4. Use the following steps from [237] to show that $\ker H = \{\mathbf{0}\}$.

(a) Suppose $(g_n)_{n=0}^{\infty} \in \ell^2$. Prove that $g(z) = \sum_{n=0}^{\infty} g_n z^n$ defines an analytic function on $\mathbb{D}$. These types of analytic functions are discussed in greater detail in Chapter 5.

(b) Prove that if $(g_n)_{n=0}^{\infty}$ belongs to $\ker H$, then $\displaystyle\int_0^1 \frac{g(t)}{1-tz}\,dt = 0$ for all $z \in \mathbb{D}$.

(c) If $G(z) = \displaystyle\sum_{n=0}^{\infty} \frac{g_n}{n+1} z^{n+1}$, prove that G is continuous on $[0,1]$.

(d) Use (b) to prove that $\displaystyle\int_0^1 g(t)t^n\,dt = G(1) - n\int_0^1 G(t)t^{n-1}\,dt = 0$ for all $n \geqslant 0$.

(e) Use the Weierstrass approximation theorem and (d) to prove that $(g_n)_{n=0}^{\infty}$ is the zero sequence.

Exercise 3.6.18. This is a continuation of Exercise 3.6.17. Prove that H is not surjective as follows. Prove that if $\mathbf{b} = (b_n)_{n=0}^{\infty} \in \ell^2$ and $H\mathbf{b} = \mathbf{e}_0$, then $H\mathbf{u} = \mathbf{0}$, where $\mathbf{u} = (0, b_0, b_1, b_2, \ldots)$.

Exercise 3.6.19. Prove that the *exponential Hilbert matrix* $E = [e_{ij}]_{i,j=0}^{\infty}$, defined by $e_{ij} = 2^{-(i+j+1)}$, is bounded on ℓ^2 and $\|E\| = \frac{2}{3}$.
Remark: See [169] for more on this.

Exercise 3.6.20. If $V = [v_{ij}]_{i,j=0}^{\infty}$ is defined by

$$v_{ij} = \frac{1}{i+j+\frac{1}{2}} \quad \text{for } i, j \geqslant 0,$$

prove that V is bounded on ℓ^2.

Exercise 3.6.21. For a sequence of complex numbers $(\alpha_n)_{n=0}^{\infty}$, consider the corresponding *weighted shift* operator on ℓ^2, defined by the matrix

$$W = \begin{bmatrix} 0 & 0 & 0 & 0 & \cdots \\ \alpha_0 & 0 & 0 & 0 & \cdots \\ 0 & \alpha_1 & 0 & 0 & \cdots \\ 0 & 0 & \alpha_2 & 0 & \cdots \\ \vdots & \vdots & \vdots & \vdots & \ddots \end{bmatrix}.$$

(a) Prove that $W\mathbf{e}_n = \alpha_n \mathbf{e}_{n+1}$ for all $n \geqslant 0$.

(b) Prove that W is a bounded operator on ℓ^2 if and only if $(\alpha_n)_{n=0}^{\infty}$ is a bounded sequence. Furthermore, prove that $\|W\| = \sup_{n \geqslant 0} |\alpha_n|$.

Remark: This discussion of weighted shift operators continues in Exercises 3.6.22 and 3.6.23. See [342] for more on weighted shift operators and connections to complex function theory.

Exercise 3.6.22. Consider the weighted shift from Exercise 3.6.21.

(a) For $\xi \in \mathbb{T}$, prove that W is unitarily equivalent to ξW.

(b) What does this say about $\sigma(W)$?

Remark: This discussion continues in Exercise 8.10.38.

Exercise 3.6.23. If W is a weighted shift operator, prove that the *self commutator* $WW^* - W^*W$ is a diagonal operator.

Exercise 3.6.24. Let

$$W = \begin{bmatrix} 0 & 0 & 0 & 0 & \cdots \\ 1 & 0 & 0 & 0 & \cdots \\ 0 & \frac{1}{2} & 0 & 0 & \cdots \\ 0 & 0 & \frac{1}{3} & 0 & \cdots \\ \vdots & \vdots & \vdots & \vdots & \ddots \end{bmatrix}.$$

(a) Prove that W is a compact operator on ℓ^2.

(b) Prove that $\sigma(W) = \{0\}$.

Exercise 3.6.25. Let

$$S = \begin{bmatrix} 0 & 0 & 0 & 0 & \cdots \\ 1 & 0 & 0 & 0 & \cdots \\ 0 & 1 & 0 & 0 & \cdots \\ 0 & 0 & 1 & 0 & \cdots \\ \vdots & \vdots & \vdots & \vdots & \ddots \end{bmatrix}$$

and $\sum_{n=0}^{\infty} |a_n| < \infty$. Prove that

$$A = \begin{bmatrix} a_0 & 0 & 0 & 0 & \cdots \\ a_1 & a_0 & 0 & 0 & \cdots \\ a_2 & a_1 & a_0 & 0 & \cdots \\ a_3 & a_2 & a_1 & a_0 & \cdots \\ \vdots & \vdots & \vdots & \vdots & \ddots \end{bmatrix}$$

is bounded on ℓ^2 and $AS = SA$.

Remark: The matrix S from this exercise and the matrix W from Exercise 3.6.24 have the following curious properties which distinguish them from finite matrices. Both S and W are lower triangular with zeros along the main diagonal. However, $\sigma(S) = \mathbb{D}^-$ (Proposition 5.1.4) and $\sigma(W) = \{0\}$.

Exercise 3.6.26. Let

$$A = \begin{bmatrix} 1 & 1 & 0 & 0 & 0 & \cdots \\ 0 & 1 & 1 & 0 & 0 & \cdots \\ 0 & 0 & 1 & 1 & 0 & \cdots \\ 0 & 0 & 0 & 1 & 1 & \cdots \\ 0 & 0 & 0 & 0 & 1 & \cdots \\ \vdots & \vdots & \vdots & \vdots & \vdots & \ddots \end{bmatrix}.$$

(a) Prove that A defines a bounded operator on ℓ^2.

(b) Compute $\|A\|$.

(c) Prove that $\sigma_p(A) = \{1 + z : |z| < 1\}$.

Exercise 3.6.27. Suppose $K(x, y) \geq 0$ is measurable on $\mathbb{R}^2$ and that there are measurable functions $p(x), q(x) > 0$ and constants $\alpha, \beta > 0$ such that

$$\int K(x, y)q(y)\, dy \leq \alpha p(x) \quad \text{and} \quad \int K(x, y)p(x)\, dx \leq \beta q(y).$$

Prove that

$$(Tf)(x) = \int K(x, y)f(y)\, dy$$

is bounded on $L^2(\mathbb{R})$ and $\|T\| \leq \sqrt{\alpha\beta}$.

Remark: This is the integral version of Schur's test.

Exercise 3.6.28. Prove that the Volterra operator $(Vf)(x) = \int_0^x f(t)\,dt$ is bounded on $L^2[0,1]$ and $\|V\| \leqslant 1$.

Remark: See Chapter 7 for more about the Volterra operator. In particular, $\|V\| = \frac{2}{\pi}$.

Exercise 3.6.29. Prove that the operator $A : L^2(0,\infty) \to L^2(0,\infty)$ defined by

$$(Af)(x) = \int_0^\infty \frac{f(y)}{x+y}\,dy$$

is bounded.

Remark: This operator appears in Exercise 17.10.22 in connection with Hankel operators.

Exercise 3.6.30. Suppose $\mathbf{b} = (b_n)_{n=-\infty}^\infty$ and $\sum_{n=-\infty}^\infty |b_n| < \infty$. Define the *convolution operator* $X_\mathbf{b} : \ell^2(\mathbb{Z}) \to \ell^2(\mathbb{Z})$ by

$$X_\mathbf{b}\mathbf{a} = \Big(\sum_{m=-\infty}^\infty a_m b_{n-m}\Big)_{n=-\infty}^\infty \quad \text{for} \quad \mathbf{a} = (a_n)_{n=-\infty}^\infty.$$

(a) Prove that $X_\mathbf{b}$ is bounded.

(b) Find the matrix representation of $X_\mathbf{b}$ with respect to the basis $(\mathbf{e}_n)_{n=-\infty}^\infty$ for $\ell^2(\mathbb{Z})$.

Exercise 3.6.31. An important class of operators is the *Hilbert–Schmidt* operators. A closely related class of operators, the trace-class operators, is covered in Exercise 14.11.30.

(a) If $(\mathbf{u}_n)_{n=1}^\infty$ is an orthonormal basis for $\mathcal{H}$ and $A \in \mathcal{B}(\mathcal{H})$, prove that the (possibly infinite) quantity $\sum_{n=1}^\infty \|A\mathbf{u}_n\|^2$ is independent of the choice of orthonormal basis $(\mathbf{u}_n)_{n=1}^\infty$.

(b) $A \in \mathcal{B}(\mathcal{H})$ is a *Hilbert–Schmidt* operator if

$$\|A\|_{HS} := \Big(\sum_{n=1}^\infty \|A\mathbf{u}_n\|^2\Big)^{\frac{1}{2}}$$

is finite for some, and hence all, orthonormal bases for $\mathcal{H}$. Prove that if $[a_{ij}]_{i,j=1}^\infty$ is the matrix representation of A with respect to an orthonormal basis $(\mathbf{u}_n)_{n=1}^\infty$, then A is Hilbert–Schmidt if and only if

$$\sum_{i,j=1}^\infty |a_{ij}|^2 < \infty.$$

In this case, the above equals $\|A\|_{HS}^2$.

(c) Prove that $\|A\| \leqslant \|A\|_{HS}$.

(d) Prove that every Hilbert–Schmidt operator is compact.

Exercise 3.6.32. Suppose $A, T \in \mathcal{B}(\mathcal{H})$ and T is Hilbert–Schmidt. Prove the following.

(a) $\|T\|_{HS} = \|T^*\|_{HS}$.

(b) $\|AT\|_{HS} \leqslant \|A\|\|T\|_{HS}$.

(c) $\|TA\|_{HS} \leqslant \|T\|_{HS}\|A\|$.

(d) The Hilbert–Schmidt operators form a two-sided ideal in $\mathcal{B}(\mathcal{H})$.

Exercise 3.6.33. A *Jacobi matrix* is an infinite matrix of the form

$$
J = \begin{bmatrix}
a_0 & b_0 & 0 & 0 & 0 & \cdots \\
c_0 & a_1 & b_1 & 0 & 0 & \cdots \\
0 & c_1 & a_2 & b_2 & 0 & \cdots \\
0 & 0 & c_2 & a_3 & b_3 & \cdots \\
\vdots & \vdots & \vdots & \vdots & \vdots & \ddots
\end{bmatrix},
$$

in which $a_n, b_n, c_n \in \mathbb{C}$. Prove that J is a compact operator on ℓ^2 if and only if $a_n \to 0$, $b_n \to 0$, and $c_n \to 0$.

Exercise 3.6.34. Extend Exercise 3.6.33 to infinite matrices $A = [a_{ij}]_{i,j=0}^{\infty}$ such that $a_{ij} = 0$ if $|i - j| > r$.

Exercise 3.6.35. Suppose that

$$
S = \begin{bmatrix}
0 & 0 & 0 & 0 & \cdots \\
1 & 0 & 0 & 0 & \cdots \\
0 & 1 & 0 & 0 & \cdots \\
0 & 0 & 1 & 0 & \cdots \\
\vdots & \vdots & \vdots & \vdots & \ddots
\end{bmatrix}
\quad \text{and} \quad
D = \begin{bmatrix}
\frac{1}{2} & 0 & 0 & 0 & \cdots \\
0 & \frac{1}{4} & 0 & 0 & \cdots \\
0 & 0 & \frac{1}{8} & 0 & \cdots \\
0 & 0 & 0 & \frac{1}{16} & \cdots \\
\vdots & \vdots & \vdots & \vdots & \ddots
\end{bmatrix}.
$$

Prove that $A = SD$ is compact and $\sigma(A) = \{0\}$.

Exercise 3.6.36. Let $(P_n)_{n=1}^{\infty}$ be a sequence of orthogonal projections in $\mathcal{B}(\mathcal{H})$ such that $P_n \mathbf{x} \to \mathbf{x}$ for each $\mathbf{x} \in \mathcal{H}$. If $A \in \mathcal{B}(\mathcal{H})$ and $A_n = P_n A P_n$ is a contraction for each n, prove that A is a contraction.

Exercise 3.6.37. Prove that if $A \in \mathcal{B}(\mathcal{H})$, there is a sequence $(A_n)_{n=1}^{\infty}$ in $\mathcal{B}(\mathcal{H})$, each of finite rank, such that $A_n \mathbf{x} \to A\mathbf{x}$ for each $\mathbf{x} \in \mathcal{H}$. In other words, $A_n \to A$ in the *strong operator topology* (Exercise 4.5.23).

Exercise 3.6.38. For a sequence $(a_n)_{n=0}^{\infty}$ of complex numbers,

$$
A = \begin{bmatrix}
\begin{array}{c|ccc}
a_0 & a_1 & a_2 & a_3 & \cdots \\
\hline
a_1 & a_1 & a_2 & a_3 & \cdots \\
\hline
a_2 & a_2 & a_2 & a_3 & \cdots \\
\hline
a_3 & a_3 & a_3 & a_3 & \cdots \\
\vdots & \vdots & \vdots & \vdots & \ddots
\end{array}
\end{bmatrix}
$$

is an *L-shaped* matrix. Notice that the (m, n) entry of A is $a_{\max\{m,n\}}$.

(a) Prove that if A is a bounded operator on ℓ^2, then $a_n = O(n^{-\frac{1}{2}})$.

(b) Let $a_{4^n} = 2^{-n}$ and $a_j = 0$ otherwise. Show that A is bounded on ℓ^2. Conclude that the condition $O(n^{-\frac{1}{2}})$ is best possible.

Remark: These matrices appear in the study of the Cesàro operator (Chapter 6). See [61, 62] for more on L-shaped matrices.

Exercise 3.6.39. Let $(a_n)_{n=0}^\infty$ satisfy $\sum_{n=0}^\infty |a_n| < \infty$ and let

$$
A = \begin{bmatrix} a_0 & a_1 & a_2 & \cdots \\ a_1 & a_2 & a_3 & \cdots \\ a_2 & a_3 & a_4 & \cdots \\ \vdots & \vdots & \vdots & \ddots \end{bmatrix}.
$$

Prove that A is a bounded operator on ℓ^2.

Exercise 3.6.40. Suppose that $A = [a_{ij}]_{i,j\in\mathbb{Z}}$ has only a finite number of nonzero diagonals. By this we mean there are $m, n \in \mathbb{Z}$ such that

$$
a_{ij} = 0 \quad \text{for } i - j > n \text{ or } i - j < -m.
$$

(a) Prove that A is a bounded operator on $\ell^2(\mathbb{Z})$ if and only if $\{a_{ij} : i, j \in \mathbb{Z}\}$ is a bounded set.

(b) Prove that

$$
\|A\| \leqslant \sum_{k=-m}^{n} \left(\sup_{i \geqslant \max\{0,k\}} |a_{i,i-k}| \right) = \sum_{k=-m}^{n} \|\mathbf{d}_k\|_\infty,
$$

where $\mathbf{d}_k$ represents the kth diagonal of A and $\|\mathbf{d}_k\|_\infty$ is the supremum norm of a sequence as defined in Example 1.8.2.

Remark: These matrices are often called *banded matrices*

Exercise 3.6.41. This exercise completes the proof of Theorem 2.6.7 (the spectral theorem for compact selfadjoint operators) in the case where $\ker A \neq \{0\}$. Suppose A is a compact, selfadjoint operator on an infinite-dimensional separable Hilbert space $\mathcal{H}$.

(a) Prove that $\mathcal{H}_1 = \ker A$ and $\mathcal{H}_2 = (\ker A)^\perp$ are invariant subspaces for A.

(b) Prove that $A_2 = A|_{\mathcal{H}_2} \in \mathcal{B}(\mathcal{H}_2)$ is selfadjoint and injective.

(c) Use Theorem 2.6.7 to prove that if $\dim \mathcal{H}_2 = \infty$, there is a sequence $(\lambda_n)_{n=1}^\infty$ of nonzero real numbers with $\lambda_n \to 0$ and an orthonormal basis $(\mathbf{x}_n)_{n=1}^\infty$ for $\mathcal{H}_2$ such that $A_2 = \sum_{n=1}^\infty \lambda_n(\mathbf{x}_n \otimes \mathbf{x}_n)$.

(d) Prove that there is an orthonormal basis $(\mathbf{z}_n)_{n=1}^\infty$ for $\mathcal{H}$ and a sequence $(\beta_n)_{n=1}^\infty$ of real numbers with $\beta_n \to 0$ such that $A = \sum_{n=1}^\infty \beta_n(\mathbf{z}_n \otimes \mathbf{z}_n)$.

3.7 Hints for the Exercises

Hint for Ex. 3.6.12. Consult Exercises 3.6.9 and 3.6.10.

Hint for Ex. 3.6.13. For part (a), observe that A is an isometry if and only if $\langle A^*Ax, x \rangle = \langle x, x \rangle$ for all $x \in \mathcal{H}$. Now consult Exercise 2.8.29.

Hint for Ex. 3.6.17: For (a), use term-by-term integration.

Hint for Ex. 3.6.19: If x_0 is the first column of E, prove that $E = 2x_0 \otimes x_0$. Now use Exercise 3.6.3.

Hint for Ex. 3.6.24: Use Theorem 2.6.9. For (a), consult Exercise 3.6.12.

Hint for Ex. 3.6.31: For (a), note that $Au_n = \sum_{k=1}^{\infty} \langle Au_n, u_k \rangle u_k$ and use Parseval's theorem. For (d), consider the operators

$$A_N x = \sum_{j=1}^{N} \langle u_j, Au_j \rangle \langle u_j, x \rangle u_j.$$

Hint for Ex. 3.6.33: Let $D_a = \operatorname{diag}(a_1, a_2, \dots)$. Similarly define D_b and D_c. Consider $D_a + SD_c + D_bS^*$, where

$$S = \begin{bmatrix} 0 & 0 & 0 & 0 & \cdots \\ 1 & 0 & 0 & 0 & \cdots \\ 0 & 1 & 0 & 0 & \cdots \\ 0 & 0 & 1 & 0 & \cdots \\ \vdots & \vdots & \vdots & \vdots & \ddots \end{bmatrix}.$$

For the other direction, consider $\langle Je_n, e_n \rangle$, $\langle Je_n, e_{n+1} \rangle$, $\langle Je_n, e_{n-1} \rangle$, and Exercise 3.6.9.

Hint for Ex. 3.6.37: Consider the matrix representation of A with respect to an orthonormal basis $(u_n)_{n=1}^{\infty}$ for $\mathcal{H}$.

Hint for Ex. 3.6.38: For (a), use the fact that each column of A belongs to ℓ^2.

Hint for Ex. 3.6.41: For (d), suppose $K = \dim \ker A$ and choose an orthonormal basis $(w_n)_{n=1}^{K}$ for $\ker A$.

4

· · **·** · ·

Two Multiplication Operators

Key Concepts: Multiplication operator, invariant subspace, Fourier series, Hardy space.

Outline: This chapter concerns the multiplication operators $M_x : L^2[0,1] \rightarrow L^2[0,1]$, defined by $(M_x f)(x) = xf(x)$, and $M_\xi : L^2(\mathbb{T}) \rightarrow L^2(\mathbb{T})$, defined by $(M_\xi g)(\xi) = \xi g(\xi)$. We discuss their spectra and invariant subspaces. This requires the introduction of Fourier series and the Hardy space H^2.

4.1 M_x on $L^2[0,1]$

From Chapter 1, recall the Lebesgue space $L^2[0,1]$ which is a Hilbert space with inner product

$$\langle f,g \rangle = \int_0^1 f(x)\overline{g(x)}\,dx$$

and corresponding norm $\|f\| = \sqrt{\langle f,f \rangle}$. In this section, we study the norm, spectrum, and invariant subspaces of the operator $M_x : L^2[0,1] \rightarrow L^2[0,1]$ defined by

$$(M_x f)(x) = xf(x). \tag{4.1.1}$$

This is a particular example of a multiplication operator (see Chapter 8).

Proposition 4.1.2. *The operator M_x is bounded on $L^2[0,1]$ and $\|M_x\| = 1$.*

Proof For $g \in L^2[0,1]$, the estimate

$$\|M_x g\|^2 = \int_0^1 |xg(x)|^2 dx \leqslant \left(\sup_{x \in [0,1]} |x^2| \right) \int_0^1 |g(x)|^2 dx = \|g\|^2$$

shows that

$$\|M_x\| = \sup_{\|g\|=1} \|M_x g\| \leqslant 1.$$

For $n \geqslant 1$, let $g_n(x) = (2n+1)^{\frac{1}{2}} x^n$. Then $\|g_n\| = 1$ and

$$\|M_x g_n\|^2 = (2n+1) \int_0^1 |x^{n+1}|^2 dx = (2n+1) \cdot \frac{1}{2n+3} \to 1 \quad \text{as } n \to \infty.$$

Thus,

$$1 = \sup_{n \geqslant 1} \|M_x g_n\| \leqslant \sup_{\|g\|=1} \|M_x g\| = \|M_x\| \leqslant 1$$

and hence $\|M_x\| = 1$. ∎

The following result determines the spectrum and point spectrum of M_x. As is standard in Lebesgue theory, we indulge in a slight abuse of language and identify functions that are equal almost everywhere (a.e.).

Proposition 4.1.3.

(a) $\sigma_p(M_x) = \varnothing$.

(b) $\sigma(M_x) = [0, 1]$.

Proof (a) Suppose $\lambda \in \mathbb{C}$ and $f \in L^2[0,1]$ satisfies $(M_x - \lambda I)f = 0$ a.e. Then $(x - \lambda)f = 0$ a.e. and hence $f = 0$ a.e. Thus, $\sigma_p(M_x) = \varnothing$.
(b) For a fixed $\lambda \notin [0, 1]$, the quantity

$$c_\lambda = \sup_{0 \leqslant x \leqslant 1} \left| \frac{1}{x - \lambda} \right|$$

is finite. For any $f \in L^2[0,1]$,

$$\int_0^1 |(x - \lambda)^{-1} f(x)|^2 dx \leqslant c_\lambda^2 \int_0^1 |f(x)|^2 \, dx = c_\lambda^2 \|f\|^2,$$

and hence the operator

$$(T_\lambda f)(x) = \frac{1}{x - \lambda} f(x)$$

is bounded on $L^2[0, 1]$. Moreover,

$$T_\lambda (M_x - \lambda I)f = f \quad \text{and} \quad (M_x - \lambda I)T_\lambda f = f \quad \text{for all } f \in L^2[0,1].$$

Therefore, $M_x - \lambda I$ is invertible and hence $\sigma(M_x) \subseteq [0, 1]$. For the reverse containment, suppose $\lambda \in [0, 1]$. If $M_x - \lambda I$ is invertible, then, given any $g \in L^2[0, 1]$, there is an $f \in L^2[0, 1]$ such that $(M_x - \lambda I)f = g$. Apply this to the constant function $g \equiv 1$ and obtain $f(x) = (x - \lambda)^{-1}$, which is not in $L^2[0, 1]$ since $\lambda \in [0, 1]$. This contradiction shows that $[0, 1] \subseteq \sigma(M_x)$ and hence $\sigma(M_x) = [0, 1]$. ∎

Proposition 1.3.6 says that $C[0, 1]$ is dense in $L^2[0, 1]$. The following refinement plays an important role in this chapter.

Proposition 4.1.4. *If $f \in L^2[0,1]$, there is a sequence $(p_n)_{n=1}^{\infty}$ of polynomials such that $\|p_n - f\| \to 0$. Consequently, if $g \in L^2[0,1]$ and*

$$\int_0^1 g(x)\overline{p(x)}\,dx = 0$$

for all polynomials p, then $g = 0$ a.e.

Proof Let $f \in L^2[0,1]$ and $\varepsilon > 0$. Proposition 1.3.6 furnishes a function $g \in C[0,1]$ such that $\|f - g\| < \varepsilon/2$. The Weierstrass approximation theorem provides a polynomial p such that $\|g - p\|_{\infty} < \varepsilon/2$. Therefore,

$$\|f - p\| \leqslant \|f - g\| + \|g - p\| \leqslant \|f - g\| + \|g - p\|_{\infty} \leqslant \frac{\varepsilon}{2} + \frac{\varepsilon}{2} = \varepsilon.$$

The second statement of the theorem follows from Corollary 1.7.4. ∎

The next result from measure theory [319, Thm. 3.12] is important for what follows.

Proposition 4.1.5. *If $f_n \to f$ in $L^2[0,1]$, then there is a subsequence $(f_{n_k})_{k=1}^{\infty}$ such that $f_{n_k} \to f$ a.e.*

For a Lebesgue-measurable set $E \subseteq [0,1]$, consider

$$\mathcal{Z}_E = \{f \in L^2[0,1] : f|_E = 0 \text{ a.e.}\}.$$

We use this definition and the previous proposition to prove the following.

Proposition 4.1.6. *$\mathcal{Z}_E$ is an M_x-invariant subspace for each measurable subset $E \subseteq [0,1]$.*

Proof First, observe that $\mathcal{Z}_E$ is a vector subspace of $L^2[0,1]$ since it is closed under addition and scalar multiplication. Second, Proposition 4.1.5 implies that if $(f_n)_{n=1}^{\infty}$ is a sequence in $\mathcal{Z}_E$ that converges to f in the norm of $L^2[0,1]$, then $f|_E = 0$ a.e. Consequently, $\mathcal{Z}_E$ is norm closed and is therefore a subspace of $L^2[0,1]$. Finally, if $f \in \mathcal{Z}_E$, then $M_x f$ has the same zeros as f almost everywhere, and hence $M_x \mathcal{Z}_E \subseteq \mathcal{Z}_E$. ∎

Are the invariant subspaces $\mathcal{Z}_E$, described in the previous proposition, all of the invariant subspaces for M_x? The answer is yes.

Theorem 4.1.7 (Wiener [376]). *Let $\mathcal{Z} \subseteq L^2[0,1]$ be an invariant subspace for M_x. Then there is a measurable set $E \subseteq [0,1]$ such that $\mathcal{Z} = \mathcal{Z}_E$.*

Proof We follow a proof from Helson [192]. Let $\mathcal{Z}$ be an invariant subspace for M_x and let q be the orthogonal projection of the constant function 1 onto $\mathcal{Z}$. This is the unique $q \in \mathcal{Z}$ such that $1 - q$ is orthogonal to $\mathcal{Z}$; that is, $\langle 1 - q, h \rangle = 0$ for every $h \in \mathcal{Z}$ (Proposition 1.7.3).

Since $M_x^n \mathcal{Z} = x^n \mathcal{Z} \subseteq \mathcal{Z}$ for all $n \geqslant 0$ and $1 - q$ is orthogonal to $\mathcal{Z}$,

$$0 = \langle pq, 1 - q \rangle = \int_0^1 pq\overline{(1 - q)}\,dx \quad \text{for all } p \in \mathbb{C}[x].$$

By Proposition 4.1.4, $q(1 - \bar{q}) = 0$ a.e. and hence $q = |q|^2$ a.e. Therefore, q assumes only the values 0 and 1, which implies that $q = \chi_F$ for some measurable $F \subseteq [0, 1]$. The claim is that $\mathcal{Z} = \chi_F L^2[0, 1]$ and hence $\mathcal{Z} = \mathcal{Z}_E$, where $E = [0, 1] \backslash F$.

To prove this, first note that

$$\bigvee\{x^n \chi_F \, : \, n \geqslant 0\} = \chi_F L^2[0, 1].$$

For the $\subseteq$ inclusion, observe that $\chi_F L^2[0, 1]$ is closed. Since $x^n \chi_F \in \chi_F L^2[0, 1]$ for all $n \geqslant 0$, it follows that $\bigvee\{x^n \chi_F \, : \, n \geqslant 0\} \subseteq \chi_F L^2[0, 1]$. For the $\supseteq$ inclusion, pick any $\chi_F f \in \chi_F L^2[0, 1]$ and use Proposition 4.1.4 to find a sequence of polynomials p_n such that $p_n \to f$ in $L^2[0, 1]$. Then, $\chi_F p_n \to \chi_F f$ in $L^2[0, 1]$ and hence $\chi_F f \in \bigvee\{x^n \chi_F \, : \, n \geqslant 0\}$.

Since $\chi_F \in \mathcal{Z}$ and $\mathcal{Z}$ is M_x-invariant,

$$\bigvee\{x^n \chi_F \, : \, n \geqslant 0\} = \chi_F L^2[0, 1] \subseteq \mathcal{Z}.$$

To show equality, suppose there is an $f \in \mathcal{Z}$ such that $f \perp \chi_F L^2[0, 1]$. In particular,

$$0 = \langle f, \chi_F p \rangle = \int_0^1 f \overline{p \chi_F} \, dx \quad \text{for all } p \in \mathbb{C}[x]$$

and hence, $f \chi_F = 0$ (Proposition 4.1.4). Since $1 - \chi_F = 1 - q$ is orthogonal to $\mathcal{Z}$, it follows that

$$0 = \langle 1 - \chi_F, pf \rangle = \int_0^1 (1 - \chi_F) \overline{pf} \, dx \quad \text{for all } p \in \mathbb{C}[x].$$

Proposition 4.1.4, implies that $(1 - \chi_F)\overline{f} = 0$ a.e. Conjugation shows that $(1 - \chi_F)f = 0$ a.e. In summary, both $f \chi_F$ and $f \chi_{F^c}$ are zero and thus $f = 0$ a.e. Therefore, $\mathcal{Z} = \chi_F L^2[0, 1]$ (Corollary 1.7.4). ∎

A simpler, but more sophisticated, proof of Theorem 4.1.7 is in Chapter 8. Although longer, the proof above is direct and uses only Lebesgue integration and Hilbert-space geometry.

4.2 Fourier Analysis

Let m denote normalized Lebesgue measure on the unit circle $\mathbb{T}$ and consider the Lebesgue space $L^2(\mathbb{T})$ of measurable $f : \mathbb{T} \to \mathbb{C}$ such that

$$\int_{\mathbb{T}} |f|^2 dm$$

is finite. When endowed with the inner product

$$\langle f, g \rangle = \int_{\mathbb{T}} f \bar{g} \, dm$$

and corresponding norm $\|f\| = \sqrt{\langle f, f \rangle}$, it turns out that $L^2(\mathbb{T})$ is a Hilbert space. Indeed, Proposition 1.3.4 can be adapted to prove the completeness of $L^2(\mathbb{T})$.

If $\xi = e^{it}$, then $(\xi^n)_{n=-\infty}^{\infty}$ is an orthonormal set in $L^2(\mathbb{T})$. To see this, compute

$$\langle \xi^j, \xi^k \rangle = \int_0^{2\pi} e^{ijt} e^{-ikt} \frac{dt}{2\pi} = \int_0^{2\pi} e^{i(j-k)t} \frac{dt}{2\pi} = \delta_{jk} \quad \text{for all } j, k \in \mathbb{Z}.$$

Since $\bigvee \{\xi^n : n \in \mathbb{Z}\} = L^2(\mathbb{T})$ (Theorem 1.3.9), $(\xi^n)_{n=-\infty}^{\infty}$ is an orthonormal basis for $L^2(\mathbb{T})$ (Proposition 1.4.15). Thus, each $f \in L^2(\mathbb{T})$ is of the form

$$f = \sum_{n=-\infty}^{\infty} \langle f, \xi^n \rangle \xi^n, \tag{4.2.1}$$

in which the series above converges in $L^2(\mathbb{T})$ norm. Furthermore, Parseval's theorem (Theorem 1.4.9b) yields

$$\|f\|^2 = \sum_{n=-\infty}^{\infty} |\langle f, \xi^n \rangle|^2. \tag{4.2.2}$$

The series in (4.2.1) is the *Fourier series* of f and the complex numbers

$$\widehat{f}(n) := \langle f, \xi^n \rangle = \int_{\mathbb{T}} f(\xi) \overline{\xi^n} \, dm(\xi) \quad \text{for } n \in \mathbb{Z},$$

are the *Fourier coefficients* of f. In general, norm convergence of a series of functions in $L^2(\mathbb{T})$ does not imply pointwise convergence almost everywhere, only that some subsequence of the partial sums converges almost everywhere (Proposition 4.1.5). However, a deep theorem of Carleson [76] says that the Fourier series of an $f \in L^2(\mathbb{T})$ converges pointwise almost everywhere to f.

Fourier series provide a natural orthogonal decomposition of $L^2(\mathbb{T})$. First write the Fourier expansion of $f \in L^2(\mathbb{T})$ as

$$f = \sum_{n=0}^{\infty} \widehat{f}(n) \xi^n + \sum_{n=-\infty}^{-1} \widehat{f}(n) \xi^n.$$

If

$$g = \sum_{n=0}^{\infty} \widehat{f}(n) \xi^n \quad \text{and} \quad h = \sum_{n=-\infty}^{-1} \widehat{f}(n) \xi^n,$$

then $g \perp h$, $f = g + h$, and g belongs to

$$H^2(\mathbb{T}) := \{f \in L^2(\mathbb{T}) : \widehat{f}(n) = 0 \text{ for all } n < 0\}. \tag{4.2.3}$$

Moreover,

$$\overline{h} = \sum_{n=-\infty}^{-1} \overline{\widehat{f}(n)} \overline{\xi}^{-n} = \sum_{n=1}^{\infty} \overline{\widehat{f}(-n)} \xi^n \in \xi H^2(\mathbb{T})$$

and hence $h \in \overline{\xi H^2(\mathbb{T})}$. This says that

$$L^2(\mathbb{T}) = H^2(\mathbb{T}) \oplus \overline{\xi H^2(\mathbb{T})}. \tag{4.2.4}$$

In (4.2.4) the notation $\oplus$ denotes the orthogonal direct sum, meaning that each $f \in L^2(\mathbb{T})$ can be written as $f = g + h$, where $g \in H^2$, $h \in \overline{\xi H^2(\mathbb{T})}$, and $g \perp h$. The space $H^2(\mathbb{T})$ is the *Hardy space* and appears again in Chapter 5. We use the following definition and proposition to make a connection between $H^2(\mathbb{T})$ and the sequence space ℓ^2.

Definition 4.2.5. Let $\mathcal{H}, \mathcal{K}$ be Hilbert spaces. Then $T \in \mathcal{B}(\mathcal{H}, \mathcal{K})$ is *unitary* if $T^* = T^{-1}$.

For $T \in \mathcal{B}(\mathcal{H}, \mathcal{K})$, the statement $T^* = T^{-1}$ means that $T^*T = I_{\mathcal{H}}$ and $TT^* = I_{\mathcal{K}}$.

Proposition 4.2.6. *For $T \in \mathcal{B}(\mathcal{H}, \mathcal{K})$, the following are equivalent.*

(a) *T is unitary.*

(b) *T is surjective and isometric.*

Proof (a) $\Rightarrow$ (b) If T is unitary, it is invertible and hence surjective. Since T is unitary, it follows that $T^*T\mathbf{x} = \mathbf{x}$ for all $\mathbf{x} \in \mathcal{H}$. Therefore,

$$\|\mathbf{x}\|_{\mathcal{H}}^2 = \langle T^*T\mathbf{x}, \mathbf{x}\rangle_{\mathcal{H}} = \langle T\mathbf{x}, T\mathbf{x}\rangle_{\mathcal{K}} = \|T\mathbf{x}\|_{\mathcal{K}}^2$$

and hence T is an isometry.

(b) $\Rightarrow$ (a) The polarization identity

$$\langle \mathbf{x}, \mathbf{y}\rangle = \frac{1}{4}(\|\mathbf{x} + \mathbf{y}\|^2 - \|\mathbf{x} - \mathbf{y}\|^2 - i\|\mathbf{x} - i\mathbf{y}\|^2 + i\|\mathbf{x} + i\mathbf{y}\|^2)$$

from Exercise 1.10.17, and the fact that $\|T\mathbf{x}\|_{\mathcal{K}} = \|\mathbf{x}\|_{\mathcal{H}}$ for all $\mathbf{x} \in \mathcal{H}$, shows that $\langle T\mathbf{x}, T\mathbf{y}\rangle_{\mathcal{K}} = \langle \mathbf{x}, \mathbf{y}\rangle_{\mathcal{H}}$ for all $\mathbf{x}, \mathbf{y} \in \mathcal{H}$. The definition of the adjoint yields $\langle T^*T\mathbf{x}, \mathbf{y}\rangle_{\mathcal{H}} = \langle \mathbf{x}, \mathbf{y}\rangle_{\mathcal{H}}$ and hence $T^*T = I_{\mathcal{H}}$ (Exercise 3.6.1). Since T is isometric, it is injective. Since T is surjective, we conclude that T is bijective, and hence has an inverse $T^{-1} \in \mathcal{B}(\mathcal{K}, \mathcal{H})$ (Lemma 2.3.5). Thus, $T^* = T^{-1}$. ∎

The following proposition permits the identification of $\ell^2(\mathbb{Z})$ and $L^2(\mathbb{T})$ in a natural and explicit manner. Along the same lines, we can also identify ℓ^2 and $H^2(\mathbb{T})$. A proof is requested in Exercise 4.5.11.

Proposition 4.2.7.

(a) *The linear transformation*

$$(a_n)_{n=-\infty}^{\infty} \mapsto \sum_{n=-\infty}^{\infty} a_n \xi^n$$

is a unitary operator from $\ell^2(\mathbb{Z})$ onto $L^2(\mathbb{T})$.

(b) *The linear transformation*

$$(a_n)_{n=0}^\infty \mapsto \sum_{n=0}^\infty a_n \xi^n$$

is a unitary operator from ℓ^2 onto $H^2(\mathbb{T})$.

The following definition is used many times in this book.

Definition 4.2.8. $A \in \mathcal{B}(\mathcal{H})$ is *unitarily equivalent* to $B \in \mathcal{B}(\mathcal{K})$ if there is a unitary operator $U \in \mathcal{B}(\mathcal{H}, \mathcal{K})$ such that $A = U^*BU$,

Exercise 4.5.17 shows that unitarily equivalent operators have the same norm, eigenvalues, and spectrum.

4.3 M_ξ on $L^2(\mathbb{T})$

This section focuses on the multiplication operator $M_\xi \in \mathcal{B}(L^2(\mathbb{T}))$ defined by $(M_\xi g)(\xi) = \xi g(\xi)$. Chapter 8, which concerns more general multiplication operators, builds upon the material below. Observe that

$$\|M_\xi g\|^2 = \int_\mathbb{T} |\xi g(\xi)|^2 dm(\xi) = \int_\mathbb{T} |g(\xi)|^2 dm(\xi) \quad \text{for all } g \in L^2(\mathbb{T})$$

and so M_ξ is bounded on $L^2(\mathbb{T})$.

Proposition 4.3.1.

(a) $M_\xi^* = M_{\bar\xi}$.

(b) M_ξ is unitary.

Proof (a) For $f, g \in L^2(\mathbb{T})$,

$$\langle M_\xi f, g \rangle = \int_\mathbb{T} \xi f(\xi)\overline{g(\xi)}\, dm(\xi) = \int_\mathbb{T} f(\xi)\overline{\bar\xi g(\xi)}\, dm(\xi) = \langle f, M_{\bar\xi} g \rangle.$$

Hence $M_\xi^* = M_{\bar\xi}$, the operator of multiplication by $\bar\xi$.

(b) Observe that $\xi\bar\xi = 1$ on $\mathbb{T}$, and hence $M_\xi M_\xi^* = M_\xi^* M_\xi = M_{\bar\xi\xi} = M_1 = I$. Therefore, M_ξ is unitary. ∎

An adaptation of the proof of Proposition 4.1.3 (see Exercise 4.5.15) yields the following.

Proposition 4.3.2.

(a) $\sigma(M_\xi) = \mathbb{T}$.

(b) $\sigma_p(M_\xi) = \emptyset$.

The invariant subspaces for M_ξ are more complicated than those for M_x on $L^2[0,1]$. There are two types of subspaces $\mathcal{S} \subseteq L^2(\mathbb{T})$ such that $M_\xi \mathcal{S} \subseteq \mathcal{S}$; those for which $M_\xi \mathcal{S} = \mathcal{S}$ and those for which $M_\xi \mathcal{S} \subsetneq \mathcal{S}$. We address the first type with a theorem of Wiener [376].

Theorem 4.3.3 (Wiener). *If $\mathcal{S} \subseteq L^2(\mathbb{T})$ is a subspace such that $M_\xi \mathcal{S} = \mathcal{S}$, then there is a measurable set $E \subseteq \mathbb{T}$ such that $\mathcal{S} = \{g \in L^2(\mathbb{T}) : g|_E = 0 \text{ a.e.}\}$.*

Proof Suppose $M_\xi \mathcal{S} = \mathcal{S}$. Since $M_\xi^* M_\xi = I$ and $M_\xi^* = M_{\overline{\xi}}$ (Proposition 4.3.1), it follows

that $M_{\overline{\xi}} \mathcal{S} = \mathcal{S}$. In other words, $p(\xi, \overline{\xi})\mathcal{S} \subseteq \mathcal{S}$ for any polynomial p in ξ and $\overline{\xi}$. The set

of such polynomials $p(\xi, \overline{\xi})$ is dense in $L^2(\mathbb{T})$ (Weierstrass approximation theorem – see Theorem 8.1.2). Consequently, an adaptation of the proof of Theorem 4.1.7 shows that $\mathcal{S}$ has the desired form. ∎

Definition 4.3.4. For $A \in \mathcal{B}(\mathcal{H})$, a subspace $\mathcal{M}$ of $\mathcal{H}$ such that $A\mathcal{M} \subseteq \mathcal{M}$ and $A^*\mathcal{M} \subseteq \mathcal{M}$ is a *reducing subspace* for A.

Theorem 4.3.3 characterizes the reducing subspaces of M_ξ. There are other invariant subspaces for M_ξ that are not the zero-based ones described in Theorem 4.3.3.

Example 4.3.5. The subspace $\mathcal{S} = H^2(\mathbb{T})$, where $H^2(\mathbb{T})$ is the Hardy space from (4.2.3), satisfies $M_\xi H^2(\mathbb{T}) \subsetneq H^2(\mathbb{T})$.

Example 4.3.6. Let q be a Lebesgue-measurable *unimodular function* on $\mathbb{T}$, that is, q is measurable with $|q| = 1$ a.e. on $\mathbb{T}$. Note that $q \in L^\infty(\mathbb{T})$. Then $\|qf\| = \|f\|$ for all $f \in H^2(\mathbb{T})$, and hence $\mathcal{S} = qH^2(\mathbb{T})$ is a proper subspace of $L^2(\mathbb{T})$ such that $M_\xi \mathcal{S} \subsetneq \mathcal{S}$.

The next result characterizes the nonreducing subspaces of M_ξ.

Theorem 4.3.7 (Helson [192]). *If $\mathcal{S} \subseteq L^2(\mathbb{T})$ is a subspace such that $M_\xi \mathcal{S} \subsetneq \mathcal{S}$, then there is a Lebesgue-measurable unimodular function q on $\mathbb{T}$ such that $\mathcal{S} = qH^2(\mathbb{T})$.*

Proof Since $M_\xi \mathcal{S}$ is a proper subspace of $\mathcal{S}$, there is a unit vector $q \in \mathcal{S}$ that is orthogonal to $M_\xi \mathcal{S}$. In particular, $q \perp M_\xi^n q$ for all $n \geq 1$. Write this orthogonality in integral form

$$0 = \langle q, M_\xi^n q \rangle = \int_{\mathbb{T}} |q(\xi)|^2 \overline{\xi}^n \, dm(\xi) \quad \text{for } n \geq 1.$$

Take complex conjugates and obtain

$$\int_{\mathbb{T}} |q(\xi)|^2 \xi^n \, dm(\xi) = 0 \quad \text{for } n \in \mathbb{Z} \setminus \{0\}.$$

The equation above says that the Fourier coefficients of $|q|^2$ are all zero, except for the zeroth one, and hence $|q|$ is constant (see (4.2.1)). Furthermore, since q is a unit vector, the constant is 1. Consequently, q is a unimodular function on $\mathbb{T}$.

It suffices to show that $\mathcal{S} = qH^2(\mathbb{T})$. Observe that $(\xi^n q)_{n=-\infty}^\infty$ is an orthonormal sequence in $L^2(\mathbb{T})$ such that $\mathcal{F} = \bigvee\{\xi^n q : n \in \mathbb{Z}\}$ satisfies $M_\xi \mathcal{F} = \mathcal{F}$. Theorem 4.3.3 ensures that $\mathcal{F} = \{f \in L^2(\mathbb{T}) : f|_E = 0 \text{ a.e.}\}$ for some measurable $E \subseteq \mathbb{T}$. However,

q is unimodular and hence $\mathcal{F} = L^2(\mathbb{T})$. Consequently, $(\xi^n q)_{n=-\infty}^{\infty}$ is an orthonormal basis for $L^2(\mathbb{T})$. Also notice that

$$qH^2(\mathbb{T}) = \bigvee\{\xi^n q : n \geq 0\} \subseteq \mathcal{S} \quad \text{and} \quad q\overline{\xi H^2(\mathbb{T})} = \bigvee\{q\xi^n : n < 0\}.$$

Moreover,

$$L^2(\mathbb{T}) = \mathcal{F} = qH^2(\mathbb{T}) \oplus q\overline{\xi H^2(\mathbb{T})}.$$

Recall that q is orthogonal to $M_\xi \mathcal{S}$ and thus orthogonal to $\xi^n \mathcal{S}$ for all $n \geq 1$. This implies $q\xi^n$ is orthogonal to $\mathcal{S}$ for all $n \leq -1$. Thus,

$$(qH^2(\mathbb{T}))^\perp = q\overline{\xi H^2(\mathbb{T})} \subseteq \mathcal{S}^\perp,$$

and hence (see Exercise 3.6.7 and Proposition 3.1.2d)

$$\mathcal{S} = (\mathcal{S}^\perp)^\perp \subseteq ((qH^2(\mathbb{T}))^\perp)^\perp = qH^2(\mathbb{T}).$$

It follows that $\mathcal{S} = qH^2(\mathbb{T})$. ∎

Complex analysis permits a more detailed study of the subspaces $qH^2(\mathbb{T})$ (Chapter 5). The key observation is that $H^2(\mathbb{T})$ can be identified with a space of analytic functions on the open unit disk. Other aspects of multiplication operators, such as cyclic vectors and commutants, are explored in Chapters 5 and 8.

4.4 Notes

From Theorem 1.3.9, the Fourier series of an $f \in L^2(\mathbb{T})$ converges in norm, meaning that

$$\lim_{N \to \infty} \left\| \sum_{n=-N}^{N} \hat{f}(n)\xi^n - f \right\| = 0.$$

On the other hand, pointwise convergence,

$$f(\xi) = \lim_{N \to \infty} \sum_{n=-N}^{N} \hat{f}(n)\xi^n,$$

is a tricky issue. If f is continuously differentiable on $\mathbb{T}$, the Fourier series converges uniformly to f [380, Ch. 2]. For continuous functions, this need not be the case. Indeed, in 1873, de Bois-Reymond [117] exhibited a continuous function whose Fourier series diverges at a point. A striking theorem of Kahane and Katznelson [209] says that given any closed set $E \subseteq \mathbb{T}$ of Lebesgue measure zero, there is a continuous function whose Fourier series diverges precisely on E. Even more dramatic is a result of Kolmogorov [219] which produces an $L^1(\mathbb{T})$ function whose Fourier series diverges everywhere. A thorough account of this is Zygmund's book [380]. On the positive side, a celebrated theorem of Carleson [76] says that the Fourier series of an $f \in L^2(\mathbb{T})$ converges to f almost everywhere.

Hilbert considered bounded linear functionals on ℓ^2 [196]. F. Riesz [299] and Fréchet [132] independently proved what is known as the Riesz representation theorem (Theorem 3.1.3): every bounded linear functional on $L^2[a, b]$ is of the form

$$f \mapsto \int_a^b f(x)g(x)\,dx$$

for some unique $g \in L^2[a, b]$. Löwig [231] studied versions of this theorem for nonseparable Hilbert spaces and Riesz [306] studied versions for abstract Hilbert spaces.

There are other representation theorems that describe the dual spaces (the set of bounded linear functionals) of certain Banach spaces. For example, Riesz [302] proved that every bounded linear functional on $C[0, 1]$ is of the form

$$f \mapsto \int_0^1 f\,dF$$

for some function F of bounded variation on $[0, 1]$. One can reformulate this using the more modern language of measure theory on $[0, 1]$.

Wiener's theorem (Theorems 4.1.7 and 4.3.3) was originally stated in terms of shifts of the Fourier transform [376, Ch. 2]. The proof presented here is due to Srinivasan and follows the presentation in Helson's book [192].

Helson's theorem (Theorem 4.3.7) generalizes to $L^2(\mu)$ spaces, where μ is a finite positive Borel measure on $\mathbb{T}$. Write $\mu = w\,dm + \mu_s$, where $w \in L^1(\mathbb{T})$ and μ_s is singular with respect to m. Suppose $\mathcal{S}$ is an M_ξ-invariant subspace of $L^2(\mu)$. If $M_\xi \mathcal{S} = \mathcal{S}$, then there is a Borel set $E \subseteq \mathbb{T}$ such that $\mathcal{S} = \chi_E L^2(\mu)$. If $M_\xi \mathcal{S} \subsetneq \mathcal{S}$, then there is a Borel set $E \subseteq \mathbb{T}$ and a function q such that $|q|^2 w = 1$ m-almost everywhere and $\mathcal{S} = qH^2(\mathbb{T}) \oplus \chi_E L^2(\mu_s)$. Exercise 4.5.4 requests a description of the M_x-invariant subspaces of $L^2(\mu)$ when μ is a measure on $[0, 1]$. See [253] for a thorough exposition.

The commutant of M_x, the set of operators $A \in \mathcal{B}(L^2[0, 1])$ such that $AM_x = M_x A$, is described in Corollary 8.3.3 (similarly for the commutant of M_ξ).

There is an interesting version of $M_x \in \mathcal{B}(L^2[0, 1])$ on the Hilbert space $\mathcal{W}$ of absolutely continuous functions on $[0, 1]$ whose derivative belongs to $L^2[0, 1]$ (see Exercise 1.10.15). In this case $\mathcal{W}$ is not only a Hilbert space-but also an algebra of continuous functions. Furthermore, the M_x-invariant subspaces of $\mathcal{W}$ are the ideals of $\mathcal{W}$ and are of the form $\{f \in \mathcal{W} : f|_E = 0\}$, where E is a closed subset of $[0, 1]$ [326].

The multiplication operator M_x plays an important role in the study of Bishop operators in Chapter 13.

4.5 Exercises

Exercise 4.5.1. For $A \in \mathcal{B}(\mathcal{H})$ and a subspace $\mathcal{M}$ of $\mathcal{H}$, prove that $A\mathcal{M} \subseteq \mathcal{M}$ if and only if $A^*\mathcal{M}^\perp \subseteq \mathcal{M}^\perp$.

Exercise 4.5.2. Let $\mathcal{M}_1$ and $\mathcal{M}_2$ be subspaces of a Hilbert space $\mathcal{H}$.

(a) Prove that $(\mathcal{M}_1 + \mathcal{M}_2)^\perp = \mathcal{M}_1^\perp \cap \mathcal{M}_2^\perp$.

(b) Prove that $(\mathcal{M}_1 \cap \mathcal{M}_2)^\perp = (\mathcal{M}_1^\perp + \mathcal{M}_2^\perp)^-$.

Exercise 4.5.3. Let $\mathcal{M}$ be a subspace of a Hilbert space $\mathcal{H}$ and let $\mathbf{x} \in \mathcal{H}$.

(a) Prove that

$$\text{dist}(\mathbf{x}, \mathcal{M}) = \sup_{\substack{\mathbf{y} \in \mathcal{M}^\perp \\ \|\mathbf{y}\|=1}} |\langle \mathbf{x}, \mathbf{y} \rangle|.$$

(b) Prove that the supremum above is attained.

Exercise 4.5.4. Let μ be a finite positive Borel measure on $[0, 1]$. Describe the invariant subspaces of M_x on $L^2(\mu)$.
Remark: See Chapter 8 for more on this.

Exercise 4.5.5. This exercise highlights a difference between the two multiplication operators considered in this chapter.

(a) Prove that M_x is unitarily equivalent to M_x^2 on $L^2[0, 1]$.

(b) Prove that M_ξ is not unitarily equivalent to M_ξ^2 on $L^2(\mathbb{T})$.

Exercise 4.5.6. This exercise highlights another difference between the two multiplication operators considered in this chapter.

(a) Prove that M_x^2 in $L^2[0, 1]$ has the same invariant subspaces as M_x.

(b) Find an M_ξ^2-invariant subspace of $L^2(\mathbb{T})$ that is not M_ξ-invariant.

Exercise 4.5.7. Suppose $f \in L^2[0, 1]$ and $1/f \in L^2[0, 1]$. Prove that

$$\bigvee \{M_x^n f : n \geqslant 0\} = L^2[0, 1].$$

Exercise 4.5.8. For M_{x^2} on $L^2[-1, 1]$, prove there is no $f \in L^2[-1, 1]$ such that $\bigvee \{M_{x^2}^n f : n \geqslant 0\} = L^2[-1, 1]$.

Exercise 4.5.9. Prove that if $A \in \mathcal{B}(\mathcal{H})$ is an isometry, then ran A is closed.

Exercise 4.5.10. Let $dA = r \, dr \, d\theta$ be area measure on $\mathbb{D}^-$ and $z = re^{i\theta}$.

(a) Prove that the invariant subspaces for $M_z f = zf$ on $L^2(dA)$ are not all of the form $\chi_E L^2(dA)$ for some measurable set $E \subseteq \mathbb{D}^-$.

(b) Prove that a subspace $\mathcal{S} \subseteq L^2(dA)$ is invariant for both M_z and $M_{\bar{z}}$ if and only if $\mathcal{M} = \chi_E L^2(dA)$ for some measurable set $E \subseteq \mathbb{D}^-$.

Remark: This topic is explored further in Chapter 8.

Exercise 4.5.11. Show that the linear transformation $(a_n)_{n=0}^\infty \mapsto \sum_{n=0}^\infty a_n \xi^n$ is a unitary operator from ℓ^2 onto $H^2(\mathbb{T})$.

Exercise 4.5.12. The *Fourier coefficients* of an $f \in L^1(\mathbb{T})$ are

$$\hat{f}(n) = \int_{\mathbb{T}} f(\xi)\overline{\xi}^n \, dm(\xi) \quad \text{for } n \in \mathbb{Z}.$$

Using the following steps, prove the *Riemann–Lebesgue lemma*: for $f \in L^1(\mathbb{T})$, $\lim_{|n|\to\infty} \hat{f}(n) = 0$.

(a) Prove the Riemann–Lebesgue lemma for $f \in L^2(\mathbb{T})$.

(b) Prove that $L^2(\mathbb{T})$ is dense in $L^1(\mathbb{T})$.

(c) Use this to prove the Riemann–Lebesgue lemma for $L^1(\mathbb{T})$.

Exercise 4.5.13. Exercise 1.10.37 introduced the class Ω of measurable functions f on $\mathbb{R}$ such that

$$\|f\|^2 = \lim_{R\to\infty} \frac{1}{2R} \int_{-R}^{R} |f(x)|^2 dx < \infty.$$

(a) If $\lambda \in \mathbb{R}$ and $e_\lambda(x) = e^{i\lambda x}$, prove that $\langle e_\alpha, e_\beta \rangle = \begin{cases} 0 & \text{if } \alpha \neq \beta, \\ 1 & \text{if } \alpha = \beta. \end{cases}$

(b) If $f = \sum_{n=0}^{N} c_n e_{\lambda_n}$, prove that $\langle f, e_{\lambda_n} \rangle = c_n$ and $\|f\|^2 = \sum_{n=0}^{N} |c_n|^2$.

Remark: As one can see, this is a version of Fourier series for Ω.

Exercise 4.5.14. Show that $(H^2(\mathbb{T}))^\perp = \overline{M_{\xi} H^2(\mathbb{T})}$.

Exercise 4.5.15. For M_{ξ} on $L^2(\mathbb{T})$, prove the following.

(a) $\sigma(M_{\xi}) = \mathbb{T}$.

(b) $\sigma_p(M_{\xi}) = \emptyset$.

Exercise 4.5.16. For $A \in \mathcal{B}(\mathcal{H})$ and a subspace $\mathcal{M} \subseteq \mathcal{H}$, let P denote the orthogonal projection of $\mathcal{H}$ onto $\mathcal{M}$.

(a) Prove that $\mathcal{M}$ is an invariant subspace for A if and only if $PAP = AP$.

(b) Prove that $\mathcal{M}$ is a reducing subspace for A if and only if $PA = AP$.

Exercise 4.5.17. Suppose $A \in \mathcal{B}(\mathcal{H})$ is unitarily equivalent to $B \in \mathcal{B}(\mathcal{K})$. Prove the following.

(a) $\|A\| = \|B\|$.

(b) $\sigma(A) = \sigma(B)$.

(c) $\sigma_p(A) = \sigma_p(B)$.

(d) $\sigma_{ap}(A) = \sigma_{ap}(B)$.

Remark: Recall the parts of the spectrum from Definition 2.4.5.

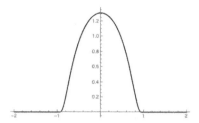

Figure 4.5.1 The "bump" function ψ.

Exercise 4.5.18. Show that M_ξ on $L^2(\mathbb{T})$ is unitarily equivalent to $M_{\overline{\xi}}$.

Exercise 4.5.19. Let T be a bounded operator on a separable Hilbert space $\mathcal{H}$.

(a) Prove that T is unitary if and only if there are orthonormal bases $(\mathbf{x}_n)_{n=1}^\infty$ and $(\mathbf{y}_n)_{n=1}^\infty$ for $\mathcal{H}$ such that $T = \sum_{n=1}^\infty \mathbf{x}_n \otimes \mathbf{y}_n$.

(b) For the unitary operator $T = M_\xi$ on $L^2(\mathbb{T})$, compute the decomposition of T as described in (a).

(c) For any orthonormal basis $(\mathbf{x}_n)_{n=1}^\infty$ for $\mathcal{H}$ and any sequence $(\xi_n)_{n=1}^\infty$ of unimodular constants, prove that $T = \sum_{n=1}^\infty \xi_n(\mathbf{x}_n \otimes \mathbf{x}_n)$ is unitary.

(d) Can every unitary $T \in \mathcal{B}(\mathcal{H})$ be written as in (c)?

Exercise 4.5.20. The discussion of the invariant subspaces for M_x on $L^2[0, 1]$ examined orthogonal projections of the form $f \mapsto \chi_E f$. This exercise develops a version of this orthogonal projection that involves C^∞ functions.

(a) Let $\varepsilon > 0$ and let ψ be a positive, even, C^∞ function on $\mathbb{R}$ whose support lies in $[-\varepsilon, \varepsilon]$ and which satisfies

$$\int_{-\infty}^\infty \psi(x)\, dx = \frac{\pi}{2}.$$

Thus, ψ is a C^∞ bump function centered at the origin (see Figure 4.5.1). Define

$$\theta(x) = \int_{-\infty}^x \psi(t)\, dt$$

and prove that $\theta(x) + \theta(-x) = \frac{\pi}{2}$ for all $x \in \mathbb{R}$.

(b) Let $s(x) = \sin(\theta(x))$ and $c(x) = \cos(\theta(x))$, where θ is the function introduced in (a). Note that s and c depend on ε. Prove that s and c are C^∞ functions such that

$$s(-x) = c(x),$$

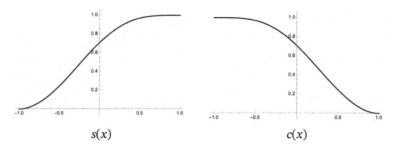

$s(x)$ $c(x)$

Figure 4.5.2 The functions $s(x)$ and $c(x)$.

$$s^2(x) + s^2(-x) = 1, \quad \text{and}$$
$$c^2(x) + c^2(-x) = 1,$$

for all $x \in \mathbb{R}$ (Figure 4.5.2).

(c) Let ω be a real-valued C^∞ function such that $\omega^2(x) + \omega^2(-x) = 1$ for all $x \in \mathbb{R}$. Define the operators P and Q on $L^2(\mathbb{R})$ by

$$(Pf)(x) = \omega^2(x)f(x) + \omega(x)\omega(-x)f(-x)$$

and

$$(Qf)(x) = \omega^2(x)f(x) - \omega(x)\omega(-x)f(-x).$$

Construct P and Q using the function $\omega = s$ introduced in (b) and prove that P and Q are orthogonal projections on $L^2(\mathbb{R})$.

(d) Prove that P and Q are smooth versions of the orthogonal projection $f \mapsto \chi_{(0,\infty)}f$, in the following sense.

   (i) if ess-sup$f \subseteq (-\infty, -\varepsilon]$, then $Pf = Qf = 0$.

   (ii) if ess-sup$f \subseteq [\varepsilon, \infty)$, then $Pf = Qf = f$.

(e) Construct P and Q using the function $\omega = c$ introduced in (b) and prove that both P and Q are smooth versions of the orthogonal projection $f \mapsto \chi_{(-\infty,0)}f$, in the following sense.

   (i) if ess-sup$f \subseteq (-\infty, -\varepsilon]$, then $Pf = Qf = f$.

   (ii) if ess-sup$f \subseteq [\varepsilon, \infty)$, then $Pf = Qf = 0$.

Exercise 4.5.21. This is a complex version of Exercise 4.5.20.

(a) Let ω be a C^∞ complex-valued function such that

$$|\omega(x)|^2 + |\omega(-x)|^2 = 1 \quad \text{for } x \in \mathbb{R}.$$

Define the operators P and Q on $L^2(\mathbb{R})$ by

$$(Pf)(x) = \overline{\omega(x)}\big(\omega(x)f(x) + \omega(-x)f(-x)\big)$$

and

$$(Qf)(x) = \overline{\omega(x)}\big(\omega(x)f(x) - \omega(-x)f(-x)\big)$$

Prove that P and Q are orthogonal projections on $L^2(\mathbb{R})$.

(b) Let $[a, b] \subseteq \mathbb{R}$. Apply the construction in the previous problem to give a smooth version of the projection $f \mapsto \chi_{[a,b]}f$.

Exercise 4.5.22. Prove that if $A_n \to A$ in the norm of $\mathcal{B}(\mathcal{H})$, then $A_n^* \to A^*$ in norm.

Exercise 4.5.23. A sequence $(A_n)_{n=1}^\infty$ in $\mathcal{B}(\mathcal{H})$ converges to $A \in \mathcal{B}(\mathcal{H})$ in the *strong operator topology* (SOT) if $\|A_n\mathbf{x} - A\mathbf{x}\| \to 0$ for each $\mathbf{x} \in \mathcal{H}$.

(a) Prove that if $A_n \to 0$ in norm, then $A_n \to 0$ (SOT).

(b) Define A on ℓ^2 by

$$A = \begin{bmatrix} 0 & 1 & 0 & 0 & \cdots \\ 0 & 0 & 1 & 0 & \cdots \\ 0 & 0 & 0 & 1 & \cdots \\ 0 & 0 & 0 & 0 & \cdots \\ \vdots & \vdots & \vdots & \vdots & \ddots \end{bmatrix}.$$

Prove that $A^n \to 0$ (SOT) but $A^{*n} \nrightarrow 0$ (SOT).

(c) If $A_n \to 0$ (SOT), does $A_n \to 0$ in norm?

Remark: One can endow $\mathcal{B}(\mathcal{H})$ with a topology that makes it a topological vector space such that a sequence $(A_n)_{n=1}^\infty$ converges to A with respect to this topology precisely when $\|A_n\mathbf{x} - A\mathbf{x}\| \to 0$ for all $\mathbf{x} \in \mathcal{H}$. See [94] for details.

Exercise 4.5.24. A sequence $(A_n)_{n=1}^\infty$ in $\mathcal{B}(\mathcal{H})$ converges to $A \in \mathcal{B}(\mathcal{H})$ in the *weak operator topology* (WOT) if $\langle A_n\mathbf{x}, \mathbf{y} \rangle \to \langle A\mathbf{x}, \mathbf{y} \rangle$ for each $\mathbf{x}, \mathbf{y} \in \mathcal{H}$.

(a) Prove that if $A_n \to 0$ in norm, then $A_n \to 0$ (WOT).

(b) Prove that if $A_n \to 0$ (WOT), then $A_n^* \to 0$ (WOT).

(c) Let $A_n = M_\xi^n|_{H^2(\mathbb{T})}$. Prove that $A_n \to 0$ (WOT), but not (SOT) or in norm.

Remark: One can endow $\mathcal{B}(\mathcal{H})$ with a topology that makes it a topological vector space such that a sequence $(A_n)_{n=1}^\infty$ converges to A with respect to this topology precisely when $\langle A_n\mathbf{x}, \mathbf{y} \rangle \to \langle A\mathbf{x}, \mathbf{y} \rangle$ for every $\mathbf{x}, \mathbf{y} \in \mathcal{H}$. See [94] for details.

4.6 Hints for the Exercises

Hint for Ex. 4.5.3: Consider the orthogonal projection onto $\mathcal{M}$.

Hint for Ex. 4.5.4: Examine the proof of Theorem 4.1.7.

Hint for Ex. 4.5.5: For (b), prove that if U is unitary and $UM_\xi = M_{\xi^2}U$, then $U\xi^n = \xi^{2n}(U1)$ for all $n \in \mathbb{Z}$. Now consider the subspace $\mathcal{M} = \bigvee\{\xi^{2k} : k \in \mathbb{Z}\}$.

Hint for Ex. 4.5.8: Given $f \in L^2[-1,1]$, find a $g \in L^2[-1,1]$ such that $\langle M_{x^2}^n f, g\rangle = 0$ for all $n \geqslant 0$.

Hint for Ex. 4.5.10: For (a), consider $\mathcal{M} = \bigvee\{z^n : n \geqslant 0\}$.

Hint for Ex. 4.5.18: Consider the operator U on $L^2(\mathbb{T})$ defined by $(Uf)(\xi) = f(\bar{\xi})$.

Hint for Ex. 4.5.20: For (a), use the fact that

$$\int_{-\infty}^{x} \psi(t)\,dt + \int_{x}^{\infty} \psi(t)\,dt = \frac{\pi}{2}.$$

For (d), note that $s(x) = 0$ for $x \leqslant -\varepsilon$ and $s(x) = 1$ for $x \geqslant \varepsilon$.

Hint for Ex. 4.5.21: For (b), choose $\varepsilon > 0$ and $\varepsilon' > 0$ such that $\varepsilon + \varepsilon' < b - a$. Construct P_1 according to the recipe in (d) with $s = s_\varepsilon$ (either with + or with −) and translate it by a. Similarly, construct P_2 according to the recipe in (e) with $c = c_{\varepsilon'}$ (again either with + or with −) and translate it by b. Then consider the orthogonal projection $P = P_1 P_2 = P_2 P_1$.

Hint for Ex. 4.5.24: For (c), see Exercise 4.5.12.

5

$\cdot$ $\cdot$ $\bullet$ $\cdot$ $\cdot$

The Unilateral Shift

Key Concepts: Shift operator (norm, adjoint, spectral properties, invariant subspaces, commutant, cyclic vectors), Hardy space, inner function, multipliers of the Hardy space.

Outline: The *unilateral shift* operator S on ℓ^2, defined by $Se_n = e_{n+1}$ for $n \geq 0$, is of supreme importance in operator theory. Despite its simple appearance in the setting of ℓ^2, this operator is best understood with complex analysis. For example, the lattice of S-invariant subspaces is described by the foundational work of Riesz, Smirnov, and Beurling on inner functions.

5.1 The Shift on ℓ^2

Definition 5.1.1. The *shift operator* is the linear transformation $S : \ell^2 \to \ell^2$ defined by

$$S(a_0, a_1, a_2,\ldots) = (0, a_0, a_1, a_2,\ldots).$$

To distinguish the shift operator S from the bilateral shift M_ξ on $L^2(\mathbb{T})$ discussed in Chapter 4, some authors use the term *unilateral shift* for emphasis. The unilateral shift is an example of a *weighted shift* on ℓ^2, as seen in Exercise 3.6.21.

Observe that S is an isometry on ℓ^2 since $\|Sa\| = \|a\|$ for all $a \in \ell^2$. In particular,

$$\|S\| = \sup_{\|a\|=1} \|Sa\| = 1 \tag{5.1.2}$$

and thus S is a bounded operator on ℓ^2 with norm one. With respect to the standard orthonormal basis $(e_n)_{n=0}^\infty$ for ℓ^2 from (1.2.8), it follows from Exercise 5.9.1 that S has the matrix representation

$$\begin{bmatrix} 0 & 0 & 0 & 0 & \cdots \\ 1 & 0 & 0 & 0 & \cdots \\ 0 & 1 & 0 & 0 & \cdots \\ 0 & 0 & 1 & 0 & \cdots \\ \vdots & \vdots & \vdots & \vdots & \ddots \end{bmatrix}. \tag{5.1.3}$$

By this we mean that the (i, j) entry of the matrix above is $\langle Se_j, e_i \rangle$.

Theorem 2.4.9a says that $\sigma(S) \subseteq \mathbb{D}^-$. Here is a more precise description of the spectrum.

Proposition 5.1.4.

(a) $\sigma(S) = \mathbb{D}^-$.

(b) $\sigma_p(S) = \emptyset$.

(c) $\sigma_{ap}(S) = \mathbb{T}$.

Proof (a) The containment $\sigma(S) \subseteq \mathbb{D}^-$ follows from Theorem 2.4.9a. Let $\lambda \in \mathbb{D}^-$ and suppose that $(S - \lambda I)\mathbf{a} = \mathbf{e}_0$ for some $\mathbf{a} = (a_n)_{n=0}^\infty \in \ell^2$. Then

$$(-\lambda a_0, a_0 - \lambda a_1, a_1 - \lambda a_2, a_2 - \lambda a_3, ...) = (1, 0, 0, ...).$$

Comparing entries yields $-\lambda a_0 = 1$. Moreover, induction provides

$$a_n = -\frac{1}{\lambda^{n+1}} \quad \text{for all } n \geqslant 0.$$

However, the sequence

$$(a_0, a_1, a_2, ...) = \left(-\frac{1}{\lambda}, -\frac{1}{\lambda^2}, -\frac{1}{\lambda^3}, ... \right)$$

does not belong to ℓ^2 since $|\lambda| \leqslant 1$. This shows that $S - \lambda I$ is not invertible and hence $\sigma(S) = \mathbb{D}^-$.

(b) Suppose $\lambda \in \mathbb{C}$ and $\mathbf{a} \in \ell^2$ satisfies $(S - \lambda I)\mathbf{a} = \mathbf{0}$. Then

$$\begin{aligned}
(0, 0, 0, 0, ...) &= (S - \lambda I)(a_0, a_1, a_2, ...) \\
&= (0, a_0, a_1, a_2, ...) - (\lambda a_0, \lambda a_1, \lambda a_2, ...) \\
&= (-\lambda a_0, a_0 - \lambda a_1, a_1 - \lambda a_2, ...).
\end{aligned}$$

Compare entries and use induction to deduce that $a_j = 0$ for every $j \geqslant 0$. Thus, $\mathbf{a} = \mathbf{0}$ and hence λ is not an eigenvalue of S. This proves that $\sigma_p(S) = \emptyset$.

(c) Fix $\xi \in \mathbb{T}$ and define the sequence of unit vectors

$$\mathbf{x}_n = \frac{1}{\sqrt{n}}(1, \bar{\xi}, \bar{\xi}^2, ..., \bar{\xi}^{n-1}, 0, 0, ...) \quad \text{for } n \geqslant 1.$$

Then

$$(S - \xi I)\mathbf{x}_n = \frac{1}{\sqrt{n}}(-\xi, 0, 0, ..., 0, \bar{\xi}^{n-1}, 0, 0, ...),$$

and hence

$$\|(S - \xi I)\mathbf{x}_n\| = \sqrt{\frac{2}{n}} \to 0.$$

Thus, $\mathbb{T} \subseteq \sigma_{ap}(S)$. Since $\sigma_{ap}(S) \subseteq \mathbb{D}^-$ (Proposition 2.4.6), it suffices to show that $\sigma_{ap}(S) \cap \mathbb{D} = \emptyset$. For $\lambda \in \mathbb{D}$ and a unit vector $\mathbf{x} \in \ell^2$, observe that

$$\|(S - \lambda I)\mathbf{x}\| \geqslant |\|S\mathbf{x}\| - |\lambda|\|\mathbf{x}\|| = (1 - |\lambda|)\|\mathbf{x}\| = 1 - |\lambda| > 0.$$

Therefore, $\inf_{\|\mathbf{x}\|=1} \|(S - \lambda I)\mathbf{x}\| > 0$, and hence $\lambda \notin \sigma_{ap}(S)$. ∎

5.2 Adjoint of the Shift

The adjoint of S is the unique $S^* \in \mathcal{B}(\ell^2)$ that satisfies $\langle S\mathbf{a}, \mathbf{b} \rangle = \langle \mathbf{a}, S^*\mathbf{b} \rangle$ for all $\mathbf{a}, \mathbf{b} \in \ell^2$. The following proposition indicates why S^* is often called the *backward shift*.

Proposition 5.2.1. *For* $\mathbf{b} = (b_n)_{n=0}^{\infty} \in \ell^2$,

$$S^*(b_0, b_1, b_2, b_3, \ldots) = (b_1, b_2, b_3, \ldots).$$

Proof The linear transformation B on ℓ^2 defined by

$$B(b_0, b_1, b_2, b_3, \ldots) = (b_1, b_2, b_3, \ldots)$$

is bounded on ℓ^2 since

$$\|B\mathbf{b}\|^2 = \sum_{n=1}^{\infty} |b_n|^2 \leqslant \sum_{n=0}^{\infty} |b_n|^2 = \|\mathbf{b}\|^2. \tag{5.2.2}$$

Since

$$
\begin{aligned}
\langle S\mathbf{a}, \mathbf{b} \rangle &= \langle (0, a_0, a_1, a_2, \ldots), (b_0, b_1, b_2, b_2, \ldots) \rangle \\
&= a_0 \overline{b_1} + a_1 \overline{b_2} + a_2 \overline{b_3} + \cdots \\
&= \langle (a_0, a_1, a_2, a_3, \ldots), (b_1, b_2, b_3, \ldots) \rangle \\
&= \langle \mathbf{a}, B\mathbf{b} \rangle
\end{aligned}
$$

for all $\mathbf{a}, \mathbf{b} \in \ell^2$, the uniqueness of the adjoint ensures that $S^* = B$. ∎

The matrix representation of S^* with respect to the standard basis $(\mathbf{e}_n)_{n=0}^{\infty}$ for ℓ^2 is

$$\begin{bmatrix} 0 & 1 & 0 & 0 & \cdots \\ 0 & 0 & 1 & 0 & \cdots \\ 0 & 0 & 0 & 1 & \cdots \\ 0 & 0 & 0 & 0 & \cdots \\ \vdots & \vdots & \vdots & \vdots & \ddots \end{bmatrix}. \tag{5.2.3}$$

As expected, it is the conjugate transpose of the matrix representation (5.1.3) of S. Compare the following description of the parts of the spectrum of S^* to that of S (Proposition 5.1.4).

Proposition 5.2.4.

(a) $\sigma(S^*) = \mathbb{D}^-$.

(b) $\sigma_p(S^*) = \mathbb{D}$.

(c) $\sigma_{ap}(S^*) = \mathbb{D}^-$.

Proof (a) Note that $\sigma(S^*) = \overline{\sigma(S)} = \overline{\mathbb{D}^-} = \mathbb{D}^-$ (Exercise 3.6.5).

(b) For $\lambda \in \mathbb{D}$, the sequence

$$\mathbf{a}_\lambda = (1, \lambda, \lambda^2, \lambda^3, \ldots) \tag{5.2.5}$$

belongs to ℓ^2 since $\|\mathbf{a}_\lambda\| = (1 - |\lambda|^2)^{-1/2}$. Moreover,

$$S^* \mathbf{a}_\lambda = (\lambda, \lambda^2, \lambda^3, \ldots) = \lambda(1, \lambda, \lambda^2, \ldots) = \lambda \mathbf{a}_\lambda$$

and hence $\mathbb{D} \subseteq \sigma_p(S^*)$. Since $\sigma_p(S^*) \subseteq \mathbb{D}^-$, it suffices to show that no $\xi \in \mathbb{T}$ belongs to $\sigma_p(S^*)$. Suppose toward a contradiction that $\xi \in \mathbb{T}$ and that $\mathbf{a} = (a_n)_{n=0}^\infty \in \ell^2 \backslash \{0\}$ with $S^* \mathbf{a} = \xi \mathbf{a}$. In other words,

$$(a_1, a_2, a_3, a_4, \ldots) = (\xi a_0, \xi a_1, \xi a_2, \xi a_3, \ldots).$$

Equating coefficients and solving a recurrence yields $a_k = \xi^k a_0$ for all $k \geqslant 0$. Since $\xi \in \mathbb{T}$, it follows that $\mathbf{a} \notin \ell^2$ unless $a_0 = 0$. Therefore, $\xi \notin \sigma_p(S^*)$.

(c) Since $\mathbb{D} = \sigma_p(S^*) \subseteq \sigma_{ap}(S^*) \subseteq \sigma(S^*) = \mathbb{D}^-$ (Proposition 2.4.6), it suffices to show that $\mathbb{T} \subseteq \sigma_{ap}(S^*)$. Given $\xi \in \mathbb{T}$, choose a sequence $(\lambda_n)_{n=1}^\infty$ in $\mathbb{D}$ such that $\lambda_n \to \xi$. Let

$$\widetilde{\mathbf{a}}_{\lambda_n} = \mathbf{a}_{\lambda_n} \sqrt{1 - |\lambda_n|^2}$$

denote the normalized eigenvectors for S^* from (5.2.5) and observe that

$$\begin{aligned}
\|(S^* - \xi I) \widetilde{\mathbf{a}}_{\lambda_n}\| &= \|(S^* - \lambda_n)\widetilde{\mathbf{a}}_{\lambda_n} + (\lambda_n - \xi)\widetilde{\mathbf{a}}_{\lambda_n}\| \\
&= \|\mathbf{0} + (\lambda_n - \xi)\widetilde{\mathbf{a}}_{\lambda_n}\| \\
&= |\lambda_n - \xi| \|\widetilde{\mathbf{a}}_{\lambda_n}\| \\
&= |\lambda_n - \xi| \to 0.
\end{aligned}$$

Thus, $\xi \in \sigma_{ap}(S^*)$. ∎

5.3 The Hardy Space

To gain a deeper understanding of the unilateral shift and its adjoint, one must view them as linear transformations on a certain Hilbert space of analytic functions on $\mathbb{D}$. We begin with the following proposition which shows that S is unitarily equivalent to the multiplication operator $M_\xi|_{H^2(\mathbb{T})}$ from Chapter 4. Recall that $H^2(\mathbb{T}) = \{f \in L^2(\mathbb{T}) : \widehat{f}(n) = 0 \text{ for all } n < 0\}$.

Proposition 5.3.1. *The operator* $U : \ell^2 \to H^2(\mathbb{T})$ *defined by*

$$U((a_0, a_1, a_2, \ldots)) = \sum_{n=0}^\infty a_n \xi^n$$

is unitary. Moreover, $U^* M_\xi|_{H^2(\mathbb{T})} U = S$.

Proof The fact that U is unitary comes from Proposition 4.2.7. For the second part, observe that for each $n \geqslant 0$,

$$U^*M_\xi U e_n = U^*M_\xi \xi^n = U^* \xi^{n+1} = e_{n+1} = S e_n.$$

The identity above extends linearly to all of ℓ^2. ∎

Below is a diagram that illustrates Proposition 5.3.1:

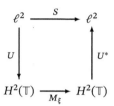

As is traditional in this subject, we identify S with $M_\xi|_{H^2(\mathbb{T})}$ and use the less cumbersome symbol S for both operators. Theorem 4.3.7 describes the S-invariant subspaces.

Theorem 5.3.2. *If $q \in H^2(\mathbb{T})$ and $|q| = 1$ almost everywhere, then $qH^2(\mathbb{T})$ is an S-invariant subspace of $H^2(\mathbb{T})$. Furthermore, every nonzero S-invariant subspace is of this form.*

Proof We first show that $qH^2(\mathbb{T}) \subseteq H^2(\mathbb{T})$. For $f \in H^2(\mathbb{T})$, note that $qf \in L^2(\mathbb{T})$ (since q is bounded). If $(p_n)_{n=1}^\infty$ is a sequence of polynomials that converges to f in $L^2(\mathbb{T})$, then $qp_n \in H^2(\mathbb{T})$ (since $q \in H^2(\mathbb{T})$) and $qp_n \to qf$ in $L^2(\mathbb{T})$. Since $H^2(\mathbb{T})$ is closed, we confirm that $qf \in H^2(\mathbb{T})$. This also shows that $qH^2(\mathbb{T})$ is closed and S-invariant.

Let $\mathcal{M}$ be a nonzero S-invariant subspace of $H^2(\mathbb{T})$. We claim that $S\mathcal{M} \subsetneq \mathcal{M}$. If otherwise, then induction confirms that $S^n \mathcal{M} = \mathcal{M}$ for $n \geqslant 0$. Consequently, each $f \in \mathcal{M}$ is of the form $f = \xi^n f_n$ for some $f_n \in H^2(\mathbb{T})$. Thus, for every $k \geqslant 0$, the first k Fourier coefficients of f are zero, which implies that f is the zero function. Since $\mathcal{M} \neq \{0\}$, it follows that $S\mathcal{M} \subsetneq \mathcal{M}$. Helson's theorem (Theorem 4.3.7) implies that $\mathcal{M} = qH^2(\mathbb{T})$. Since $\mathcal{M} \subseteq H^2(\mathbb{T})$ it follows that $q = q \cdot 1 \in H^2(\mathbb{T})$. ∎

The functions $q \in H^2(\mathbb{T})$ with $|q| = 1$ everywhere are *inner functions*. Their exact description is specified after $H^2(\mathbb{T})$ is identified with a space of analytic functions on $\mathbb{D}$.

Definition 5.3.3. The *Hardy space* H^2 is the set of power series

$$f(z) = \sum_{n=0}^\infty a_n z^n, \quad \text{where } \mathbf{a} = (a_n)_{n=0}^\infty \in \ell^2.$$

As it stands now, the Hardy space is a set of formal power series. The minimal disk of convergence of an $f \in H^2$ is addressed by the following.

Proposition 5.3.4. *Every*

$$f(z) = \sum_{n=0}^\infty a_n z^n \in H^2$$

has a radius of convergence at least 1 and thus defines an analytic function on $\mathbb{D}$.

Proof For each $\lambda \in \mathbb{D}$, the Cauchy–Schwarz inequality implies that

$$\sum_{n=0}^{\infty} |a_n||\lambda|^n \leqslant \left(\sum_{n=0}^{\infty} |a_n|^2 \right)^{\frac{1}{2}} \left(\sum_{n=0}^{\infty} |\lambda|^{2n} \right)^{\frac{1}{2}}$$

$$= \|\mathbf{a}\| \left(\frac{1}{1 - |\lambda|^2} \right)^{\frac{1}{2}}. \tag{5.3.5}$$

Therefore, the power series converges absolutely at every point in $\mathbb{D}$ and so its radius of convergence is at least 1. Now use the fact that a power series defines an analytic function on the interior of its disk of convergence [9, p. 38]. ∎

Define an inner product on H^2 by

$$\langle f, g \rangle = \sum_{n=0}^{\infty} a_n \overline{b_n},$$

where $(a_n)_{n=0}^{\infty}$ and $(b_n)_{n=0}^{\infty}$ are the sequence of Taylor coefficients of f and g, respectively. Using the fact that ℓ^2 is complete (Proposition 1.2.5) and that the bijective map

$$\mathbf{a} \mapsto \sum_{n=0}^{\infty} a_n z^n \tag{5.3.6}$$

from ℓ^2 to H^2 is an isometry, it follows that H^2 is complete and hence is a Hilbert space. Furthermore, for $f \in H^2$, the inequality (5.3.5) can be rephrased as

$$|f(\lambda)| \leqslant \|f\| \frac{1}{\sqrt{1 - |\lambda|^2}} \quad \text{for } \lambda \in \mathbb{D}. \tag{5.3.7}$$

This yields the following.

Proposition 5.3.8. *If $f_n \to f$ in H^2 norm, then $f_n \to f$ uniformly on compact subsets of $\mathbb{D}$.*

Proof Let $0 < r < 1$. For $|\lambda| \leqslant r$, (5.3.7) implies that

$$|f_n(\lambda) - f(\lambda)| \leqslant \|f_n - f\| \frac{1}{\sqrt{1 - |\lambda|^2}} \leqslant \|f_n - f\| \frac{1}{\sqrt{1 - r^2}}.$$

Thus, $f_n \to f$ uniformly on $|z| \leqslant r$. ∎

The identification (5.3.6) of ℓ^2 and H^2 provides an important connection between Fourier analysis and complex analysis.

Proposition 5.3.9. *If*

$$f(z) = \sum_{n=0}^{\infty} a_n z^n$$

is an analytic function on $\mathbb{D}$, then

$$\sup_{0 < r < 1} \int_0^{2\pi} |f(re^{i\theta})|^2 \frac{d\theta}{2\pi} = \sum_{n=0}^{\infty} |a_n|^2.$$

Proof For a fixed $0 < r < 1$,

$$
\int_0^{2\pi} |f(re^{i\theta})|^2 \frac{d\theta}{2\pi} = \int_0^{2\pi} f(re^{i\theta})\overline{f(re^{i\theta})} \frac{d\theta}{2\pi}
$$

$$
= \int_0^{2\pi} \Big(\sum_{k=0}^{\infty} a_k r^k e^{ik\theta} \Big)\Big(\sum_{\ell=0}^{\infty} \overline{a_\ell} r^\ell e^{-i\ell\theta} \Big) \frac{d\theta}{2\pi}
$$

$$
= \sum_{k,\ell=0}^{\infty} a_k \overline{a_\ell} r^{k+\ell} \int_0^{2\pi} e^{i(k-\ell)\theta} \frac{d\theta}{2\pi}
$$

$$
= \sum_{k,\ell=0}^{\infty} a_k \overline{a_\ell} r^{k+\ell} \delta_{k\ell}
$$

$$
= \sum_{k=0}^{\infty} |a_k|^2 r^{2k}.
$$

The series above increases with r and hence

$$
\sup_{0 < r < 1} \int_0^{2\pi} |f(re^{i\theta})|^2 \frac{d\theta}{2\pi} = \lim_{r \to 1^-} \int_0^{2\pi} |f(re^{i\theta})|^2 \frac{d\theta}{2\pi}
$$

$$
= \lim_{r \to 1^-} \sum_{n=0}^{\infty} r^{2n} |a_n|^2.
$$

If $\sum_{n=0}^{\infty} |a_n|^2 = \infty$, Fatou's lemma shows that

$$
\lim_{r \to 1^-} \sum_{n=0}^{\infty} r^{2n} |a_n|^2 = \infty.
$$

If $\sum_{n=0}^{\infty} |a_n|^2 < \infty$, the dominated convergence theorem implies that

$$
\lim_{r \to 1^-} \sum_{n=0}^{\infty} r^{2n} |a_n|^2 = \sum_{n=0}^{\infty} |a_n|^2,
$$

which completes the proof. ∎

Recall normalized Lebesgue measure m on $\mathbb{T}$ from Chapter 4.

Corollary 5.3.10. *An analytic function f on $\mathbb{D}$ belongs to H^2 if and only if*

$$
\sup_{0 < r < 1} \int_{\mathbb{T}} |f(r\xi)|^2 \, dm(\xi) < \infty.
$$

If $f \in H^2$, then the supremum above equals $\|f\|^2$.

Corollary 5.3.11. *The set of bounded analytic functions on $\mathbb{D}$ is dense in H^2. In particular, the polynomials are dense in H^2.*

Proof If f is a bounded analytic function on $\mathbb{D}$, then

$$\|f\|^2 = \sup_{0<r<1} \int_{\mathbb{T}} |f(r\xi)|^2 dm(\xi) \leqslant \sup_{z \in \mathbb{D}} |f(z)|^2 < \infty$$

and hence $f \in H^2$. For any $g(z) = \sum_{n=0}^{\infty} b_n z^n \in H^2$, the Nth Taylor polynomial $g_N(z) = \sum_{n=0}^{N} b_n z^n$ satisfies $\|g - g_N\|^2 = \sum_{n=N+1}^{\infty} |b_n|^2 \to 0$ as $N \to \infty$. Thus, the polynomials, and hence the bounded analytic functions, are dense in H^2. ∎

The next result, whose proof would draw us too far afield, permits the identification of H^2 and $H^2(\mathbb{T})$ via boundary values [202].

Proposition 5.3.12. *If*

$$f(z) = \sum_{n=0}^{\infty} a_n z^n \in H^2,$$

then for almost every $\xi \in \mathbb{T}$,

$$f(\xi) := \lim_{r \to 1^-} f(r\xi)$$

exists and defines a function in $H^2(\mathbb{T})$. Furthermore, $\widehat{f}(n) = a_n$ for $n \geqslant 0$ and

$$\int_{\mathbb{T}} |f|^2 dm = \sum_{n=0}^{\infty} |a_n|^2.$$

A version of the Cauchy integral formula recovers the values of $f \in H^2$ in $\mathbb{D}$ from its boundary values on $\mathbb{T}$.

Proposition 5.3.13. *If $f \in H^2(\mathbb{T})$, then the analytic function on $\mathbb{D}$ defined by*

$$f(z) = \int_{\mathbb{T}} \frac{f(\xi)}{1 - \bar{\xi}z} dm(\xi) \tag{5.3.14}$$

belongs to H^2. Furthermore,

$$f(z) = \sum_{n=0}^{\infty} \widehat{f}(n) z^n.$$

Proof Fix $z \in \mathbb{D}$. For $\xi \in \mathbb{T}$,

$$\frac{1}{1 - \bar{\xi}z} = \sum_{n=0}^{\infty} \bar{\xi}^n z^n$$

and the series above converges uniformly in ξ. Thus, the following exchange of sum and integral is justified:

$$\int_{\mathbb{T}} \frac{f(\xi)}{1 - \bar{\xi}z} dm(\xi) = \int_{\mathbb{T}} f(\xi) \left(\sum_{n=0}^{\infty} \bar{\xi}^n z^n \right) dm(\xi)$$

$$= \sum_{n=0}^{\infty} z^n \int_{\mathbb{T}} f(\xi)\overline{\xi}^{-n} \, dm(\xi)$$

$$= \sum_{n=0}^{\infty} \widehat{f}(n)z^n,$$

which completes the proof. ∎

The previous two propositions yield the following (Exercises 5.9.6 and 5.9.7).

Corollary 5.3.15. *For fixed $\lambda \in \mathbb{D}$, the function*

$$k_\lambda(z) = \frac{1}{1 - \overline{\lambda}z}, \quad \text{where } z \in \mathbb{D},$$

satisfies the following.

(a) $k_\lambda \in H^2$.

(b) $\|k_\lambda\| = \dfrac{1}{\sqrt{1 - |\lambda|^2}}$.

(c) $\langle f, k_\lambda \rangle = f(\lambda)$ *for all $f \in H^2$.*

The function k_λ is the *reproducing kernel* for H^2. Reproducing kernels for other Hilbert spaces of analytic functions appear in Chapters 9 and 10. One can show directly that Proposition 5.3.13 implies that for each $\lambda \in \mathbb{D}$, the corresponding evaluation functional $f \mapsto f(\lambda)$ is bounded on H^2. Item (c) is an expression of the Riesz representation theorem in this setting.

There is a natural unitary operator

$$\sum_{n=0}^{\infty} a_n z^n \mapsto \sum_{n=0}^{\infty} a_n \xi^n$$

from H^2, the space of analytic functions with square-summable power series coefficients, and $H^2(\mathbb{T})$, the subspace of $L^2(\mathbb{T})$ whose negatively indexed Fourier coefficients vanish. It is traditional in Hardy-space theory to dispense with the difference in notation between H^2 and $H^2(\mathbb{T})$, and to use H^2 for both spaces.

5.4 Bounded Analytic Functions

What makes the characterization of the S-invariant subspaces of H^2 from Theorem 5.3.2 more interesting is the fact that a complete description of inner functions is available. We refer the reader to [202] for the details of the discussion below.

Definition 5.4.1. Let H^∞ denote the set of bounded analytic functions on $\mathbb{D}$. For $f \in H^\infty$, define

$$\|f\|_\infty = \sup_{z \in \mathbb{D}} |f(z)|.$$

One can show that H^∞ is a normed linear space (Definition 1.8.8). A normal family argument (Exercise 5.9.5) verifies the following.

Proposition 5.4.2. *H^∞ is a Banach space.*

As with H^2, there are results concerning the boundary values of H^∞ functions. For $g \in L^\infty(\mathbb{T})$, the *essential supremum* of g is

$$\text{ess-sup}_\mathbb{T} |g| = \sup\{a \geqslant 0 : m(\{\xi \in \mathbb{T} : |g(\xi)| > a\}) > 0\}.$$

Theorem 5.4.3 (Fatou–Smirnov).

(a) *If $f \in H^\infty$, then*

$$f(\xi) = \lim_{r \to 1-} f(r\xi)$$

exists for almost every $\xi \in \mathbb{T}$. Furthermore, this boundary function is an essentially bounded function on $\mathbb{T}$ and

$$\text{ess-sup}_\mathbb{T} |f| = \|f\|_\infty.$$

(b) *If $f \in H^2(\mathbb{T})$ is an essentially bounded function on $\mathbb{T}$, then its Cauchy integral*

$$f(z) = \int_\mathbb{T} \frac{f(\xi)}{1 - \bar{\xi}z} \, dm(\xi)$$

defines an H^∞ function whose radial boundary values coincide with f almost everywhere on $\mathbb{T}$.

So far, an inner function is a $q \in H^2(\mathbb{T})$ such that $|q| = 1$ almost everywhere. The theorem above proves the following.

Corollary 5.4.4. *If $q \in H^2(\mathbb{T})$ is inner, then*

$$z \mapsto \int_\mathbb{T} \frac{q(\xi)}{1 - \bar{\xi}z} \, dm(\xi)$$

defines an H^∞ function whose radial boundary values equal those of q almost everywhere. If $q \in H^\infty$ and q has unimodular boundary values, then this boundary function is an inner function.

Example 5.4.5. For fixed $a \in \mathbb{D}$, consider the Möbius transformation

$$B(z) = \frac{a - z}{1 - \bar{a}z}.$$

For any $\xi \in \mathbb{T}$,

$$|B(\xi)| = \left| \frac{a - \xi}{1 - \bar{a}\xi} \right| = \left| \xi \frac{1 - a\bar{\xi}}{1 - \bar{a}\xi} \right| = \left| \frac{1 - \bar{\xi}a}{1 - \bar{a}\xi} \right| = 1,$$

and hence the maximum modulus theorem implies that $B(\mathbb{D}) \subseteq \mathbb{D}$. Thus, B belongs to H^2 (Corollary 5.3.11) and is an inner function. A calculation reveals that $B(B(z)) = z$ for all $z \in \mathbb{D}$, so the range of B contains $\mathbb{D}$. Thus, $B(\mathbb{D}) = \mathbb{D}$, which implies that B is also a disk automorphism.

Example 5.4.6. Building on the previous example, any finite product

$$B(z) = \gamma z^N \prod_{i=1}^{n} \frac{z - a_i}{1 - \overline{a_i} z}, \tag{5.4.7}$$

where $(a_i)_{i=1}^{n}$ is a finite sequence in $\mathbb{D}\backslash\{0\}$ (repetitions allowed), $\gamma \in \mathbb{T}$, and $N \in \mathbb{N} \cup \{0\}$, is an inner function. These are the *finite Blaschke products* [144].

For an infinite number of zeros, there are convergence issues that can be resolved with the following theorem [202, Ch. 5].

Theorem 5.4.8 (Blaschke). *Suppose that* $(a_i)_{i=1}^{\infty}$ *is a sequence of points in* $\mathbb{D}\backslash\{0\}$, *repetitions allowed, such that* $\sum_{i=1}^{\infty}(1 - |a_i|) < \infty$. *Then*

$$B(z) = \prod_{i=1}^{\infty} \frac{\overline{a_i}}{|a_i|} \frac{a_i - z}{1 - \overline{a_i} z}$$

converges for every $z \in \mathbb{D}$ *and defines an inner function.*

Example 5.4.9. A nonconstant inner function need not have zeros in $\mathbb{D}$. For example, consider

$$f(z) = \exp\left(-\frac{1+z}{1-z}\right).$$

For $z \in \mathbb{D}$,

$$|f(z)| = \exp\left(-\operatorname{Re}\frac{1+z}{1-z}\right) = \exp\left(-\frac{1-|z|^2}{|1-z|^2}\right) < 1,$$

and therefore $f \in H^2$ (Corollary 5.3.11). For every $\xi = e^{it} \in \mathbb{T}\backslash\{1\}$,

$$\lim_{r\to 1^-} \frac{1+r\xi}{1-r\xi} = \frac{1+\xi}{1-\xi} = i\cot(t/2)$$

so $|f(\xi)| = 1$. Thus, f is an inner function without zeros on $\mathbb{D}$. This particular inner function is important in Chapter 20.

The example above can be generalized to the following [202, Ch. 5].

Theorem 5.4.10. *Let* μ *be a positive finite Borel measure on* $\mathbb{T}$ *that is singular with respect to* m. *Then*

$$f(z) = \exp\left(-\int_{\mathbb{T}} \frac{\xi + z}{\xi - z} d\mu(\xi)\right)$$

is an inner function without zeros in $\mathbb{D}$.

The reader needing a review of singular measures should consult [317]. An inner function of this form is a *singular inner function*. The classification of inner functions is a combination of the two theorems above [202, Ch. 5].

Theorem 5.4.11. *Every inner function is of the form $B(z)S_\mu(z)$ with*

$$B(z) = \gamma z^N \prod_{i=1}^{\infty} \frac{\overline{a_i}}{|a_i|} \frac{a_i - z}{1 - \overline{a_i} z}$$

and

$$S_\mu(z) = \exp\left(-\int_{\mathbb{T}} \frac{\xi + z}{\xi - z} d\mu(\xi)\right),$$

where $\gamma \in \mathbb{T}$, $N \geqslant 0$, $(a_i)_{i=1}^{\infty}$ is a sequence in $\mathbb{D}\backslash\{0\}$ such that $\sum_{i=1}^{\infty}(1 - |a_i|) < \infty$, and μ is a finite positive Borel measure on $\mathbb{T}$ that is singular with respect to m.

Unimodular functions on $\mathbb{T}$ arose in Chapter 4 in the study of the M_ξ-invariant subspaces of $L^2(\mathbb{T})$. The quotient of two inner functions is unimodular on $\mathbb{T}$. However, there are unimodular functions that are not the quotient of two inner functions (Exercise 5.9.25).

Theorems 5.3.2 and 5.4.11 yield a fundamental theorem of Beurling that concretely describes the invariant subspaces of S on H^2.

Theorem 5.4.12 (Beurling [53]). *Let $\mathcal{M}$ be a nonzero S-invariant subspace of H^2.*

(a) *There is an inner function q, uniquely determined up to a unimodular constant factor, such that $\mathcal{M} \cap (S\mathcal{M})^{\perp} = \text{span}\{q\}$.*

(b) *$\mathcal{M} = qH^2$.*

(c) *The smallest S-invariant subspace containing q is $\mathcal{M} = qH^2$.*

5.5 Multipliers of H^2

An analytic function $\varphi : \mathbb{D} \to \mathbb{C}$ is a *multiplier* of H^2 if $\varphi H^2 \subseteq H^2$. Note that the set of multipliers of H^2 forms an algebra in the sense that this set is closed under addition, scalar multiplication, and multiplication.

Proposition 5.5.1. *If φ is a multiplier of H^2, then the operator $M_\varphi : H^2 \to H^2$ defined by*

$$M_\varphi f = \varphi f$$

is bounded.

Proof Suppose $f_n \to f$ and $M_\varphi f_n \to g$ in H^2. Then Proposition 5.3.8 implies that $f_n \to f$ and $\varphi f_n \to g$ pointwise in $\mathbb{D}$, and hence $M_\varphi f = g$. The closed graph theorem (Theorem 2.2.2) ensures that M_φ is bounded. ∎

Before proceeding to the description of the multiplier algebra of H^2, we need the following result from harmonic analysis [202, p. 34].

Theorem 5.5.2 (Fatou). *If $f \in L^1(\mathbb{T})$ and*

$$\mathscr{P}(f)(\lambda) = \int_{\mathbb{T}} \frac{1 - |\lambda|^2}{|\xi - \lambda|^2} f(\xi) \, dm(\xi) \quad \text{for } \lambda \in \mathbb{D}, \tag{5.5.3}$$

then, for almost every $\zeta \in \mathbb{T}$,

$$\lim_{r \to 1^-} \mathscr{P}(f)(r\zeta) = f(\zeta).$$

The function $\mathscr{P}(f)$ is the *Poisson integral* of f and it is harmonic on $\mathbb{D}$ (see Chapter 12). If f is continuous on $\mathbb{T}$, then $\mathscr{P}(f)$ is the solution to the *Dirichlet problem* (given a continuous function φ on $\mathbb{T}$, find a Φ that is continuous on $\mathbb{D}^-$ and harmonic on $\mathbb{D}$ such that $\Phi|_{\mathbb{T}} = \varphi$).

Proposition 5.5.4. H^∞ *is the multiplier algebra of H^2. Moreover, if $\varphi \in H^\infty$, then $\|M_\varphi\| = \|\varphi\|_\infty$.*

Proof If $\varphi \in H^\infty$ and $f \in H^2$, Corollary 5.3.10 implies that

$$
\begin{aligned}
\|\varphi f\|^2 &= \sup_{0 < r < 1} \int_{\mathbb{T}} |\varphi(r\xi) f(r\xi)|^2 dm(\xi) \\
&\leqslant \|\varphi\|_\infty^2 \sup_{0 < r < 1} \int_{\mathbb{T}} |f(r\xi)|^2 dm(\xi) \\
&= \|\varphi\|_\infty^2 \|f\|^2.
\end{aligned}
$$

Thus, φ is a multiplier of H^2 and $\|M_\varphi\| \leqslant \|\varphi\|_\infty$.

For $\lambda \in \mathbb{D}$, define the normalized reproducing kernel $\widetilde{k}_\lambda = k_\lambda / \|k_\lambda\|$, where k_λ is defined in Corollary 5.3.15. The Cauchy–Schwarz inequality implies that

$$|\langle M_\varphi \widetilde{k}_\lambda, \widetilde{k}_\lambda \rangle| \leqslant \|M_\varphi \widetilde{k}_\lambda\| \|\widetilde{k}_\lambda\| \leqslant \|M_\varphi\| \|\widetilde{k}_\lambda\| \|\widetilde{k}_\lambda\| = \|M_\varphi\|.$$

Therefore,

$$\|M_\varphi\| \geqslant |\langle \varphi \widetilde{k}_\lambda, \widetilde{k}_\lambda \rangle| = \left| \int_{\mathbb{T}} \frac{1 - |\lambda|^2}{|\xi - \lambda|^2} \varphi(\zeta) \, dm(\zeta) \right| = |\mathscr{P}(\varphi)(\lambda)|.$$

Now let $\lambda = r\xi$, where $\xi \in \mathbb{T}$ and $r \in (0, 1)$, and use Theorem 5.5.2 to see that $\|M_\varphi\| \geqslant |\varphi(\xi)|$ for almost every $\xi \in \mathbb{T}$. Theorem 5.4.3 yields the desired lower bound $\|M_\varphi\| \geqslant \|\varphi\|_\infty$, and hence establishes equality.

For the converse, let φ be a multiplier of H^2. Proposition 5.5.1 says that M_φ is a bounded operator. For $\lambda, w \in \mathbb{D}$,

$$
\begin{aligned}
(M_\varphi^* k_\lambda)(w) = \langle M_\varphi^* k_\lambda, k_w \rangle &= \langle k_\lambda, M_\varphi k_w \rangle \\
&= \overline{\langle \varphi k_w, k_\lambda \rangle} = \overline{\varphi(\lambda) k_w(\lambda)} = \overline{\varphi(\lambda)} k_\lambda(w). \tag{5.5.5}
\end{aligned}
$$

Thus, $\overline{\varphi(\mathbb{D})} = \{\overline{\varphi(\lambda)} : \lambda \in \mathbb{D}\} \subseteq \sigma_p(M_\varphi^*) \subseteq \sigma(M_\varphi^*)$. Since $\sigma(M_\varphi^*)$ is compact (Theorem 2.4.9b), $\varphi(\mathbb{D})$ is a bounded set and hence φ belongs to H^∞. ∎

5.6 Commutant of the Shift

For many of the operators $A \in \mathcal{B}(\mathcal{H})$ considered in this book, we will explore the set of operators $B \in \mathcal{B}(\mathcal{H})$ that commute with A.

Definition 5.6.1. For $A \in \mathcal{B}(\mathcal{H})$, the *commutant*, written $\{A\}'$, is the set of all $B \in \mathcal{B}(\mathcal{H})$ such that $AB = BA$.

For the shift operator S, the commutant can be identified in a natural way with H^∞, the multiplier algebra of H^2.

Corollary 5.6.2. $\{S\}' = \{M_\varphi : \varphi \in H^\infty\}$.

Proof For $\varphi \in H^\infty$ and $f \in H^2$,

$$(SM_\varphi f)(z) = z\varphi(z)f(z) = \varphi(z)(zf(z)) = (M_\varphi Sf)(z),$$

and hence $M_\varphi \in \{S\}'$. Now suppose that $A \in \mathcal{B}(H^2)$ and $AS = SA$. Then $AS^n = S^n A$ for any $n \geqslant 0$ and thus $AM_p = M_p A$ for any $p \in \mathbb{C}[z]$. Apply this to the constant function 1 and deduce that $Ap = pA1$. Given $f \in H^2$, Corollary 5.3.11 provides a sequence $(p_n)_{n=1}^\infty$ of polynomials such that $p_n \to f$ in H^2 norm. By Proposition 5.3.8, $p_n \to f$ pointwise in $\mathbb{D}$. Therefore, $Af = fA1$ and hence $A1$ is a multiplier of H^2. Proposition 5.5.4 implies that $A1 \in H^\infty$ and $A = M_{A1}$. ∎

An alternate, in fact the original, proof of this is in [68].

5.7 Cyclic Vectors

An important class of vectors $\mathbf{x} \in \mathcal{H}$ associated with an $A \in \mathcal{B}(\mathcal{H})$ are those whose *orbit* $\{A^n \mathbf{x} : n \geqslant 0\}$ has dense linear span in $\mathcal{H}$.

Definition 5.7.1. For an $A \in \mathcal{B}(\mathcal{H})$, a vector $\mathbf{x} \in \mathcal{H}$ is a *cyclic vector* for A if

$$\bigvee \{A^n \mathbf{x} : n \geqslant 0\} = \mathcal{H}.$$

A complete description of the cyclic vectors for S is known (the outer functions – see the endnotes for this chapter) but a detailed discussion of this would take us too far off course. In this section we cover a few examples of cyclic vectors for S.

Example 5.7.2. If $f(\lambda) = 0$ for some $\lambda \in \mathbb{D}$, then Proposition 5.3.8 ensures that every function in $\bigvee\{z^n f : n \geqslant 0\}$ vanishes at λ. Thus, the constant function 1 is not in $\bigvee\{S^n f : n \geqslant 0\}$ and hence f is not a cyclic vector for S.

The following lemma is useful in the next few examples.

Lemma 5.7.3. *If $\mathcal{M}$ is an S-invariant subspace of H^2 containing the constant function 1, then $\mathcal{M} = H^2$.*

Proof If $1 \in \mathcal{M}$, then the S-invariance of $\mathcal{M}$ implies that $\mathcal{M}$ contains every polynomial. Since $\mathcal{M}$ is closed and the polynomials are dense in H^2 (Corollary 5.3.11), it follows that $\mathcal{M} = H^2$. ■

Example 5.7.4. If $f \in H^2$ and

$$\inf_{z \in \mathbb{D}} |f(z)| = \delta > 0, \tag{5.7.5}$$

then f is a cyclic vector for S. First notice that $1/f \in H^2$. Corollary 5.3.11 provides a sequence of polynomials such that $p_n \to 1/f$ in the norm of H^2. By Proposition 5.3.12,

$$\int_{\mathbb{T}} \left| p_n - \frac{1}{f} \right|^2 dm \to 0,$$

and hence

$$\int_{\mathbb{T}} |p_n f - 1|^2 dm = \int_{\mathbb{T}} \frac{1}{|f|^2} \left| p_n - \frac{1}{f} \right|^2 dm$$

$$\leqslant \frac{1}{\delta^2} \int_{\mathbb{T}} \left| p_n - \frac{1}{f} \right|^2 dm \to 0.$$

Since the constant function 1 belongs to $\bigvee\{S^n f : n \geqslant 0\}$, Lemma 5.7.3 shows this subspace must be H^2. Therefore, f is a cyclic vector for S.

Example 5.7.6. The condition (5.7.5) is sufficient, but not necessary for the cyclicity of f. Consider the case when $f(z) = 1 - z$, for which the infimum in (5.7.5) is zero. Nevertheless, $f(z) = 1 - z$ is a cyclic vector for S. To see why, define

$$p_n(z) = \sum_{j=0}^{n-1} \frac{n-j}{n} z^j \quad \text{for } n \geqslant 1, \tag{5.7.7}$$

and perform a calculation to obtain

$$1 - (1-z)p_n(z) = \frac{1}{n} \sum_{j=1}^{n} z^j.$$

Thus,

$$\|1 - (1-z)p_n(z)\|^2 = \left\| \frac{1}{n} \sum_{j=1}^{n} z^j \right\|^2 = \frac{1}{n^2} n = \frac{1}{n} \to 0.$$

Thus, 1 belongs to $\bigvee\{S^n f : n \geqslant 0\}$. By Lemma 5.7.3, f is a cyclic vector. The papers [38, 336] find approximating polynomials p_n for (5.7.7) when $f(z) = (1-z)^k$ for $k \in \mathbb{N}$ and $k \geqslant 2$. See Exercise 5.9.22 for another proof.

5.8 Notes

This chapter introduced the Hardy space H^2. More generally, the Hardy spaces H^p for $0 < p \leqslant \infty$ have a long and storied history. The study began with Hardy's 1914 paper [174] which proved that for $f = \sum_{n=0}^{\infty} a_n z^n$ analytic on $\mathbb{D}$ and $p > 0$,

$$M_p(f, r) = \int_{\mathbb{T}} |f(r\xi)|^p dm(\xi)$$

is an increasing function of $0 \leqslant r < 1$. For $p = 2$, this follows from the identity (Proposition 5.3.9)

$$\int_{\mathbb{T}} |f(r\xi)|^2 dm = \sum_{n=0}^{\infty} r^{2n} |a_n|^2.$$

Furthermore, he proved that $\log M_p(f, r)$ is a convex function of $\log r$. Standard results from the theory of Poisson integrals show that if $\varphi \in L^1(\mathbb{T})$ and $\widehat{\varphi}(n) = 0$ for all $n < 0$, then there is an analytic function f on $\mathbb{D}$ such that

$$\sup_{0<r<1} M_1(f, r) < \infty$$

and whose radial boundary values are equal to φ almost everywhere. In 1916, the Riesz brothers [307] proved the converse of this. F. Riesz coined the term H_p as well as "Hardy space" to describe what we call H^p [305]. The Hardy spaces H^p are well understood and texts such as [118, 149, 202, 220, 239] contain all of the details from various perspectives.

Beurling's landmark 1948 paper [53] connected operator theory with complex function theory. In this paper he studied the $T \in \mathcal{B}(\mathcal{H})$ which satisfy the following four properties.

(a) The eigenvectors of T have dense linear span in $\mathcal{H}$.

(b) $\|Tx\| \leqslant \|x\|$ and $\|T^n x\| \to 0$ for every $x \in \mathcal{H}$.

(c) $\|T^* x\| = \|x\|$ for all $x \in \mathcal{H}$.

(d) At least one eigenvalue of T is simple.

These four properties show the existence of an orthonormal basis $(\mathbf{u}_n)_{n=0}^{\infty}$ for $\mathcal{H}$ satisfying $T\mathbf{u}_0 = 0$, $T\mathbf{u}_n = \mathbf{u}_{n-1}$ for all $n \geqslant 1$, and $T^* \mathbf{u}_n = \mathbf{u}_{n+1}$ for all $n \geqslant 0$. Furthermore, every eigenvalue λ lies in $\mathbb{D}$, is simple, and its corresponding eigenvector φ_λ is of the form

$$\varphi_\lambda = \sum_{n=0}^{\infty} \lambda^n \mathbf{u}_n.$$

Since $(\mathbf{u}_n)_{n=0}^{\infty}$ is an orthonormal basis for $\mathcal{H}$, every $x \in \mathcal{H}$ is of the form

$$x = \sum_{n=0}^{\infty} \langle x, \mathbf{u}_n \rangle \mathbf{u}_n.$$

Beurling defined $F_{\mathbf{x}}(\lambda) = \langle \varphi_\lambda, \mathbf{x} \rangle$ for $\lambda \in \mathbb{D}$ and observed that

$$F_{\mathbf{x}}(\lambda) = \sum_{n=0}^{\infty} \lambda^n \langle \mathbf{u}_n, \mathbf{x} \rangle$$

and $F_{\mathbf{x}} \in H^2$. This gives rise to the unitary operator $U : \mathcal{H} \to H^2$ defined as the linear extension of $U\mathbf{u}_n = z^n$. Via U, the operator T corresponds to

$$f \mapsto \frac{f - f(0)}{z},$$

the backward shift on H^2, while T^* corresponds to $f \mapsto zf$, the forward shift S. We encounter constructions like this again in Chapters 6 and 9, where an abstract operator is revealed to be unitarily equivalent to the concrete operator of multiplication by z on a Hilbert space of analytic functions.

Beurling coined the terms "inner" and "outer" and showed that the cyclic vectors for S are precisely the outer functions and that the nonzero S-invariant subspaces are uH^2, where u is inner. He also discussed the lattice of invariant subspaces for S and showed, for two inner functions u_1, u_2, that $u_1/u_2 \in H^2$ if and only if $u_1 H^2 \subseteq u_2 H^2$. It is known that $u_1 H^2 \cap u_2 H^2 = vH^2$, where v is the "least common multiple" of u_1 and u_2, and that $u_1 H^2 \bigvee u_2 H^2 = wH^2$, where w is the "greatest common divisor" of u_1 and u_2.

Helson's theorem (Theorem 4.3.7) described the invariant subspaces for the bilateral shift M_ξ on $L^2(\mathbb{T})$. The cyclic vectors for M_ξ are those functions $f \in L^2(\mathbb{T})$ such that $|f| > 0$ almost everywhere and $\log |f| \notin L^1(m)$. See [202, Ch. 4] for the details.

The backward shift S^* on H^2 is an influential operator that we discuss in Chapter 20 in terms of model spaces. The cyclic vectors and invariant subspaces for S^* are known, but too technical to describe in this early chapter. Several good sources for this are [143, 250, 251, 252].

There are other invariant-subspace results for M_z on spaces related to the Hardy space. Korenblum [221] considered the space H_1^2, the space of analytic functions f on $\mathbb{D}$ such that $f' \in H^2$. The inner product on H_1^2 is $\langle f, g \rangle = \langle f, g \rangle_{H^2} + \langle f', g' \rangle_{H^2}$. This space is an algebra of continuous functions on $\mathbb{D}^-$ and the M_z-invariant subspaces of H_1^2 are closed ideals. These ideals can be described using inner functions, as in Beurling's theorem, but they also depend on possible zeros on $\mathbb{T}$.

There are also shifts of higher multiplicity. Let

$$(H^2)^{(n)} = \{\mathbf{f} = (f_1, f_2, ..., f_n) : f_j \in H^2\}$$

and endow this space with the norm $\|\mathbf{f}\|^2 = \sum_{j=1}^{n} \|f_j\|^2$. Define the shift operator $S^{(n)}$ on this space by

$$S^{(n)}\mathbf{f} = (Sf_1, Sf_2, ..., Sf_n).$$

The invariant subspaces of $S^{(n)}$ are completely characterized. The inner functions from Beurling's theorem are replaced by matrix-valued analogues of inner functions [202, Ch. 7].

There are Hardy spaces $H^2(\Omega)$ of general domains $\Omega \subseteq \mathbb{C}$. Chapter 11 concerns the Hardy space of the upper half plane. For a bounded domain Ω, the operator $M_z f = zf$ is bounded on $H^2(\Omega)$ and has an associated lattice of invariant subspaces. The invariant subspaces were described by Hitt [201] when Ω is an annulus; Aleman and Richter [15] when Ω is a multiply connected domain; Aleman and Olin [14] when Ω is a crescent domain; and Aleman, Feldman, and Ross [11] when Ω is a slit domain. The invariant subspaces can be complicated and are not always described by inner functions.

5.9 Exercises

Exercise 5.9.1. Prove that with respect to the standard orthonormal basis $(\mathbf{e}_n)_{n=0}^{\infty}$ for ℓ^2, the unilateral shift S has the matrix representation

$$\begin{bmatrix} 0 & 0 & 0 & 0 & \cdots \\ 1 & 0 & 0 & 0 & \cdots \\ 0 & 1 & 0 & 0 & \cdots \\ 0 & 0 & 1 & 0 & \cdots \\ \vdots & \vdots & \vdots & \vdots & \ddots \end{bmatrix}.$$

Exercise 5.9.2. Verify that the unilateral shift S on ℓ^2 satisfies the following identities.

(a) $S^*S = I$.

(b) $SS^* = I - \mathbf{e}_0 \otimes \mathbf{e}_0$.

Exercise 5.9.3. For any $A \in \mathcal{B}(\mathcal{H})$, prove that $\sigma_{ap}(A)$ is closed.

Exercise 5.9.4. For any $A \in \mathcal{B}(\mathcal{H})$, prove that $\sigma_p(A)^- \subseteq \sigma_{ap}(A)$.

Exercise 5.9.5. Prove that H^∞ is a Banach space.

Exercise 5.9.6. Consider the following functions in $L^2(\mathbb{T})$:

$$f_\alpha(\xi) = \frac{1}{\alpha - \xi} \quad \text{for } \alpha \in \mathbb{C} \backslash \mathbb{T}.$$

Evaluate $\langle f_\alpha, f_\beta \rangle$, $\|f_\alpha\|$, and $\|f_\alpha f_\beta\|$.

Exercise 5.9.7. For each $\lambda \in \mathbb{D}$, prove the following.

(a) $k_\lambda(z) = \dfrac{1}{1 - \bar{\lambda}z}$ belongs to H^2.

(b) $\langle f, k_\lambda \rangle = f(\lambda)$ for all $f \in H^2$.

Exercise 5.9.8. Prove that $\bigvee \{k_\lambda : \lambda \in \mathbb{D}\} = H^2$.

Exercise 5.9.9. Prove that the inequality in (5.3.7) can be improved to say that if $g \in H^2$, then $|g(\lambda)| = o\left(\dfrac{1}{\sqrt{1 - |\lambda|^2}} \right)$ for $\lambda \in \mathbb{D}$.

Exercise 5.9.10. Let $\lambda \in \mathbb{D}$.

(a) Consider the operator $A_\lambda : H^2 \to H^2$ defined by $A_\lambda = k_{\bar{\lambda}} \otimes k_\lambda$. Prove that the matrix representation of A_λ with respect to the orthonormal basis $(z^n)_{n=0}^\infty$ is

$$\begin{bmatrix} 1 & \lambda & \lambda^2 & \lambda^3 & \cdots \\ \lambda & \lambda^2 & \lambda^3 & \lambda^4 & \cdots \\ \lambda^2 & \lambda^3 & \lambda^4 & \lambda^5 & \cdots \\ \lambda^3 & \lambda^4 & \lambda^5 & \lambda^6 & \cdots \\ \vdots & \vdots & \vdots & \vdots & \ddots \end{bmatrix}.$$

(b) Prove that $A_\lambda^2 = (1 - \lambda^2)^{-1} A_\lambda$.

Exercise 5.9.11. Let $(\lambda_n)_{n=1}^\infty$ be a sequence in $\mathbb{D}$ such that

$$(1 - |\lambda_{n+1}|) \leqslant c(1 - |\lambda_n|) \quad \text{for all } n \geqslant 1$$

for some constant $0 < c < 1$. If

$$f_n(z) = \frac{\sqrt{1 - |\lambda_n|^2}}{1 - \bar{\lambda}_n z} \quad \text{for } z \in \mathbb{D},$$

prove that $(f_n)_{n=1}^\infty$ satisfies

$$\sum_{n=1}^\infty |\langle f, f_n \rangle|^2 \leqslant \|f\|^2 \quad \text{for all } f \in H^2.$$

Exercise 5.9.12. For any $f(z) = \sum_{j=0}^\infty a_j z^j \in H^2$ and $n \geqslant 1$, define $\Lambda_n : H^2 \to \mathbb{C}$ by $\Lambda_n(f) = a_n$.

(a) Prove that Λ_n is a bounded linear functional on H^2.

(b) The Riesz representation theorem (Theorem 3.1.3) provides a function $k_n \in H^2$ such that $\Lambda_n(f) = \langle f, k_n \rangle$ for all $f \in H^2$. Use the Cauchy integral formula to find k_n.

(c) Compute $\|\Lambda_n\|$.

Exercise 5.9.13. The *self commutator* of $A \in \mathcal{B}(\mathcal{H})$ is $A^*A - AA^*$. Compute the self commutator of S.

Exercise 5.9.14. Prove that the backward shift operator S^* on H^2 satisfies

$$(S^* f)(z) = \frac{f(z) - f(0)}{z} \quad \text{for all } f \in H^2.$$

Exercise 5.9.15. For each $\lambda \in \mathbb{C} \setminus \mathbb{D}^-$, prove that

$$((S^* - \lambda I)^{-1} f)(z) = \frac{zf - \frac{1}{\lambda} f(\frac{1}{\lambda})}{1 - \lambda z} \quad \text{for all } f \in H^2.$$

Exercise 5.9.16. For each $\lambda \in \mathbb{D}$, prove that the difference-quotient operator

$$(Q_\lambda f)(z) = \frac{f(z) - f(\lambda)}{z - \lambda}$$

is bounded on H^2.

Exercise 5.9.17. Let $a_1, a_2, \ldots, a_n \in \mathbb{D}$ be distinct and

$$B(z) = \prod_{i=1}^{n} \frac{z - a_i}{1 - \overline{a_i}z}.$$

Prove that $(BH^2)^\perp$ is an S^*-invariant subspace of H^2 and

$$(BH^2)^\perp = \bigvee \left\{ \frac{1}{1 - \overline{a_i}z} : 1 \leqslant i \leqslant n \right\}.$$

Exercise 5.9.18. Let $A = S^*|_{(BH^2)^\perp}$, where $(BH^2)^\perp$ is the S^*-invariant subspace from Exercise 5.9.17.

(a) Compute $\sigma(A)$.

(b) Compute $\|A\|$.

Exercise 5.9.19. Prove that the nth power of the unilateral shift on H^2, that is, $(S^n f)(z) = z^n f(z)$ for $f \in H^2$, is unitarily equivalent to the shift of multiplicity n, defined by

$$S^{(n)}\mathbf{f} = (Sf_1, Sf_2, \ldots, Sf_n) \quad \text{for } \mathbf{f} = (f_1, f_2, \ldots, f_n) \in (H^2)^{(n)}.$$

Exercise 5.9.20. For two nonzero S-invariant subspaces $\mathcal{M}_1, \mathcal{M}_2$ of H^2, prove that $S|_{\mathcal{M}_1}$ is unitarily equivalent to $S|_{\mathcal{M}_2}$.

Exercise 5.9.21. For $a \in \mathbb{D}$, recall from Example 5.4.5 the function

$$u_a(z) = \frac{a - z}{1 - \overline{a}z}.$$

Prove that M_{u_a} (multiplication by u_a on H^2) is unitarily equivalent to S.

Exercise 5.9.22. Example 5.7.6 reveals that $f(z) = 1 - z$ is a cyclic vector for the shift on H^2. Here is another proof.

(a) Prove that $f \perp (1 - z)z^n$ if and only if $\widehat{f}(n) = \widehat{f}(n + 1)$.

(b) Use this to prove that if f is orthogonal to $\bigvee\{(1 - z)z^n : n \geqslant 0\}$, then $f \equiv 0$.

Exercise 5.9.23. Let p be a polynomial whose roots lie in $|z| \geqslant 1$. Prove that pH^2 is dense in H^2. When is $pH^2 = H^2$?

Exercise 5.9.24. If $A \in \mathcal{M}_n$ has n distinct eigenvalues, prove that A is a cyclic operator on $\mathbb{C}^n$ in the sense that there is an $\mathbf{x} \in \mathbb{C}^n$ such that $\bigvee\{A^n\mathbf{x} : n \geqslant 0\} = \mathbb{C}^n$.

Exercise 5.9.25. This problem requires knowledge of the Smirnov class N^+ of functions $f = \varphi/\psi$, where $\varphi, \psi \in H^\infty$ and ψ is an outer function. The unfamiliar reader should consult [118, Ch. 2]. One important fact needed here is that if $f \in N^+\backslash\{0\}$, then

$$f(\xi) = \lim_{r\to 1^-} f(r\xi)$$

exists and is nonzero for almost every $\xi \in \mathbb{T}$. The goal here is to produce a bounded measurable function f on $\mathbb{T}$ such that $|f| = 1$ almost everywhere on $\mathbb{T}$, but which is not the quotient of two inner functions.

(a) For a measurable set $E \subseteq \mathbb{T}$ with $m(E) > 0$, define $N^+(E) = \{f|_E : f \in N^+\}$. Prove that if f, g are inner functions and $f/g \in N^+(E)$, then f/g is an inner function.

(b) Define h on $\mathbb{T}$ by

$$h(e^{i\theta}) = \begin{cases} 1 & \text{if } 0 \leqslant \theta \leqslant \pi, \\ -1 & \text{if } \pi < \theta < 2\pi. \end{cases}$$

Prove there are no inner functions f, g such that $h = f/g$ almost everywhere on $\mathbb{T}$.

Exercise 5.9.26. Let K be a compact operator on H^2. Show that $KS^n \to 0$ in operator norm as follows.

(a) Given $\varepsilon > 0$, prove that there is a finite-rank operator T on H^2 such that $\|KS^n\| \leqslant \|TS^n\| + \varepsilon$.

(b) Since T has finite rank, there are $f_i, g_i \in H^2$ for $1 \leqslant i \leqslant N$ such that $T = \sum_{i=1}^{N} f_i \otimes g_i$. Prove that $TS^n = \sum_{i=1}^{N} f_i \otimes (S^{*n}g_i)$.

(c) Prove that $\|S^{*n}g\| \to 0$ as $n \to \infty$.

Exercise 5.9.27. Use these steps from [169] to see that $\inf_K \|S - K\| = 1$, where the infimum is taken over the compact operators on H^2.

(a) Prove that $\inf_K \|S - K\| \leqslant 1$.

(b) If K is compact, prove that $1 \in \sigma((S - K)^*(S - K))$.

(c) Prove that $\|(S - K)^*(S - K)\| \geqslant 1$.

Exercise 5.9.28. This problem explores the commutant of S^2 on H^2.

(a) Prove that every $f \in H^2$ has an orthogonal decomposition

$$f(z) = g(z^2) + zh(z^2), \quad \text{where } g, h \in H^2.$$

(b) For $\varphi, \psi \in H^\infty$, prove that $A_{\varphi,\psi} : H^2 \to H^2$ defined by

$$A_{\varphi,\psi}(g(z^2) + zh(z^2)) = \varphi(z)g(z^2) + \psi(z)h(z^2)$$

is bounded on H^2 and commutes with S^2.

(c) Prove that any $A \in \mathcal{B}(H^2)$ that commutes with S^2 is of the form $A = A_{\varphi,\psi}$.

(d) Prove that $f(z) \mapsto f(-z)$ is a bounded operator on H^2 that commutes with S^2, but is not of the form M_{φ} on H^2 for some $\varphi \in H^{\infty}$.

(e) How is the operator $f(z) \mapsto f(-z)$ realized as $A_{\varphi,\psi}$?

Exercise 5.9.29. If $U \in \mathcal{B}(H^2)$ is unitary and S is the unilateral shift, follow these steps from [169] to see that $\|S - U\| = 2$.

(a) Prove that $\|S - U\| \leqslant 2$.

(b) Prove that $\|S - U\| = \|U^*S - I\|$.

(c) Prove that $-1 \in \sigma(U^*S)$.

(d) Prove that $\|U^*S - I\| \geqslant 2$.

Exercise 5.9.30. Follow this idea from [246] to prove that the unilateral shift S is *irreducible*: there is no subspace $\mathcal{M}$ of ℓ^2 with $\mathcal{M} \neq \{0\}$ and $\mathcal{M} \neq \ell^2$ such that $\mathcal{M}$ and $\mathcal{M}^{\perp}$ are invariant for S (equivalently, $S\mathcal{M} \subseteq \mathcal{M}$ and $S^*\mathcal{M} \subseteq \mathcal{M}$ by Exercise 4.5.1).

(a) Suppose $\mathcal{M} \neq \{0\}$ with $S\mathcal{M} \subseteq \mathcal{M}$ and $S^*\mathcal{M} \subseteq \mathcal{M}$. Let $\mathbf{x} = (x_n)_{n=0}^{\infty} \in \mathcal{M}$ and let n be the smallest index such that $x_n \neq 0$. Define $\mathbf{y} = S^{n+1}(S^{*(n+1)}\mathbf{x})$ and prove that

$$y_k = \begin{cases} x_k & \text{if } k > n, \\ 0 & \text{if } k \leqslant n. \end{cases}$$

(b) Prove that $\mathbf{x} - \mathbf{y} = (0, 0, 0, \ldots, 0, x_n, 0, 0, \ldots)$, in which x_n occurs in the nth position.

(c) Use the above to prove that $\mathcal{M} = \ell^2$.

Exercise 5.9.31. Here is another proof that the unilateral shift S on H^2 is irreducible. Suppose $\mathcal{M}$ is a subspace of H^2 such that $S\mathcal{M} \subseteq \mathcal{M}$ and $S^*\mathcal{M} \subseteq \mathcal{M}$. If P is the orthogonal projection of H^2 onto $\mathcal{M}$, recall from Exercise 4.5.16 that $SP = PS$. Consult Corollary 5.6.2 and use the fact that P satisfies $P^2 = P$ to deduce that $P = I$ or $P = 0$.

Exercise 5.9.32. Prove that S^2 has proper nonzero reducing subspaces $\mathcal{M}$. That is, find a proper nonzero subspace $\mathcal{M}$ of H^2 such that $S^2\mathcal{M} \subseteq \mathcal{M}$ and $S^{*2}\mathcal{M} \subseteq \mathcal{M}$.

Exercise 5.9.33. Follow this idea from [168] to prove that there is no bounded operator on H^2 whose square is the unilateral shift.

(a) Prove that it suffices to show there is no $B \in \mathcal{B}(H^2)$ such that $B^2 = S^*$.

(b) Suppose toward a contradiction that $B^2 = S^*$. Prove that $\ker B = \ker S^*$ and $\ker B$ is the subspace of constant functions.

(c) Prove that the constant functions belong to $\operatorname{ran} B$.

(d) Let $f \in H^2$ be such that $Bf = 1$. Prove that 1 and f are linearly independent.

(e) Obtain a contradiction and conclude that no such B exists.

Exercise 5.9.34. Here is another proof that the unilateral shift S does not have a square root.

(a) Prove that if $A \in \mathcal{B}(H^2)$ and $A^2 = S$, then $A \in \{S\}'$.

(b) Corollary 5.6.2 says that $A = M_\varphi$ for some $\varphi \in H^\infty$. Derive a contradiction from this.

Exercise 5.9.35.

(a) Prove that the bilateral shift $(M_\xi f)(\xi) = \xi f(\xi)$ on $L^2(\mathbb{T})$ has a square root. In other words, find a $T \in \mathcal{B}(L^2(\mathbb{T}))$ such that $T^2 = M_\xi$.

(b) Prove that there are infinitely many square roots of M_ξ.

(c) For one of the square roots produced in (b), find the matrix representation of T with respect to the orthonormal basis $(\xi^n)_{n=-\infty}^{\infty}$ for $L^2(\mathbb{T})$.

Exercise 5.9.36. The *numerical range* of $A \in \mathcal{B}(\mathcal{H})$ is $W(A) = \{\langle Ax, x \rangle : \|x\| = 1\}$. We explore the numerical range of several operators in this book such as the Cesàro and Volterra operators. Prove that $W(S^*) = \mathbb{D}$ as follows.

(a) Prove that $W(S^*) \subseteq \mathbb{D}^-$.

(b) Prove that $\mathbb{D} \subseteq W(S^*)$.

(c) If $\xi \in W(S^*) \cap \mathbb{T}$, then there is an $x \in \ell^2$ with $\langle S^*x, x \rangle = \xi$. Prove that $S^*x = \lambda x$ for some $\lambda \in \mathbb{T}$, which contradicts Proposition 5.2.4.

(d) Prove that $W(S^*) = \mathbb{D}$.

Exercise 5.9.37. Here is a useful function-theoretic result with a geometric proof [192]. Proposition 5.3.12 says that every $f \in H^2 \backslash \{0\}$ has finite radial limits almost everywhere on $\mathbb{T}$. Use the following steps to show that the radial-limit function for f cannot be zero on any set of positive measure.

(a) If $f \in H^2 \backslash \{0\}$, divide by a suitable power of ξ to assume, without loss of generality, that f has the Fourier series $f(\xi) = a_0 + a_1\xi + a_2\xi^2 + \cdots$ with $a_0 \neq 0$. Prove that

$$\{f(\xi)(1 + b_1\xi + b_2\xi^2 + \cdots + b_n\xi^n) : b_j \in \mathbb{C}, n \geq 1\},$$

and its closure $\mathcal{X}$ is convex and that the leading coefficient of any element of $\mathcal{X}$ is a_0.

(b) From Exercise 1.10.31, there is a unique $g \in \mathcal{X}$ such that $\|g\| \leq \|h\|$ for all $h \in \mathcal{X}$. For all $\lambda \in \mathbb{C}$ and $n \geq 1$, prove that

$$\|g + \lambda \xi^n g\|^2 = \|g\|^2(1 + |\lambda|^2) + 2 \operatorname{Re} \lambda \int_{\mathbb{T}} |g(\xi)|^2 \xi^n dm(\xi).$$

(c) Use (b) to prove that $\widehat{|g|^2}(n) = 0$ for all $n \leq -1$.

(d) Prove that $\widehat{|g|^2}(n) = 0$ for all $n \geqslant 1$.

(e) Prove that $|g|$ is a nonzero constant function on $\mathbb{T}$.

(f) Use a contradiction argument to show that $f \in H^2 \backslash \{0\}$ cannot have radial limits equal to zero on a set of positive measure.

5.10 Hints for the Exercises

Hint for Ex. 5.9.3: Mimic the proof of Proposition 5.2.4.

Hint for Ex. 5.9.5: Use a normal-family argument to argue completeness.

Hint for Ex. 5.9.6: Consider the $|\alpha| < 1$ and $|\alpha| > 1$ cases separately (similarly for β).

Hint for Ex. 5.9.7: For $\xi \in \mathbb{T}$, write $k_\lambda(\xi) = \sum_{n=0}^{\infty} (\xi \overline{\lambda})^n$.

Hint for Ex. 5.9.8: Suppose $f \in H^2$ annihilates k_λ for all $\lambda \in \mathbb{D}$.

Hint for Ex. 5.9.9: The result is true for a polynomial. Approximate $g \in H^2$ with a suitable polynomial and use the estimate from (5.3.7).

Hint for Ex. 5.9.11: Use Schur's test (Theorem 3.3.1) to show that the Gram matrix $[\langle f_j, f_i \rangle]_{i,j=1}^{\infty}$ is a bounded operator on $\ell^2(\mathbb{N})$.

Hint for Ex. 5.9.17: Show that

$$\left\langle \frac{f(z) - f(0)}{z}, z^k B(z) \right\rangle = 0$$

for all $f \in (BH^2)^\perp$ and $k \geqslant 0$.

Hint for Ex. 5.9.21: Recall that $u_a(u_a(z)) = z$ and show that $(Ug)(z) = g(u_a(z))u_a'(z)$ defines a unitary operator on H^2.

Hint for Ex. 5.9.26: Consult Exercise 3.6.3.

Hint for Ex. 5.9.28: For (c), use Exercise 5.9.19 and consult the proof of Corollary 5.6.2.

Hint for Ex. 5.9.33: For (c), prove that $\ker S^* \subseteq \operatorname{ran} B$.

Hint for Ex. 5.9.36: For (b), consider the vectors from (5.2.5).

Hint for Ex. 5.9.37: For (f) use the fact that g is the limit of sequence of functions of the form $f(\xi)(1 + \sum_{j=1}^{n} b_j \xi^j)$. Now use Proposition 4.1.5.

6

. . • . .

The Cesàro Operator

Key Concepts: Cesàro summation, properties of the Cesàro operator (matrix representation, norm, adjoint, spectrum, numerical range), hyponormal operator, subnormal operator, subnormality of the Cesàro operator, operators related to the Cesàro operator.

Outline: The Cesàro operator on ℓ^2, which originates in summability theory, opens the door to subnormal operators. This connection appears when the Cesàro operator is viewed as a multiplication operator on a Hilbert space of analytic functions.

6.1 Cesàro Summability

Let $\sum_{n=0}^{\infty} a_n$ be an infinite series of complex numbers. For $n \geqslant 0$, let $s_n = \sum_{j=0}^{n} a_j$ denote the corresponding sequence of partial sums. The series *converges* to $L \in \mathbb{C}$, written $\sum_{n=0}^{\infty} a_n = L$, if $s_n \to L$. If no such L exists, the series *diverges*. To deal with divergent series, Cesàro proposed the following method of summation. The nth *Cesàro mean* is

$$\sigma_n = \frac{1}{n+1} \sum_{j=0}^{n} s_j \quad \text{for } n \geqslant 0,$$

the average of the first $n + 1$ partial sums. Then $\sum_{n=0}^{\infty} a_n$ is *Cesàro summable* to L if

$$\lim_{n \to \infty} \sigma_n = L.$$

For example, the *Grandi series* $1 - 1 + 1 - 1 + 1 - 1 + 1 - 1 + \cdots$ diverges since

$$s_n = \begin{cases} 1 & n \text{ even,} \\ 0 & n \text{ odd.} \end{cases}$$

However,

$$\sigma_n = \begin{cases} \dfrac{n/2+1}{n+1} & n \text{ even,} \\ \dfrac{1}{2} & n \text{ odd} \end{cases}$$

which converges to 1/2. Thus, the Grandi series is Cesàro summable to 1/2 even though it diverges in the usual sense. Exercise 6.7.1 shows that if a series converges in the traditional sense, then it converges in the Cesàro sense to the same value. An excellent survey of summability methods for infinite series is Hardy's classic text [178]. Cesàro summability connects to the problem of pointwise convergence of Fourier series [352, 380].

6.2 The Cesàro Operator

Although Cesàro summability was studied by others (see [178] for a survey), the initial paper on the Cesàro operator by Brown, Halmos, and Shields appeared in 1965 [69]. We largely follow their presentation. For $\mathbf{a} = (a_n)_{n=0}^{\infty} \in \ell^2$, define the sequence $C\mathbf{a}$ whose nth term is

$$(C\mathbf{a})(n) = \frac{1}{n+1} \sum_{j=0}^{n} a_j. \tag{6.2.1}$$

In other words,

$$C((a_0, a_1, a_2, ...)) = \left(a_0, \frac{a_0 + a_1}{2}, \frac{a_0 + a_1 + a_2}{3}, ...\right).$$

The *Cesàro operator* is the linear transformation $\mathbf{a} \mapsto C\mathbf{a}$, also denoted by C, from ℓ^2 to the vector space of complex sequences. In fact, C maps ℓ^2 to itself.

Proposition 6.2.2. *The Cesàro operator is bounded on ℓ^2.*

This proposition is a consequence of the following.

Lemma 6.2.3 (Hardy [175]). *If $b_n \geq 0$ for all $n \geq 1$, then*

$$\sum_{n=1}^{N} \left(\frac{b_1 + b_2 + \cdots + b_n}{n}\right)^2 \leq 16 \sum_{n=1}^{N} b_n^2 \quad \text{for all } N \geq 1.$$

Proof Although there are more general modern proofs of this result (see [69] and Example 3.3.3), Hardy's original argument is the most direct. Define

$$B_0 = 0 \quad \text{and} \quad B_n = b_1 + b_2 + \cdots + b_n \quad \text{for } n \geq 1,$$

and

$$\Phi_n = \frac{1}{n^2} + \frac{1}{(n+1)^2} + \frac{1}{(n+2)^2} + \cdots \quad \text{for } n \geq 1.$$

Observe that

$$\Phi_n < \frac{1}{n^2} + \int_n^{\infty} \frac{dx}{x^2} = \frac{1}{n^2} + \frac{1}{n} \leq \frac{2}{n}. \tag{6.2.4}$$

For each $N \geq 1$, it follows that

$$\sum_{n=1}^{N} \left(\frac{B_n}{n}\right)^2 = \sum_{n=1}^{N} B_n^2 (\Phi_n - \Phi_{n+1})$$

$$= \sum_{n=1}^{N} (B_n^2 - B_{n-1}^2)\Phi_n - B_N^2 \Phi_{N+1} \qquad \text{(summation by parts)}$$

$$\leqslant \sum_{n=1}^{N} (B_n^2 - B_{n-1}^2)\Phi_n$$

$$\leqslant 2 \sum_{n=1}^{N} b_n B_n \Phi_n.$$

$$\leqslant 4 \sum_{n=1}^{N} b_n \frac{B_n}{n} \qquad \text{(by (6.2.4))}$$

$$\leqslant 4 \Big(\sum_{n=1}^{N} b_n^2 \Big)^{\frac{1}{2}} \Big(\sum_{n=1}^{N} \Big(\frac{B_n}{n}\Big)^2 \Big)^{\frac{1}{2}}.$$

Rearrange the terms above and square the result to obtain

$$\sum_{n=1}^{N} \Big(\frac{B_n}{n}\Big)^2 \leqslant 16 \sum_{n=1}^{N} b_n^2,$$

which completes the proof. ∎

As we will see in a moment, the constant 16 on the right side of Hardy's inequality can be improved. In fact, Hardy mentions this in his paper [175].

For the standard basis $(e_n)_{n=0}^{\infty}$ for ℓ^2, obverse that

$$Ce_0 = 1e_0 + \frac{1}{2}e_1 + \frac{1}{3}e_2 + \frac{1}{4}e_3 + \cdots,$$

$$Ce_1 = 0e_0 + \frac{1}{2}e_1 + \frac{1}{3}e_2 + \frac{1}{4}e_3 + \cdots,$$

$$Ce_2 = 0e_0 + 0e_1 + \frac{1}{3}e_2 + \frac{1}{4}e_3 + \cdots,$$

$$\vdots$$

Thus, with respect to this basis, the Cesàro operator has the matrix representation

$$[C] = \begin{bmatrix} 1 & 0 & 0 & 0 & 0 & \cdots \\ \frac{1}{2} & \frac{1}{2} & 0 & 0 & 0 & \cdots \\ \frac{1}{3} & \frac{1}{3} & \frac{1}{3} & 0 & 0 & \cdots \\ \frac{1}{4} & \frac{1}{4} & \frac{1}{4} & \frac{1}{4} & 0 & \cdots \\ \frac{1}{5} & \frac{1}{5} & \frac{1}{5} & \frac{1}{5} & \frac{1}{5} & \cdots \\ \vdots & \vdots & \vdots & \vdots & \vdots & \ddots \end{bmatrix}. \qquad (6.2.5)$$

Consequently, the matrix representation of C^* is

$$[C^*] = \begin{bmatrix} 1 & \frac{1}{2} & \frac{1}{3} & \frac{1}{4} & \frac{1}{5} & \cdots \\ 0 & \frac{1}{2} & \frac{1}{3} & \frac{1}{4} & \frac{1}{5} & \cdots \\ 0 & 0 & \frac{1}{3} & \frac{1}{4} & \frac{1}{5} & \cdots \\ 0 & 0 & 0 & \frac{1}{4} & \frac{1}{5} & \cdots \\ 0 & 0 & 0 & 0 & \frac{1}{5} & \cdots \\ \vdots & \vdots & \vdots & \vdots & \vdots & \ddots \end{bmatrix}$$

and hence

$$(C^*\mathbf{a})(n) = \sum_{j=n}^{\infty} \frac{a_j}{j+1} \quad \text{for } n \geqslant 0. \tag{6.2.6}$$

A matrix computation (see Exercise 6.7.2) shows that

$$[(I - C)(I - C)^*] = \begin{bmatrix} 0 & 0 & 0 & 0 & \cdots \\ 0 & \frac{1}{2} & 0 & 0 & \cdots \\ 0 & 0 & \frac{2}{3} & 0 & \cdots \\ 0 & 0 & 0 & \frac{3}{4} & \cdots \\ \vdots & \vdots & \vdots & \vdots & \ddots \end{bmatrix}, \tag{6.2.7}$$

which is a diagonal operator on ℓ^2. By Proposition 3.1.5 and Example 2.1.7,

$$\|I - C\|^2 = \|(I - C)(I - C)^*\| = 1. \tag{6.2.8}$$

Thus, $\|C\| = \|I - (I - C)\| \leqslant \|I\| + \|I - C\| = 1 + 1 = 2$. In fact, equality holds above.

Proposition 6.2.9. $\|C\| = 2$.

Proof First observe that $\|C^*\| = \|C\| \leqslant 2$ (Proposition 3.1.5d). For each $\alpha > \frac{1}{2}$, define the ℓ^2 sequence

$$\mathbf{a}_\alpha = \left(\frac{1}{(n+1)^\alpha} \right)_{n=0}^{\infty}.$$

Then

$$\|C^*\mathbf{a}_\alpha\|^2 = \sum_{m=0}^{\infty} \left(\sum_{n=m}^{\infty} \frac{1}{(n+1)^{\alpha+1}} \right)^2 \qquad \text{(by (6.2.6))}$$

$$\geqslant \sum_{m=0}^{\infty} \left(\int_{m+1}^{\infty} \frac{dx}{x^{\alpha+1}} \right)^2$$

$$= \sum_{m=0}^{\infty} \left(\frac{1}{\alpha(m+1)^\alpha} \right)^2$$

$$= \frac{1}{\alpha^2} \sum_{m=0}^{\infty} \frac{1}{(m+1)^{2\alpha}}$$

$$= \frac{1}{\alpha^2} \|\mathbf{a}_\alpha\|^2.$$

Thus, $\|C^*\mathbf{a}_\alpha\| \geqslant \alpha^{-1}\|\mathbf{a}_\alpha\|$ for all $\alpha > \frac{1}{2}$ and hence $\|C^*\| \geqslant \alpha^{-1}$. Letting $\alpha \to \frac{1}{2}$ gives us $\|C^*\| \geqslant 2$. Another application of Proposition 3.1.5d shows that $\|C\| = 2$. ∎

6.3 Spectral Properties

We now use the results from the previous section to determine the spectrum of the Cesàro operator.

Theorem 6.3.1. *The following hold for the Cesàro operator C on ℓ^2.*

(a) $\sigma_p(C) = \varnothing$.

(b) $\sigma_p(C^*) = \{z : |z - 1| < 1\}$.

(c) $\sigma(C) = \{z : |z - 1| \leqslant 1\}$.

Proof (a) Suppose $\mathbf{a} = (a_n)_{n=0}^{\infty} \in \ell^2 \setminus \{0\}$ and $C\mathbf{a} = \lambda\mathbf{a}$. Then $a_0 = (C\mathbf{a})(0) = \lambda a_0$. If $n = 1$, then

$$\frac{a_0 + a_1}{2} = (C\mathbf{a})(1) = \lambda a_1.$$

Consequently, $a_1 = 2\lambda a_1 - \lambda a_0$. Induction confirms that

$$a_n = \lambda((n+1)a_n - na_{n-1}) \quad \text{for } n \geqslant 1. \tag{6.3.2}$$

Now suppose $m \geqslant 0$ is the smallest integer such that $a_m \neq 0$. Then (6.3.2) yields

$$\lambda = \frac{1}{m+1}$$

and hence $0 < \lambda \leqslant 1$. Solve for a_n in (6.3.2) and deduce

$$a_n = \frac{\lambda n}{\lambda n - (1 - \lambda)}a_{n-1} = \frac{n}{n - m}a_{n-1} \quad \text{for } n \geqslant m + 1,$$

which implies that $|a_n| \geqslant |a_{n-1}|$ for all $n \geqslant m + 1$. This contradicts the fact that $\mathbf{a} \in \ell^2 \setminus \{0\}$. Therefore, $\sigma_p(C) = \varnothing$. See Exercise 6.7.3 for another proof of this.

(b) The first step is to prove $\sigma_p(C^*) \supseteq \{z : |z - 1| < 1\}$. An analysis similar to (a) shows that for $\lambda \neq 0$,

$$C^*\mathbf{a} = \lambda\mathbf{a} \iff a_{n+1} = \left(1 - \frac{1}{\lambda(n+1)}\right)a_n \quad \text{for } n \geqslant 0.$$

Solving the recursion yields

$$a_n = a_0 \prod_{j=1}^{n} \left(1 - \frac{1}{j\lambda}\right) \quad \text{for } n \geqslant 1. \tag{6.3.3}$$

Now suppose that $|\lambda - 1| < 1$, or equivalently,

$$2\,\mathrm{Re}\,\frac{1}{\lambda} = 1 + \varepsilon \quad \text{for some } \varepsilon > 0.$$

An estimate in [69] (see Exercise 6.7.6) shows that

$$|a_n|^2 = O\left(\frac{1}{n^{1+\varepsilon}}\right)$$

and hence $\mathbf{a} \in \ell^2$. Therefore,

$$\{z : |z - 1| < 1\} \subseteq \sigma_p(C^*). \tag{6.3.4}$$

The next step is to show the reverse containment. By (6.2.8) and Theorem 2.4.9,

$$\sigma(I - C) \subseteq \{z : |z| \leqslant 1\}$$

and hence

$$\sigma(C) \subseteq \{z : |z - 1| \leqslant 1\}. \tag{6.3.5}$$

Since $\sigma_p(C^*) \subseteq \sigma(C^*) = \overline{\sigma(C)}$, it follows that

$$\{z : |z - 1| < 1\} \subseteq \sigma_p(C^*) \subseteq \{z : |z - 1| \leqslant 1\}.$$

It suffices to show that no λ satisfying $|1 - \lambda| = 1$ lies in $\sigma_p(C^*)$. By (6.2.7), $\|(I - C)(I - C^*)\mathbf{a}\| < \|\mathbf{a}\|$ for $\mathbf{a} \in \ell^2 \setminus \{0\}$, and hence

$$
\begin{aligned}
\|(I - C^*)\mathbf{a}\|^2 &= \langle (I - C^*)\mathbf{a}, (I - C^*)\mathbf{a} \rangle \\
&= \langle (I - C)(I - C^*)\mathbf{a}, \mathbf{a} \rangle \\
&\leqslant \|(I - C)(I - C^*)\mathbf{a}\|\,\|\mathbf{a}\| \\
&< \|\mathbf{a}\|^2. \tag{6.3.6}
\end{aligned}
$$

Suppose $\lambda \in \sigma_p(C^*)$ and $|1 - \lambda| = 1$. Then $1 - \lambda \in \sigma_p(I - C^*)$ and thus there is a corresponding unit eigenvector $\mathbf{b}$. But then, by (6.3.6),

$$1 = |1 - \lambda| = \|(I - C^*)\mathbf{b}\| < \|\mathbf{b}\| = 1,$$

which is a contradiction. This completes the proof of (b).
(c) Use the containments (6.3.4) and (6.3.5). ∎

Lemma 6.4.14 (below) computes eigenvectors for C^*.

6.4 Other Properties of the Cesàro Operator

An operator $A \in \mathcal{B}(\mathcal{H})$ is *normal* if $A^*A = AA^*$. In particular, $A^*A - AA^* = 0$ for any normal operator A. We say that $A \in \mathcal{B}(\mathcal{H})$ is *hyponormal* if

$$\langle (A^*A - AA^*)\mathbf{x}, \mathbf{x}\rangle \geqslant 0 \quad \text{for all } \mathbf{x} \in \mathcal{H}, \tag{6.4.1}$$

or equivalently, if $\|A\mathbf{x}\| \geqslant \|A^*\mathbf{x}\|$ for all $\mathbf{x} \in \mathcal{H}$. For $T \in \mathcal{B}(\mathcal{H})$, we use the notation $T \geqslant 0$ to denote the condition $\langle T\mathbf{x}, \mathbf{x}\rangle \geqslant 0$ for all $\mathbf{x} \in \mathcal{H}$. Hyponormal operators are one of several types of operators (subnormal, seminormal, posinormal) that are generalizations of the well-understood class of normal operators. We will formally study normal operators in Chapter 8 and subnormal operators in Chapter 19.

Theorem 6.4.2. *The Cesàro operator C is hyponormal.*

Proof Exercise 6.7.7 shows that with respect to the standard basis $(\mathbf{e}_n)_{n=0}^{\infty}$ for ℓ^2, the matrix representation of $C^*C - CC^*$ is the "L-shaped" matrix

$$T = \begin{bmatrix} \alpha_0 - \beta_0 & \alpha_1 - \beta_1 & \alpha_2 - \beta_2 & \alpha_3 - \beta_3 & \cdots \\ \alpha_1 - \beta_1 & \alpha_1 - \beta_1 & \alpha_2 - \beta_2 & \alpha_3 - \beta_3 & \cdots \\ \alpha_2 - \beta_2 & \alpha_2 - \beta_2 & \alpha_2 - \beta_2 & \alpha_3 - \beta_3 & \cdots \\ \alpha_3 - \beta_3 & \alpha_3 - \beta_3 & \alpha_3 - \beta_3 & \alpha_3 - \beta_3 & \cdots \\ \vdots & \vdots & \vdots & \vdots & \ddots \end{bmatrix}, \tag{6.4.3}$$

where

$$\alpha_n = \sum_{j=n}^{\infty} \frac{1}{(j+1)^2} \quad \text{and} \quad \beta_n = \frac{1}{n+1}. \tag{6.4.4}$$

We must prove that $T \geqslant 0$. Let

$$T_n = P_n(C^*C - CC^*)P_n,$$

where P_n is the orthogonal projection of ℓ^2 onto $\text{span}\{\mathbf{e}_j : 0 \leqslant j \leqslant n\}$. Since $P_n\mathbf{x} \to \mathbf{x}$ as $n \to \infty$, the Cauchy–Schwarz inequality implies that $\langle T_n\mathbf{x}, \mathbf{x}\rangle \to \langle T\mathbf{x}, \mathbf{x}\rangle$ for all $\mathbf{x} \in \ell^2$. Thus, it suffices to show that $T_n \geqslant 0$ for all n. We must verify that the finite L-shaped matrices

$$\begin{bmatrix} z_0 & z_1 & z_2 & z_3 & \cdots & z_n \\ z_1 & z_1 & z_2 & z_3 & \cdots & z_n \\ z_2 & z_2 & z_2 & z_3 & \cdots & z_n \\ z_3 & z_3 & z_3 & z_3 & \cdots & z_n \\ \vdots & \vdots & \vdots & \vdots & \ddots & \vdots \\ z_n & z_n & z_n & z_n & \cdots & z_n \end{bmatrix}, \tag{6.4.5}$$

where $z_k = \alpha_k - \beta_k$, are positive semidefinite. By Sylvester's criterion for positive definiteness [141, Thm. 16.4.3], this can be done by showing that the determinant of each matrix (6.4.5) is positive. Subtract the second column from the first, then subtract the third column from the second, and continue this way through the columns. The determinant of (6.4.5) is unchanged but the resulting matrix is upper triangular:

$$
\begin{bmatrix}
z_0 - z_1 & \star & \star & \star & \cdots & \star \\
0 & z_1 - z_2 & \star & \star & \cdots & \star \\
0 & 0 & z_2 - z_3 & \star & \cdots & \star \\
0 & 0 & 0 & z_3 - z_4 & \cdots & \star \\
\vdots & \vdots & \vdots & \vdots & \ddots & \vdots \\
0 & 0 & 0 & 0 & \cdots & z_n
\end{bmatrix}.
$$

In the above, $\star$ denotes an entry whose exact value does not concern us. Thus, we just need to check that

$$(z_0 - z_1)(z_1 - z_2)(z_2 - z_3) \cdots (z_{n-1} - z_n)z_n > 0. \tag{6.4.6}$$

To do this, recall the definitions of α_n and β_n from (6.4.4) and observe that for $k = 0, 1, 2, \ldots, n - 1$,

$$
\begin{aligned}
z_k - z_{k+1} &= (\alpha_k - \beta_k) - (\alpha_{k+1} - \beta_{k+1}) \\
&= \left(\sum_{j=k}^{\infty} \frac{1}{(j+1)^2} - \frac{1}{k+1} \right) - \left(\sum_{j=k+1}^{\infty} \frac{1}{(j+1)^2} - \frac{1}{k+2} \right) \\
&= \frac{1}{(k+1)^2} - \frac{1}{(k+1)(k+2)} > 0
\end{aligned}
$$

and

$$z_n = \sum_{j=n}^{\infty} \frac{1}{(j+1)^2} - \frac{1}{n+1} > \int_n^{\infty} \frac{dx}{(x+1)^2} - \frac{1}{n+1} = 0.$$

This verifies (6.4.6). ∎

See Exercise 6.7.19 for another proof of the hyponormality of C.

An $S \in \mathcal{B}(\mathcal{H})$ is *subnormal* if there is a Hilbert space $\mathcal{K} \supseteq \mathcal{H}$ and a normal operator $N \in \mathcal{B}(\mathcal{K})$ such that $N\mathcal{H} \subseteq \mathcal{H}$ and $N|_{\mathcal{H}} = S$. Equivalently, a subnormal operator is the restriction of a normal operator to one of its invariant subspaces. Note that every normal operator is subnormal. One example of a subnormal operator is M_z on $H^2(\mu)$, where μ is a finite positive compactly supported Borel measure on $\mathbb{C}$, and $H^2(\mu)$ is the closure of the polynomials in $L^2(\mu)$. Furthermore, a subnormal operator is hyponormal (Exercise 19.6.9) but not vice versa [95, p. 47]. The next result is one of the gems in the study of the Cesàro operator.

Theorem 6.4.7 (Kriete–Trutt [223]). *The Cesàro operator is subnormal.*

The next few results set up an outline of the proof. Recall that $\mathbf{a} = (a_n)_{n=0}^{\infty} \in \ell^2$ is identified with $f(z) = \sum_{n=0}^{\infty} a_n z^n \in H^2$ (Proposition 4.2.7). Consequently, one can view the Cesàro operator as an operator on H^2 defined by

$$(Cf)(z) = \frac{1}{z} \int_0^z \frac{f(\xi)}{1 - \xi} d\xi \quad \text{for } z \in \mathbb{D}. \tag{6.4.8}$$

Since $f(\xi)(1 - \xi)^{-1}$ is analytic on the simply connected domain $\mathbb{D}$, the value of the antiderivative above is independent of the path from 0 to z. A calculation with power series (see Exercise 6.7.8) shows that if $f(z) = \sum_{j=0}^{\infty} a_j z^j$, then

$$(Cf)(z) = \sum_{n=0}^{\infty} \left(\frac{1}{n+1} \sum_{j=0}^{n} a_j \right) z^n. \tag{6.4.9}$$

In other words, the nth Taylor coefficient of Cf is the nth term of the Cesàro sequence from (6.2.1) corresponding to $(a_n)_{n=0}^{\infty}$.

Here is an adjoint formula for C, related to the one in (6.2.6).

Proposition 6.4.10. *Let $f \in H^2$ and let*

$$F(\lambda) = \int_0^{\lambda} f(w)\,dw \quad \text{for } \lambda \in \mathbb{D},$$

denote the antiderivative of f that vanishes at $\lambda = 0$. Then F extends continuously to $\mathbb{D}^-$, and hence $F(1)$ is well defined. Moreover,

$$(C^* f)(\lambda) = \frac{F(1) - F(\lambda)}{1 - \lambda} = \frac{1}{1 - \lambda} \int_{\lambda}^{1} f(w)\,dw.$$

Proof If $f(z) = \sum_{n=0}^{\infty} a_n z^n \in H^2$, then

$$F(\lambda) = \sum_{n=0}^{\infty} \frac{a_n}{n+1} \lambda^{n+1} \quad \text{for all } \lambda \in \mathbb{D}.$$

The Cauchy–Schwarz inequality shows that the series above converges absolutely and uniformly on $\mathbb{D}^-$ and thus F extends continuously to $\mathbb{D}^-$. Therefore,

$$F(1) = \sum_{n=0}^{\infty} \frac{a_n}{n+1}$$

is well defined.

Recall from Chapter 5 that

$$k_{\lambda}(z) = \frac{1}{1 - \bar{\lambda} z} = \sum_{n=0}^{\infty} \bar{\lambda}^n z^n,$$

the reproducing kernel for H^2, satisfies $k_{\lambda} \in H^2$ and $\langle f, k_{\lambda} \rangle = f(\lambda)$ for all $\lambda \in \mathbb{D}$ and $f \in H^2$. Thus,

$$(C^* f)(\lambda) = \langle C^* f, k_{\lambda} \rangle = \langle f, C k_{\lambda} \rangle.$$

Identify k_{λ} with the ℓ^2 sequence $(\bar{\lambda}^n)_{n=0}^{\infty}$ and use (6.2.5) to deduce that

$$(C k_{\lambda})(z) = 1 + \frac{1 + \bar{\lambda}}{2} z + \frac{1 + \bar{\lambda} + \bar{\lambda}^2}{3} z^2 + \frac{1 + \bar{\lambda} + \bar{\lambda}^2 + \bar{\lambda}^3}{4} z^3 + \cdots$$

$$= \sum_{n=0}^{\infty} \frac{1}{n+1} \left(\frac{1 - \overline{\lambda}^{n+1}}{1 - \overline{\lambda}} \right) z^n.$$

Therefore,

$$(C^* f)(\lambda) = \langle f, C k_\lambda \rangle$$

$$= \left\langle \sum_{n=0}^{\infty} a_n z^n, \sum_{n=0}^{\infty} \frac{1}{n+1} \left(\frac{1 - \overline{\lambda}^{n+1}}{1 - \overline{\lambda}} \right) z^n \right\rangle$$

$$= \sum_{n=0}^{\infty} a_n \frac{1}{n+1} \left(\frac{1 - \lambda^{n+1}}{1 - \lambda} \right)$$

$$= \frac{1}{1 - \lambda} \left(\sum_{n=0}^{\infty} \frac{a_n}{n+1} - \sum_{n=0}^{\infty} \frac{a_n}{n+1} \lambda^{n+1} \right)$$

$$= \frac{F(1) - F(\lambda)}{1 - \lambda}$$

$$= \frac{1}{1 - \lambda} \int_{\lambda}^{1} f(w) \, dw,$$

which completes the proof. ∎

By Theorem 6.3.1, $\sigma_p(C^*) = \{ z : |z - 1| < 1 \}$. We now compute the corresponding eigenvectors for C^*. Observe that $1 - z$ is nonzero on $\mathbb{D}$, and hence a branch of the logarithm of $1 - z$ exists on $\mathbb{D}$ and can be chosen so that $\log 1 = 0$. For each $w \in \mathbb{D}$, define the analytic function φ_w on $\mathbb{D}$ by

$$\varphi_w(z) = (1 - z)^{w/(1-w)}$$

and observe that $\varphi_w(0) = 1$.

Remark 6.4.11. Note that as w runs through $\mathbb{D}$, the exponent $w/(1 - w)$ in the formula for φ_w runs over the right half plane $\{ z : \operatorname{Re} z > -\frac{1}{2} \}$. This simple observation is used below.

Lemma 6.4.12. $\varphi_w \in H^2$ for all $w \in \mathbb{D}$.

Proof From the definition of powers of complex numbers, observe that

$$\varphi_w(z) = \exp \left(\frac{w}{1 - w} \log(1 - z) \right).$$

Also notice that

$$\frac{w}{1 - w} \log(1 - z) = \left(\operatorname{Re} \frac{w}{1 - w} + i \operatorname{Im} \frac{w}{1 - w} \right) \left(\log|1 - z| + i \arg(1 - z) \right)$$

$$= \operatorname{Re} \left(\frac{w}{1 - w} \right) \log|1 - z| - \operatorname{Im} \left(\frac{w}{1 - w} \right) \arg(1 - z) + i\star,$$

where $\star$ denotes a real number whose exact value is unimportant. It follows that

$$|\varphi_w(z)| = \exp \left(\operatorname{Re} \left(\frac{w}{1 - w} \log(1 - z) \right) \right)$$

$$= |1 - z|^{\operatorname{Re}(w/1-w)} \exp \left(- \operatorname{Im} \left(\frac{w}{1 - w} \right) \arg(1 - z) \right).$$

Since $\arg(1 - z)$ is bounded on $\mathbb{D}$, it follows that $\varphi_w \in H^2$ by Remark 6.4.11. ∎

Lemma 6.4.13. $\bigvee\{\varphi_w : w \in \mathbb{D}\} = H^2$.

Proof For $n \geqslant 0$, observe that

$$\varphi_{\frac{n}{n+1}}(z) = (1 - z)^n,$$

so $\bigvee\{\varphi_w : w \in \mathbb{D}\}$ contains the polynomials (Exercise 6.7.9). Now use the density of the polynomials in H^2. ∎

Lemma 6.4.14. $(I - C^*)\varphi_w = w\varphi_w$ for all $w \in \mathbb{D}$.

Proof Proposition 6.4.10 implies that

$$
\begin{aligned}
((I - C^*)\varphi_w)(z) &= \varphi_w(z) - (C^*\varphi_w)(z) \\
&= \varphi_w(z) - \frac{1}{1-z}\left(\int_z^1 \varphi_w(t)\, dt\right) \\
&= (1-z)^{w/(1-w)} - \frac{1}{1-z}(1-z)^{w/(1-w)+1}(1-w) \\
&= (1-z)^{w/(1-w)} - (1-z)^{w/(1-w)}(1-w) \\
&= w(1-z)^{w/(1-w)} \\
&= w\,\varphi_w(z),
\end{aligned}
$$

which completes the proof. ∎

Here is a sketch of the proof of Theorem 6.4.7.

Proof Let $\mathcal{H}$ denote the space of analytic functions on $\mathbb{D}$ of the form

$$F(z) = \langle f, \varphi_{\bar{z}}\rangle_{H^2} \quad \text{for } f \in H^2.$$

Lemma 6.4.13 implies that $F(z) = 0$ for all $z \in \mathbb{D}$ if and only if $f \equiv 0$. Define a norm on $\mathcal{H}$ by $\|F\|_{\mathcal{H}} = \|f\|_{H^2}$. By the definition of $\mathcal{H}$, the operator $U : H^2 \to \mathcal{H}$ given by $(Uf)(z) = \langle f, \varphi_{\bar{z}}\rangle$ is unitary.

If

$$\psi_n(z) = \frac{1}{(z-1)^n}\left(z - \frac{1}{2}\right)\cdots\left(z - \frac{n-1}{n}\right) \quad \text{for } n \geqslant 1, \qquad (6.4.15)$$

then $(\psi_n)_{n=1}^{\infty}$ is an orthonormal basis for $\mathcal{H}$. This follows from the fact that $(z^n)_{n=0}^{\infty}$ is an orthonormal basis for H^2 and hence, since U is unitary, $\psi_n = Uz^n$ is an orthonormal basis for $\mathcal{H}$. Since

$$\psi_n(z) = \langle \xi^n, \varphi_{\bar{z}}\rangle_{H^2} = \int_0^{2\pi} e^{in\theta}(1 - e^{-i\theta})^{z/(1-z)}\frac{d\theta}{2\pi},$$

one can use an integral formula from [125, p. 12] to obtain (6.4.15). Observe that the function $K_\lambda(z)$ defined by

$$K_\lambda(z) = \langle \varphi_{\bar{\lambda}}, \varphi_{\bar{z}}\rangle_{H^2} = (U\varphi_{\bar{\lambda}})(z)$$

is analytic in $z \in \mathbb{D}$ and coanalytic in $\lambda \in \mathbb{D}$. Moreover, for $F = Uf \in \mathcal{H}$ and $\lambda \in \mathbb{D}$,

$$\langle F, K_\lambda \rangle_{\mathcal{H}} = \langle Uf, U\varphi_{\overline{\lambda}} \rangle_{\mathcal{H}} = \langle f, \varphi_{\overline{\lambda}} \rangle_{H^2} = (Uf)(\lambda) = F(\lambda).$$

Thus, $\mathcal{H}$ is a reproducing kernel Hilbert space. One can show [125, p. 12] that

$$K_\lambda(z) = \Gamma\left(\frac{\overline{\lambda}}{1-\overline{\lambda}} + \frac{z}{1-z} + 1\right) \bigg/ \Gamma\left(\frac{\overline{\lambda}}{1-\overline{\lambda}} + 1\right)\Gamma\left(\frac{z}{1-z} + 1\right), \qquad (6.4.16)$$

where

$$\Gamma(z) := \int_0^\infty x^{z-1} e^{-x} dx$$

is the *Gamma function*. Note that $\Gamma(z)$ is an analytic function initially defined on $\{z : \operatorname{Re} z > 0\}$ that extends to a meromorphic function on $\mathbb{C}$ with simple poles at $\{\ldots, -3, -2, -1, 0\}$ [351, Ch. 6].

The unitary operator U above satisfies

$$\begin{aligned}
(U(I - C)f)(z) &= \langle (I - C)f, \varphi_{\overline{z}} \rangle_{H^2} \\
&= \langle f, (I - C^*)\varphi_{\overline{z}} \rangle_{H^2} \\
&= \langle f, \overline{z}\varphi_{\overline{z}} \rangle_{H^2} \\
&= z\langle f, \varphi_{\overline{z}} \rangle_{H^2} \\
&= z(Uf)(z)
\end{aligned}$$

for all $f \in H^2$. Thus, $U(I - C) = M_z U$ on $\mathcal{H}$. In particular, this shows that M_z (multiplication by z) is a bounded operator on $\mathcal{H}$.

Since M_z is bounded on $\mathcal{H}$ and $\psi_1 = (z - 1)^{-1} \in \mathcal{H}$ by (6.4.15), the polynomials are contained in $\mathcal{H}$. It can be shown that the polynomials are actually dense in $\mathcal{H}$. The heart of the Kriete–Trutt paper [223] is the construction of a finite positive Borel measure μ on $\mathbb{D}^-$ such that

$$\|p\|_{\mathcal{H}}^2 = \int_{\mathbb{D}^-} |p|^2 d\mu \quad \text{for all } p \in \mathbb{C}[z].$$

This measure μ is supported on the sequence of circles

$$\gamma_n = \left\{z : \left|z - \frac{n}{n+1}\right| = \frac{1}{n+1}\right\} \quad \text{for } n \geqslant 0;$$

(see Figure 6.4.1). Furthermore, $\mu(\gamma_n) = 2^{-n-1}$ and $\mu|_{\gamma_n}$ is mutually absolutely continuous with respect to arc length measure on γ_n.

Let $H^2(\mu)$ denote the closure of the polynomials in $L^2(\mu)$. For $p \in \mathbb{C}[z]$, define $Qp = p$ and extend Q unitarily to all of $\mathcal{H}$ using the density of the polynomials in $\mathcal{H}$ and in $H^2(\mu)$. The composition $QU : H^2 \rightarrow H^2(\mu)$ defines a unitary operator that

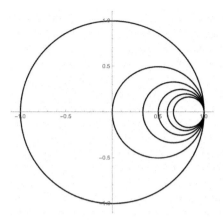

Figure 6.4.1 The circles (from left to right) $\gamma_0, \gamma_1, \gamma_2, \gamma_3, \gamma_4, \gamma_5$.

intertwines $I - C$ and M_z (multiplication by z) on $H^2(\mu)$. This is summarized in the following commutative diagram:

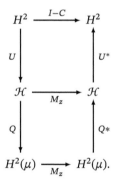

Since M_z on $H^2(\mu)$ has a normal extension to $L^2(\mu)$, M_z on $H^2(\mu)$ is subnormal. It follows (using the fact that an operator unitarily equivalent to a subnormal operator is subnormal; see Proposition 19.1.7) that $I - C$, and hence C, is subnormal. ∎

As a reminder, we study subnormal operators more thoroughly in Chapter 19.

6.5 Other Versions of the Cesàro Operator

For $f \in L^2[0, 1]$, define

$$(C_1 f)(x) = \frac{1}{x} \int_0^x f(t)\, dt \quad \text{for } x \in [0, 1], \tag{6.5.1}$$

and for $h \in L^2(0, \infty)$, define

$$(C_\infty h)(t) = \frac{1}{t} \int_0^t h(s) \, ds \quad \text{for } t \in (0, \infty).$$

Both C_1 and C_∞ define linear transformations. The operator C_1 is the *finite continuous Cesàro operator* and C_∞ is the *infinite continuous Cesàro operator*.

Proposition 6.5.2. *The operators C_1 and C_∞ are bounded on $L^2[0, 1]$ and $L^2(0, \infty)$, respectively.*

The proof uses a continuous version of Hardy's inequality (Exercise 6.7.5). One can also use an integral version of Schur's theorem (see [69] and the endnotes of Chapter 3). Although these two operators initially seem mysterious, they are, up to unitary equivalence, well-known operators [69]. Recall the unilateral shift $Sf = zf$ on H^2 (Chapter 5) and the bilateral shift $M_\xi g = \xi g$ on $L^2(\mathbb{T})$ (Chapter 4).

Theorem 6.5.3. *$I - C_1^*$ is unitarily equivalent to the unilateral shift and $I - C_\infty^*$ is unitarily equivalent to the bilateral shift.*

Proof We only outline the proof of the first statement. Let $Q = I - C_1^*$ and

$$f_\alpha(x) = x^\alpha \quad \text{for } \operatorname{Re} \alpha > -\frac{1}{2}.$$

Exercise 6.7.12 ensures that

$$Q^* f_\alpha = \frac{\alpha}{\alpha + 1} f_\alpha.$$

Changing parameters, define $\beta = \bar{\alpha} + \frac{1}{2}$ and $g_\beta = f_{\bar{\beta} - \frac{1}{2}}$ for $\operatorname{Re} \beta > 0$. Then,

$$Q^* g_\beta = \overline{\varphi(\beta)} g_\beta, \quad \text{where} \quad \varphi(\beta) = \frac{\beta - \frac{1}{2}}{\beta + \frac{1}{2}}.$$

For $f \in L^2[0, 1]$, define

$$\widetilde{f}(\beta) = \int_0^1 f(t) \overline{g_\beta(t)} \, dt.$$

The change of variables $t = e^{-u}$ for $u \geqslant 0$ yields

$$\widetilde{f}(\beta) = \int_0^\infty f(e^{-u}) e^{-u/2} e^{-u\beta} \, du.$$

Since $L^2(0, \infty) = \{f(e^{-u/2}) e^{-u/2} : f \in L^2[0, 1]\}$, it follows that $\{\widetilde{f} : f \in L^2[0, 1]\}$ is the set of Laplace transforms of functions in $L^2(0, \infty)$. By the Paley–Wiener theorem (11.8.2), this set is precisely $H^2(\operatorname{Re} z > 0)$, the Hardy space of the right half plane. Furthermore,

$$\widetilde{Qf}(\beta) = \langle Qf, g_\beta \rangle_{L^2[0,1]} = \langle f, Q^* g_\beta \rangle_{L^2[0,1]} = \varphi(\beta) \widetilde{f}(\beta).$$

Since the operator $f \mapsto \tilde{f}$ is unitary, Q is unitarily equivalent to multiplication by φ on $H^2(\mathrm{Re}\, z > 0)$. The change of variables $w = \varphi(z)$ (which maps $\{\mathrm{Re}\, z > 0\}$ to $\mathbb{D}$), shows that Q is unitarily equivalent to multiplication by z on $H^2(\mathbb{D})$. ∎

Corollary 6.5.4. $I - C_\infty$ *is unitarily equivalent to the bilateral shift.*

Proof By Exercise 4.5.18, M_ξ is unitarily equivalent to $M_{\bar{\xi}}$. Now apply Theorem 6.5.3. ∎

Recall that the Cesàro operator acts on H^2 by

$$(Cf)(z) = \frac{1}{z} \int_0^z \frac{f(\xi)}{1 - \xi} \, d\xi \quad \text{for } z \in \mathbb{D}.$$

This suggests a generalization of C defined as

$$(C_g f)(z) = \frac{1}{z} \int_0^z f(\xi) g'(\xi) \, d\xi \quad \text{for } z \in \mathbb{D},$$

where g is an analytic function on $\mathbb{D}$. Pommerenke [272] proved that C_g is bounded on H^2 if and only if g is of bounded mean oscillation. Results of Aleman and Cima [10] extend the boundedness of C_g to other Banach spaces of analytic functions on $\mathbb{D}$.

6.6 Notes

Cesàro summation initially appears in an 1890 paper of Cesàro [78]. The boundedness of the three Cesàro operators mentioned in this chapter have been known since the 1950s. Indeed, Hardy, Littlewood, and Pólya discuss this in their book [179, Ch. IX]. Brown, Halmos, and Shields [69] were the first to explore the spectrum and norm of the Cesàro operator.

Kriete and Trutt [224] proved that $I - C$ is unitarily equivalent to M_z on $H^2(\mu)$. They also showed that the invariant subspace structure of M_z on $H^2(\mu)$ is complicated. Since C has the same invariant subspaces as $I - C$, this complexity carries over to C. This same paper also identifies the commutant (recall Definition 5.6.1) of M_z with H^∞, in the sense that $\{M_z\}' = \{M_\varphi : \varphi \in H^\infty\}$. Since C and $I - C$ have the same commutant, one can use the unitary operator $QU : H^2 \to H^2(\mu)$, defined in the proof of Theorem 6.4.7 to show that $\{C\}' = \{(QU)^* M_\varphi (QU) : \varphi \in H^\infty\}$. Although this description of $\{C\}'$ is specific, it is hidden behind several unitary operators. Shields and Wallen [343] proved that $\{C\}'$ is the closure of $\{p(C) : p \in \mathbb{C}[z]\}$ in the weak operator topology. In fact, Shields and Wallen start with the fact that $I - C$ is unitarily equivalent to M_z on the Hilbert space $\mathcal{H}$ of analytic functions that appears in the proof of Theorem 6.4.7. They go on to show that the commutant of M_z is identified with H^∞, in that $A \in \{M_z\}'$ if and only if $A = M_\varphi$ for some $\varphi \in H^\infty$.

Versions of the Cesàro operator were explored on many function spaces beyond those discussed in this chapter [249, 268]. These operators often enjoy some of the properties they do in the H^2 setting. For example, Hardy's inequality can be extended to ℓ^p [179, p. 239] to show that the Cesàro matrix defines a bounded operator on ℓ^p with

$$\|C\|_{\ell^p \to \ell^p} = \frac{p}{p-1} \quad \text{for } 1 < p < \infty.$$

Furthermore, (6.2.8) shows that $\|I - C\|_{\ell^2 \to \ell^2} = 1$. For the Cesàro operator on ℓ^p, the paper [348] resolves a question posed in [41] and shows that

$$\|I - C\|_{\ell^p \to \ell^p} = \begin{cases} \dfrac{1}{p-1} & \text{if } 1 < p \leqslant 2, \\[2mm] m_p^{-\frac{1}{p}} & \text{if } 2 < p < \infty, \end{cases}$$

where $m_p = \min\{pt^{p-1} + (1-t)^p - t^p : 0 \leqslant t \leqslant \frac{1}{2}\}$.

Kriete and Trutt [223] showed that the Cesàro operator is related to M_z on a space $\mathcal{H}$ of analytic functions and went on in [224] to explore the mysterious properties of $\mathcal{H}$. In particular, the zero sets and the M_z-invariant subspaces of $\mathcal{H}$ are particularly complicated and are far from being understood. Cowen [99] gave an alternate proof of the subnormality of the Cesàro operator using composition operators. Related Cesàro operators with associated spaces of analytic functions are found in [212, 291].

A result from [97], using the fact that the Cesàro operator is subnormal, shows that a branch of $\sqrt{z}$ is analytic on $\sigma(C)$ and $\sqrt{C}$ is a bounded operator on ℓ^2. In other words, the Cesàro operator has a square root. One can also obtain the same result using the theory of accretive operators [211]. See Exercise 6.7.25 for more on this.

Hyponormal operators were originally studied by Halmos in 1950 [166] and developed into a cohesive theory by Clancey [86]. The term "hyponormal" first appeared in a 1961 book by Berberian [42]. See [238] for a good survey of this material.

6.7 Exercises

Exercise 6.7.1. If a series of complex numbers converges to L, prove it is Cesàro summable to L.

Exercise 6.7.2. Prove that

$$(I - C)(I - C)^* = \begin{bmatrix} 0 & 0 & 0 & 0 & \cdots \\ 0 & \frac{1}{2} & 0 & 0 & \cdots \\ 0 & 0 & \frac{2}{3} & 0 & \cdots \\ 0 & 0 & 0 & \frac{3}{4} & \cdots \\ \vdots & \vdots & \vdots & \vdots & \ddots \end{bmatrix}.$$

Exercise 6.7.3. Use the definition of C from (6.4.8) to show that $\sigma_p(C) = \varnothing$.

Exercise 6.7.4. For each integer $n \geqslant 1$, prove that $z^{n-1} \notin \mathrm{ran}(n^{-1}I - C)$.

Exercise 6.7.5. If f is a nonnegative measurable function on $[0, \infty)$, prove that

$$\int_0^\infty \left(\frac{1}{x} \int_0^x f(t)\,dt\right)^2 dx \leqslant 4 \int_0^\infty f(x)^2\,dx.$$

Remark: This proves that the operator C_∞ mentioned in this chapter is bounded.

Exercise 6.7.6. Prove that if $2\,\mathrm{Re}\,\frac{1}{\lambda} = 1 + \varepsilon$ for some $\varepsilon > 0$ and

$$a_{n+1} = \left(1 - \frac{1}{\lambda(n+1)}\right)a_n \quad \text{for all } n \geqslant 0,$$

then $|a_n|^2 = O(1/n^{1+\varepsilon})$.

Exercise 6.7.7. Prove that with respect to the standard basis $(\mathbf{e}_n)_{n=0}^\infty$ for ℓ^2, the matrix representation of $C^*C - CC^*$ is

$$\begin{bmatrix}
\alpha_0 - \beta_0 & \alpha_1 - \beta_1 & \alpha_2 - \beta_2 & \alpha_3 - \beta_3 & \cdots \\
\alpha_1 - \beta_1 & \alpha_1 - \beta_1 & \alpha_2 - \beta_2 & \alpha_3 - \beta_3 & \cdots \\
\alpha_2 - \beta_2 & \alpha_2 - \beta_2 & \alpha_2 - \beta_2 & \alpha_3 - \beta_3 & \cdots \\
\alpha_3 - \beta_3 & \alpha_3 - \beta_3 & \alpha_3 - \beta_3 & \alpha_3 - \beta_3 & \cdots \\
\vdots & \vdots & \vdots & \vdots & \ddots
\end{bmatrix},$$

where

$$\alpha_n = \sum_{j=n}^\infty \frac{1}{(j+1)^2} \quad \text{and} \quad \beta_n = \frac{1}{n+1} \quad \text{for } n \geqslant 0.$$

Exercise 6.7.8. Prove that if $f(z) = \sum_{j=0}^\infty a_j z^j \in H^2$, then the Cesàro operator C from (6.4.8) can be written as

$$(Cf)(z) = \sum_{n=0}^\infty \left(\frac{1}{n+1} \sum_{j=0}^n a_j\right) z^n.$$

Exercise 6.7.9. Prove that the linear span of $\{(1-z)^n : n \geqslant 0\}$ contains every polynomial. *Remark:* This detail is important in the proof of Lemma 6.4.13.

Exercise 6.7.10. Prove that C is not compact.

Exercise 6.7.11. Follow these steps from [268] to compute the resolvent of the Cesàro operator C.

(a) Prove that for $\lambda \in \mathbb{C}\backslash\{0\}$ and $h \in H^2$, a solution f to $(\lambda I - C)f = h$ satisfies the differential equation

$$f'(z) + \left(\frac{1}{z} - \frac{1}{\lambda z(1-z)}\right)f(z) = \frac{1}{\lambda z}\frac{d}{dz}(zh(z)) \quad \text{for all } z \in \mathbb{D}.$$

(b) Prove that when multiplied by $z^{1-\frac{1}{\lambda}}(1-z)^{\frac{1}{\lambda}}$, this differential equation becomes

$$\frac{d}{dz}\left(z^{1-\frac{1}{\lambda}}(1-z)^{\frac{1}{\lambda}}f(z)\right) = \frac{z^{-\frac{1}{\lambda}}}{\lambda}(1-z)^{\frac{1}{\lambda}}\frac{d}{dz}(zh(z)).$$

(c) Prove that

$$f(z) = \frac{h(z)}{\lambda} + \frac{1}{\lambda^2} z^{\frac{1}{\lambda}-1}(1-z)^{-\frac{1}{\lambda}} \int_0^z w^{-\frac{1}{\lambda}}(1-w)^{\frac{1}{\lambda}-1} h(w)\, dw.$$

Exercise 6.7.12. Find a formula for the adjoint of the operator

$$(C_1 f)(x) = \frac{1}{x} \int_0^x f(t)\, dt$$

on $L^2[0, 1]$ from (6.5.1).

Exercise 6.7.13. Consider the following generalization of the Cesàro operator on ℓ^2 explored in [288]. For $p \in \mathbb{R}$ and $n \geqslant 0$, define

$$(C_p \mathbf{x})(n) = \frac{1}{(n+1)^p} \sum_{j=0}^n x_j \quad \text{for } \mathbf{x} = (x_n)_{n=0}^\infty \in \ell^2,$$

and note that C_1 is the standard Cesàro operator C.

(a) Compute the matrix representation of C_p with respect to the standard basis for ℓ^2.

(b) Prove that C_p is bounded on ℓ^2 if $p > 1$.

(c) Prove that C_p does not map ℓ^2 to ℓ^2 if $p < \frac{1}{2}$.

Exercise 6.7.14. This is a continuation of Exercise 6.7.13.

(a) For $a > 0$, let

$$D_a = \begin{bmatrix} \frac{1}{1^a} & 0 & 0 & 0 & \cdots \\ 0 & \frac{1}{2^a} & 0 & 0 & \cdots \\ 0 & 0 & \frac{1}{3^a} & 0 & \cdots \\ 0 & 0 & 0 & \frac{1}{4^a} & \cdots \\ \vdots & \vdots & \vdots & \vdots & \ddots \end{bmatrix}.$$

Prove that $C_p = D_{p-1} C_1$.

(b) Prove that C_p is compact for $p > 1$.

(c) Prove that for $p > 1$, the operator C_p has Hilbert–Schmidt (see Exercise 3.6.31) norm $\|C_p\|_{HS}^2 = \zeta(2p-1)$, where $\zeta(z) = \sum_{n=1}^\infty n^{-z}$ is the Riemann zeta function.

(d) Prove that C_p is not a bounded operator on ℓ^2 if $0 < p < 1$.

Remark: C_p is not hyponormal when $p > 1$ (and thus not subnormal).

Exercise 6.7.15. Here is another interesting class of operators from [285] that are related to the Cesàro operator. For $0 < |\lambda| \leqslant 1$, define $A_\lambda : \ell^2 \to \ell^2$ by

$$(A_\lambda x)(n) = \frac{1}{n+1} \sum_{j=0}^{n} \overline{\lambda}^{n-j} x_j \quad \text{for } n \geqslant 0.$$

Notice that A_1 is the Cesàro operator C.

(a) Prove that A_λ is bounded and $\|A_\lambda\| \leqslant 2$.

(b) Compute the matrix representation of A_λ with respect to the standard basis $(\mathbf{e}_n)_{n=0}^{\infty}$ for ℓ^2.

(c) Prove that A_λ is unitarily equivalent to $A_{|\lambda|}$.

Exercise 6.7.16. This is a continuation of Exercise 6.7.15.

(a) Identify ℓ^2 with H^2 in the usual way. If $0 < \lambda < 1$, prove that A_λ is represented on H^2 by

$$(A_\lambda f)(z) = \frac{1}{z} \int_0^z \frac{f(\xi)}{1 - \lambda\xi} \, d\xi.$$

(b) Prove that

$$(A_\lambda^* f)(z) = \frac{1}{z - \lambda} \int_\lambda^z f(\xi) \, d\xi.$$

(c) If $0 < \lambda < 1$ and $n \geqslant 1$ is an integer, prove that $1/n$ is an eigenvalue of A_λ with corresponding eigenvector $f_n(z) = z^{n-1}(1 - \lambda z)^{-n}$.

(d) If $0 < \lambda < 1$ and $n \geqslant 1$ is an integer, prove that each $1/n$ is an eigenvalue of A_λ^* with corresponding eigenvector $f_n(z) = (z - \lambda)^{n-1}$.

Remark: The operators A_λ for $0 < \lambda < 1$ are compact but not hyponormal.

Exercise 6.7.17. Prove that the Cesàro operator C is not bounded below on ℓ^2, that is, $\inf_{\|\mathbf{x}\|=1} \|C\mathbf{x}\| = 0$.

Exercise 6.7.18. This is a continuation of Exercise 6.7.17 and follows [234]. Prove that if $x_0 \geqslant x_1 \geqslant x_2 \geqslant \cdots \geqslant 0$ and $\mathbf{x} = (x_n)_{n=0}^{\infty} \in \ell^2$, then

$$\|C\mathbf{x}\|^2 \geqslant \frac{\pi^2}{6} \|\mathbf{x}\|^2.$$

Use the following steps.

(a) Use an integral estimate to prove that

$$2n \sum_{k=n+1}^{\infty} \frac{1}{k^2} > \sum_{k=1}^{n} \frac{1}{k^2}.$$

(b) Prove that

$$\|C\mathbf{x}\|^2 = \frac{\pi^2}{6}\|\mathbf{x}\|^2 - \sum_{n=1}^{\infty}\Big(\sum_{k=1}^{n}\frac{1}{k^2}\Big)x_n^2 + \sum_{n=1}^{\infty}\Big(2\sum_{k=n+1}^{\infty}\frac{1}{k^2}\Big)\sum_{j=0}^{n-1}x_j x_n.$$

(c) Use the fact that x_k is a positive decreasing sequence to obtain the estimate.

Remark: The paper [40] proves the following ℓ^p version of this result: if $1 < p < \infty$ and $x_0 \geqslant x_1 \geqslant x_2 \geqslant \cdots \geqslant 0$ with $\mathbf{x} = (x_n)_{n=0}^{\infty} \in \ell^p$, then $\|C\mathbf{x}\|_{\ell^p \to \ell^p}^p \geqslant \zeta(p)\|\mathbf{x}\|_{\ell^p}^p$. where $\zeta(z)$ is the Riemann zeta function.

Exercise 6.7.19. Here is a proof from [290] that C is hyponormal.

(a) Prove that $CC^* = C^*DC$, where

$$D = \begin{bmatrix} \frac{1}{2} & 0 & 0 & 0 & \cdots \\ 0 & \frac{2}{3} & 0 & 0 & \cdots \\ 0 & 0 & \frac{3}{4} & 0 & \cdots \\ 0 & 0 & 0 & \frac{4}{5} & \cdots \\ \vdots & \vdots & \vdots & \vdots & \ddots \end{bmatrix}.$$

(b) Prove that $C^*C - CC^* \geqslant 0$.

Remark: $A \in \mathcal{B}(\mathcal{H})$ is *posinormal* if $AA^* = A^*PA$ for some positive operator P. This concept is discussed further in Exercise 19.6.26. Note that the Cesàro operator is posinormal.

Exercise 6.7.20. Let S denote the unilateral shift on ℓ^2.

(a) Prove that $C = (C - S^*)C^*$.

(b) Prove that C is contained in the left ideal generated by C^*.

Exercise 6.7.21. Continuing with the notation from Exercises 6.7.19 and 6.7.20, show that $C^*C = CPC^*$, where $P = (C^* - S)(C - S^*)$, and hence that C^* is also posinormal.

Exercise 6.7.22. Let $S \in \mathcal{B}(\ell^2)$ denote the unilateral shift and C the Cesàro matrix. Prove that $S^*CS - C$ is a Hilbert–Schmidt operator.

Exercise 6.7.23. For the Cesàro operator C, follow this idea from [287] to prove that the numerical range (see Exercise 2.8.41) $W(C)$ of C is $\{z : |z - 1| < 1\}$ as follows.

(a) Prove that $\lambda \in W(C)$ if and only if $\bar{\lambda} \in W(C^*)$.

(b) Prove that the eigenvalues of C^* belong to $W(C^*)$.

(c) For any $A \in \mathcal{B}(\mathcal{H})$, prove that if $\|A\| \in W(A)$, then $\|A\|$ is an eigenvalue of A.

(d) Show that $W(C)$ does not contain any z such that $|z - 1| = 1$.

(e) Prove that $W(C)$ does not contain any z such that $|z - 1| > 1$.

Exercise 6.7.24. Let $\mathcal{V}$ denote the vector space of all complex sequences $\mathbf{a} = (a_n)_{n=0}^{\infty}$. The Cesàro operator C is a linear transformation on $\mathcal{V}$.

(a) For each $m \geqslant 0$, prove that

$$C\mathbf{a}^{(m)} = \frac{1}{m+1}\mathbf{a}^{(m)}, \quad \text{where} \quad a_n^{(m)} = \begin{cases} 0 & \text{for } 0 \leqslant n < m, \\ \binom{m+k}{m} & \text{if } n = m+k, k \geqslant 0. \end{cases}$$

(b) Prove that $\left\{\frac{1}{m+1} : m \geqslant 0\right\}$ is the set of all eigenvalues of C on $\mathcal{V}$.

(c) Prove that each eigenspace is one dimensional.

Remark: See [228] for more on this result.

Exercise 6.7.25. This result of Hausdorff [185] continues the discussion in Exercise 6.7.24 of the Cesàro operator on the space of all sequences.

(a) Consider the matrix

$$W = \begin{bmatrix} 1 & 0 & 0 & 0 & 0 & \cdots \\ 1 & -1 & 0 & 0 & 0 & \cdots \\ 1 & -2 & 1 & 0 & 0 & \cdots \\ 1 & -3 & 3 & -1 & 0 & \cdots \\ 1 & -4 & 6 & -4 & 1 & \cdots \\ \vdots & \vdots & \vdots & \vdots & \vdots & \ddots \end{bmatrix}.$$

Notice how the rows of W consist of alternating binomial coefficients. Prove that $W^2 = I$.

(b) For $\alpha \in \mathbb{R}$, define

$$H_\alpha = W \begin{bmatrix} 1 & 0 & 0 & 0 & \cdots \\ 0 & \frac{1}{2^\alpha} & 0 & 0 & \cdots \\ 0 & 0 & \frac{1}{3^\alpha} & 0 & \cdots \\ 0 & 0 & 0 & \frac{1}{4^\alpha} & \cdots \\ \vdots & \vdots & \vdots & \vdots & \ddots \end{bmatrix} W.$$

(c) Prove that $H_{\alpha+\beta} = H_\alpha H_\beta$ for all $\alpha, \beta \in \mathbb{R}$.

(d) Prove that H_1 is the Cesàro matrix C.

(e) Prove that $(H_{1/2})^2 = C$.

Remark: See [203] for more on this.

Exercise 6.7.26. Use the fact that $I - C$ is unitarily equivalent to M_z on a certain $H^2(\mu)$ space on $\mathbb{D}$ to prove that the Cesàro operator has a bounded square root.

Exercise 6.7.27. This exercise makes a connection between the Cesàro operator C on ℓ^2 and composition operators on the Hardy space H^2.

(a) For $0 < \alpha < 1$, use Schur's test (Theorem 3.3.1) to prove that

$$
A_\alpha = \begin{bmatrix}
1 & \alpha & \alpha^2 & \alpha^3 & \cdots \\
0 & (1-\alpha) & 2\alpha(1-\alpha) & 3\alpha^2(1-\alpha) & \cdots \\
0 & 0 & (1-\alpha)^2 & 3\alpha(1-\alpha)^2 & \cdots \\
0 & 0 & 0 & (1-\alpha)^3 & \cdots \\
\vdots & \vdots & \vdots & \vdots & \ddots
\end{bmatrix},
$$

that is

$$
(A_\alpha)_{i,j} = \begin{cases} \binom{j}{i} \alpha^{j-i}(1-\alpha)^i & \text{for } j \geqslant i, \\ 0 & \text{for } j < i, \end{cases}
$$

defines a bounded operator on ℓ^2.

(b) Prove that A_α commutes with C^*.

(c) Prove that A_α is the matrix representation of the composition operator $f(z) \mapsto f(\alpha + (1-\alpha)z)$ on H^2 with respect to the orthonormal basis $(z^n)_{n=0}^\infty$.

Remark: See [108, 343] for more on the commutant of C. Composition operators are explored in Chapter 18.

6.8 Hints for the Exercises

Hint for Ex. 6.7.4: If $\left(\frac{1}{n}I - C\right)f = z^{n-1}$, then f satisfies a certain first-order linear differential equation.

Hint for Ex. 6.7.6: Use (6.3.3) and the estimate $1 + x \leqslant e^x$ for $x \geqslant 0$.

Hint for Ex. 6.7.10: Use Theorem 2.6.9 and the spectral properties of C.

Hint for Ex. 6.7.14: For (d), observe that $C_1 = D_{1-p}C_p$ and consult Exercise 6.7.10.

Hint for Ex. 6.7.15: For (a), verify that $\|A_\lambda(x_n)_{n=0}^\infty\| \leqslant \|C(|x_n|)_{n=0}^\infty\|$ for each $(x_n)_{n=0}^\infty \in \ell^2$. For (c), conjugate A_λ with a suitable diagonal operator to obtain $A_{|\lambda|}$.

Hint for Ex. 6.7.22: Consult Exercise 3.6.31.

Hint for Ex. 6.7.23: For (d), recall that $\|C - I\| = 1$. For (e), use the convexity of $W(C)$.

Hint for Ex. 6.7.26: The Taylor coefficients of $\sqrt{1-z}$ form an ℓ^1 sequence.

7

·   ·   ●   ·   ·

The Volterra Operator

Key Concepts: Properties of the Volterra operator (adjoint, norm, spectrum, invariant subspaces, commutant), nilpotent operator, quasinilpotent operator, complex symmetric operator.

Outline: This chapter covers the Volterra operator

$$f(x) \mapsto \int_0^x f(t)\, dt$$

on $L^2[0,1]$. This operator is compact and quasinilpotent with no eigenvalues. The Volterra operator makes deep connections with function theory, in particular with the compression of the shift operator to a model space.

7.1 Basic Facts

Recall from Chapter 1 that the norm and inner product on $L^2[0,1]$ are

$$\|f\| = \left(\int_0^1 |f(x)|^2 dx \right)^{\frac{1}{2}} \quad \text{and} \quad \langle f, g \rangle = \int_0^1 f(x)\overline{g(x)}\, dx,$$

respectively. For $f \in L^2[0,1]$, define

$$(Vf)(x) = \int_0^x f(t)\, dt \quad \text{for } x \in [0,1]. \tag{7.1.1}$$

For $0 \leqslant x \leqslant y \leqslant 1$, the Cauchy–Schwarz inequality implies that

$$
\begin{aligned}
|(Vf)(x) - (Vf)(y)| &= \left| \int_x^y f(t)\, dt \right| \\
&\leqslant \int_x^y |f(t)|\, dt \\
&= \int_0^1 \chi_{[x,y]}(t)|f(t)|\, dt
\end{aligned}
$$

$$\leqslant \left(\int_0^1 \chi^2_{[x,y]}(t)\, dt \right)^{\frac{1}{2}} \left(\int_0^1 |f(t)|^2\, dt \right)^{\frac{1}{2}}$$

$$= |x - y|^{\frac{1}{2}} \|f\| \tag{7.1.2}$$

and hence $VL^2[0,1]$ is contained in $C[0,1] \subseteq L^2[0,1]$. Thus, (7.1.1) defines a linear transformation $V : L^2[0,1] \to L^2[0,1]$.

Proposition 7.1.3. *V is bounded on $L^2[0,1]$.*

Proof For $f \in L^2[0,1]$, set $x = 0$ in (7.1.2) to see that

$$|(Vf)(y)|^2 \leqslant y\|f\|^2 \quad \text{for all } y \in [0,1].$$

Integrate both sides and obtain

$$\|Vf\|^2 \leqslant \|f\|^2 \int_0^1 y\, dy = \frac{1}{2}\|f\|^2.$$

Therefore, V is a bounded operator on $L^2[0,1]$ and

$$\|V\| = \sup_{\|f\|=1} \|Vf\| \leqslant \frac{1}{\sqrt{2}},$$

which completes the proof. ∎

The linear transformation V is the *Volterra operator*. The previous proposition yields the estimate $\|V\| \leqslant 1/\sqrt{2}$. We compute the exact value of $\|V\|$ in Proposition 7.2.1 below. For the next result, recall the discussion of compact operators from Chapter 2.

Proposition 7.1.4. *The Volterra operator is compact.*

Proof To prove that V is compact, it suffices to show that $(Vf_n)_{n=1}^{\infty}$ has a convergent subsequence for every bounded sequence $(f_n)_{n=1}^{\infty}$ in $L^2[0,1]$. By (7.1.2), $(Vf_n)_{n=1}^{\infty}$ is a uniformly bounded and equicontinuous sequence of continuous functions on $[0,1]$. The Arzelà–Ascoli theorem [94, p. 179] provides a subsequence $(Vf_{n_k})_{k=1}^{\infty}$ that converges uniformly to a continuous function. Since uniform convergence implies $L^2[0,1]$ convergence (see (1.3.5)), $(Vf_{n_k})_{k=1}^{\infty}$ converges in $L^2[0,1]$. Thus, V is compact. ∎

One can also prove the compactness of the Volterra operator with Exercise 3.6.12 and Proposition 7.2.1 (below).

Proposition 7.1.5. *The adjoint of the Volterra operator is*

$$(V^*f)(x) = \int_x^1 f(t)\, dt \quad \text{for } x \in [0,1].$$

Proof For $f, g \in L^2$, let

$$u(x) = \int_0^x f(t)\,dt \quad \text{and} \quad v(x) = \int_0^x \overline{g(t)}\,dt.$$

Then $du = f(x)\,dx$ and $dv = \overline{g(x)}\,dx$. The integration by parts formula yields

$$\langle f, V^*g \rangle = \langle Vf, g \rangle$$

$$= \int_0^1 \left(\int_0^x f(t)\,dt \right)\overline{g(x)}\,dx$$

$$= \left(\int_0^x f(t)\,dt \right)\left(\int_0^x \overline{g(t)}\,dt \right)\Big|_{x=0}^{x=1} - \int_0^1 \left(\int_0^x \overline{g(t)}\,dt \right)f(x)\,dx$$

$$= \left(\int_0^1 f(x)\,dx \right)\left(\int_0^1 \overline{g(t)}\,dt \right) - \int_0^1 \left(\int_0^x \overline{g(t)}\,dt \right)f(x)\,dx$$

$$= \int_0^1 f(x)\left(\int_x^1 \overline{g(t)}\,dt \right)dx$$

$$= \int_0^1 f(x)\overline{\left(\int_x^1 g(t)\,dt \right)}\,dx,$$

which verifies the desired adjoint formula. ∎

7.2 Norm, Spectrum, and Resolvent

Proposition 7.1.3 yields the estimate $\|V\| \leqslant 1/\sqrt{2}$. The next proposition gives the exact value of $\|V\|$.

Proposition 7.2.1. *The Volterra operator V satisfies the following properties.*

(a) V^*V *is a compact operator with eigenvalues*

$$\lambda_n = \frac{4}{(2n+1)^2\pi^2} \quad \text{for } n \geqslant 0,$$

and corresponding unit eigenvectors

$$f_n(x) = \sqrt{2}\cos\left(\frac{2n+1}{2}\pi x\right).$$

(b) $\|V\| = \dfrac{2}{\pi}$.

Proof (a) Since V is compact (Proposition 7.1.4), and the space of compact operators is closed under adjoints (Exercise 3.6.11) and products (Proposition 2.5.5), it follows that V^*V is compact.

If $\lambda > 0$ and $V^*Vf = \lambda f$ for some $f \in L^2[0,1]\backslash\{0\}$, then Proposition 7.1.5 says that

$$\int_x^1 \int_0^t f(s)\,ds\,dt = \lambda f(x) \quad \text{for all } x \in [0,1]. \tag{7.2.2}$$

Differentiate the previous equation twice and obtain $-f(x) = \lambda f''(x)$. The solutions of this differential equation are of the form

$$f(x) = ae^{i\omega x} + be^{-i\omega x}, \tag{7.2.3}$$

. where $a, b \in \mathbb{C}$ and $\omega^2 = 1/\lambda$. Set $x = 1$ in (7.2.2) and deduce that $f(1) = 0$. Differentiating both sides of (7.2.2) shows that

$$-\int_0^x f(s)\,ds = \lambda f'(x).$$

Evaluate the previous equation at $x = 0$ and obtain $f'(0) = 0$. The conditions

$$f(1) = 0, \quad f'(0) = 0, \quad \text{and} \quad f(x) = ae^{i\omega x} + be^{-i\omega x},$$

say that

$$\omega = \frac{2n+1}{2}\pi$$

for some integer $n \geqslant 0$ and that f is of the form

$$f(x) = c\cos\left(\frac{2n+1}{2}\pi x\right)$$

for some constant c. Thus, since $\lambda = 1/\omega^2$, the eigenvalues of V^*V are

$$\lambda_n = \frac{4}{(2n+1)^2\pi^2},$$

and the corresponding (unit) eigenvectors f_n are

$$f_n(x) = \sqrt{2}\cos\left(\frac{2n+1}{2}\pi x\right). \tag{7.2.4}$$

(b) The eigenvectors $(f_n)_{n=0}^\infty$ form an orthonormal basis for $L^2[0,1]$ (Exercise 7.7.2) and the matrix representation of V^*V with respect to this basis is the diagonal matrix

$$\begin{bmatrix} \lambda_0 & 0 & 0 & 0 & \cdots \\ 0 & \lambda_1 & 0 & 0 & \cdots \\ 0 & 0 & \lambda_2 & 0 & \cdots \\ 0 & 0 & 0 & \lambda_3 & \cdots \\ \vdots & \vdots & \vdots & \vdots & \ddots \end{bmatrix}.$$

Proposition 2.1.1 reveals that

$$\|V^*V\| = \sup_{n\geqslant 0} \lambda_n = \sup_{n\geqslant 0} \frac{4}{(2n+1)^2\pi^2} = \frac{4}{\pi^2}.$$

Finally, Proposition 3.1.5e yields

$$\|V\| = \sqrt{\|V^*V\|} = \sqrt{\frac{4}{\pi^2}} = \frac{2}{\pi},$$

which completes the proof. ∎

One can also write V^*V using the tensor notation from (2.6.2):

$$V^*V = \sum_{n=0}^{\infty} \frac{8}{(2n+1)^2\pi^2} \cos\left(\frac{2n+1}{2}\pi x\right) \otimes \cos\left(\frac{2n+1}{2}\pi x\right).$$

The next proposition summarizes the spectral properties of V.

Proposition 7.2.5. *The Volterra operator V satisfies the following.*

(a) $\sigma(V) = \{0\}$.

(b) $\sigma_p(V) = \emptyset$.

(c) $\sigma_{ap}(V) = \{0\}$.

Proof (a) By induction (Exercise 7.7.3),

$$(V^n f)(x) = \int_0^x f(t) \frac{(x-t)^{n-1}}{(n-1)!} dt \quad \text{for all } n \geqslant 1. \tag{7.2.6}$$

Exercise 7.7.4 ensures that

$$\|V^n\| \leqslant \frac{1}{(n-1)!}. \tag{7.2.7}$$

Thus, for any $z \in \mathbb{C}$,

$$(I - zV)^{-1} = \sum_{n=0}^{\infty} z^n V^n,$$

where the series above converges in the operator norm (Proposition 2.3.9). Since

$$(zI - V)^{-1} = \frac{1}{z}\left(I - \frac{1}{z}V\right)^{-1} \quad \text{for all } z \neq 0,$$

it follows that $\sigma(V) \subseteq \{0\}$. Thus, $\sigma(V) = \{0\}$ since the spectrum of a bounded operator is nonempty (Theorem 2.4.9c). See Exercise 8.10.8 for another proof of (a).

(b) The proof of this is requested in Exercise 7.7.5.

(c) Note that $\sigma_{ap}(V) \subseteq \sigma(V) = \{0\}$ (Proposition 2.4.6). To show that $0 \in \sigma_{ap}(V)$, we proceed as follows. Observe that

$$h_n(x) = e^{2\pi i n x}, \quad \text{where } n \in \mathbb{Z},$$

is an orthonormal basis for $L^2[0, 1]$ (Theorem 1.3.9) and, for $n \geqslant 1$,

$$(Vh_n)(x) = \frac{(e^{2i\pi n x} - 1)}{2\pi i n} = \frac{1}{2\pi i n}(h_n(x) - h_0(x)).$$

By orthogonality (Proposition 1.4.6), $\|h_n - h_0\|^2 = \|h_n\|^2 + \|h_0\|^2 = 2$ and hence

$$\|Vh_n\|^2 = \frac{1}{4\pi^2 n^2}\|h_n - h_0\|^2 = \frac{1}{4\pi^2 n^2} \cdot 2 = \frac{1}{2\pi^2 n^2}.$$

Thus, $\|Vh_n\| \to 0$ as $n \to \infty$, so $0 \in \sigma_{ap}(V)$. ∎

Recall the definition of a Hilbert–Schmidt operator from Exercise 3.6.31.

Proposition 7.2.8. *The Volterra operator is a Hilbert–Schmidt operator with Hilbert–Schmidt norm $1/\sqrt{2}$.*

Proof From Proposition 7.2.1, the eigenvalues and corresponding unit eigenvectors of V^*V are

$$\lambda_n = \frac{4}{(2n+1)^2\pi^2} \quad \text{and} \quad f_n(x) = \sqrt{2}\cos\left(\frac{2n+1}{2}\pi x\right) \quad \text{for } n \geq 0.$$

Moreover, $(f_n)_{n=0}^{\infty}$ is an orthonormal basis for $L^2[0,1]$. The Hilbert–Schmidt norm of V satisfies

$$\|V\|_{HS}^2 = \sum_{n=0}^{\infty} \|Vf_n\|^2$$

$$= \sum_{n=0}^{\infty} \langle Vf_n, Vf_n \rangle$$

$$= \sum_{n=0}^{\infty} \langle V^*Vf_n, f_n \rangle$$

$$= \sum_{n=0}^{\infty} \lambda_n$$

$$= \frac{4}{\pi^2} \sum_{n=0}^{\infty} \frac{1}{(2n+1)^2}$$

$$= \frac{4}{\pi^2} \cdot \frac{\pi^2}{8}$$

$$= \frac{1}{2}.$$

Thus, $\|V\|_{HS} = 1/\sqrt{2}$. ∎

The fact that every Hilbert–Schmidt operator is compact (Exercise 3.6.31) yields another proof that the Volterra operator is compact.

Below we produce an explicit formula for the *resolvent* $(zI-V)^{-1}$ for $z \neq 0$. Since $\sigma(V) = \{0\}$, it follows that $(zI - V)^{-1}$ is a bounded operator. From (7.2.7), the series $\sum_{n=0}^{\infty} z^n V^n$ converges in operator norm.

Proposition 7.2.9. *For $z \neq 0$ and $f \in L^2[0,1]$,*

$$((zI - V)^{-1}f)(x) = \frac{1}{z}\left(f(x) + \frac{1}{z}\int_0^x \exp\left(\frac{x-y}{z}\right)f(y)\,dy\right).$$

Proof For $z \neq 0$ and $f \in L^2[0, 1]$,

$$(I - zV)^{-1}f = \sum_{n=0}^{\infty} z^n V^n f = f + z \sum_{n=1}^{\infty} z^{n-1} V^n f.$$

Thus,

$$((I - zV)^{-1}f)(x) = f(x) + z \sum_{n=1}^{\infty} z^{n-1} \int_0^x \frac{(x-y)^{n-1}}{(n-1)!} f(y)\, dy \qquad \text{(by (7.2.6))}$$

$$= f(x) + z \int_0^x \left(\sum_{n=1}^{\infty} \frac{(z(x-y))^{n-1}}{(n-1)!} \right) f(y)\, dy$$

$$= f(x) + z \int_0^x e^{z(x-y)} f(y)\, dy.$$

Now replace z with $1/z$ and then multiply by $1/z$. ∎

7.3 Other Properties of the Volterra Operator

Each $T \in \mathcal{B}(\mathcal{H})$ has a Cartesian decomposition $T = \operatorname{Re} T + i \operatorname{Im} T$, where

$$\operatorname{Re} T = \frac{1}{2}(T + T^*) \quad \text{and} \quad \operatorname{Im} T = \frac{1}{2i}(T - T^*) \tag{7.3.1}$$

are selfadjoint operators. Apply this to the Volterra operator V to see that

$$(V + V^*)f = \int_0^1 f(t)\, dt = (1 \otimes 1)f, \tag{7.3.2}$$

and hence $\operatorname{Re} V$ is a rank-one operator, namely $1/2$ times the orthogonal projection of $L^2[0, 1]$ onto the subspace of constant functions. Furthermore (Exercise 7.7.7),

$$(\operatorname{Im} Vf)(x) = \frac{1}{2i} \int_0^1 \operatorname{sgn}(t - x) f(t)\, dt.$$

Definition 7.3.3. $T \in \mathcal{B}(\mathcal{H})$ is *nilpotent* if $T^n = 0$ for some $n \geqslant 0$ and *quasinilpotent* if $\sigma(T) = \{0\}$.

A nilpotent operator is quasinilpotent, but the converse does not hold. From Theorem 8.4.4 below, it follows that T is quasinilpotent if and only if $\|T^n\|^{1/n} \to 0$.

Proposition 7.3.4. *The Volterra operator is quasinilpotent but not nilpotent.*

Proof Proposition 7.2.5 implies that V is quasinilpotent. Any nilpotent operator has zero as an eigenvalue (Exercise 7.7.9) which, for the Volterra operator, is not the case (Proposition 7.2.5b). ∎

Definition 7.3.5. $T \in \mathcal{B}(\mathcal{H})$ is *complex symmetric* if there is a conjugation C (a conjugate-linear, isometric, involution) on $\mathcal{H}$ such that $T = CT^*C$.

Exercise 7.7.15 yields the following.

Proposition 7.3.6. *If* $(Cf)(x) = \overline{f(1-x)}$, *then C is a conjugation on* $L^2[0,1]$ *and* $V = CV^*C$. *Thus, V is a complex symmetric operator.*

We can take advantage of the complex-symmetric nature of V to obtain an elegant matrix representation of V [146]. Consider the functions

$$w_n(x) = e^{2\pi i n(x-\frac{1}{2})} \quad \text{for } n \in \mathbb{Z}.$$

Exercise 7.7.19 shows that $(w_n)_{n=-\infty}^{\infty}$ is a *C-real* orthonormal basis for $L^2[0,1]$, in that it is an orthonormal basis and $Cw_n = w_n$ for all $n \in \mathbb{Z}$. Furthermore (Exercise 7.7.19), with respect to the basis $(w_n)_{n=-\infty}^{\infty}$, the Volterra operator has the matrix representation

$$[V] = \begin{bmatrix}
\ddots & \vdots & \vdots & \vdots & \vdots & \vdots & \vdots & \vdots & \iddots \\
\cdots & \frac{i}{6\pi} & 0 & 0 & \frac{i}{6\pi} & 0 & 0 & 0 & \cdots \\
\cdots & 0 & \frac{i}{4\pi} & 0 & -\frac{i}{4\pi} & 0 & 0 & 0 & \cdots \\
\cdots & 0 & 0 & \frac{i}{2\pi} & \frac{i}{2\pi} & 0 & 0 & 0 & \cdots \\
\cdots & \frac{i}{6\pi} & -\frac{i}{4\pi} & \frac{i}{2\pi} & \boxed{\frac{1}{2}} & -\frac{i}{2\pi} & \frac{i}{4\pi} & -\frac{i}{6\pi} & \cdots \\
\cdots & 0 & 0 & 0 & -\frac{i}{2\pi} & -\frac{i}{2\pi} & 0 & 0 & \cdots \\
\cdots & 0 & 0 & 0 & \frac{i}{4\pi} & 0 & -\frac{i}{4\pi} & 0 & \cdots \\
\cdots & 0 & 0 & 0 & -\frac{i}{6\pi} & 0 & 0 & -\frac{i}{6\pi} & \cdots \\
\iddots & \vdots & \vdots & \vdots & \vdots & \vdots & \vdots & \vdots & \ddots
\end{bmatrix}, \quad (7.3.7)$$

where the box denotes the $(0,0)$ entry. The matrix above is self-transpose since the matrix representation of any complex symmetric operator with respect to a *C*-real orthonormal basis is self transpose (Exercise 7.7.18).

A result from [148] gives the singular value decomposition of any compact *C*-symmetric operator T as

$$T = \sum_{n=0}^{\infty} \sigma_n(Ce_n \otimes e_n),$$

where σ_n are the eigenvalues of $|T|$ (the singular values of T) and e_n are certain eigenvectors of $|T| = (T^*T)^{1/2}$. See Chapter 14 for a formal presentation of the operator $|T|$. For the Volterra operator V, this becomes

$$V = \sum_{n=0}^{\infty} \frac{2}{(n+\frac{1}{2})\pi} \sin\left(\frac{2n+1}{2}\pi x\right) \otimes \cos\left(\frac{2n+1}{2}\pi x\right),$$

and thus

$$(Vf)(x) = \int_0^1 \left(\sum_{n=0}^{\infty} \frac{2}{(n+\frac{1}{2})\pi} \sin\left(\frac{2n+1}{2}\pi x\right) \cos\left(\frac{2n+1}{2}\pi y\right) \right) f(y)\, dy.$$

Observe that

$$\sum_{n=0}^{\infty} \frac{2}{(n + \frac{1}{2})\pi} \sin\left(\frac{2n+1}{2}\pi x\right) \cos\left(\frac{2n+1}{2}\pi y\right)$$

is a double Fourier expansion of the kernel $k(x, y) = \chi_\Delta(x, y)$, where Δ is the triangle $\{(x, y) : 0 \leqslant y \leqslant x, 0 \leqslant x \leqslant 1\}$. Therefore,

$$(Vf)(x) = \int_0^1 k(x, y)f(y)\, dy.$$

7.4 Invariant Subspaces

With any operator, one is always interested in a description of its invariant subspaces. The Volterra operator is no different. From Proposition 4.1.5, it follows that

$$\mathcal{F}_a = \chi_{[a,1]}L^2[0, 1], \quad \text{where } a \in [0, 1],$$

the set of functions in $L^2[0, 1]$ which vanish almost everywhere on $[0, a]$, is a subspace of $L^2[0, 1]$. Another way to see that $\mathcal{F}_a$ is closed is to observe that the operator $f \mapsto \chi_{[a,1]}f$ is the orthogonal projection of $L^2[0, 1]$ onto $\mathcal{F}_a$. One can also see that $V\mathcal{F}_a \subseteq \mathcal{F}_a$, and hence $\mathcal{F}_a$ is an invariant subspace for V. The Gelfand problem, posed in 1938 by I. Gelfand [150], asks whether these are all of the V-invariant subspaces of $L^2[0, 1]$. This was resolved in 1949 by Agmon [8].

Theorem 7.4.1 (Agmon). *For each $a \in [0, 1]$, $\mathcal{F}_a$ is an invariant subspace for V. Moreover, every invariant subspace for V is equal to $\mathcal{F}_a$ for some $a \in [0, 1]$.*

Proof The proof we outline here is due to Sarason [323]. Further details of the discussion below are covered in Chapter 20 once we know more about compressed shift operators. Let

$$\Theta(z) = \exp\left(\frac{z+1}{z-1}\right) \quad \text{for } z \in \mathbb{D}.$$

By Theorem 5.4.10, Θ is an inner function and ΘH^2 is an invariant subspace for the shift operator $(Sf)(z) = zf(z)$ on H^2. By Exercise 4.5.1, $(\Theta H^2)^\perp$ is an invariant subspace for S^*. Let

$$T = PS|_{(\Theta H^2)^\perp},$$

where P is the orthogonal projection of $L^2(\mathbb{T})$ onto $(\Theta H^2)^\perp$. This is the compression of the shift S to $(\Theta H^2)^\perp$. By results in Chapter 20 the operators $\frac{1}{2}(T + I)$ and $(I + V)^{-1}$ are unitarily equivalent via a unitary operator $W : L^2[0, 1]$ onto $(\Theta H^2)^\perp$. Also observe that T and $\frac{1}{2}(T + I)$ have the same invariant subspaces.

We claim that V and $(I + V)^{-1}$ also have the same invariant subspaces. This follows from the fact that $\sigma(V) = \{0\}$ and so $\|V^n\|^{\frac{1}{n}} \to 0$ (via the spectral radius formula – see

Theorem 8.4.4) and thus the series

$$\sum_{n=0}^{\infty}(-1)^n V^n$$

converges in operator norm to $(I + V)^{-1}$ (Exercise 7.7.20). By the spectral mapping theorem (Lemma 8.4.1), we have $\sigma((I+V)^{-1}-I) = \{0\}$ and thus, again by the spectral radius formula,

$$\|((I + V)^{-1} - I)^n\|^{\frac{1}{n}} \to 0.$$

This shows that

$$\sum_{n=1}^{\infty}(-1)^n((I + V)^{-1} - I)^n$$

converges in operator norm to V (Exercise 7.7.20). Since V and $(I + V)^{-1}$ can be approximated in operator norm by polynomials in the other, they have the same invariant subspaces. An invariant subspace of V corresponds (via the unitary operator W) to a T-invariant subspace $\mathcal{K}$ of $(\Theta H^2)^{\perp}$. By [143, p. 193], $\mathcal{K} = (\Theta H^2)^{\perp} \cap \Theta^a H^2$ for some $a \in [0,1]$. A final computation shows that $(\Theta H^2)^{\perp} \cap \Theta^a H^2$ corresponds to $\chi_{[a,1]}L^2[0,1]$. ∎

There are no interesting reducing subspaces for V.

Corollary 7.4.2. *If $\mathcal{F}$ is invariant for V and V^*, then $\mathcal{F} = L^2[0,1]$ or $\mathcal{F} = \{0\}$.*

Proof By Exercise 4.5.1, $\mathcal{F}$ is invariant for both V and V^* if and only if $\mathcal{F}$ and $\mathcal{F}^{\perp}$ are invariant for V. From Theorem 7.4.1, $\mathcal{F} = \chi_{[a,1]}L^2[0,1]$. One can show that $(\chi_{[a,1]}L^2[0,1])^{\perp} = \chi_{[0,a]}L^2[0,1]$ which, by Theorem 7.4.1 again, is V-invariant only when $a = 0$ or $a = 1$. In other words, $\mathcal{F} = L^2[0,1]$ or $\mathcal{F} = \{0\}$. ∎

7.5 Commutant

The commutant $\{V\}' = \{T \in \mathcal{B}(L^2[0,1]) : TV = VT\}$ of the Volterra operator is difficult to understand completely. Certainly $p(V) \in \{V\}'$ for all $p \in \mathbb{C}[z]$. Furthermore, the strong operator closure of $\{p(V) : p \in \mathbb{C}[z]\}$ is contained in $\{V\}'$. The following shows that this is precisely the commutant. In the above, recall from Exercise 4.5.23 that a sequence $A_n \in \mathcal{B}(\mathcal{H})$ converges to $A \in \mathcal{B}(\mathcal{H})$ in the *strong operator topology* (SOT) if $A_n\mathbf{x} \to A\mathbf{x}$ for each $\mathbf{x} \in \mathcal{H}$.

Theorem 7.5.1 (J. Erdos [126]). *The commutant of V is the strong operator closure of $\{p(V) : p \in \mathbb{C}[z]\}$.*

Proof We follow the original proof from [126]. Let $\mathcal{V}$ denote the strong operator closure of $\{p(V) : p \in \mathbb{C}[z]\}$. Observe that $\mathcal{V} \subseteq \{V\}'$.

To prove the reverse containment, we need a few details. For $f, g \in L^2[0,1]$, define

$$(T_g f)(x) = \int_0^x g(x-t)f(t)\,dt.$$

This is the convolution of f and g. The Cauchy–Schwarz inequality shows that T_g is a bounded operator on $L^2[0,1]$ with

$$\|T_g\| \leqslant \|g\| \quad \text{for all } g \in L^2[0,1]. \tag{7.5.2}$$

A short integral substitution reveals that

$$(T_g f)(x) = (T_f g)(x) \tag{7.5.3}$$

and

$$T_f 1 = V f \quad \text{for all } f \in L^2[0,1], \tag{7.5.4}$$

where 1 denotes the function whose value is identically 1.

It follows from (7.2.6) that if $p \in \mathbb{C}[x]$, then T_p is a polynomial in V with zero constant term. Furthermore, for $g \in L^2[0,1]$, Proposition 4.1.4 provides a sequence of polynomials $(p_n)_{n=1}^\infty$ such that $p_n \to g$ in $L^2[0,1]$. Now observe that (7.5.2) implies that $T_{p_n} \to T_g$ in operator norm (and hence strongly) and thus

$$T_g \in \mathcal{V} \quad \text{for all } g \in L^2[0,1]. \tag{7.5.5}$$

Now suppose that $A \in \{V\}'$. Then A commutes with $p(V)$ for all $p \in \mathbb{C}[x]$. Since $T_f \in \mathcal{V}$ for any $f \in L^2[0,1]$, it follows that A commutes with T_f. Therefore,

$$
\begin{aligned}
AVf &= AT_f 1 && \text{(by (7.5.4))} \\
&= T_f A1 && (A \text{ commutes with } T_f) \\
&= T_{A1} f && \text{(by (7.5.3)).}
\end{aligned}
$$

Consequently,

$$VA = AV = T_{A1} \in \mathcal{V}. \tag{7.5.6}$$

To prove that $A \in \mathcal{V}$, it suffices to show that given $\varepsilon > 0$ and $f_1, f_2, \dots, f_n \in L^2[0,1]$, there is a polynomial p such that $p(0) = 0$ and

$$\|(A - p(V))f_j\| < \varepsilon \quad \text{for all } 1 \leqslant j \leqslant n.$$

For $k \geqslant 1$, let

$$(V^k)^{(n)} = \bigoplus_{j=1}^n V^k : \bigoplus_{j=1}^n L^2[0,1] \to \bigoplus_{j=1}^n L^2[0,1]$$

be defined by

$$(V^k)^{(n)}(g_1, g_2, \dots, g_n) = (V^k g_1, V^k g_2, \dots, V^k g_n).$$

See Chapter 14 for a more formal treatment of direct sums of Hilbert spaces and operators. Let $\mathbf{f} = (f_1, f_2, \ldots, f_n)$ and

$$\mathcal{M} = \bigvee \{(V^k)^{(n)}\mathbf{f} : k \geqslant 1\}.$$

It is sufficient to show that $A^{(n)}\mathbf{f} \in \mathcal{M}$, where

$$A^{(n)} = \bigoplus_{j=1}^{n} A : \bigoplus_{j=1}^{n} L^2[0,1] \to \bigoplus_{j=1}^{n} L^2[0,1]$$

is defined by

$$A^{(n)}(g_1, g_2, \ldots, g_n) = (Ag_1, Ag_2, \ldots, Ag_n).$$

Suppose toward a contradiction that $A^{(n)}\mathbf{f} \notin \mathcal{M}$. Then there exists an $\mathbf{m} \in \mathcal{M}$ such that $\mathbf{g} = A^{(n)}\mathbf{f} - \mathbf{m}$ is orthogonal to $\mathcal{M}$. By (7.5.6) $AV \in \mathscr{V}$, and hence

$$(V^k)^{(n)}A^{(n)}\mathbf{f} \in \mathcal{M} \quad \text{for all } k \geqslant 1.$$

Thus,

$$(V^k)^{(n)}A^{(n)}\mathbf{g} \in \mathcal{M} \quad \text{and} \quad \langle (V^k)^n\mathbf{g}, \mathbf{g} \rangle = 0 \quad \text{for all } k \geqslant 1.$$

Let $\mathbf{h}_\lambda = V^{(n)}\mathbf{g} - \lambda\mathbf{g}$, where $\lambda > 0$. Then

$$\langle V^{(n)}\mathbf{h}_\lambda, \mathbf{h}_\lambda \rangle = \langle V^{(n)}(V^{(n)}\mathbf{g} - \lambda\mathbf{g}), V^{(n)}\mathbf{g} - \lambda\mathbf{g} \rangle$$
$$= \langle (V^2)^{(n)}\mathbf{g}, V^{(n)}\mathbf{g} \rangle - \lambda \langle V^{(n)}\mathbf{g}, V^{(n)}\mathbf{g} \rangle.$$

Since V, and hence $V^{(n)}$, is injective (Proposition 7.2.5), deduce that $\mathrm{Re}\langle V^{(n)}\mathbf{h}_\lambda, \mathbf{h}_\lambda \rangle < 0$ for sufficiently large $\lambda > 0$. However, (7.3.2) says that $\mathrm{Re}\, V = \frac{1}{2}(1 \otimes 1)$, which is a positive operator. Thus, $\mathrm{Re}\, V^{(n)}$ is also a positive operator, and hence $\mathrm{Re}\langle V^{(n)}\mathbf{h}_\lambda, \mathbf{h}_\lambda \rangle \geqslant 0$, which is a contradiction. ∎

In another sense, when looking for a function-theoretic understanding of the commutant, the situation becomes more involved. By our discussion of the invariant subspaces of V, it follows that $\{V\}' = W^*\{T\}'W$. As part of an interpolation result of Sarason [324] (see also [143]), $\{T\}' = \{PM_g|_{(\Theta H^2)^\perp} : g \in H^\infty\}$, where P is the orthogonal projection of H^2 onto $(\Theta H^2)^\perp$. Notice how each element of $\{T\}'$ is a compression of the multiplication operator M_g on H^2 to $(\Theta H^2)^\perp$. The following theorem summarizes this discussion.

Theorem 7.5.7 (Sarason [324]). $\{V\}' = \{W^*(PM_g|_{(\Theta H^2)^\perp})W : g \in H^\infty\}$.

Even though the Sarason result implies Theorem 7.5.1, we included an independent proof of Theorem 7.5.1 since it relied on elegant elementary techniques, while Sarason's theorem relies on some deep operator-interpolation results.

7.6 Notes

The Volterra operator was initially studied by Volterra [367, 368] and generalized in various directions by Gohberg and Kreĭn [154]. In particular, the Volterra operator belongs to a general class of integral operators

$$f(x) \mapsto \int_X k(x,y)f(y)\,dy$$

and these were some of the first operators studied. A well-done and readable introductory treatment of the Volterra operator on the space of continuous functions on $[0,1]$ is Shapiro's text [340]. A source for general integral operators on L^2 spaces is the text of Halmos and Sunder [172].

The initial papers on complex symmetric operators are due to Garcia and Putinar [147, 148]. These two authors, along with Prodan [146], studied various connections of complex symmetric operators to mathematical physics.

Our presentation of $\sigma(V) = \{0\}$ (Proposition 7.2.5) used the fact that

$$\lim_{n \to \infty} \|V^n\|^{\frac{1}{n}} = 0.$$

Kershaw [213] refined this decay result to

$$\lim_{n \to \infty} n!\|V^n\| = \frac{1}{2}.$$

The Gelfand problem asks for a description of the invariant subspaces for the Volterra operator. Although posed by I. Gelfand [150] and solved by Agmon [8], there are various related problems explored by Sakhnovich [322], Brodskiĭ [65] (the cyclic vectors for V), Donoghue [113] (solution of the Gelfand problem for $L^p[0,1]$), Kalisch [210], and, as we discussed in this chapter, Sarason [323].

Aleman and Korenblum examined a class of Volterra operators

$$(V_a f)(z) = \int_a^z f(w)\,dw \quad \text{for } a \in \mathbb{D}^-,$$

on the Hardy space H^2 and classified their invariant subspaces [12]. The book of Gohberg and Kreĭn [154] explores many other types of Volterra operators.

One can also study operators of the form

$$(Tf)(x) = xf(x) + \int_0^x f(t)\,dt$$

on $L^2[0,1]$. In other words, $T = M_x + V$ is multiplication by x plus the Volterra operator. Sarason [326] characterized the T-invariant subspaces by relating them to the ideals of the Sobolev space of absolutely continuous functions on $[0,1]$ whose derivative belongs to $L^2[0,1]$. Ong [256, 257] studied the invariant subspaces of the operator

$$f \mapsto xf(x) + n\int_0^x f(t)\,dt,$$

where $n \geqslant 1$ is an integer. Čučković and Paudyal [103] studied the analytic version of $M_x + V$, namely the operator on H^2 defined by

$$f(z) \mapsto zf(z) + \int_0^z f(w)\, dw.$$

7.7 Exercises

Exercise 7.7.1. Prove that $f \in \operatorname{ran} V$ if and only if f is absolutely continuous on $[0, 1]$, $f' \in L^2[0, 1]$, and $f(0) = 0$.

Exercise 7.7.2. Prove that the functions

$$f_n(x) = \sqrt{2} \cos\left(\frac{2n+1}{2}\pi x\right) \quad \text{for } n \geqslant 0$$

form an orthonormal basis for $L^2[0, 1]$.

Exercise 7.7.3. For $n \geqslant 1$ and $f \in L^2[0, 1]$, prove that

$$(V^n f)(x) = \int_0^x f(t)\frac{(x-t)^{n-1}}{(n-1)!}\, dt.$$

Exercise 7.7.4. For $n \geqslant 1$, prove that $\|V^n\| \leqslant \dfrac{1}{(n-1)!}$.

Exercise 7.7.5. Prove directly that the Volterra operator V has no eigenvalues.

Exercise 7.7.6. Prove that the Volterra operator is not similar to a weighted shift on ℓ^2. In other words, show there is no invertible operator $T : L^2[0, 1] \to \ell^2$ such that $TVT^{-1} = W$ for some weighted shift W (see Exercise 3.6.21).
Remark: See [168] for more on operators similar to a weighted shift.

Exercise 7.7.7.

(a) Prove that

$$(\operatorname{Im} V f)(x) = \frac{1}{2i}\int_0^1 \operatorname{sgn}(t-x)f(t)\, dt \quad \text{for } f \in L^2[0, 1].$$

(b) Compute $\sigma(\operatorname{Im} V)$.

Exercise 7.7.8. Prove that $\|\operatorname{Re} V\| = \dfrac{1}{2}$ and $\|\operatorname{Im} V\| = \dfrac{1}{\pi}$.
Remark: See [214] for more on this.

Exercise 7.7.9.

(a) Prove that 0 is an eigenvalue of every nilpotent operator.

(b) Prove that 0 is the only eigenvalue of a nilpotent operator.

Exercise 7.7.10. If $T \in \mathcal{B}(\mathcal{H})$ is nilpotent, prove that $I - T$ is invertible and find a formula for its inverse as a polynomial function of T.

Exercise 7.7.11. Let T be a linear transformation on a finite-dimensional vector space $\mathcal{V}$ such that for each $\mathbf{v} \in \mathcal{V}$, there is an integer $m(\mathbf{v}) \geqslant 1$ such that $T^{m(\mathbf{v})}(\mathbf{v}) = \mathbf{0}$. Prove that T is nilpotent.

Exercise 7.7.12. Prove that the bounded operator $A : L^2(0, \infty) \to L^2(0, \infty)$ defined by $(Af)(x) = f(x + 1)$ is quasinilpotent but not nilpotent.

Exercise 7.7.13. Prove that the operator $A \in \mathcal{B}(\ell^2)$ defined by

$$A(a_0, a_1, a_2, a_3, \ldots) = \left(0, \frac{a_0}{2^1}, \frac{a_1}{2^2}, \frac{a_2}{2^3}, \frac{a_3}{2^4}, \ldots\right)$$

is quasinilpotent but not nilpotent.

Exercise 7.7.14. Prove that the operator $T : H^2 \to H^2$ defined by $(Tf)(z) = zf(z/2)$ is compact and quasinilpotent. This exercise continues in Exercise 18.8.35.
Remark: This operator was explored in [113].

Exercise 7.7.15. Define $(Cf)(x) = \overline{f(1 - x)}$ for $f \in L^2$.

(a) Prove that $C(af + bg) = \overline{a}Cf + \overline{b}Cg$ for all $f, g \in L^2[0, 1]$ and $a, b \in \mathbb{C}$.

(b) Prove that $C^2 = I$.

(c) Prove that $\|Cf\| = \|f\|$ for every $f \in L^2[0, 1]$.

(d) Prove that $V = CV^*C$.

Exercise 7.7.16. Prove that

$$VV^* = \sum_{n=0}^{\infty} \frac{4}{(2n+1)^2\pi^2} g_n \otimes g_n, \quad \text{where} \quad g_n(x) = \sin\left(\frac{2n+1}{2}\pi x\right).$$

Exercise 7.7.17. Let $f \in L^2[0, 1]$.

(a) Prove that $(VV^*f)(x) = \displaystyle\int_0^1 \min(x, y)f(y)\, dy$.

(b) Prove that $(V^*Vf)(x) = \displaystyle\int_0^1 (1 - \max(x, y))f(y)\, dy$.

Exercise 7.7.18. Suppose $T \in \mathcal{B}(\mathcal{H})$ and C is a conjugation on $\mathcal{H}$ such that $T = CT^*C$. If $(\mathbf{u}_n)_{n=1}^{\infty}$ is a C-real orthonormal basis, that is, $C\mathbf{u}_n = \mathbf{u}_n$ for all $n \geqslant 1$, show that $[\langle T\mathbf{u}_j, \mathbf{u}_i \rangle]_{i,j=1}^{\infty}$, the matrix representation of T with respect to this basis, is self-transpose.

Exercise 7.7.19. Let $(Cf)(x) = \overline{f(1 - x)}$ on $L^2[0, 1]$ and

$$w_n(x) = e^{2\pi i n(x - \frac{1}{2})} \quad \text{for } n \in \mathbb{Z}.$$

Prove the following.

(a) $(w_n)_{n=-\infty}^{\infty}$ is an orthonormal basis for $L^2[0,1]$.

(b) $Cw_n = w_n$ for all $n \in \mathbb{Z}$.

(c) With respect to the basis $(w_n)_{n=-\infty}^{\infty}$, the matrix representation of V is

$$[V] = \begin{bmatrix}
\ddots & \vdots & \vdots & \vdots & \vdots & \vdots & \vdots & \vdots & \iddots \\
\cdots & \dfrac{i}{6\pi} & 0 & 0 & \dfrac{i}{6\pi} & 0 & 0 & 0 & \cdots \\
\cdots & 0 & \dfrac{i}{4\pi} & 0 & -\dfrac{i}{4\pi} & 0 & 0 & 0 & \cdots \\
\cdots & 0 & 0 & \dfrac{i}{2\pi} & \dfrac{i}{2\pi} & 0 & 0 & 0 & \cdots \\
\cdots & \dfrac{i}{6\pi} & -\dfrac{i}{4\pi} & \dfrac{i}{2\pi} & \boxed{\dfrac{1}{2}} & -\dfrac{i}{2\pi} & \dfrac{i}{4\pi} & -\dfrac{i}{6\pi} & \cdots \\
\cdots & 0 & 0 & 0 & -\dfrac{i}{2\pi} & -\dfrac{i}{2\pi} & 0 & 0 & \cdots \\
\cdots & 0 & 0 & 0 & \dfrac{i}{4\pi} & 0 & -\dfrac{i}{4\pi} & 0 & \cdots \\
\cdots & 0 & 0 & 0 & -\dfrac{i}{6\pi} & 0 & 0 & -\dfrac{i}{6\pi} & \cdots \\
\iddots & \vdots & \vdots & \vdots & \vdots & \vdots & \vdots & \vdots & \ddots
\end{bmatrix}.$$

Remark: The box denotes the $(0,0)$ entry.

Exercise 7.7.20.

(a) Prove that the series $\sum_{n=0}^{\infty}(-1)^n V^n$ converges in operator norm to $(V+I)^{-1}$.

(b) Prove that the series $\sum_{n=1}^{\infty}(-1)^n((V+I)^{-1}-I)^n$ converges in operator norm to V.

Exercise 7.7.21. The endnotes of this chapter mentioned how Sarason characterized the invariant subspaces of $M_x + V$ on $L^2[0,1]$, where $M_x f = xf$ on $L^2[0,1]$, by relating this problem to the ideals of $\mathcal{W}$, the algebra of absolutely continuous functions on $[0,1]$ whose derivative belongs to $L^2[0,1]$ [326]. Follow the steps in these next three problems to see how this works.

(a) Exercise 1.10.15 asserts that $\mathcal{W}$ is a Hilbert space with norm

$$\|f\|^2 = \int_0^1 |f(x)|^2\, dx + \int_0^1 |f'(x)|^2\, dx.$$

Prove there exists a $c > 0$ such that $|f(x) - f(y)| \leqslant c|x-y|^{\frac{1}{2}}\|f\|$ for all $x, y \in [0,1]$ and all $f \in \mathcal{W}$.

(b) Prove that there exists a $k > 0$ such that $\|fg\| \leqslant k\|f\|\|g\|$ for all $f, g \in \mathcal{W}$.

(c) Prove that $(Mf)(x) = xf(x)$ is a bounded linear operator on $\mathcal{W}$.

(d) Prove that a subspace $\mathcal{K}$ of $\mathcal{W}$ is M-invariant if and only if $\mathcal{K}$ is a (topologically closed) ideal of $\mathcal{W}$.

Exercise 7.7.22. This exercise continues Exercise 7.7.21.

(a) Define $Q : L^2[0,1] \to W$ by $(Qf)(x) = f(0) + \int_0^x f(t)\,dt$. Prove that Q is bounded, invertible, and satisfies $M_x + V = Q^{-1}MQ$.

(b) Show that the invariant subspaces of $M_x + V$ on $L^2[0,1]$ are in bijective and order-preserving correspondence with the closed ideals of W.

Exercise 7.7.23. This is a continuation of Exercise 7.7.22. The closed ideals of W are of the form $W(E) = \{f \in W : f|_E = 0\}$, where E is a closed subset of $[0,1]$. Assuming this fact, describe the invariant subspaces for $M_x + V$ on $L^2[0,1]$.

Exercise 7.7.24. Define $k : [0,1] \times [0,1] \to \mathbb{R}$ by

$$k(s,t) = \begin{cases} (1-s)t & \text{if } s \geqslant t, \\ (1-t)s & \text{if } s < t. \end{cases}$$

(a) Prove that $A : L^2[0,1] \to L^2[0,1]$ defined by $(Af)(s) = \int_0^1 k(s,t)f(t)\,dt$, is compact.

(b) Compute the eigenvalues and eigenvectors of A.

Exercise 7.7.25. Prove that the operator $A : L^2[0,1] \to L^2[0,1]$ defined by

$$(Af)(x) = \frac{1}{x}\int_0^x f(t)\,dt$$

is bounded but not compact.

Exercise 7.7.26. Prove that the Volterra operator V has a square root (first discovered in [280]) as follows.

(a) Prove that

$$(Wf)(x) = \frac{1}{\sqrt{\pi}}\int_0^x \frac{f(t)}{\sqrt{x-t}}\,dt$$

defines a bounded linear operator on $L^2[0,1]$.

(b) Prove that $W^2 = V$.

Remark: A result from [324] shows that $\pm W$ are the only two (bounded) square roots of V.

Exercise 7.7.27. Prove that the operator W defined in Exercise 7.7.26 is compact.

Exercise 7.7.28. Give an example of an $A \in \mathcal{B}(\mathcal{H})$ that is not compact, but such that A^2 is compact.

Exercise 7.7.29. Show that $\{V\}'$ is the closure of $\{p(V) : p \in \mathbb{C}[z]\}$ in the weak operator topology.

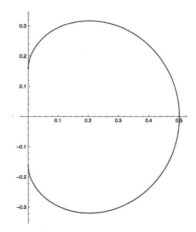

Figure 7.7.1 The numerical range of the Volterra operator is the region inside the curve.

Exercise 7.7.30. This problem discusses $W(V) = \{\langle Vf, f \rangle : \|f\| = 1\}$, the numerical range of the Volterra operator V.

(a) For $\|f\| = 1$, prove that $0 \leqslant \mathrm{Re}\langle Vf, f \rangle \leqslant \frac{1}{2}$ and thus $W(V) \subseteq \{z : 0 \leqslant \mathrm{Re}\, z \leqslant \frac{1}{2}\}$.

(b) Prove that if $z \in W(V)$, then $\bar{z} \in W(V)$.

(c) For each $\theta \in [0, 2\pi]$, define $f_\theta(x) = e^{i\theta x}$ and prove that

$$\langle Vf_\theta, f_\theta \rangle = -\frac{1}{\theta^2}(i\theta - 1 + e^{-i\theta}).$$

(d) Prove that $W(V)$ contains the region bounded by the curves

$$\theta \mapsto \frac{1 - \cos\theta}{\theta^2} \pm i\frac{\theta - \sin\theta}{\theta^2}.$$

See Figure 7.7.1.

Remark: A more technical discussion shows that $W(V)$ is equal to the set in (d) [215]. The numerical ranges of other operators are computed in [107, Ch. 9].

7.8 Hints for the Exercises

Hint for Ex. 7.7.4: Consult Proposition 7.1.3.
Hint for Ex. 7.7.5: Differentiate both sides of $Vf = \lambda f$ and consider the resulting boundary value problem.
Hint for Ex. 7.7.6: Look at $\ker V^*$ and $\ker W^*$.
Hint for Ex. 7.7.16: Use the conjugation C from Exercise 7.7.15 and the formula for V^*V proved in this chapter.

Hint for Ex. 7.7.21: For (a), note that $f(x) = f(0) + \int_0^x f'(t)\,dt$.

Hint for Ex. 7.7.25: Consult Theorem 6.5.3.

Hint for Ex. 7.7.26: Use the integral version of Schur's test from Exercise 3.6.27.

Hint for Ex. 7.7.29: Use Theorem 7.5.1 and the fact that $\{V\}'$ is closed in the weak operator topology.

8

. . **.** . .

Multiplication Operators

Key Concepts: Multiplication operator, normal operator, spectral theorem, continuous functional calculus, cyclic vector, ∗-cyclic vector, Bram's theorem, commutant of a normal operator, Fuglede–Putnam theorem.

Outline: We discuss multiplication operators $M_\varphi f = \varphi f$ on $L^2(\mu)$, where μ is a finite positive Borel measure on a compact set in $\mathbb{C}$ and φ is a μ-essentially bounded function. These operators represent normal operators on Hilbert spaces via the spectral theorem.

8.1 Multipliers of Lebesgue Spaces

Let $(X, \mathcal{A}, \mu)$ be a measure space, where $X \subseteq \mathbb{C}$ is compact, $\mathcal{A}$ is the collection of all Borel subsets of X, and μ is a finite positive Borel measure on X. By this we mean that $\mu(X) < \infty$ and $\mu(E) \geqslant 0$ for every Borel subset $E \subseteq X$. The space $L^2(\mu) := L^2(X, \mathcal{A}, \mu)$ is the set of all complex-valued μ-measurable functions f on X such that the integral

$$\int |f(z)|^2 d\mu(z)$$

is finite. Use a similar version of the proof of Proposition 1.3.1 to define an inner product and corresponding norm on $L^2(\mu)$ by

$$\langle f, g \rangle = \int f(z)\overline{g(z)}\, d\mu(z) \quad \text{and} \quad \|f\| = \left(\int |f(z)|^2\, d\mu(z) \right)^{\frac{1}{2}},$$

respectively. Note that $\|f\| = 0$ if and only if $f = 0$ μ-almost everywhere and so, as was done with $L^2[0, 1]$ and $L^2(\mathbb{T})$ in Chapter 4, we equate functions that are equal μ-almost everywhere. The inner-product spaces $L^2(\mu)$ are complete hence they are Hilbert spaces [319, Thm. 3.11]. A suitable modification of the proof of the Riesz–Fischer theorem (Proposition 1.3.4) also verifies this.

Let supp(μ) denote the *support* of μ (the complement of the union of all open sets with zero μ-measure). The space $L^\infty(\mu)$ of μ-essentially bounded functions on X is the set of all complex-valued μ-measurable functions f on X such that

$$\|f\|_\infty = \sup\{a \geq 0 : \mu(\{z \in X : |f(z)| > a\}) > 0\}$$

is finite. The quantity $\|f\|_\infty$ is the μ-*essential supremum* of f. Since μ is a finite measure, it follows that $L^\infty(\mu) \subseteq L^2(\mu)$.

Remark 8.1.1. Since the definition of the μ-essential supremum depends on the measure μ, one might argue that we should write $\|f\|_{\infty,\mu}$ to denote this dependence. We find this notation cumbersome and opt for the simpler $\|f\|_\infty$. The dependence on μ will always be clear from context.

Let $C(X)$ denote the set of complex-valued continuous functions on X endowed with the supremum norm

$$\|f\|_\infty = \sup_{z \in X} |f(z)|.$$

The analogue of Proposition 1.3.6 says that $C(X)$ is a dense subset of $L^2(\mu)$ (see also [319, Ch. 3]). To discuss another dense set in $L^2(\mu)$, we need the following theorem.

Theorem 8.1.2 (Weierstrass approximation theorem). *For a compact set $X \subseteq \mathbb{C}$ the following hold.*

(a) $\mathbb{C}[z, \overline{z}]$, *the set of polynomials in the complex variables z and $\overline{z}$, is dense in $C(X)$.*

(b) *If $X \subseteq \mathbb{R}$, then $\mathbb{C}[x]$, the set of polynomials in the real variable x, is dense in $C(X)$.*

These two forms of the Weierstrass approximation theorem yield useful dense subsets of $L^2(\mu)$.

Corollary 8.1.3. *Let μ be a finite positive Borel measure with compact support $X \subseteq \mathbb{C}$.*

(a) $\mathbb{C}[z, \overline{z}]$ *is dense in $L^2(\mu)$.*

(b) *If $X \subseteq \mathbb{R}$, then $\mathbb{C}[x]$ is dense in $L^2(\mu)$.*

Remark 8.1.4. Some examples of $L^2(\mu)$ spaces explored in this chapter are the following:

(a) $L^2[0, 1] = L^2(\lambda)$, where λ is Lebesgue measure on $[0, 1]$,

(b) $L^2(\mathbb{T}) = L^2(m)$, where m is normalized Lebesgue measure on $\mathbb{T}$,

(c) $L^2(dA)$, where dA is planar Lebesgue measure on $\mathbb{D}$, and

(d) $L^2(\sigma)$, where σ is the discrete measure

$$d\sigma = \sum_{n=1}^\infty c_n \delta_{\lambda_n}.$$

Here $(c_n)_{n=1}^\infty$ is a sequence of positive numbers with $\sum_{n=1}^\infty c_n < \infty$, $(\lambda_n)_{n=1}^\infty$ is a bounded sequence of complex numbers, and δ_{λ_n} is the *Dirac measure* defined on all subsets E of $\mathbb{C}$ by

$$\delta_{\lambda_n}(E) = \begin{cases} 1 & \text{if } \lambda_n \in E, \\ 0 & \text{if } \lambda_n \notin E. \end{cases}$$

This next proposition identifies the multiplier algebra of $L^2(\mu)$.

Proposition 8.1.5. *If φ is a μ-measurable function, then $\varphi L^2(\mu) \subseteq L^2(\mu)$ if and only if $\varphi \in L^\infty(\mu)$.*

Proof If $\varphi \in L^\infty(\mu)$, then

$$\|\varphi f\|^2 = \int |\varphi f|^2 \, d\mu \leqslant \|\varphi\|_\infty^2 \int |f|^2 \, d\mu = \|\varphi\|_\infty^2 \|f\|^2.$$

Thus, $\varphi L^2(\mu) \subseteq L^2(\mu)$ and hence the multiplication operator $M_\varphi : L^2(\mu) \to L^2(\mu)$ defined by

$$M_\varphi f = \varphi f$$

is well defined and bounded with

$$\|M_\varphi\| \leqslant \|\varphi\|_\infty. \tag{8.1.6}$$

Conversely, suppose that $\varphi L^2(\mu) \subseteq L^2(\mu)$. Then M_φ defines a linear transformation from $L^2(\mu)$ to itself. If $\|f_n - f\| \to 0$ and $\|\varphi f_n - g\| \to 0$, there is a subsequence $(f_{n_k})_{k=1}^\infty$ such that $f_{n_k} \to f$ and $\varphi f_{n_k} \to g$ μ-almost everywhere (Proposition 4.1.5). Thus, $g = \varphi f$ and hence the graph of M_φ is closed. The closed graph theorem (Theorem 2.2.2) ensures that M_φ is bounded on $L^2(\mu)$. Let χ denote the characteristic function of X. Since $\mu(X) < \infty$, it follows that $\chi \in L^2(\mu)$ and for all $n \geqslant 1$,

$$\int |\varphi|^{2n} d\mu = \|M_\varphi^n \chi\|^2 \leqslant \|M_\varphi^n\|^2 \|\chi\|^2 \leqslant \|M_\varphi\|^{2n} \mu(X).$$

Thus,

$$\|\varphi\|_{L^{2n}(\mu)} = \left(\int |\varphi|^{2n} d\mu \right)^{\frac{1}{2n}} \leqslant \|M_\varphi\| \mu(X)^{\frac{1}{2n}}.$$

Let $n \to \infty$ and observe that $\|\varphi\|_{L^{2n}(\mu)} \to \|\varphi\|_\infty$ [319, Ch. 3] and hence

$$\|\varphi\|_\infty \leqslant \|M_\varphi\|. \tag{8.1.7}$$

Therefore, $\varphi \in L^\infty(\mu)$. ∎

Equations (8.1.6) and (8.1.7) yield the following.

Corollary 8.1.8. *If $\varphi \in L^\infty(\mu)$, then $\|M_\varphi\| = \|\varphi\|_\infty$.*

A direct consequence of Corollary 8.1.8 is that the symbol φ for a multiplication operator M_φ is (essentially) unique.

Corollary 8.1.9. *For $\varphi, \psi \in L^\infty(\mu)$, the following are equivalent.*

(a) $M_\varphi = M_\psi$,

(b) $\varphi = \psi$ μ-almost everywhere.

The set of multiplication operators $\{M_\varphi : \varphi \in L^\infty(\mu)\}$ on $L^2(\mu)$ is a commutative algebra. In particular,

$$M_\varphi M_\psi = M_{\varphi\psi} = M_\psi M_\varphi \quad \text{for all } \varphi, \psi \in L^\infty(\mu). \tag{8.1.10}$$

The description of the spectrum of M_φ requires the following definition.

Definition 8.1.11. The *essential range* of a μ-measurable $\varphi : X \to \mathbb{C}$ is

$$\mathcal{R}_\varphi := \bigcap \{\varphi(E)^- : E \text{ is } \mu\text{-measurable and } \mu(E^c) = 0\}.$$

Note that $\mathcal{R}_\varphi$ depends on μ but we suppress the μ since it is clear from context. One also sees that $\mathcal{R}_\varphi$ is closed and that $\mathcal{R}_\varphi \subseteq \varphi(X)^-$. Exercise 8.10.1 shows that $w \in \mathcal{R}_\varphi$ if and only if

$$\mu\big(\varphi^{-1}(\{z : |z - w| < r\})\big) > 0 \quad \text{for all } r > 0.$$

Moreover, $\|\varphi\|_\infty = \max\{|w| : w \in \mathcal{R}_\varphi\}$. This definition enables us to discuss the parts of the spectrum of M_φ.

Proposition 8.1.12. *Let* $\varphi \in L^\infty(\mu)$.

(a) $\sigma(M_\varphi) = \mathcal{R}_\varphi$.

(b) $\sigma_p(M_\varphi) = \{\lambda \in \mathbb{C} : \mu(\{z : \varphi(z) = \lambda\}) > 0\}$.

(c) $\sigma_{ap}(M_\varphi) = \mathcal{R}_\varphi$.

Proof (a) Notice that $\lambda \notin \sigma(M_\varphi)$ if and only if $\lambda I - M_\varphi$ is an invertible operator on $L^2(\mu)$. If $f, g \in L^2(\mu)$ and $(\lambda I - M_\varphi)f = g$, then $f = (\lambda - \varphi)^{-1}g$ μ-almost everywhere. Thus, $\lambda \notin \sigma(M_\varphi)$ if and only if $M_{(\lambda-\varphi)^{-1}}$ is a bounded operator on $L^2(\mu)$. By Proposition 8.1.5, this is true if and only if there exists an $\varepsilon > 0$ such that $|\varphi - \lambda| \geq \varepsilon$ μ-almost everywhere. This last statement holds precisely when $\lambda \notin \mathcal{R}_\varphi$.

(b) Observe that $\lambda I - M_\varphi$ is injective if and only if the conditions $f \in L^2(\mu)$ and $(\lambda - \varphi)f = 0$ μ-almost everywhere imply that $f = 0$ μ-almost everywhere. This holds if and only if $\lambda - \varphi \neq 0$ μ-almost everywhere. Thus, $\lambda I - M_\varphi$ is injective if and only if $\mu(\{z : \varphi(z) = \lambda\}) = 0$.

(c) See Exercise 8.10.2. ∎

Recall that $A \in \mathcal{B}(\mathcal{H})$ is normal if $AA^* = A^*A$ and selfadjoint if $A = A^*$. Every selfadjoint operator is normal, although the converse is false. The importance of multiplication operators stems from the fact that they serve as models for certain normal operators. We begin with the following results.

Proposition 8.1.13. $M_\varphi^* = M_{\bar\varphi}$ *for every* $\varphi \in L^\infty(\mu)$.

Proof For $f, g \in L^2(\mu)$,

$$\langle M_\varphi f, g \rangle = \int \varphi f \bar{g} \, d\mu = \int f \overline{\bar\varphi g} \, d\mu = \langle f, M_{\bar\varphi} g \rangle.$$

Thus, by the definition of the adjoint, $M_\varphi^* = M_{\bar\varphi}$. ∎

Proposition 8.1.14. M_φ on $L^2(\mu)$ is normal for every $\varphi \in L^\infty(\mu)$. Moreover, M_φ is selfadjoint if and only if φ is real valued μ-almost everywhere.

Proof From Proposition 8.1.13 and (8.1.10),

$$M_\varphi^* M_\varphi - M_\varphi M_\varphi^* = M_{\overline{\varphi}} M_\varphi - M_\varphi M_{\overline{\varphi}} = M_{|\varphi|^2} - M_{|\varphi|^2} = 0.$$

Thus, M_φ is normal. Next, observe that $M_\varphi^* = M_\varphi$ if and only if $M_{\overline{\varphi}} = M_\varphi$, which is true if and only if $\varphi = \overline{\varphi}$ μ-almost everywhere (Corollary 8.1.9). ∎

8.2 Cyclic Vectors

We begin our discussion with the following definition.

Definition 8.2.1. An operator $A \in \mathcal{B}(\mathcal{H})$ is *cyclic* if there exists an $\mathbf{x} \in \mathcal{H}$ such that

$$\bigvee \{A^n \mathbf{x} : n \geq 0\} = \mathcal{H}.$$

Such a vector $\mathbf{x}$ is a *cyclic vector* for A.

Exercise 13.9.5 shows that if A has a cyclic vector, it has a set of cyclic vectors whose linear span is dense.

A thorough discussion of the cyclicity of multiplication operators requires the Riesz representation theorem for bounded linear functionals on $C(X)$. If μ is a finite complex Borel measure with $\mathrm{supp}(\mu) \subseteq X$, then the linear functional $\Lambda_\mu : C(X) \to \mathbb{C}$ defined by

$$\Lambda_\mu(f) = \int_X f \, d\mu$$

is well defined on $C(X)$. Furthermore,

$$|\Lambda_\mu(f)| = \left| \int_X f \, d\mu \right| \leq \int_X |f| \, d|\mu| \leq \|f\|_\infty |\mu|(X) \quad \text{for all } f \in C(X).$$

In the above, the *total variation* of μ is defined by

$$|\mu|(X) = \sup \sum_{n=1}^\infty |\mu(A_n)|,$$

where the supremum is taken over all countable Borel partitions $X = \bigcup \{A_n : n \geq 1\}$ of X. It follows that Λ_μ is a bounded linear functional on $C(X)$. Furthermore,

$$\|\Lambda_\mu\| = \sup_{\|f\|_\infty = 1} |\Lambda_\mu(f)| = |\mu|(X)$$

and such Λ_μ comprise all of the bounded linear functionals on $C(X)$ [94, p. 78].

Theorem 8.2.2 (Riesz representation theorem). *If Λ is a bounded linear functional on $C(X)$, then there is a unique finite complex Borel measure μ on X such that $\Lambda = \Lambda_\mu$.*

Corollary 8.2.3. *If μ is a finite complex Borel measure on X, the following are equivalent.*

(a) $\int f \, d\mu = 0$ *for all* $f \in C(X)$.

(b) $\mu = 0$.

The next result characterizes the cyclic vectors of a certain multiplication operator.

Proposition 8.2.4. *Suppose μ is a finite positive Borel measure with compact support in $\mathbb{R}$.*

(a) M_x *is a cyclic operator on $L^2(\mu)$.*

(b) $f \in L^2(\mu)$ *is a cyclic vector for M_x if and only if $|f| > 0$ μ-almost everywhere.*

Proof Observe that $f \in L^2(\mu)$ is cyclic if and only if $\{pf \ : \ p \in \mathbb{C}[x]\}$ is dense in $L^2(\mu)$. This holds if and only if the conditions $g \in L^2(\mu)$ and $\langle pf, g \rangle = 0$ for all $p \in \mathbb{C}[x]$ imply $g = 0$ μ-almost everywhere. Suppose that $|f| > 0$ μ-almost everywhere and

$$\int p f \overline{g} \, d\mu = 0 \quad \text{for all } p \in \mathbb{C}[x]. \tag{8.2.5}$$

The Weierstrass approximation theorem implies that (8.2.5) also holds for any $p \in C(X)$. By Corollary 8.2.3, $d\nu = f\overline{g} \, d\mu$ is the zero measure, in other words $f\overline{g} = 0$ μ-almost everywhere. Since $|f| > 0$ μ-almost everywhere, it follows that $g = 0$ μ-almost everywhere.

Conversely, if $f = 0$ on a μ-measurable set A with $\mu(A) > 0$, then $g = \chi_A \in L^2(\mu)\backslash\{0\}$ and $f\overline{g} = 0$ μ-almost everywhere. Therefore,

$$\langle pf, g \rangle = \int p f \overline{g} \, d\mu = 0 \quad \text{for all } p \in \mathbb{C}[x],$$

and hence f is not cyclic for M_x. ∎

If $\text{supp}(\mu) \not\subset \mathbb{R}$, then Proposition 8.2.4 requires significant modification, since $\mathbb{C}[z]$ may not be dense in $L^2(\mu)$. For example, if m is Lebesgue measure on $\mathbb{T}$, then the closure of $\mathbb{C}[z]$ in $L^2(m)$ is the Hardy space H^2 and not $L^2(m)$ (see Example 8.2.9 below).

Definition 8.2.6. M_z on $L^2(\mu)$ is *$*$-cyclic* if there is an $f \in L^2(\mu)$ such that

$$\bigvee \{ M_z^j M_{\overline{z}}^k f \ : \ j, k \geqslant 0 \} = L^2(\mu).$$

The function f is a *$*$-cyclic vector for M_z.*

The normality of M_z ensures that $M_z^j M_{\overline{z}}^k = M_{\overline{z}}^k M_z^j$ for any integers $j, k \geqslant 0$ (Proposition 8.1.14).

Proposition 8.2.7. *Suppose μ is a finite positive Borel measure with compact support in $\mathbb{C}$.*

(a) M_z *on $L^2(\mu)$ is $*$-cyclic.*

(b) $f \in L^2(\mu)$ *is a $*$-cyclic vector if and only if $|f| > 0$ μ-almost everywhere.*

Proof Follow the proof of Proposition 8.2.4 and use the density of $\mathbb{C}[z, \overline{z}]$ in $L^2(\mu)$ (Theorem 8.1.3a). ∎

When is M_φ cyclic? The following theorem, whose proof is in [337, 378], answers this question.

Theorem 8.2.8. *For $\varphi \in L^2(\mu)$, the following are equivalent.*

(a) M_φ *is cyclic.*

(b) φ *is injective on a set of full measure.*

We encountered an example of this result in Exercise 4.5.8.

If $\mathrm{supp}(\mu) \subseteq \mathbb{R}$, the cyclic and $*$-cyclic vectors for M_z are the same (since $M_z = M_z^*$). If $\mathrm{supp}(\mu) \not\subset \mathbb{R}$, they can be different.

Example 8.2.9. The constant function 1 on $\mathbb{T}$ is a $*$-cyclic vector for M_ξ on $L^2(\mathbb{T})$ since

$$\bigvee\{M_\xi^j M_\xi^k 1 \, : \, j, k \geqslant 0\} = \bigvee\{\xi^n \, : \, n \in \mathbb{Z}\} = L^2(\mathbb{T})$$

(Theorem 1.3.9). However,

$$\bigvee\{M_\xi^n 1 \, : \, n \geqslant 0\} = \bigvee\{\xi^n \, : \, n \geqslant 0\} = H^2,$$

which is a proper subspace of $L^2(\mathbb{T})$. Thus, 1 is a $*$-cyclic vector for M_ξ but not a cyclic vector for M_ξ.

Example 8.2.10. Continuing with Example 8.2.9, observe that Theorem 8.2.8 guarantees that M_ξ on $L^2(\mathbb{T})$ is cyclic. However, the cyclic vectors for M_ξ are not so obvious. For $f \in L^2(\mathbb{T})$, Szegő's formula [202, p. 49] says that

$$\inf_{p \in \mathbb{C}[\xi]} \int_\mathbb{T} |\bar{\xi} f - pf|^2 dm = \exp\Big(\int_\mathbb{T} \log |f| dm\Big).$$

Select an $f \in L^2(\mathbb{T}) \backslash \{0\}$ such that $\log |f|$ is not integrable; for example

$$f(\xi) = \exp\Big(-\frac{1}{|\xi - 1|}\Big) \quad \text{for } \xi \in \mathbb{T} \backslash \{1\}.$$

Consequently,

$$\bar{\xi} f \in \bigvee\{M_\xi^n f \, : \, n \geqslant 0\}.$$

Repeated applications of Szegő's formula yield

$$\bar{\xi}^k f \in \bigvee\{M_\xi^n f \, : \, n \geqslant 0\} \quad \text{for all } k \geqslant 0.$$

This means that

$$\bigvee\{M_\xi^n f \, : \, n \geqslant 0\} = \bigvee\{\xi^m f \, : \, m \in \mathbb{Z}\}.$$

Given any $g \in C(\mathbb{T})$, the Weierstrass approximation theorem (Theorem 8.1.2) produces a sequence $(p_n(\xi, \bar{\xi}))_{n=1}^\infty$ of polynomials in ξ and $\bar{\xi}$ such that $\|p_n - g\|_\infty \to 0$. It follows that $\|p_n f - gf\| \leqslant \|p_n - g\|_\infty \|f\| \to 0$. Since $|f| > 0$, the set $\{gf \, : \, g \in C(\mathbb{T})\}$ is dense in $L^2(\mathbb{T})$. Thus, f is cyclic.

We summarize Proposition 8.2.7 and Example 8.2.10 as follows.

Proposition 8.2.11. *For $f \in L^2(\mathbb{T})$, the following are equivalent.*

(a) *f is a cyclic vector for M_ξ.*

(b) *$|f| > 0$ m-almost everywhere and $\log|f| \notin L^1(\mathbb{T})$.*

One can show that M_z on $L^2(dA)$, where dA is area measure on $\mathbb{D}$, does not have the constant function 1 as a cyclic vector (Exercise 8.10.19). In fact, examples like these lead the reader to wonder if M_z on $L^2(\mu)$ has any cyclic vectors. A surprising theorem of Bram puts this issue to rest.

Theorem 8.2.12 (Bram [63]). *If M_φ on $L^2(\mu)$ is $*$-cyclic, then it is cyclic.*

Here are some of the ingredients used to prove Bram's theorem. The hypothesis that M_φ is $*$-cyclic allows the use of a version of the spectral theorem for normal operators (see Theorem 19.2.3) to prove that M_φ is unitarily equivalent to M_z on $L^2(\nu)$, where ν is a finite positive Borel measure with compact support in $\mathbb{C}$. Bram's rather elaborate construction (see also [95, p. 232]) produces a finite positive Borel measure ν_1 on $\mathbb{C}$ such that

(a) $\nu \ll \nu_1$ and $\nu_1 \ll \nu$,

(b) $H^2(\nu_1) = L^2(\nu_1)$,

(c) $\psi = \sqrt{\dfrac{d\nu_1}{d\nu}}$ is bounded.

For any $f \in L^2(\nu)$, it follows from the definition of ψ that $f/\psi \in L^2(\nu_1)$ and thus there is a sequence of polynomials $(p_n)_{n=1}^\infty$ such that

$$\int \left| p_n - \frac{f}{\psi} \right|^2 d\nu_1 \to 0.$$

Then

$$\int |p_n\psi - f|^2 d\nu = \int |p_n\psi - f|^2 \frac{1}{\psi^2} d\nu_1 = \int \left| p_n - \frac{f}{\psi} \right|^2 d\nu_1 \to 0.$$

Thus, ψ is a cyclic vector for M_z on $L^2(\nu)$. Since M_z on $L^2(\nu)$ is unitarily equivalent to M_φ on $L^2(\mu)$, Exercise 8.10.30 shows that M_φ has a cyclic vector.

Example 8.2.13. Exercise 8.10.19 shows that the constant function 1 is not a cyclic vector for M_z on $L^2(dA)$. However, it is a $*$-cyclic vector. Thus, Bram's theorem implies that M_z has a cyclic vector.

8.3 Commutant

What is the commutant $\{M_\varphi\}' = \{A \in \mathcal{B}(L^2(\mu)) : AM_\varphi = M_\varphi A\}$ of M_φ? The following theorem aids the analysis.

Theorem 8.3.1 (Fuglede [136]). *If $\varphi \in L^\infty(\mu)$ and $A \in \mathcal{B}(L^2(\mu))$ with $AM_\varphi = M_\varphi A$, then $AM_{\overline{\varphi}} = M_{\overline{\varphi}}A$.*

Proof We follow a wonderful proof by Rosenblum [309]. The proof uses operator-valued analytic functions (see Exercise 8.10.22). The hypotheses imply that $AM_\varphi^k = M_\varphi^k A$ for all $k \geqslant 0$. Using operator-valued power series, it follows that

$$Ae^{\overline{\lambda}M_\varphi} = e^{\overline{\lambda}M_\varphi}A \quad \text{for all } \lambda \in \mathbb{C}.$$

Define the operator-valued entire function by

$$F(\lambda) = e^{\lambda M_{\overline{\varphi}}}Ae^{-\lambda M_{\overline{\varphi}}}. \tag{8.3.2}$$

Since $A = e^{-\overline{\lambda}M_\varphi}Ae^{\overline{\lambda}M_\varphi}$ and M_φ commutes with $M_{\overline{\varphi}}$, one concludes that

$$F(\lambda) = e^{\lambda M_{\overline{\varphi}}}\left(e^{-\overline{\lambda}M_\varphi}Ae^{\overline{\lambda}M_\varphi}\right)e^{-\lambda M_{\overline{\varphi}}} = e^{\lambda M_{\overline{\varphi}}-\overline{\lambda}M_\varphi}Ae^{\overline{\lambda}M_\varphi-\lambda M_{\overline{\varphi}}}.$$

Exercise 8.10.23 shows that the operators

$$U = e^{\lambda M_{\overline{\varphi}}-\overline{\lambda}M_\varphi} \quad \text{and} \quad V = e^{\overline{\lambda}M_\varphi-\lambda M_{\overline{\varphi}}}$$

are unitary, and hence

$$
\begin{aligned}
\|F(\lambda)\| &= \|UAV\| \\
&\leqslant \|U\|\|A\|\|V\| && \text{(by Proposition 2.1.12b)} \\
&= \|A\| && (U \text{ and } V \text{ are isometric}).
\end{aligned}
$$

This means that F is a bounded entire operator-valued function, so, via a similar argument used to prove that the spectrum of an operator is nonempty (Theorem 2.4.9), $F(\lambda) = F(0) = A$ for all $\lambda \in \mathbb{C}$. Take derivatives of F (using (8.3.2)) and set $\lambda = 0$ to obtain $AM_{\overline{\varphi}} = M_{\overline{\varphi}}A$. ∎

Putman [276] generalized Fuglede's theorem as follows. If $T, M, N \in \mathcal{B}(\mathcal{H})$ with M, N normal and $MT = TN$, then $M^*T = TN^*$ (Exercise 8.10.24). Fuglede's theorem also yields a characterization of the commutant of M_z on $L^2(\mu)$.

Corollary 8.3.3. $\{M_z\}' = \{M_\varphi : \varphi \in L^\infty(\mu)\}$.

Proof The $\supseteq$ containment follows from (8.1.10). For the reverse containment, suppose that $A \in \mathcal{B}(L^2(\mu))$ and $AM_z = M_z A$. We need to produce a $\varphi \in L^\infty(\mu)$ such that $A = M_\varphi$. Fuglede's theorem (Theorem 8.3.1) implies that $AM_{\overline{z}} = M_{\overline{z}}A$. These two facts say that $AM_{p(z,\overline{z})} = M_{p(z,\overline{z})}A$ for all $p \in \mathbb{C}[z,\overline{z}]$. Apply this to $\chi \in L^2(\mu)$, the characteristic function of supp(μ), and conclude that $Ap\chi = pA\chi$ for all $p \in \mathbb{C}[z, \bar{z}]$. The density of $\mathbb{C}[z, \bar{z}]$ in $L^2(\mu)$ (Corollary 8.1.3) says that given $f \in L^2(\mu)$, there is a sequence $(p_n)_{n=1}^{\infty}$ in $\mathbb{C}[z, \bar{z}]$ such that $p_n \to f$ in $L^2(\mu)$ norm. Passing to a subsequence and relabeling, we may assume that $p_n \to f$ pointwise μ-almost everywhere. Consequently, $Af = fA\chi$ for all $f \in L^2(\mu)$, and hence $A\chi \in L^{\infty}(\mu)$ (Proposition 8.1.5). Furthermore, $A = M_{A\chi}$. ∎

8.4 Spectral Radius

For $p(z) = \sum_{k=0}^{n} c_k z^k \in \mathbb{C}[z]$ and $T \in \mathcal{B}(\mathcal{H})$, define

$$p(T) = \sum_{k=0}^{n} c_k T^k.$$

By convention, $T^0 = I$, the identity operator on $\mathcal{H}$. The following important lemma says that this *polynomial functional calculus* respects the spectrum.

Lemma 8.4.1 (Polynomial spectral mapping theorem). *If $A \in \mathcal{B}(\mathcal{H})$ and $p \in \mathbb{C}[z]$, then* $\sigma(p(A)) = p(\sigma(A)) = \{p(\lambda) : \lambda \in \sigma(A)\}$.

Proof Without loss of generality, suppose p is nonconstant.

($\supseteq$) If $\lambda \in \sigma(A)$, the polynomial $p(z) - p(\lambda)$ vanishes at $z = \lambda$ and hence $p(z) - p(\lambda) = (z - \lambda)q(z)$ for some $q \in \mathbb{C}[z]$. Since $\lambda \in \sigma(A)$, it follows that the operator $p(A) - p(\lambda)I = (A - \lambda I)q(A)$ is not invertible. Thus, $p(\lambda) \in \sigma(p(A))$ and hence $p(\sigma(A)) \subseteq \sigma(p(A))$.

($\subseteq$) Let $\xi \in \sigma(p(A))$ and let $\lambda_1, \lambda_2, ..., \lambda_n$ be the zeros of $p(z) - \xi$, repeated according to multiplicity. Then

$$p(z) - \xi = c(z - \lambda_1)(z - \lambda_2) \cdots (z - \lambda_n), \tag{8.4.2}$$

where $c \neq 0$. If $\lambda_1, \lambda_2, ..., \lambda_n \notin \sigma(A)$, then

$$p(A) - \xi I = c(A - \lambda_1 I)(A - \lambda_2 I) \cdots (A - \lambda_n I)$$

is a product of invertible operators and hence invertible. This is impossible since $\xi \in \sigma(p(A))$. Therefore, $\lambda_i \in \sigma(A)$ for some $1 \leqslant i \leqslant n$, and hence $\xi = p(\lambda_i)$ by (8.4.2). Consequently, $\sigma(p(A)) \subseteq p(\sigma(A))$. ∎

Definition 8.4.3. The *spectral radius* of $A \in \mathcal{B}(\mathcal{H})$ is

$$r(A) = \sup_{\lambda \in \sigma(A)} |\lambda|.$$

Since $\sigma(A)$ is a compact subset of $\mathbb{C}$ the supremum above is attained. Moreover, Theorem 2.4.9 yields $r(A) \leqslant \|A\|$. Equality is attained for a diagonal operator, although other examples exist. Strict inequality occurs for

$$A = \begin{bmatrix} 0 & 1 \\ 0 & 0 \end{bmatrix},$$

for which $r(A) = 0$ and $\|A\| = 1$. The next result of Beurling [51] and Gelfand [151] is known as the *spectral radius formula*.

Theorem 8.4.4 (Beurling–Gelfand). *If $A \in \mathcal{B}(\mathcal{H})$, then*

$$r(A) = \lim_{n \to \infty} \|A^n\|^{\frac{1}{n}} .$$

The proof of this theorem requires some preliminaries. For $A \in \mathcal{B}(\mathcal{H})$, observe that $\mathcal{A} = \bigvee\{A^k : k \geq 0\}$ is a Banach space and an algebra. Recall that $\mathcal{A}^*$ denotes the dual space of $\mathcal{A}$, the set of bounded linear functionals φ on $\mathcal{A}$ (Definition 2.2.4). With respect to the norm

$$\|\varphi\| = \sup_{\substack{a \in \mathcal{A}, \\ \|a\|=1}} |\varphi(a)|,$$

$\mathcal{A}^*$ is itself a Banach space and one can consider $\mathcal{A}^{**} = (\mathcal{A}^*)^*$, the space of bounded linear functionals on $\mathcal{A}^*$, that is, the space of all $\ell : \mathcal{A}^* \to \mathbb{C}$ with finite norm

$$\|\ell\| = \sup_{\substack{\varphi \in \mathcal{A}^* \\ \|\varphi\|=1}} |\ell(\varphi)|.$$

For $a \in \mathcal{A}$, one has $\hat{a} \in \mathcal{A}^{**}$ defined by $\hat{a}(\varphi) = \varphi(a)$. The Hahn–Banach extension theorem (Theorem 2.2.5) implies that the map $a \mapsto \hat{a}$ is a linear isometry from $\mathcal{A}$ into $\mathcal{A}^{**}$. We are now ready for the proof of Theorem 8.4.4.

Proof Lemma 8.4.1 ensures that $\lambda \in \sigma(A)$ implies that $\lambda^n \in \sigma(A^n)$, so $|\lambda^n| \leq \|A^n\|$ (Theorem 2.4.9). For $\lambda \in \sigma(A)$, it follows that

$$|\lambda| \leq \liminf_{n \to \infty} \|A^n\|^{\frac{1}{n}} ,$$

and hence

$$r(A) \leq \liminf_{n \to \infty} \|A^n\|^{\frac{1}{n}} .$$

To conclude the proof, it suffices to show that

$$\limsup_{n \to \infty} \|A^n\|^{\frac{1}{n}} \leq r(A). \tag{8.4.5}$$

For $\varphi \in \mathcal{A}^*$, consider

$$f(z) = \varphi\big((zI - A)^{-1}\big), \tag{8.4.6}$$

which is analytic on $\mathbb{C}\backslash\sigma(A)$. For $|z| > \|A\|$, consider the series development

$$\varphi\big((zI - A)^{-1}\big) = \varphi\big(z^{-1}(I - z^{-1}A)^{-1}\big)$$

$$= \varphi\Big(z^{-1} \sum_{n=0}^{\infty} (z^{-1}A)^n\Big)$$

$$= z^{-1} \sum_{n=0}^{\infty} \varphi(z^{-n}A^n)$$

$$= z^{-1} \sum_{n=0}^{\infty} z^{-n}\varphi(A^n). \qquad (8.4.7)$$

This series agrees with f on $|z| > \|A\|$. Since f, as defined in (8.4.6), is analytic on $|z| > r(A)$, the series in (8.4.7) converges for all $|z| > r(A)$.

For each fixed $|z| > r(A)$, the convergence of (8.4.7) implies that the evaluation functionals

$$\widehat{z^{-n}A^n}(\varphi) = \varphi(z^{-n}A^n) \quad \text{for } n \geqslant 0,$$

which belong to $\mathcal{A}^{**}$, satisfy

$$\sup_{n \geqslant 0} |\widehat{z^{-n}A^n}(\varphi)| < \infty \quad \text{for all } \varphi \in \mathcal{A}^*.$$

Since $a \mapsto \hat{a}$ is a linear isometry from $\mathcal{A}$ to $\mathcal{A}^{**}$, the principle of uniform boundedness (Theorem 2.2.3) provides constants $C(z) > 0$ such that

$$\|z^{-n}A^n\|_{\mathcal{A}} = \big\|\widehat{z^{-n}A^n}\big\|_{\mathcal{A}^{**}} \leqslant C(z) \quad \text{for all } n \geqslant 0.$$

This yields $\|A^n\|^{\frac{1}{n}} \leqslant |z|C(z)^{\frac{1}{n}}$, so $\limsup_{n\to\infty} \|A^n\|^{\frac{1}{n}} \leqslant |z|$. Since this holds for all $|z| > r(A)$, the desired inequality (8.4.5) follows. ∎

8.5 Selfadjoint and Positive Operators

Recall that $A \in \mathcal{B}(\mathcal{H})$ is selfadjoint if $A = A^*$. Since real numbers are characterized by the condition $z = \bar{z}$, one often thinks of selfadjoint operators as the operator analogues of real numbers. We explore this connection further, along with positive operators, the analogues of nonnegative numbers. We begin with a few facts about selfadjoint operators.

Theorem 8.5.1. Let $A \in \mathcal{B}(\mathcal{H})$ be selfadjoint.

(a) $\sigma(A) \subseteq \mathbb{R}$.

(b) $\sigma(A) = \sigma_{ap}(A)$.

(c) $\|A\| = r(A)$.

(d) Eigenvectors of A corresponding to distinct eigenvalues are orthogonal.

Proof (a) Let $\lambda = \alpha + i\beta$, in which $\alpha, \beta \in \mathbb{R}$ and $\beta \neq 0$. Then, $A - \lambda I = B - i\beta I$, in which $B = A - \alpha I$ is selfadjoint. Thus,

$$
\begin{aligned}
\|(A - \lambda I)\mathbf{x}\|^2 &= \|B\mathbf{x} - i\beta\mathbf{x}\|^2 \\
&= \langle B\mathbf{x} - i\beta\mathbf{x}, B\mathbf{x} - i\beta\mathbf{x} \rangle \\
&= \langle B\mathbf{x}, B\mathbf{x} \rangle - i\beta \langle \mathbf{x}, B\mathbf{x} \rangle + i\beta \langle B\mathbf{x}, \mathbf{x} \rangle + |\beta|^2 \langle \mathbf{x}, \mathbf{x} \rangle \\
&= \|B\mathbf{x}\|^2 + i\beta \big(\langle B\mathbf{x}, \mathbf{x} \rangle - \langle B\mathbf{x}, \mathbf{x} \rangle \big) + |\beta|^2 \|\mathbf{x}\|^2 \\
&= \|(A - \alpha I)\mathbf{x}\|^2 + |\beta|^2 \|\mathbf{x}\|^2 \\
&\geqslant |\beta|^2 \|\mathbf{x}\|^2,
\end{aligned}
$$

and hence $A - \lambda I$ is bounded below. In particular, $\ker(A - \lambda I) = \{\mathbf{0}\}$ if $\operatorname{Im} \lambda \neq 0$. From Proposition 3.1.7 observe that

$$
(\operatorname{ran}(A - \lambda I))^- = (\ker(A - \lambda I)^*)^\perp = (\ker(A - \bar{\lambda} I))^\perp = \{\mathbf{0}\}^\perp = \mathcal{H}.
$$

Then $A - \lambda I$ is bounded below and has dense range, so it is invertible (Lemma 2.3.5). Therefore, $\sigma(A) \cap \{z : \operatorname{Im} z \neq 0\} = \varnothing$ and hence $\sigma(A) \subseteq \mathbb{R}$.

(b) It suffices to prove that $\sigma(A) \subseteq \sigma_{ap}(A)$; the reverse inclusion always holds (Proposition 2.4.6). Suppose that $\alpha \in \sigma(A)$ (note that $\alpha \in \mathbb{R}$ by (a)). Lemma 2.3.5 implies that $A - \alpha I$ is not bounded below or $A - \alpha I$ does not have dense range. If $A - \alpha I$ is not bounded below, then $\alpha \in \sigma_{ap}(A)$ by definition. If $A - \alpha I$ does not have dense range, then

$$
\mathcal{H} \neq (\operatorname{ran}(A - \alpha I))^- = (\ker(A - \alpha I)^*)^\perp = (\ker(A - \alpha I))^\perp,
$$

and thus $\ker(A - \alpha I) \neq \{\mathbf{0}\}$. In other words, $\alpha \in \sigma_p(A) \subseteq \sigma_{ap}(A)$.

(c) Observe that $\|A^2\| = \|A^*A\| = \|A\|^2$ (recall Proposition 3.1.5). Similarly,

$$
\|A^4\| = \|A^2 A^2\| = \|(A^2)^* A^2\| = \|A^2\|^2 = (\|A\|^2)^2 = \|A\|^4. \tag{8.5.2}
$$

Induction confirms that

$$
\|A^{2^k}\| = \|A\|^{2^k} \quad \text{for all } k \geqslant 0, \tag{8.5.3}
$$

and the spectral radius formula yields

$$
r(A) = \lim_{n \to \infty} \|A^n\|^{\frac{1}{n}} = \lim_{k \to \infty} \|A^{2^k}\|^{\frac{1}{2^k}} = \lim_{k \to \infty} \|A\| = \|A\|.
$$

(d) Suppose that $A\mathbf{x} = \lambda\mathbf{x}$ and $A\mathbf{y} = \mu\mathbf{y}$, in which $\mathbf{x}, \mathbf{y} \neq \mathbf{0}$ and $\lambda \neq \mu$ are real. Then,

$$
\lambda \langle \mathbf{x}, \mathbf{y} \rangle = \langle \lambda\mathbf{x}, \mathbf{y} \rangle = \langle A\mathbf{x}, \mathbf{y} \rangle = \langle \mathbf{x}, A\mathbf{y} \rangle = \langle \mathbf{x}, \mu\mathbf{y} \rangle = \mu \langle \mathbf{x}, \mathbf{y} \rangle,
$$

and hence $\langle \mathbf{x}, \mathbf{y} \rangle = 0$ because $\lambda \neq \mu$. ∎

Definition 8.5.4. Let $A \in \mathcal{B}(\mathcal{H})$.

(a) A is *positive*, denoted $A \geqslant 0$, if $\langle A\mathbf{x}, \mathbf{x} \rangle \geqslant 0$ for all $\mathbf{x} \in \mathcal{H}$.

(b) A is *strictly positive*, denoted $A > 0$, if $\langle A\mathbf{x}, \mathbf{x} \rangle > 0$ for all $\mathbf{x} \in \mathcal{H} \backslash \{\mathbf{0}\}$.

In linear algebra, the corresponding terms for matrices are *positive semidefinite* and *positive definite*, respectively. It is usually assumed, as part of the definition of positivity, that $A = A^*$. Theorem 8.5.8 (below) shows this assumption is unnecessary in a complex inner product space. The set of positive operators on $\mathcal{H}$ is closed under nonnegative linear combinations. That is, if $A, B \geqslant 0$ and $\alpha, \beta \geqslant 0$, then $\alpha A + \beta B \geqslant 0$ (Exercise 8.10.13).

Example 8.5.5. A diagonal operator D_Λ on ℓ^2 is positive if and only if $\Lambda \subseteq [0, \infty)$.

Proposition 8.5.6. *If $A, B \in \mathcal{B}(\mathcal{H})$ and $A \geqslant 0$, then $B^*AB \geqslant 0$. In particular, $T^*T \geqslant 0$ for any $T \in \mathcal{B}(\mathcal{H})$.*

Proof See Exercise 8.10.14. ∎

The next lemma is a uniqueness result that employs a polarization-type identity.

Lemma 8.5.7. *If $A \in \mathcal{B}(\mathcal{H})$ is selfadjoint and $\langle A\mathbf{x}, \mathbf{x} \rangle = 0$ for all $\mathbf{x} \in \mathcal{H}$, then $A = 0$.*

Proof Suppose that $A = A^*$ and $\langle A\mathbf{x}, \mathbf{x} \rangle = 0$ for all $\mathbf{x} \in \mathcal{H}$. Then

$$
\begin{aligned}
0 &= \langle A(\mathbf{x} + \mathbf{y}), \mathbf{x} + \mathbf{y} \rangle - \langle A(\mathbf{x} - \mathbf{y}), \mathbf{x} - \mathbf{y} \rangle \\
&= (\langle A\mathbf{x}, \mathbf{x} \rangle + \langle A\mathbf{x}, \mathbf{y} \rangle + \langle A\mathbf{y}, \mathbf{x} \rangle + \langle A\mathbf{y}, \mathbf{y} \rangle) \\
&\quad - (\langle A\mathbf{x}, \mathbf{x} \rangle - \langle A\mathbf{x}, \mathbf{y} \rangle - \langle A\mathbf{y}, \mathbf{x} \rangle + \langle A\mathbf{y}, \mathbf{y} \rangle) \\
&= 2(\langle A\mathbf{x}, \mathbf{y} \rangle + \langle A\mathbf{y}, \mathbf{x} \rangle) \\
&= 4 \operatorname{Re} \langle A\mathbf{x}, \mathbf{y} \rangle
\end{aligned}
$$

for all $\mathbf{x}, \mathbf{y} \in \mathcal{H}$. Let $\mathbf{y} = A\mathbf{x}$ and conclude that $\|A\mathbf{x}\|^2 = 0$ for all $\mathbf{x} \in \mathcal{H}$. ∎

The following selfadjointness criterion only holds for complex Hilbert spaces. Recall that each $T \in \mathcal{B}(\mathcal{H})$ has a Cartesian decomposition

$$
T = A + iB,
$$

in which $A, B \in \mathcal{B}(\mathcal{H})$ are selfadjoint; see (7.3.1).

Theorem 8.5.8. *If $T \in \mathcal{B}(\mathcal{H})$ and $\langle T\mathbf{x}, \mathbf{x} \rangle \in \mathbb{R}$ for all $\mathbf{x} \in \mathcal{H}$, then $T = T^*$.*

Proof If $\langle T\mathbf{x}, \mathbf{x} \rangle \in \mathbb{R}$ for all $\mathbf{x} \in \mathcal{H}$, then

$$
\begin{aligned}
\langle (T - T^*)\mathbf{x}, \mathbf{x} \rangle &= \langle T\mathbf{x}, \mathbf{x} \rangle - \langle T^*\mathbf{x}, \mathbf{x} \rangle \\
&= \langle T\mathbf{x}, \mathbf{x} \rangle - \langle \mathbf{x}, T\mathbf{x} \rangle \\
&= \langle T\mathbf{x}, \mathbf{x} \rangle - \overline{\langle T\mathbf{x}, \mathbf{x} \rangle} \\
&= 2i \operatorname{Im} \langle T\mathbf{x}, \mathbf{x} \rangle \\
&= 0.
\end{aligned}
$$

Thus, $A = \frac{1}{2i}(T - T^*)$ is selfadjoint and satisfies $\langle A\mathbf{x}, \mathbf{x} \rangle = 0$ for all $\mathbf{x} \in \mathcal{H}$. The preceding lemma ensures that $A = 0$ and hence $T = T^*$. ∎

Lemma 8.5.9. *If $A, B \in \mathcal{B}(\mathcal{H})$ are positive and $A + B = 0$, then $A = B = 0$.*

Proof For all $\mathbf{x} \in \mathcal{H}$,

$$0 \leqslant \max\{\langle A\mathbf{x}, \mathbf{x}\rangle, \langle B\mathbf{x}, \mathbf{x}\rangle\} \leqslant \langle A\mathbf{x}, \mathbf{x}\rangle + \langle B\mathbf{x}, \mathbf{x}\rangle = \langle (A + B)\mathbf{x}, \mathbf{x}\rangle = 0.$$

Lemma 8.5.7 ensures that $A = B = 0$. ∎

Theorem 8.5.10. *Let $A \in \mathcal{B}(\mathcal{H})$. The following are equivalent.*

(a) *A is positive.*

(b) *A is selfadjoint and $\sigma(A) \subseteq [0, \infty)$.*

(c) *$A = B^*B$ for some $B \in \mathcal{B}(\mathcal{H})$.*

Proof (a) $\Rightarrow$ (b) If $A \geqslant 0$, then Theorem 8.5.8 ensures that $A = A^*$, so $\sigma(A) = \sigma_{ap}(A)$ (Theorem 8.5.1). If $\lambda \in \sigma(A)$, then there exists a sequence of unit vectors $(\mathbf{x}_n)_{n=1}^{\infty}$ such that $(A - \lambda I)\mathbf{x}_n \to \mathbf{0}$. Since the $\mathbf{x}_n$ are unit vectors we see that

$$|\langle (\lambda I - A)\mathbf{x}_n, \mathbf{x}_n\rangle| \leqslant \|(\lambda I - A)\mathbf{x}_n\| \|\mathbf{x}_n\| = \|(\lambda I - A)\mathbf{x}_n\| \to 0 \quad \text{as } n \to \infty.$$

Thus,

$$\begin{aligned}
\lambda = \lambda \|\mathbf{x}_n\|^2 &= \lambda \langle \mathbf{x}_n, \mathbf{x}_n\rangle = \langle \lambda \mathbf{x}_n, \mathbf{x}_n\rangle \\
&= \langle (\lambda I - A)\mathbf{x}_n, \mathbf{x}_n\rangle + \langle A\mathbf{x}_n, \mathbf{x}_n\rangle \\
&\geqslant \langle (\lambda I - A)\mathbf{x}_n, \mathbf{x}_n\rangle + 0 \qquad &(\text{since } \langle A\mathbf{x}_n, \mathbf{x}_n\rangle \geqslant 0) \\
&= \langle (\lambda I - A)\mathbf{x}_n, \mathbf{x}_n\rangle \to 0.
\end{aligned}$$

Therefore, $\lambda \geqslant 0$.

(b) $\Rightarrow$ (c) This follows from Theorem 8.6.4 below.

(c) $\Rightarrow$ (a) If $A = B^*B$, then $\langle A\mathbf{x}, \mathbf{x}\rangle = \langle B^*B\mathbf{x}, \mathbf{x}\rangle = \langle B\mathbf{x}, B\mathbf{x}\rangle = \|B\mathbf{x}\|^2 \geqslant 0$. ∎

Example 8.5.11. Notice that the Volterra operator V is not a positive operator yet $\sigma(V) = \{0\} \subseteq [0, \infty)$ (Proposition 7.2.5). Thus, the selfadjointness hypothesis in statement (b) of Theorem 8.5.10 is crucial.

8.6 Continuous Functional Calculus

Our presentation follows [284]. Lemma 8.4.1 implies that $p(\sigma(A)) = \sigma(p(A))$ for all $p \in \mathbb{C}[z]$ and $A \in \mathcal{B}(\mathcal{H})$. For selfadjoint operators, this can be pushed much further. For $p \in \mathbb{C}[z]$, let

$$\widetilde{p}(z) = \overline{p(\bar{z})}.$$

In other words, if $p(z) = a_0 + a_1 z + a_2 z^2 + \cdots + a_k z^k$, then $\widetilde{p}(z) = \overline{a_0} + \overline{a_1}z + \overline{a_2}z^2 + \cdots + \overline{a_k}z^k$.

Lemma 8.6.1. *If $A \in \mathcal{B}(\mathcal{H})$ is selfadjoint and $p \in \mathbb{C}[z]$, then*

$$\|p(A)\| = \sup_{\lambda \in \sigma(A)} |p(\lambda)|.$$

Proof If $p(z) = \sum_{k=0}^{n} c_k z^k$ and $A = A^*$, then

$$p(A)^* = \left(\sum_{k=0}^{n} c_k A^k \right)^* = \sum_{k=0}^{n} \overline{c_k} A^k = \tilde{p}(A)$$

and

$$\tilde{p}(A)p(A) = \left(\sum_{j=0}^{n} \overline{c_j} A^j \right) \left(\sum_{k=0}^{n} c_k A^k \right) = \sum_{j,k=0}^{n} \overline{c_j} c_k A^{j+k} = (\tilde{p}p)(A),$$

in which $(\tilde{p}p)(A)$ is selfadjoint. Consequently,

$$
\begin{aligned}
\|p(A)\|^2 &= \|p(A)^* p(A)\| && \text{(by Proposition 3.1.5)} \\
&= \|\tilde{p}(A)p(A)\| \\
&= \|(\tilde{p}p)(A)\| \\
&= r((\tilde{p}p)(A)) && \text{(by Theorem 8.5.1)} \\
&= \sup_{\lambda \in \sigma(\tilde{p}p(A))} |\lambda| \\
&= \sup_{\lambda \in \sigma(A)} |(\tilde{p}p)(\lambda)| && \text{(by Lemma 8.4.1)} \\
&= \sup_{\lambda \in \sigma(A)} \left| \sum_{j,k=0}^{n} \overline{c_j} c_k \lambda^{j+k} \right| \\
&= \sup_{\lambda \in \sigma(A)} \left| \sum_{j,k=0}^{n} \overline{c_j} c_k \overline{\lambda^j} \lambda^k \right| && \text{(since } \sigma(A) \subseteq \mathbb{R}) \\
&= \sup_{\lambda \in \sigma(A)} |p(\lambda)|^2,
\end{aligned}
$$

which completes the proof. ∎

The next theorem extends the polynomial spectral mapping theorem (Lemma 8.4.1) to a much larger class of functions.

Theorem 8.6.2 (Continuous functional calculus). *Let $A \in \mathcal{B}(\mathcal{H})$ be selfadjoint and let $X = \sigma(A)$. There is a unique map $\Phi : C(X) \to \mathcal{B}(\mathcal{H})$ satisfying the following.*

(a) *Φ is a $*$-homomorphism, in the sense that for all $f, g \in C(X)$, the following hold.*

    (i) *Φ is linear.*

    (ii) *$\Phi(fg) = \Phi(f)\Phi(g)$.*

    (iii) *$\Phi(1) = I$.*

    (iv) *$\Phi(\overline{f}) = \Phi(f)^*$.*

(b) Φ *is an isometry.*

(c) $\Phi(z) = A$, *where z denotes the identity function on X.*

For $f \in C(X)$, define $f(A) = \Phi(f)$.

(d) *If $f \geq 0$, then $f(A) \geq 0$.*

(e) $\sigma(f(A)) = f(\sigma(A))$.

(f) $\|f(A)\| = \sup\limits_{\lambda \in \sigma(A)} |f(\lambda)|$.

Proof Since $\sigma(A) \subseteq \mathbb{R}$, the Weierstrass approximation theorem (Theorem 8.1.2) ensures that $\mathbb{C}[z]$ is dense in $C(X)$. If $\Phi, \Phi' : C(X) \to \mathcal{B}(\mathcal{H})$ are linear maps that satisfy (a), (b), and (c), then Φ and Φ' agree on the polynomials and hence on $C(X)$ by continuity. This takes care of uniqueness. We now prove existence.

Recall the notation

$$\|g\|_\infty = \sup_{z \in X} |g(z)| \quad \text{for } g \in C(X).$$

For $p \in \mathbb{C}[z]$, define $\Phi(p) = p(A)$ and note that $\Phi(pq) = \Phi(p)\Phi(q)$ for all $p, q \in \mathbb{C}[z]$. Lemma 8.6.1 implies that

$$\|\Phi(p)\|_{\mathcal{B}(\mathcal{H})} = \|p(A)\|_{\mathcal{B}(\mathcal{H})} = \|p\|_\infty. \tag{8.6.3}$$

The next step is to extend Φ to a map $\Phi : C(X) \to \mathcal{B}(\mathcal{H})$ that satisfies (a), (b), (c). To do this, suppose $(p_n)_{n=1}^\infty$ is a sequence of polynomials such that $p_n \to f$ in $C(X)$. Then $(p_n)_{n=1}^\infty$ is a Cauchy sequence in $C(X)$. Thus, $(\Phi(p_n))_{n=1}^\infty$ is a Cauchy sequence in $\mathcal{B}(\mathcal{H})$ by (8.6.3). By the completeness of $\mathcal{B}(\mathcal{H})$ (Proposition 2.1.14), $\Phi(p_n) \to T$ for some $T \in \mathcal{B}(\mathcal{H})$. Suppose that $(q_n)_{n=1}^\infty$ is another sequence of polynomials that converges in $C(X)$ to f, so that $\Phi(q_n) \to T'$ for some $T' \in \mathcal{B}(\mathcal{H})$. For all $n \geq 1$,

$$\|T - T'\| \leq \|T - \Phi(p_n)\| + \|\Phi(p_n) - \Phi(q_n)\| + \|\Phi(q_n) - T'\|.$$

Then

$$\limsup_{n \to \infty} (\|T - \Phi(p_n)\| + \|\Phi(p_n) - \Phi(q_n)\| + \|\Phi(q_n) - T'\|)$$

$$= \lim_{n \to \infty} \|T - \Phi(p_n)\| + \limsup_{n \to \infty} \|\Phi(p_n) - \Phi(q_n)\| + \lim_{n \to \infty} \|\Phi(q_n) - T'\|$$

$$= 0 + \limsup_{n \to \infty} \|p_n - q_n\|_\infty + 0$$

$$\leq \limsup_{n \to \infty} (\|p_n - f\|_\infty + \|f - q_n\|_\infty)$$

$$= 0.$$

It follows that $T = T'$, and therefore $\Phi(f) = T$.

Now we need to verify (a), (b), and (c). Let $(p_n)_{n=1}^{\infty}$ and $(q_n)_{n=1}^{\infty}$ be sequences in $\mathbb{C}[z]$ such that $p_n \to f$ and $q_n \to g$ in $C(X)$. Then, $p_n q_n \to fg$ in $C(X)$, so the continuity of Φ implies that

$$\Phi(fg) = \lim_{n\to\infty} \Phi(p_n q_n) = \lim_{n\to\infty} \Phi(p_n)\Phi(q_n) = \Phi(f)\Phi(g).$$

This verifies (a). The proofs of (b) and (c) are requested in Exercise 8.10.16. If $f \geqslant 0$, then $f = g^2$ for some real valued $g \in C(X)$. Then

$$f(A) = \Phi(f) = \Phi(g^2) = \Phi(g)\Phi(g) = \Phi(\overline{g})\Phi(g) = \Phi(g)^*\Phi(g) \geqslant 0,$$

which verifies (d). The proofs of (e) and (f) are requested in Exercise 8.10.17. ∎

A nonnegative real number has a unique nonnegative square root. The analogue of this is true for positive operators.

Theorem 8.6.4. *If $A \in \mathcal{B}(\mathcal{H})$ and $A \geqslant 0$, then there exists a unique $B \geqslant 0$ such that $A = B^2$.*

Proof If $A \geqslant 0$, then $f(z) = \sqrt{z}$ is continuous on $\sigma(A)$ since $\sigma(A) \subseteq [0, \infty)$ (Theorem 8.5.10). The previous theorem ensures that $B = f(A) \geqslant 0$ and

$$B^2 = f(A)f(A) = \Phi(f)\Phi(f) = \Phi(f^2) = \Phi(z) = A.$$

This verifies the existence of a nonnegative square root.

For the uniqueness, observe that if $C \geqslant 0$ and $C^2 = A$, then $CA = C^3 = AC$, so C commutes with A. It follows that $p(A)C = Cp(A)$ for every $p \in \mathbb{C}[z]$. Since $\sigma(A)$ is a compact subset of $[0, \infty)$, the Weierstrass approximation theorem yields a sequence $(p_n)_{n=1}^{\infty}$ in $\mathbb{C}[z]$ that converges uniformly to $\sqrt{z}$ on $\sigma(A)$. Hence,

$$\begin{aligned}
\|CB - BC\| &= \|Cf(A) - f(A)C\| \\
&= \|Cf(A) - Cp_n(A) + p_n(A)C - f(A)C\| \\
&\leqslant \|C\| \|(f - p_n)(A)\| + \|(p_n - f)(A)\| \|C\| \\
&= 2\|C\| \|f - p_n\|_{\infty} \to 0.
\end{aligned}$$

Thus, C commutes with B and hence

$$\begin{aligned}
(B - C)^*B(B - C) &+ (B - C)^*C(B - C) \\
&= (B - C)\big[B(B - C) + C(B - C)\big] \\
&= (B - C)(B + C)(B - C) \\
&= (B^2 - C^2)(B - C) \\
&= (A - A)(B - C) = 0.
\end{aligned}$$

Since $(B - C)^*B(B - C) \geqslant 0$ and $(B - C)^*C(B - C) \geqslant 0$ (Proposition 8.5.6), Lemma 8.5.9 ensures that they are both zero. Consequently,

$$0 = (B - C)^*B(B - C) - (B - C)^*C(B - C)$$

$$= (B - C)[B(B - C) - C(B - C)]$$
$$= (B - C)(B^2 - BC - CB + C^2)$$
$$= (B - C)^3, \qquad\qquad \text{(since } BC = CB)$$

so $\|B - C\|^4 = \|(B-C)^4\| = 0$ since $B - C$ is selfadjoint (see (8.5.2)). Thus, $B = C$. ∎

8.7 The Spectral Theorem

The following theorem says that every cyclic selfadjoint operator is unitarily equivalent to a multiplication operator. It is one of the main ways in which measure theory connects with the study of Hilbert-space operators.

Theorem 8.7.1. *Let $A \in \mathcal{B}(\mathcal{H})$ be selfadjoint and let $\mathbf{x}$ be a cyclic vector for A. Then there is a finite positive Borel measure $\mu_\mathbf{x}$ with $\mathrm{supp}(\mu_\mathbf{x}) = \sigma(A)$ and a unitary $U : \mathcal{H} \to L^2(\sigma(A), \mu_\mathbf{x})$ that satisfy the following.*

(a) $(UAU^*f)(z) = zf(z)$ *for all $f \in L^2(\sigma(A), \mu_\mathbf{x})$.*

(b) *$U\mathbf{x}$ is the constant function 1 on $\sigma(A)$.*

(c) *1 is a cyclic vector for M_z on $L^2(\sigma(A), \mu_\mathbf{x})$.*

Proof For $f \in C(\sigma(A))$, the continuous functional calculus (Theorem 8.6.2) ensures that $f(A)$ is well defined and $\|f(A)\| = \|f\|_\infty$. Moreover, $f(A) \geqslant 0$ whenever $f \geqslant 0$. Thus,

$$\varphi(f) = \langle f(A)\mathbf{x}, \mathbf{x} \rangle$$

is well defined on $C(\sigma(A))$ and

$$|\varphi(f)| \leqslant \|f(A)\|\|\mathbf{x}\| \leqslant \|f\|_\infty \|\mathbf{x}\|.$$

In other words, φ is a bounded linear functional on $C(\sigma(A))$. If $f \geqslant 0$, then $f(A) \geqslant 0$ and hence $\varphi(f) \geqslant 0$, that is, φ is a positive linear functional. The Riesz representation theorem for positive bounded linear functionals on $C(\sigma(A))$ [319, p. 42] provides a unique finite positive Borel measure $\mu_\mathbf{x}$ on $\sigma(A)$ such that

$$\varphi(f) = \int f d\mu_\mathbf{x} \quad \text{for all } f \in C(\sigma(A)).$$

We claim that the map

$$f(A)\mathbf{x} \mapsto f, \qquad\qquad (8.7.2)$$

defined initially for $f \in C(\sigma(A))$, extends to a unitary operator from $\mathcal{H}$ onto $L^2(\sigma(A), d\mu_\mathbf{x})$. First note that the image of $\mathbf{x}$ under this map is the constant function 1 in $L^2(\sigma(A), \mu_\mathbf{x})$. Next recall the map $\Phi(f) = f(A)$ from Theorem 8.6.2. Since

$$\|f(A)\mathbf{x}\|^2 = \langle \Phi(f)\mathbf{x}, \Phi(f)\mathbf{x} \rangle$$
$$= \langle \Phi(f)^*\Phi(f)\mathbf{x}, \mathbf{x} \rangle$$

$$= \langle \Phi(\overline{f}) \Phi(f) \mathbf{x}, \mathbf{x} \rangle$$
$$= \langle \Phi(|f|^2) \mathbf{x}, \mathbf{x} \rangle$$
$$= \langle |f|^2 (A) \mathbf{x}, \mathbf{x} \rangle$$
$$= \int_{\sigma(A)} |f|^2 \, d\mu_{\mathbf{x}}$$
$$= \|f\|^2_{L^2(\sigma(A), \mu_{\mathbf{x}})},$$

it follows that the map from (8.7.2) is well defined and isometric on $C(\sigma(A))$. Since $\mathbf{x}$ is a cyclic vector for A, it follows that $\{f(A)\mathbf{x} : f \in C(\sigma(A))\}^- = \mathcal{H}$, so (8.7.2) extends to an isometry $U : \mathcal{H} \to L^2(\sigma(A), \mu_{\mathbf{x}})$. Note that

$$C(\sigma(A)) \subseteq \operatorname{ran} U \subseteq L^2(\sigma(A), \mu_{\mathbf{x}}).$$

The range of an isometry is closed (Exercise 4.5.9) and $C(\sigma(A))$ is dense in $L^2(\sigma(A), \mu_{\mathbf{x}})$ (Proposition 1.3.6). Therefore, $\operatorname{ran} U = L^2(\sigma(A), \mu_{\mathbf{x}})$, and hence U is a unitary operator from $\mathcal{H}$ onto $L^2(\sigma(A), \mu_{\mathbf{x}})$.

If $f \in C(\sigma(A))$, use the fact that $\Phi(z) = A$ to conclude that

$$UAU^* f = UA(f(A)\mathbf{x}) = U[(zf)(A)\mathbf{x}] = zf = M_z f,$$

the operator of multiplication by z on $L^2(\sigma(A), \mu_{\mathbf{x}})$. Since $C(\sigma(A))$ is dense in $L^2(\sigma(A), \mu_{\mathbf{x}})$, it follows that $UAU^* = M_z$, which completes the proof. ∎

Below is a commutative diagram that illustrates the spectral theorem.

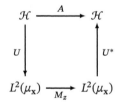

Not every selfadjoint operator is cyclic; consider the identity operator on a Hilbert space of dimension two or more. In the general setting, the *cyclic subspace* generated by a given $\mathbf{x} \in \mathcal{H}$ and $A \in \mathcal{B}(\mathcal{H})$ is

$$[\mathbf{x}] = \bigvee \{\mathbf{x}, A\mathbf{x}, A^2\mathbf{x}, \ldots\}.$$

This is a closed subspace of $\mathcal{H}$ which is A-invariant since A is bounded and

$$A[\mathbf{x}] = A\left(\bigvee \{\mathbf{x}, A\mathbf{x}, A^2\mathbf{x}, \ldots\}\right) = \bigvee \{A\mathbf{x}, A^2\mathbf{x}, A^3\mathbf{x}, \ldots\} \subseteq [\mathbf{x}].$$

The next lemma follows from the definition of selfadjointness and the definition of a cyclic subspace (Exercise 8.10.32).

Lemma 8.7.3. *Let $A \in \mathcal{B}(\mathcal{H})$ be selfadjoint and $\mathbf{x} \in \mathcal{H}$. Then the restriction of A to $[\mathbf{x}]$ is a cyclic selfadjoint operator with cyclic vector $\mathbf{x}$.*

Zorn's lemma and Theorem 8.7.1 provide us with the following theorem. See Chapter 14 for a formal treatment of direct sums of Hilbert spaces and operators.

Theorem 8.7.4. *Let $A \in \mathcal{B}(\mathcal{H})$ be a selfadjoint operator on a separable Hilbert space $\mathcal{H}$. Then there is an orthogonal decomposition*

$$\mathcal{H} = \bigoplus_{n=1}^{N} \mathcal{H}_n,$$

in which $N \in \mathbb{N} \cup \{\infty\}$, such that the following hold.

(a) *Each $\mathcal{H}_n$ is A-invariant.*

(b) *There is an $\mathbf{x}_n \in \mathcal{H}_n$ which is cyclic for $A|_{\mathcal{H}_n}$, that is, $\mathcal{H}_n = \{p(A)\mathbf{x}_n : p \in \mathbb{C}[x]\}^-$.*

(c) *There exist finite positive Borel measures μ_n on $\mathbb{R}$ with $\mathrm{supp}(\mu_n) \subseteq \sigma(A)$ and a unitary operator*

$$U : \mathcal{H} \to \bigoplus_{n=1}^{N} L^2(\mathbb{R}, \mu_n)$$

such that $(UAU^*f)_n = zf_n$, where we write $f \in \bigoplus_{n=1}^{N} L^2(\mathbb{R}, \mu_n)$ as $f = (f_j)_{j=1}^{N}$. Moreover,

$$U\mathbf{x}_n = (0, 0, ..., 0, 1, 0, ...),$$

in which $U\mathbf{x}_n$ has the constant function 1 in the nth component.

The issue of uniqueness is subtle and handled with *multiplicity theory* [24, 94], which is beyond the scope of this book. The previous theorem permits most proofs concerning selfadjoint operators to start with the phrase "without loss of generality, suppose that $A = M_z$ on $L^2(X, \mu)$, where μ is a finite positive Borel measure on a compact $X \subseteq \mathbb{R}$."

There is a more general spectral theorem for normal operators (see Theorem 19.2.3). In that theorem, one replaces "cyclic vector" with "$*$-cyclic vector," meaning that one considers orbits under nonnegative powers of the normal operator N and its adjoint N^* simultaneously. The spectrum $\sigma(N)$ of a normal operator N is a compact subset of $\mathbb{C}$; hence one considers compactly supported, finite, positive, Borel measures on $\mathbb{C}$.

8.8 Revisiting Diagonal Operators

For a bounded sequence $\Lambda = (\lambda_n)_{n=0}^{\infty}$ of complex numbers, recall from Chapter 2 the diagonal operator D_Λ acting on ℓ^2 by

$$D_\Lambda \mathbf{e}_n = \lambda_n \mathbf{e}_n \quad \text{for } n \geqslant 0.$$

Note that $D_\Lambda^* \mathbf{e}_n = \overline{\lambda_n} \mathbf{e}_n$, and hence D_Λ commutes with its adjoint, that is, D_Λ is a normal operator. Here is a spectral theorem for D_Λ.

Theorem 8.8.1. *Suppose* $\Lambda = (\lambda_n)_{n=0}^{\infty}$ *is a bounded sequence of distinct complex numbers. If* μ *is the discrete measure on* $\mathbb{C}$ *defined by*

$$\mu = \sum_{n=0}^{\infty} \frac{1}{(n+1)^2} \delta_{\lambda_n},$$

then D_{Λ} *is unitarily equivalent to* M_z *on* $L^2(\mu)$.

Proof Define $U : \ell^2 \to L^2(\mu)$ by

$$Uc = \sum_{n=0}^{\infty} (n+1)c_n \chi_{\{\lambda_n\}}, \quad \text{where } c = (c_n)_{n=0}^{\infty},$$

and note that

$$\|Uc\|_{L^2(\mu)}^2 = \sum_{n=0}^{\infty} |c_n|^2 (n+1)^2 \mu(\{\lambda_n\})$$

$$= \sum_{n=0}^{\infty} |c_n|^2 (n+1)^2 \frac{1}{(n+1)^2}$$

$$= \sum_{n=0}^{\infty} |c_n|^2$$

$$= \|c\|_{\ell^2}^2,$$

so U is isometric. If $f \in L^2(\mu)$, then

$$f = \sum_{n=0}^{\infty} f(\lambda_n)\chi_{\{\lambda_n\}} \quad \text{and} \quad \sum_{n=0}^{\infty} \frac{|f(\lambda_n)|^2}{(n+1)^2} < \infty.$$

If

$$c_n = \frac{f(\lambda_n)}{n+1} \quad \text{for } n \geq 0,$$

then $c = (c_n)_{n=0}^{\infty} \in \ell^2$ and $Uc = f$. Thus, U is isometric and surjective, hence unitary. Finally,

$$UD_{\Lambda}e_n = U(\lambda_n e_n) = \lambda_n Ue_n = \lambda_n(n+1)\chi_{\{\lambda_n\}} = M_z Ue_n.$$

Thus, D_{Λ} is unitarily equivalent to M_z on $L^2(\mu)$. ∎

If $(w_n)_{n=0}^{\infty}$ is any positive summable sequence, then

$$\mu = \sum_{n=0}^{\infty} w_n \delta_{\lambda_n}$$

is also a measure such that D_{Λ} is unitarily equivalent to M_z on $L^2(\mu)$. The theorem above has some important consequences.

Corollary 8.8.2. *If Λ is a bounded sequence of distinct complex numbers, then D_Λ is cyclic.*

Proof Theorem 8.8.1 says that D_Λ is unitarily equivalent to M_z on $L^2(\mu)$. Note that D_Λ is cyclic if and only if M_z is cyclic (Exercise 8.10.30). By Proposition 8.2.7, M_z is $*$-cyclic. Now apply Bram's theorem (Theorem 8.2.12) to see that M_z is cyclic. ∎

One can also prove the corollary above by using Theorem 8.2.8. Here are a few examples of invariant subspaces for diagonal operators.

Example 8.8.3. For any bounded sequence Λ and $E \subseteq \mathbb{N}_0$, consider

$$\ell_E^2 = \{\mathbf{a} = (a_n)_{n=0}^\infty \in \ell^2 : a_n = 0 \text{ for } n \in E\}.$$

One can show that ℓ_E^2 is a D_Λ-invariant subspace. Note that $\ell_\varnothing^2 = \ell^2$ and $\ell_{\mathbb{N}_0}^2 = \{0\}$. Also note that ℓ_E^2 is invariant for D_Λ^* and hence is a reducing subspace for D_Λ.

Is every invariant subspace for D_Λ of this form? In general, the answer is no.

Example 8.8.4. Suppose $\Lambda = (1, 1, \lambda_2, \lambda_3, ...)$, where $(\lambda_n)_{n=2}^\infty$ is a bounded sequence. Then

$$\mathcal{M} = \{(c, -c, 0, 0, ...) : c \in \mathbb{C}\}$$

is an invariant subspace for D_Λ that is not of the form ℓ_E^2 for any $E \subseteq \mathbb{N}_0$.

One might think that the invariant subspaces for D_Λ that are not of the form ℓ_E^2 arise from repetitions in the sequence Λ. In fact, one can create a bounded sequence Λ of distinct complex numbers for which there exists an invariant subspace of D_Λ that is not of the form ℓ_E^2 (see Exercise 8.10.36). Here is a positive result about reducing subspaces.

Theorem 8.8.5. *Suppose Λ is a bounded sequence of distinct complex numbers. Then every reducing subspace for D_Λ is of the form ℓ_E^2 for some $E \subseteq \mathbb{N}_0$.*

Proof If

$$U\mathbf{c} = \sum_{n=0}^\infty (n+1)c_n \chi_{\{\lambda_n\}}$$

is the unitary operator from Theorem 8.8.1 and $\mathcal{M}$ is a reducing subspace for D_Λ, then $\mathcal{F} = U\mathcal{M}$ is reducing for M_z on $L^2(\mu)$. If P is the orthogonal projection of $L^2(\mu)$ onto $\mathcal{F}$, then Exercise 4.5.16 says that $PM_z = M_zP$. Corollary 8.3.3, which describes the commutant of M_z, says that $P = M_\varphi$ for some $\varphi \in L^\infty(\mu)$.
Since P is an orthogonal projection, it is selfadjoint and idempotent. Thus, $M_\varphi^* = M_\varphi$ and hence φ is real valued (Proposition 8.1.14). Now use the fact that M_φ is idempotent to see that $M_{\varphi^2} = M_\varphi^2 = M_\varphi$. Corollary 8.1.9 now yields $\varphi^2 = \varphi$. Since μ is discrete, μ-almost everywhere (normally needed in the statements above) means everywhere. Since $\varphi^2 = \varphi$ and φ is real valued, it follows that $\varphi = \chi_W$ for some $W \subseteq \Lambda$. Thus, $\mathcal{F} = PL^2(\mu) = \chi_W L^2(\mu) = \{f \in L^2(\mu) : f|_{W^c} = 0\}$. The formula for U ensures that if $E = \{n : \lambda_n \in W^c\}$, then $\mathcal{M} = U^{-1}\mathcal{F} = \ell_E^2$. ∎

Corollary 8.8.6. *If Λ is a bounded sequence of distinct real numbers, then every invariant subspace of D_Λ is of the form ℓ_E^2 for some $E \subseteq \mathbb{N}_0$.*

The proof of Theorem 8.8.5 yields the following more general result.

Corollary 8.8.7. *If μ is a finite positive Borel measure with compact support in $\mathbb{C}$, then each reducing subspace for M_z on $L^2(\mu)$ is of the form $\chi_W L^2(\mu)$ for some $W \subseteq \mathbb{C}$. If $\mathrm{supp}(\mu) \subseteq \mathbb{R}$, then every invariant subspace for M_z is of the form $\chi_W L^2(\mu)$ for some $W \subseteq \mathbb{R}$.*

Note that the previous corollary generalizes Theorem 4.3.3.

What is special about the invariant subspaces ℓ_E^2? Observe that when $n \notin E$, then $\mathbf{e}_n \in \ell_E^2$ and $\mathbf{e}_n$ is an eigenvector for D_Λ. Furthermore, $\ell_E^2 = \bigvee \{\mathbf{e}_n : n \notin E\}$. In other words, ℓ_E^2 is an invariant subspace consisting of the closed linear span of the eigenvectors contained in it. Such invariant subspaces have the *spectral synthesis property*. In Exercise 8.10.36, we give an example of a cyclic diagonal operator that has invariant subspaces without the spectral synthesis property.

8.9 Notes

The spectral theorem for selfadjoint operators was developed by Hilbert [196], Riesz [301], and Hellinger [190].

There are other versions of the spectral theorem that correspond to the familiar one for selfadjoint matrices. For a selfadjoint $n \times n$ matrix A, list the distinct eigenvalues $\lambda_1, \lambda_2, ..., \lambda_d$ and orthogonal projections $P_1, P_2, ..., P_d$ onto the corresponding eigenspaces $\ker(A - \lambda_j I)$, for $1 \leqslant j \leqslant d$. Then

$$A = \sum_{j=1}^{d} \lambda_j P_j.$$

Here

$$\sum_{j=1}^{d} P_j = I \quad \text{and} \quad P_i P_j = 0 \text{ for } i \neq j. \tag{8.9.1}$$

There is a version of this formula for general selfadjoint operators:

$$A = \int_{\sigma(A)} \lambda \, dP_\lambda,$$

where P_λ is a certain family of orthogonal projections known by Hilbert as *Zerlegung der Einheit* (a resolution of the identity) that enjoy a continuous analogue of (8.9.1). An excellent source for the spectral theorem for normal operators is Conway's book [94, Ch. 9].

There is a version of the spectral theorem for unbounded selfadjoint operators that arises in mathematical physics and the study of Schrödinger and Sturm–Liouville operators [283].

The multiplicity theory for normal operators, due to Hahn and Hellinger, determines when two normal operators are unitarily equivalent [1, 24, 222]. For a Borel measure ν with compact support in $\mathbb{C}$, consider the space $L^2(\nu, \ell^2)$ of $f : \mathbb{C} \to \ell^2$, that is, $f(\lambda) = (f_n(\lambda))_{n=1}^{\infty}$ such that each component function f_n is Borel measurable and

$$\|f\| = \left(\int \|f(\lambda)\|_{\ell^2}^2 \, d\nu(\lambda) \right)^{\frac{1}{2}} < \infty.$$

The corresponding inner product is

$$\langle f, g \rangle = \int \langle f(\lambda), g(\lambda) \rangle_{\ell^2} \, d\nu(\lambda).$$

Let ℓ_j^2 denote the set of sequences $\mathbf{a} = (a_k)_{k=0}^{\infty}$ such that $a_k = 0$ for $k > j$ and let $n : \mathbb{C} \to \mathbb{N} \cup \{\infty\}$ be a Borel function, the *multiplicity function*. Now consider the subspace $\mathscr{D}$ of all $f \in L^2(\nu, \ell^2)$ such that $f(\lambda) \in \ell_{n(\lambda)}^2$ ν-almost everywhere. The operator $M = M(\nu, n)$ on $\mathscr{D}$ defined by $(Mf)(\lambda) = \lambda f(\lambda)$ is bounded. Furthermore, $(M^*f)(\lambda) = \bar{\lambda} f(\lambda)$ and thus M is normal. The Hahn–Hellinger theorem says that if $N \in \mathcal{B}(\mathcal{H})$ is normal, then there is a measure ν and a multiplicity function n such that N is unitarily equivalent to M on $\mathscr{D}$. Furthermore, $M(\nu_1, n_1)$ is unitarily equivalent to $M(\nu_2, n_2)$ if and only if ν_1 and ν_2 have the same sets of measure zero and $n_1 = n_2$ ν_1-almost everywhere (or ν_2-almost everywhere). The papers [1, 222] give a recipe for computing ν and n for certain multiplication operators M_φ on $L^2(\mu)$. If the multiplicity function n is identically equal to 1, then $\mathscr{D} = L^2(\nu)$ and $(Mf)(\lambda) = \lambda f(\lambda)$ becomes what was presented in this chapter.

The spectral theorem can also be stated in terms of multiplication operators. If N is a normal operator, there is a measure space $(X, \mathcal{A}, \mu)$ and a $\varphi \in L^\infty(\mu)$ such that N is unitarily equivalent to M_φ on $L^2(\mu)$. In this generality, X is a compact Hausdorff space (not necessarily a subset of $\mathbb{C}$). If N is cyclic, then X can be taken to be a subset of $\mathbb{C}$ and $\varphi(z) = z$.

A description of the commutant of M_z on $L^2(m)$, where m is Lebesgue measure on $\mathbb{T}$, is given in [68].

8.10 Exercises

Exercise 8.10.1. Let $\varphi \in L^\infty(\mu)$, where μ is a finite positive Borel measure on a compact set $X \subseteq \mathbb{C}$. Recall the definition of $\mathscr{R}_\varphi$ from 8.1.11.

(a) Prove that $w \in \mathscr{R}_\varphi$ if and only if $\mu(\varphi^{-1}(\{z : |z - w| < r\})) > 0$ for all $r > 0$.

(b) Prove that $\mathscr{R}_\varphi \subseteq \varphi(X)^-$.

(c) If φ is continuous on X, prove that $\mathscr{R}_\varphi = \varphi(X)$.

(d) If $\psi \in L^\infty(\mu)$ and $\psi = \varphi$ μ-almost everywhere, then $\mathscr{R}_\psi = \mathscr{R}_\varphi$.

(e) $\|\varphi\|_\infty = \max\{|w| : w \in \mathscr{R}_\varphi\}$.

Exercise 8.10.2. For $\varphi \in L^\infty(\mu)$, prove that $\sigma_{ap}(M_\varphi) = \mathcal{R}_\varphi$ using the following steps. Let $M = M_\varphi$ and let $\lambda \in \sigma(M) \backslash \sigma_p(M)$.

(a) Use the normality of M to prove that $\|(M - \lambda I)f\| = \|(M^* - \bar{\lambda}I)f\|$ for all $f \in L^2(\mu)$.

(b) Prove that $M - \lambda I$ is injective.

(c) Prove that $M - \lambda I$ has dense range.

(d) Prove that $M - \lambda I$ is an invertible linear transformation from $L^2(\mu)$ onto its range but its inverse does not extend to a bounded operator on $L^2(\mu)$.

(e) Prove there exists a sequence $(g_n)_{n=1}^\infty$ in $\text{ran}(M - \lambda I)$ such that their pre-images under $M - \lambda I$ have unbounded norm.

(f) Use Proposition 8.1.12 and the above to prove that $\sigma_{ap}(M) = \sigma(M)$.

Exercise 8.10.3. For $\varphi \in L^\infty(\mu)$ and $\varepsilon > 0$, let $X_{\varphi,\varepsilon} = \{z : |\varphi(z)| > \varepsilon\}$.

(a) Prove that M_φ is compact on $L^2(\mu)$ if and only if $\chi_{\mathbb{C} \backslash X_{\varphi,\varepsilon}} L^2(\mu)$ is finite dimensional for every $\varepsilon > 0$.

(b) Suppose μ is a measure which has no atoms, in other words, $\mu(\{\lambda\}) = 0$ for each $\lambda \in \mathbb{C}$. If M_φ is compact, prove that φ is zero μ-almost everywhere.

Exercise 8.10.4. Let $\varphi \in L^\infty(\mu)$. Prove there is a sequence of invertible multiplication operators on $L^2(\mu)$ that converge to M_φ.

Exercise 8.10.5. Let μ and ν be finite positive Borel measures having compact support in $\mathbb{R}$. Let A denote M_x on $L^2(\mu)$ and let B denote M_x on $L^2(\nu)$. Show that A and B are unitarily equivalent if and only if μ and ν have the same sets of measure zero.

Exercise 8.10.6. Prove that the operator A on $L^2[0,1] \oplus L^2[0,1]$ defined by $A(f,g) = (M_x f, M_x g)$ has no cyclic vectors.

Exercise 8.10.7. Let I and J be intervals in $\mathbb{R}$ and suppose $\varphi : I \to J$ is differentiable and bijective with differentiable inverse. Show that $(Uf)(x) = f(\varphi(x))|\varphi'(x)|^{1/2}$ defines a unitary operator from $L^2(I)$ onto $L^2(J)$ such that $UM_x = M_\varphi U$, where M_φ is multiplication by φ on $L^2(J)$.

Exercise 8.10.8. Let V denote the Volterra operator from Chapter 7. Use Theorem 8.4.4 to prove that $\sigma(V) = \{0\}$.
Remark: See Proposition 7.2.5 for another proof.

Exercise 8.10.9. Prove that $A \in \mathcal{B}(\mathcal{H})$ is normal if and only if $\|Ax\| = \|A^*x\|$ for all $x \in \mathcal{H}$.

Exercise 8.10.10. Prove that if $A \in \mathcal{B}(\mathcal{H})$ is unitary, then $\sigma(A) \subseteq \mathbb{T}$. Is the converse true?

Exercise 8.10.11. If $N \in \mathcal{B}(\mathcal{H})$ is normal, prove that $\ker N = \ker N^*$ and $\text{ran } N = \text{ran } N^*$.

Exercise 8.10.12. Let $A \in \mathcal{B}(\mathcal{H})$ be normal.

(a) Prove that A is selfadjoint if and only if $\sigma(A) \subseteq \mathbb{R}$.

(b) Prove that A is an orthogonal projection if and only if $\sigma(A) \subseteq \{0, 1\}$.

(c) Prove that A is unitary if and only if $\sigma(A) \subseteq \mathbb{T}$.

Exercise 8.10.13. Prove that if $A, B \in \mathcal{B}(\mathcal{H})$, $A, B \geqslant 0$, and $\alpha, \beta \geqslant 0$, then $\alpha A + \beta B \geqslant 0$.

Exercise 8.10.14. Prove that if $A, B \in \mathcal{B}(\mathcal{H})$ and $A \geqslant 0$, then $B^* A B \geqslant 0$.

Exercise 8.10.15. Find a real inner product space $\mathcal{H}$ and an $A \in \mathcal{B}(\mathcal{H}) \backslash \{0\}$ such that $\langle A\mathbf{x}, \mathbf{x} \rangle = 0$ for all $\mathbf{x} \in \mathcal{H}$.

Exercise 8.10.16. Verify statements (b) and (c) of Theorem 8.6.2.

Exercise 8.10.17. Verify statements (e) and (f) of Theorem 8.6.2.

Exercise 8.10.18. Give an example of a finite positive Borel measure μ with $\text{supp}(\mu) \not\subseteq \mathbb{R}$ such that $\mathbb{C}[z]$ is dense in $L^2(\mu)$.

Exercise 8.10.19. Prove that the constant function 1 is not a cyclic vector for M_z on $L^2(dA)$, where dA is area measure on $\mathbb{D}$.

Exercise 8.10.20. Revisit the convolution operator $X_\mathbf{b}$ on $\ell^2(\mathbb{Z})$ from Exercise 3.6.30.

(a) Prove that $X_\mathbf{b}$ is a normal operator.

(b) Prove that $X_\mathbf{b}$ is unitarily equivalent to a multiplication operator M_ψ on an $L^2(\mu)$ space. Identify the measure μ and the symbol ψ.

Exercise 8.10.21. As we have seen in this chapter, the cyclic and $*$-cyclic vectors for M_z on $L^2(\mu)$ are not necessarily the same. Similarly, the invariant and reducing subspaces are not necessarily the same. However, in certain circumstances, they are. Lavrentiev [95] proved that if $K \subseteq \mathbb{C}$ is compact, $\mathbb{C} \backslash K$ is connected, and the interior of K is empty, then given any continuous function f on K and an $\varepsilon > 0$, there is a polynomial p such that

$$\max_{z \in K} |f(z) - p(z)| < \varepsilon.$$

(a) Prove that if the measure μ is supported on a compact set K with the properties above, then the cyclic and $*$-cyclic vectors for M_z on $L^2(\mu)$ are the same.

(b) Similarly, prove that the invariant and reducing subspaces for M_z on $L^2(\mu)$ are the same.

Exercise 8.10.22. Certain proofs in this chapter (Theorem 8.3.1 for example) used the concept of an operator-valued analytic function. The purpose of this exercise is to complement that discussion. Let V be a Banach space and let Ω be an open subset of $\mathbb{C}$. A function $f : \Omega \to V$ is *analytic* on Ω if

$$f'(z_0) = \lim_{z \to z_0} \frac{f(z) - f(z_0)}{z - z_0}$$

exists for every $z_0 \in \Omega$, and *weakly analytic* on Ω if

$$\lim_{z \to z_0} \frac{\varphi(f(z)) - \varphi(f(z_0))}{z - z_0}$$

exists for every $z_0 \in \Omega$ and every $\varphi \in V^*$.

(a) Prove that f is analytic if and only if f is weakly analytic.

(b) Prove that if f is analytic on Ω and $K \subseteq \Omega$ is compact, then $\sup\limits_{z \in K} \|f(z)\| < \infty$.

(c) If $z_0 \in \Omega$, f is analytic on Ω, and γ is a positively oriented simple closed continuous piecewise C^2 curve that contains z_0 in its interior, then

$$f(z_0) = \frac{1}{2\pi i} \oint_\gamma \frac{f(z)}{z - z_0} dz,$$

where the integral above is defined as the norm limit of its Riemann sums.

Exercise 8.10.23. Let $A, B \in \mathcal{B}(\mathcal{H})$.

(a) If $AB = BA$, prove that $e^{A+B} = e^A e^B$.

(b) If A is selfadjoint, prove that e^{iA} is unitary.

(c) If A is normal, prove that e^{A-A^*} is unitary.

(d) Prove that e^{zA} is an operator-valued entire function.

Exercise 8.10.24. Let $T, M, N \in B(\mathcal{H})$ be such that M and N are normal and $MT = TN$. Prove that $M^*T = TN^*$.
Remark: See also Theorem 14.2.10.

Exercise 8.10.25. If $A, B \in \mathcal{B}(\mathcal{H})$ are normal and $AB = BA$, prove that AB is also normal.

Exercise 8.10.26. If $A \in \mathcal{B}(\mathcal{H})$ is normal, prove that $A^* = f(A)$ for some continuous function f.

Exercise 8.10.27. Suppose $A \in \mathcal{B}(\mathcal{H})$ is a contraction. Prove that $(I - AA^*)^{\frac{1}{2}}A = A(I - A^*A)^{\frac{1}{2}}$ as follows.

(a) Let $S = (I - AA^*)^{\frac{1}{2}}$ and $T = (I - A^*A)^{\frac{1}{2}}$. Prove that $S^2A = AT^2$.

(b) Prove that $p(S^2)A = Ap(T^2)$ for all $p \in \mathbb{C}[z]$.

(c) Prove that $f(S^2)A = Af(T^2)$ for every $f \in C[0, 1]$.

(d) Choose an appropriate f to prove the desired identity.

Remark: The identity $(I - AA^*)^{\frac{1}{2}}A = A(I - A^*A)^{\frac{1}{2}}$ appears in the proof of Lemma 14.7.3, which concerns the Julia operator.

Exercise 8.10.28. $A, B \in \mathcal{B}(\mathcal{H})$ are *similar* if there an invertible $S \in \mathcal{B}(\mathcal{H})$ such that $AS = SB$. Use the steps below to show that if A and B are normal operators that are similar, then they are unitarily equivalent.

(a) Prove that $(SS^*)A(SS^*)^{-1} = A$.

(b) Write $S^* = UP$, where U is unitary and $P = (SS^*)^{\frac{1}{2}}$ (Theorem 14.9.15). Prove that $PAP^{-1} = A$.

(c) Prove that $B = S^*A(S^*)^{-1} = UPAP^{-1}U^*$.

(d) Prove that A is unitarily equivalent to B.

Exercise 8.10.29. The proof of Bram's theorem (Theorem 8.2.12) says that M_z on $L^2(\mu)$ is cyclic with a bounded cyclic vector f.

(a) Prove that $|f| > 0$ μ-almost everywhere.

(b) Consider the set $\mathscr{P}$ of products fg, where $g \in L^\infty(\mu)$ and $|g| > 0$ μ-almost everywhere. Prove that every $fg \in \mathscr{P}$ is cyclic for M_z.

(c) Prove that $\mathscr{P}$ is dense in $L^2(\mu)$.

Remark: See [129] for more on this.

Exercise 8.10.30. Suppose $A \in \mathcal{B}(\mathcal{H})$ is unitarily equivalent to $B \in \mathcal{B}(\mathcal{K})$. Prove that A is cyclic if and only if B is cyclic.

Exercise 8.10.31. Prove the *Hellinger–Toeplitz theorem*: if A is a linear transformation on a Hilbert space $\mathcal{H}$ such that $\langle Ax, y \rangle = \langle x, Ay \rangle$ for all $x, y \in \mathcal{H}$, then A is bounded.

Exercise 8.10.32. Let $A \in \mathcal{B}(\mathcal{H})$ be selfadjoint and $x \in \mathcal{H}$. Prove that the restriction of A to $[x]$, the A-invariant subspace generated by x, is a cyclic selfadjoint operator with cyclic vector x.

Exercise 8.10.33. Let $N \in \mathcal{B}(\mathcal{H})$ be normal.

(a) Prove that $\|(N^*N)^k\| = \|N^k\|^2$ for all $k \geq 0$.

(b) Use this to prove that $r(N) = \|N\|$, where $r(A)$ denotes the spectral radius of A.

Exercise 8.10.34. If $N \in \mathcal{B}(\mathcal{H})$ is normal and $\lambda \notin \sigma(N)$, prove that

$$\|(\lambda I - N)^{-1}\| = \frac{1}{\text{dist}(\lambda, \sigma(N))}.$$

Exercise 8.10.35. The *multiplicity function* for a diagonal operator D_Λ is $m_\Lambda(\lambda) = \dim \ker(D_\Lambda - \lambda I)$. Prove that two diagonal operators are unitarily equivalent if and only if their multiplicity functions are the same.

Exercise 8.10.36. This problem shows that if the eigenvalues of D_Λ are not all real, the invariant subspaces for a diagonal operator D_Λ can be complicated. Use the steps below to prove there is a sequence Λ of distinct complex numbers such that not every D_Λ-invariant subspace is of the form $\ell_E^2 = \{(a_n)_{n=0}^\infty \in \ell^2 : a_n = 0 \text{ for all } n \in E\}$ for some $E \subseteq \mathbb{N}_0$. This construction comes from [371].

(a) A result from [71] says there exists a sequence $\Lambda = (\lambda_n)_{n=0}^\infty$ of distinct points in $\mathbb{D}$ and an absolutely summable sequence $(w_n)_{n=0}^\infty$ of nonzero complex numbers such that

$$f(z) = \sum_{n=0}^\infty \frac{w_n}{1 - \lambda_n z} = 0 \quad \text{for all } z \in \mathbb{D}.$$

Assume this result and prove there are $\mathbf{a}, \mathbf{b} \in \ell^2$ such that $a_j \overline{b}_j = w_j$ for all $j \geq 0$.

(b) For $z \in \mathbb{D}$, prove that $f(z) = \sum_{N=0}^\infty z^N \left(\sum_{n=0}^\infty w_n \lambda_n^N \right)$.

(c) Use (a) and (b) to prove that $\langle D_\Lambda^N \mathbf{a}, \mathbf{b} \rangle = 0$ for all $N \geq 0$.

(d) Prove that $\mathbf{e}_n \notin \bigvee \{D_\Lambda^N \mathbf{a} : N \geq 0\}$ for all $n \geq 0$.

(e) Prove that the invariant subspace $\bigvee \{D_\Lambda^N \mathbf{a} : N \geq 0\}$ is not of the form ℓ_E^2 for any $E \subseteq \mathbb{N}_0$.

Remark: Constructions like the one in (a) date back to Borel [57, 58].

Exercise 8.10.37. Perform the decomposition in Theorem 8.7.4 for a diagonal operator D_Λ, where repetitions in Λ are allowed.

Exercise 8.10.38. Recall the weighted shift W on ℓ^2 defined by $W\mathbf{e}_n = \alpha_n \mathbf{e}_{n+1}$ from Exercise 3.6.21. If $\alpha_n \to 0$, prove that $\sigma(W) = \{0\}$.

Exercise 8.10.39. The notation $A^{1/2}$ refers to the unique positive square root of a positive $A \in \mathcal{B}(\mathcal{H})$ (Theorem 8.6.4). This exercise illustrates that "the square root of an operator" does not make sense in general. We have already seen this phenomenon in Exercise 5.9.33.

(a) Let

$$A = \begin{bmatrix} 0 & 1 \\ 0 & 0 \end{bmatrix}.$$

Prove that there does not exist any 2×2 matrix B such that $B^2 = A$.

(b) Let

$$A = \begin{bmatrix} 1 & 0 \\ 0 & -1 \end{bmatrix},$$

which is selfadjoint but not positive. Prove that there are precisely four 2×2 matrices B, none of which are positive, such that $B^2 = A$.

(c) Prove that there are infinitely many 2×2 matrices B such that $B^2 = 0$

Exercise 8.10.40. Prove that $A \in \mathcal{M}_n$ is a square root of the identity matrix $I \in \mathcal{M}_n$ if and only if $A = PDP^{-1}$, where $P \in \mathcal{M}_n$ is invertible and $D \in \mathcal{M}_n$ is diagonal with eigenvalues contained in $\{-1, 1\}$.

Exercise 8.10.41. Prove that if $A \in \mathcal{B}(\mathcal{H})$ is selfadjoint, then there is a $B \in \mathcal{B}(\mathcal{H})$ (not necessarily selfadjoint) such that $B^2 = A$.

8.11 Hints for the Exercises

Hint for Ex. 8.10.9: Consider the selfadjoint operator $S = AA^* - A^*A$.
Hint for Ex. 8.10.25: Write $C = AB$. Then

$$CC^* = (AB)(AB)^* = (AB)(BA)^* = (AB)(A^*B^*) = A(BA^*)B^*.$$

Hint for Ex. 8.10.28: For (a), use Fuglede's theorem.
Hint for Ex. 8.10.31: For $\|x\| \leq 1$, consider the bounded linear functional $\Lambda_x(\mathbf{y}) = \langle A\mathbf{x}, \mathbf{y} \rangle$ and consult the uniform boundedness principle.
Hint for Ex. 8.10.33: For (a), look at (8.5.3) and note that N^*N is selfadjoint.
Hint for Ex. 8.10.34: Use Exercise 8.10.33.

9

· · **·** · ·

The Dirichlet Shift

Key Concepts: Properties of the Dirichlet shift: norm, adjoint, spectrum, 2-isometry, commutant, invariant subspaces, cyclic vectors.

Outline: This chapter explores the operator-theoretic aspects of the shift operator $(M_z f)(z) = z f(z)$ on the Dirichlet space. In particular, we examine the 2-isometry property and discuss the invariant subspaces, cyclic vectors, and commutant of the Dirichlet shift.

9.1 The Dirichlet Space

For an analytic function f on $\mathbb{D}$, the *Dirichlet integral* of f is

$$D(f) := \frac{1}{\pi} \int_{\mathbb{D}} |f'(z)|^2 \, dA(z), \tag{9.1.1}$$

where dA is planar Lebesgue area measure. The Dirichlet integral has an appealing geometric interpretation: it is $1/\pi$ times the area of $f(\mathbb{D})$, counting multiplicity (see Exercises 9.9.3 and 9.9.4). Polar coordinates provide the following useful formula for the Dirichlet integral (see Exercise 9.9.1).

Proposition 9.1.2. *If $f(z) = \sum_{n=0}^{\infty} a_n z^n$ is analytic on $\mathbb{D}$, then*

$$D(f) = \sum_{n=1}^{\infty} n |a_n|^2. \tag{9.1.3}$$

See Exercise 9.9.10 for another formula for $D(f)$. Observe that $D(f) = 0$ if and only if f is a constant function.

Definition 9.1.4. The *Dirichlet space* $\mathcal{D}$ is the set of analytic f on $\mathbb{D}$ with $D(f) < \infty$.

The Cauchy–Schwarz inequality shows that $\mathcal{D}$ is a vector space under function addition and scalar multiplication. Since

$$\sum_{n=1}^{\infty} |a_n|^2 \leqslant \sum_{n=1}^{\infty} n|a_n|^2 = D(f), \tag{9.1.5}$$

it follows that $\mathcal{D}$ is contained in the Hardy space H^2 (Definition 5.3.3) with proper inclusion (Exercise 9.9.5). The norm on $\mathcal{D}$ is defined by

$$\|f\| = \Big(\sum_{n=0}^{\infty} (n+1)|a_n|^2 \Big)^{\frac{1}{2}}. \tag{9.1.6}$$

The $n+1$ above ensures that $\|f\| = 0$ if and only if $f \equiv 0$. Notice that $\|f\|^2 = \|f\|_{H^2}^2 + D(f)$ (recall the Hardy space norm from Proposition 5.3.9) and that the corresponding inner product is

$$\langle f, g \rangle = \sum_{n=0}^{\infty} (n+1) a_n \overline{b_n}, \tag{9.1.7}$$

where $f(z) = \sum_{n=0}^{\infty} a_n z^n$ and $g(z) = \sum_{n=0}^{\infty} b_n z^n$ belong to $\mathcal{D}$. In fact, the following results imply that $\mathcal{D}$, equipped with the inner product (9.1.7), is a reproducing kernel Hilbert space. We first establish completeness. The following lemma is used below, and again in the next chapter, to show a given inner-product space is a Hilbert space.

Lemma 9.1.8. *Suppose $\mathcal{V}$ and $\mathcal{W}$ are normed vector spaces and that the linear transformation $U : \mathcal{V} \to \mathcal{W}$ is isometric and surjective. If $\mathcal{W}$ is complete, then so is $\mathcal{V}$.*

Proof Suppose $(\mathbf{v}_n)_{n=1}^{\infty}$ is a Cauchy sequence in $\mathcal{V}$. If $\mathbf{w}_n = U\mathbf{v}_n$, the isometric assumption implies that $(\mathbf{w}_n)_{n=1}^{\infty}$ is a Cauchy sequence in $\mathcal{W}$. The completeness of $\mathcal{W}$ implies that $\mathbf{w}_n$ converges to some $\mathbf{w} \in \mathcal{W}$. Since U is surjective, $\mathbf{w} = U\mathbf{v}$ for some $\mathbf{v} \in \mathcal{V}$. Observe that U^{-1} is also isometric and hence $\mathbf{v}_n = U^{-1}\mathbf{w}_n \to U^{-1}\mathbf{w} = \mathbf{v}$. Consequently, $\mathcal{V}$ is complete. ∎

Corollary 9.1.9. *The Dirichlet space is a Hilbert space.*

Proof The discussion above shows that $\mathcal{D}$ is an inner-product space so it suffices to show that $\mathcal{D}$ is complete. The map $\sum_{n=0}^{\infty} a_n z^n \mapsto (\sqrt{n+1}\, a_n)_{n=0}^{\infty}$ from $\mathcal{D}$ to ℓ^2 is isometric and surjective. Since ℓ^2 is complete (Proposition 1.2.5), the previous lemma implies that $\mathcal{D}$ is complete. ∎

Exercise 9.9.2 provides the following useful orthonormal basis for $\mathcal{D}$.

Proposition 9.1.10. *The sequence*

$$\Big(\frac{z^n}{\sqrt{n+1}} \Big)_{n=0}^{\infty}$$

is an orthonormal basis for $\mathcal{D}$. In particular, the polynomials are dense in $\mathcal{D}$.

Next we find the reproducing kernel for $\mathcal{D}$ (Exercise 9.9.7).

Proposition 9.1.11. *Let*

$$k_\lambda(z) = \begin{cases} \dfrac{1}{\bar\lambda z} \log\left(\dfrac{1}{1 - \bar\lambda z}\right) & \text{if } \lambda, z \in \mathbb{D}\backslash\{0\}, \\ 1 & \text{if } \lambda \text{ or } z \text{ is zero.} \end{cases}$$

For all $\lambda \in \mathbb{D}$ and $f \in \mathcal{D}$, the following hold.

(a) $k_\lambda \in \mathcal{D}$ *and*

$$\|k_\lambda\|^2 = \begin{cases} \dfrac{1}{|\lambda|^2} \log \dfrac{1}{1 - |\lambda|^2} & \text{if } \lambda \in \mathbb{D}\backslash\{0\}, \\ 1 & \text{if } \lambda = 0. \end{cases}$$

(b) $f(\lambda) = \langle f, k_\lambda \rangle$.

(c) $|f(\lambda)| \leqslant \|f\| \|k_\lambda\|$.

Although the emphasis of this chapter is on the shift operator on the Dirichlet space, the Dirichlet space itself is a fascinating class of analytic functions that is not as well understood as the closely related Hardy space (Chapter 5). For example, the Blaschke condition (Theorem 5.4.8) completely describes the zeros of Hardy space functions. For the Dirichlet space, however, the zero sets are more subtle and are not completely characterized. It is also the case that the radial limits of functions in the Dirichlet space exist more often than for typical functions in the Hardy space.

9.2 The Dirichlet Shift

The shift on the Hardy space was explored in Chapter 5. This section explores the shift $f \mapsto zf$ on the Dirichlet space.

Proposition 9.2.1. *If $f \in \mathcal{D}$, then $zf \in \mathcal{D}$. Furthermore, the operator $M_z : \mathcal{D} \to \mathcal{D}$ defined by*

$$(M_z f)(z) = zf(z)$$

is bounded and $\|M_z\| = \sqrt{2}$.

Proof For $f(z) = \sum_{n=0}^{\infty} a_n z^n \in \mathcal{D}$,

$$\|M_z f\|^2 = \sum_{n=1}^{\infty} (n+1)|a_{n-1}|^2 = \sum_{n=0}^{\infty} (n+2)|a_n|^2.$$

Since

$$\frac{n+2}{n+1} \leqslant 2 \quad \text{for all } n \geqslant 0,$$

it follows that $\|M_z f\|^2 \leq 2 \sum_{n=0}^{\infty} (n+1)|a_n|^2 = 2\|f\|^2$. This estimate shows that

$$\|M_z\| = \sup_{\|f\|=1} \|M_z f\| \leq \sqrt{2}.$$

To obtain equality above, observe that $\|1\| = 1$ and $\|z\| = \sqrt{2}$. From here it follows that $\|M_z\| \geq \|M_z 1\| = \|z\| = \sqrt{2}$, and hence $\|M_z\| = \sqrt{2}$. ■

The operator M_z is the *Dirichlet shift*. Since

$$M_z f_n = \sqrt{\frac{n+2}{n+1}} f_{n+1} \quad \text{for all } n \geq 0,$$

where

$$f_n(z) = \frac{z^n}{\sqrt{n+1}}$$

is the orthonormal basis from Proposition 9.1.10, the matrix representation of the Dirichlet shift with respect to $(f_n)_{n=0}^{\infty}$ is

$$\begin{bmatrix} 0 & 0 & 0 & 0 & \cdots \\ \sqrt{\frac{2}{1}} & 0 & 0 & 0 & \cdots \\ 0 & \sqrt{\frac{3}{2}} & 0 & 0 & \cdots \\ 0 & 0 & \sqrt{\frac{4}{3}} & 0 & \cdots \\ \vdots & \vdots & \vdots & \vdots & \ddots \end{bmatrix}. \tag{9.2.2}$$

Compare this to the matrix representation (5.1.3) of the shift on the Hardy space. In particular, the Dirichlet shift is a weighted shift in the sense of Exercise 3.6.21. This next proposition shows that the Dirichlet shift is *expansive*.

Proposition 9.2.3. $\|M_z f\| \geq \|f\|$ *for all* $f \in \mathcal{D}$.

Proof For $f(z) = \sum_{n=0}^{\infty} a_n z^n \in \mathcal{D}$,

$$\|M_z f\|^2 = \sum_{n=0}^{\infty} (n+2)|a_n|^2 \geq \sum_{n=0}^{\infty} (n+1)|a_n|^2 = \|f\|^2,$$

which completes the proof. ■

The adjoint of M_z on $\mathcal{D}$ is more complicated than the adjoint of the unilateral shift on H^2 (Exercise 5.9.14).

Proposition 9.2.4. *If* $g(z) = \sum_{n=0}^{\infty} b_n z^n \in \mathcal{D}$, *then*

$$(M_z^* g)(z) = \sum_{n=0}^{\infty} \frac{n+2}{n+1} b_{n+1} z^n. \tag{9.2.5}$$

Proof For $f(z) = \sum_{n=0}^{\infty} a_n z^n \in \mathcal{D}$, use the inner product in (9.1.7) to deduce

$$\langle f, M_z^* g \rangle = \langle M_z f, g \rangle$$

$$= \sum_{n=1}^{\infty} (n+1) a_{n-1} \overline{b_n}$$

$$= \sum_{n=0}^{\infty} (n+2) a_n \overline{b_{n+1}}$$

$$= \sum_{n=0}^{\infty} (n+1) a_n \overline{\left(\frac{n+2}{n+1} b_{n+1} \right)}$$

$$= \langle f, h \rangle,$$

where

$$h(z) = \sum_{n=0}^{\infty} \frac{n+2}{n+1} b_{n+1} z^n.$$

The definition of the Dirichlet norm from (9.1.6) shows that $h \in \mathcal{D}$. This verifies the adjoint formula in (9.2.5). ∎

Although the spectrum of the Dirichlet shift is the same as the spectrum of the unilateral shift on H^2, the proof is more complicated.

Proposition 9.2.6. *For the Dirichlet shift M_z, the following hold.*

(a) $\sigma(M_z) = \mathbb{D}^-$.

(b) $\sigma_p(M_z) = \varnothing$.

(c) $\sigma_{ap}(M_z) = \mathbb{T}$.

Proof (a) A generalization of Proposition 9.2.1 (see Exercise 9.9.12) yields

$$\|M_z^k\| = \sqrt{k+1} \quad \text{for } k \geqslant 0. \tag{9.2.7}$$

The spectral radius formula (Theorem 8.4.4) says that

$$\sup_{\lambda \in \sigma(M_z)} |\lambda| = \lim_{k \to \infty} \|M_z^k\|^{\frac{1}{k}} = \lim_{k \to \infty} (k+1)^{\frac{1}{2k}} = 1$$

and hence $\sigma(M_z) \subseteq \mathbb{D}^-$. For the reverse inclusion, let $\lambda \in \mathbb{D}$ and observe that

$$(\lambda I - M_z)\mathcal{D} \subseteq \{f \in \mathcal{D} : f(\lambda) = 0\} \subsetneq \mathcal{D}.$$

Thus, $\lambda I - M_z$ is not invertible and hence $\mathbb{D} \subseteq \sigma(M_z)$. Since the spectrum of a bounded operator is compact (Theorem 2.4.9b), the observations above imply that $\sigma(M_z) = \mathbb{D}^-$.

(b) If $(\lambda I - M_z) f = 0$, then $(z - \lambda) f(z) = 0$ for all $z \in \mathbb{D}$ and, since f is an analytic function, $f \equiv 0$. Thus, $\sigma_p(M_z) = \varnothing$.

(c) For $\xi \in \mathbb{T}$ and $n \geqslant 2$, define

$$g_n(z) = \frac{1}{c_n}(1 + \bar{\xi}z + \bar{\xi}^2 z^2 + \cdots + \bar{\xi}^{n-1} z^{n-1}),$$

where

$$c_n = (1 + 2 + 3 + \cdots + n)^{\frac{1}{2}} = \left(\frac{n(n+1)}{2}\right)^{\frac{1}{2}}.$$

Note that each g_n is a unit vector in $\mathcal{D}$ and

$$(M_z - \xi I)g_n = -\frac{\xi}{c_n}(1 - \bar{\xi}^n z^n).$$

Thus,

$$\|(M_z - \xi I)g_n\| = \frac{1}{c_n}(n+2)^{1/2} = \left(\frac{2}{n(n+1)}\right)^{\frac{1}{2}}(n+2)^{\frac{1}{2}} \to 0,$$

and hence $\xi \in \sigma_{ap}(M_z)$. It follows that $\mathbb{T} \subseteq \sigma_{ap}(M_z)$. If $\lambda \in \mathbb{D}$ and $f \in \mathcal{D}$, then

$$
\begin{aligned}
\|(M_z - \lambda I)f\| &\geqslant \|M_z f\| - |\lambda|\|f\| \\
&\geqslant \|f\| - |\lambda|\|f\| \qquad \text{(Proposition 9.2.3)} \\
&= (1 - |\lambda|)\|f\|.
\end{aligned}
$$

Consequently, $\inf_{\|f\|=1} \|(M_z - \lambda I)f\| \geqslant 1 - |\lambda| > 0$ and hence $\lambda \notin \sigma_{ap}(M_z)$. Therefore, $\sigma_{ap}(M_z) = \mathbb{T}$. ∎

Proposition 9.2.8. *For the adjoint M_z^* of the Dirichlet shift, the following hold.*

(a) $\sigma(M_z^*) = \mathbb{D}^-$.

(b) $\sigma_p(M_z^*) = \mathbb{D}$.

(c) $\sigma_{ap}(M_z^*) = \mathbb{D}^-$.

Proof (a) $\sigma(M_z^*) = \mathbb{D}^-$ follows from $\sigma(M_z) = \mathbb{D}^-$ (Proposition 3.1.6).

(b) To compute the eigenvalues of M_z^*, proceed as in (5.5.5) and use the reproducing kernels $k_\lambda(z)$ from Proposition 9.1.11 to show that $M_z^* k_\lambda = \bar{\lambda} k_\lambda$. Thus, $\mathbb{D} \subseteq \sigma_p(M_z^*)$. Since $\mathbb{D} \subseteq \sigma_p(M_z^*) \subseteq \sigma(M_z^*) = \mathbb{D}^-$, it suffices to show that no $\xi \in \mathbb{T}$ is an eigenvalue for M_z^*. To the contrary, suppose that $g(z) = \sum_{n=0}^{\infty} b_n z^n \in \mathcal{D}\setminus\{0\}$ and $\xi \in \mathbb{T}$ with $M_z^* g = \xi g$. Then (9.2.5) says that

$$\sum_{n=0}^{\infty} \frac{n+2}{n+1} b_{n+1} z^n = \xi \sum_{n=0}^{\infty} b_n z^n.$$

Compare the coefficients of z^n to obtain the recurrence

$$b_{n+1} = \xi \frac{n+1}{n+2} b_n \quad \text{for all } n \geqslant 0,$$

which implies

$$b_n = \frac{\xi^n}{n+1} b_0 \quad \text{for all } n \geqslant 1.$$

In light of (9.1.3), g cannot belong to $\mathcal{D}$ unless $b_0 = 0$, that is, $g \equiv 0$. Thus, no $\xi \in \mathbb{T}$ is an eigenvalue of M_z^*.

(c) Mimic the proof in Proposition 5.2.4. ∎

9.3 The Dirichlet Shift is a 2-isometry

Recall from Exercise 3.6.13 that $A \in \mathcal{B}(\mathcal{H})$ is an isometry if and only if $A^*A - I = 0$. Unlike the unilateral shift on the Hardy space, the Dirichlet shift M_z is not an isometry. Indeed, $\|1\| = 1$ but $\|M_z 1\| = \|z\| = \sqrt{2}$. In fact, Proposition 9.2.3 says that M_z is expansive. An operator $A \in \mathcal{B}(\mathcal{H})$ is a *2-isometry* if

$$A^{*2}A^2 - 2A^*A + I = 0,$$

or equivalently,

$$\|A^2\mathbf{x}\|^2 - 2\|A\mathbf{x}\|^2 + \|\mathbf{x}\|^2 = 0 \quad \text{for all } \mathbf{x} \in \mathcal{H}.$$

These types of operators and their generalizations are studied in [5, 6, 7]. The next result says that the Dirichlet shift is a 2-isometry. A generalization of the Dirichlet shift is used to model certain 2-isometries (see [293] and the discussion below).

Proposition 9.3.1. *The Dirichlet shift is a 2-isometry.*

Proof For any $f(z) = \sum_{n=0}^{\infty} a_n z^n \in \mathcal{D}$,

$$\|M_z^k f\|^2 = \sum_{n=0}^{\infty} (n+k+1)|a_n|^2 \quad \text{for all } k \geqslant 0.$$

Therefore,

$$\|M_z^2 f\|^2 + \|f\|^2 = \sum_{n=0}^{\infty} (2n+4)|a_n|^2 = 2\|M_z f\|^2,$$

and hence M_z is a 2-isometry. ∎

Every 2-isometry $A \in \mathcal{B}(\mathcal{H})$ is expansive in the sense that $\|A\mathbf{x}\| \geqslant \|\mathbf{x}\|$ for every $\mathbf{x} \in \mathcal{H}$ (Exercise 9.9.19). Proposition 9.2.3 verifies this directly for the Dirichlet shift. Without getting too far into the details, we briefly introduce a class of spaces related to $\mathcal{D}$ such that the shift M_z on these spaces model a wide class of 2-isometries. Given a finite positive Borel measure μ on $\mathbb{T}$, the function

$$P_\mu(z) := \int_{\mathbb{T}} \frac{1 - |z|^2}{|\zeta - z|^2} \, d\mu(\zeta) \quad \text{for } z \in \mathbb{D},$$

is the *Poisson integral of μ* on $\mathbb{D}$ (see Chapter 12). Observe that $P_\mu \geq 0$ and P_μ is harmonic on $\mathbb{D}$ (Theorem 12.1.6). Use P_μ as a weight to define the *harmonically weighted Dirichlet integral*

$$D_\mu(f) := \frac{1}{\pi} \int_{\mathbb{D}} |f'(z)|^2 P_\mu(z) \, dA(z). \tag{9.3.2}$$

The corresponding *harmonically weighted Dirichlet space* $\mathcal{D}_\mu$ is the space of analytic functions f on $\mathbb{D}$ such that $D_\mu(f) < \infty$. One defines a norm on $\mathcal{D}_\mu$ via

$$\|f\|_{\mathcal{D}_\mu}^2 := \|f\|_{H^2}^2 + D_\mu(f).$$

If μ is normalized Lebesgue measure m on $\mathbb{T}$, then $P_\mu(z) = 1$ for all $z \in \mathbb{D}$ (Exercise 12.5.2) and hence the space $\mathcal{D}_\mu$ is the classical Dirichlet space $\mathcal{D}$.

Although harder to prove, M_z on $\mathcal{D}_\mu$ is a well-defined bounded operator and a 2-isometry. Since a nonzero analytic function cannot have a zero of infinite order,

$$\bigcap_{n \geq 0} M_z^n \mathcal{D}_\mu = \{0\}. \tag{9.3.3}$$

An operator $T \in B(\mathcal{H})$ which satisfies the analogue of (9.3.3), namely

$$\bigcap_{n \geq 0} T^n \mathcal{H} = \{0\},$$

is *analytic*. The polynomials are dense in $\mathcal{D}_\mu$ [293, Cor. 3.8] and hence

$$\bigvee \{M_z^n 1 : n \geq 0\} = \mathcal{D}_\mu.$$

Therefore, M_z is a cyclic operator (recall Definition 8.2.1) with cyclic vector 1. To summarize, M_z on $\mathcal{D}_\mu$ is a cyclic analytic 2-isometry. The following theorem says that, up to unitary equivalence, these are all of the analytic cyclic 2-isometries.

Theorem 9.3.4 (Richter [293]). *Any cyclic analytic 2-isometry on a Hilbert space is unitarily equivalent to M_z on $\mathcal{D}_\mu$ for some finite positive Borel measure μ on $\mathbb{T}$.*

Notice the pattern continued by this theorem. Beurling represented a certain operator as the shift on the Hardy space (Chapter 5); Kriete and Trutt represented the Cesàro operator as M_z on a space of analytic functions (Theorem 6.4.7); and Richter represented certain 2-isometries as M_z on a Dirichlet-type space.

9.4 Multipliers and Commutant

An analytic function φ on $\mathbb{D}$ such that $\varphi \mathcal{D} \subseteq \mathcal{D}$ is a *multiplier* of $\mathcal{D}$. Proposition 5.5.4 says that the multipliers of H^2 are H^∞, the set of bounded analytic functions on $\mathbb{D}$. For the Dirichlet space, the multipliers are more complicated. The proof of the next result is similar to that of Proposition 5.5.1.

Proposition 9.4.1. *If φ is a multiplier of $\mathcal{D}$, then φ belongs to H^∞ and the corresponding multiplication operator M_φ on $\mathcal{D}$ is bounded.*

Since the constant functions belong to $\mathcal{D}$, the multipliers of $\mathcal{D}$ are a subset of $\mathcal{D}$, and indeed, a proper subset of $\mathcal{D}$ (Exercise 9.9.18). Although every multiplier of $\mathcal{D}$ belongs to H^∞, the converse is not true [121, Thm. 5.1.6]. The multipliers of $\mathcal{D}$ remain mysterious and lack a simple satisfactory description. However, if one is looking for specific examples of multipliers, Exercise 9.9.17 shows that if $\varphi' \in H^\infty$, then φ is a multiplier of $\mathcal{D}$.

Essentially the same proof that was used for the shift on the Hardy space (Corollary 5.6.2) describes the commutant of the Dirichlet shift.

Proposition 9.4.2. *For $A \in \mathcal{B}(\mathcal{D})$, the following are equivalent.*

(a) $AM_z = M_z A$.

(b) $A = M_\varphi$, *where φ is a multiplier of $\mathcal{D}$.*

9.5 Invariant Subspaces

Beurling's theorem (Theorem 5.4.12) says that if $\mathcal{M}$ is a nonzero S-invariant subspace of H^2, then $\mathcal{M} \cap (S\mathcal{M})^\perp$ is one dimensional and is spanned by an inner function. Furthermore, $[\mathcal{M} \cap (S\mathcal{M})^\perp]$, the S-invariant subspace generated by $\mathcal{M} \cap (S\mathcal{M})^\perp$, equals $\mathcal{M}$. Richter and Sundberg proved a version of Beurling's theorem for the Dirichlet space.

Theorem 9.5.1 (Richter–Sundberg [295, 296]). *Suppose $\mathcal{M}$ is a nonzero M_z-invariant subspace of $\mathcal{D}$. Then $\mathcal{M} \cap (M_z\mathcal{M})^\perp$ is one dimensional and is the span of a multiplier of $\mathcal{D}$. Furthermore, $[\mathcal{M} \cap (M_z\mathcal{M})^\perp] = \mathcal{M}$.*

A more precise description of $\mathcal{M}$ comes from the next theorem.

Theorem 9.5.2 (Richter–Sundberg [295, 296]). *Let $\mathcal{M}$ be an M_z-invariant subspace of $\mathcal{D}$. Then $\mathcal{M} \cap (M_z\mathcal{M})^\perp$ is spanned by a multiplier φ of $\mathcal{D}$ and $\mathcal{M} = \varphi\mathcal{D}_\nu$, where $d\nu = |\varphi|^2 dm$.*

9.6 Cyclic Vectors

The cyclic vectors for M_z on $\mathcal{D}$ (recall Definition 5.7.1) are somewhat mysterious and, although there is a conjecture as to what they might be (see [70, 121, 315] and the end notes for this chapter), the problem of characterizing the cyclic vectors remains open. Let us prove a few results about the cyclic vectors for M_z on $\mathcal{D}$.

Proposition 9.6.1. *If $f \in \mathcal{D}$ is cyclic for M_z, then f is cyclic for the unilateral shift S on H^2.*

Proof If $f \in \mathcal{D}$ is cyclic for M_z, then there is a sequence of polynomials $(p_n)_{n=1}^\infty$ such that $p_n f \to 1$ in the norm of $\mathcal{D}$. By (9.1.5), $\|p_n f - 1\|_\mathcal{D} \geqslant \|p_n f - 1\|_{H^2}$ and it follows that $p_n f \to 1$ in H^2. Thus, 1 belongs to $\mathcal{W}$, the closed linear span in H^2 of $\{S^n f : n \geqslant 0\}$. Due to the S-invariance of $\mathcal{W}$, it follows that every polynomial belongs to $\mathcal{W}$. Consequently, f is cyclic for S on H^2. ∎

Below is a specific example of a cyclic vector for the Dirichlet shift.

Example 9.6.2. $f(z) = 1 - z$ is a cyclic vector for M_z on $\mathcal{D}$. We follow a proof from [37] (see also [38]). Exercise 9.9.21 gives an alternate proof. For $n \geqslant 1$, let

$$h_n = \sum_{k=1}^{n} \frac{1}{k} \quad \text{and} \quad p_n(z) = \sum_{k=1}^{n} \left(1 - \frac{h_k}{h_n}\right) z^{k-1}.$$

Then,

$$
\begin{aligned}
(1 - z)p_n(z) &= (1 - z) \sum_{k=1}^{n} \left(1 - \frac{h_k}{h_n}\right) z^{k-1} \\
&= \sum_{k=1}^{n} \left(1 - \frac{h_k}{h_n}\right) z^{k-1} - \sum_{k=1}^{n} \left(1 - \frac{h_k}{h_n}\right) z^{k} \\
&= \sum_{k=1}^{n} \left(1 - \frac{h_k}{h_n}\right) z^{k-1} - \sum_{k=2}^{n+1} \left(1 - \frac{h_{k-1}}{h_n}\right) z^{k-1} \\
&= \left(1 - \frac{h_1}{h_n}\right) + \sum_{k=2}^{n} \left(\frac{h_{k-1}}{h_n} - \frac{h_k}{h_n}\right) z^{k-1} - \left(1 - \frac{h_n}{h_n}\right) z^{n} \\
&= 1 - \sum_{k=1}^{n} \frac{1}{kh_n} z^{k-1}.
\end{aligned}
$$

Therefore,

$$\|1 - (1 - z)p_n(z)\|^2 = \left\|\sum_{k=1}^{n} \frac{1}{kh_n} z^{k-1}\right\|^2 = \sum_{k=1}^{n} (k - 1 + 1)\frac{1}{k^2 h_n^2} = \frac{1}{h_n^2} \sum_{k=1}^{n} \frac{1}{k} = \frac{1}{h_n}.$$

Note that

$$h_n = \sum_{k=1}^{n} \frac{1}{k} > \int_{1}^{n} \frac{dx}{x} = \log n.$$

Hence, $(1 - z)p_n(z) \to 1$ in $\mathcal{D}$. This shows that $f(z) = 1 - z$ is a cyclic vector for M_z on $\mathcal{D}$.

Although the proof that the reader is encouraged to explore in Exercise 9.9.21 might seem easier, we included the proof above since it explicitly computes polynomials p_n such that $\|(1 - z)p_n - 1\| \to 0$.

9.7 The Bilateral Dirichlet Shift

For $f \in L^2(\mathbb{T})$ with Fourier expantion $f = \sum_{n=-\infty}^{\infty} \widehat{f}(n)\xi^n$, define

$$\|f\|_{\mathscr{D}} = \left(\sum_{n=-\infty}^{\infty} (|n| + 1)|\widehat{f}(n)|^2\right)^{\frac{1}{2}}.$$

The space $\{f \in L^2(\mathbb{T}) : \|f\|_{\mathscr{D}} < \infty\}$ is the *harmonic Dirichlet space* (not to be confused with the harmonically weighted Dirichlet space $\mathcal{D}_\mu$ mentioned earlier). There is also a shift operator $(M_\xi g)(\xi) = \xi g(\xi)$ defined on $\mathscr{D}$ that is bounded and invertible, although not unitary like M_ξ on $L^2(\mathbb{T})$. Using logarithmic capacity, one can describe the M_ξ-invariant subspaces $\mathcal{M} \subseteq \mathscr{D}$ such that $M_\xi \mathcal{M} = \mathcal{M}$ as the set of functions which vanish (quasi-everywhere) on some subset $E \subseteq \mathbb{T}$ (see [294] and the endnotes of this chapter). This is the analogue of Wiener's theorem (Theorem 4.3.3) for $\mathscr{D}$. The invariant subspaces $\mathcal{M}$ such that $M_\xi \mathcal{M} \subsetneq \mathcal{M}$ do not yield as simple a description as they did for $L^2(\mathbb{T})$ (Theorem 4.3.7). In fact, they can be very complicated [314].

9.8 Notes

The Dirichlet integral gets its name from the *Dirichlet problem*: one is given a smooth function $f : \mathbb{T} \to \mathbb{R}$ and asked to find a harmonic $u : \mathbb{D} \to \mathbb{R}$ that is continuous on $\mathbb{D}^-$ such that $u|_\mathbb{T} = f$. Here harmonic means that $\partial_x^2 u + \partial_y^2 u = 0$. We explore this problem in Chapter 12. One approach to solving this problem is via the *Dirichlet principle*: minimize the Dirichlet integral

$$\int_\mathbb{D} \left(|\partial_x v|^2 + |\partial_y v|^2 \right) dA$$

over all smooth functions v on $\mathbb{D}^-$ with $v|_\mathbb{T} = f$. In a series of lectures, Dirichlet showed three things. First, any v that is harmonic on $\mathbb{D}$ with $v|_\mathbb{T} = f$ minimizes the Dirichlet integral. Second, any u with $u|_\mathbb{T} = f$ that minimizes the Dirichlet integral is harmonic. Third, there is only one function with the prescribed boundary values that minimizes the Dirichlet integral. The paper [158] contains an excellent historical survey, along with extensions of the Dirichlet problem and principle to more general planar domains.

Good sources on the Dirichlet space are [22, 121, 315]. Much of our presentation on the Dirichlet space focuses on the Dirichlet shift and not on the function-theoretic properties of $\mathcal{D}$. The two are closely related and one cannot understand the operator theory without knowledge of the function theory of the ambient space.

In his doctoral thesis [50], Beurling explored the properties of analytic functions with finite Dirichlet integral. An important paper of Beurling from 1940 [52] shows that if $f \in \mathcal{D}$, then $\lim_{r \to 1^-} f(r\xi)$ exists for every $\xi \in \mathbb{T}$, except possibly for a set of logarithmic capacity zero. Authors sometimes use the term *quasi-everywhere* to mean "everywhere except possibly for a set of logarithmic capacity zero." The notion of logarithmic capacity used in Beurling's paper was originally developed by de la Vallée Poussin [365] who defined the *capacity* $c(E)$ of a closed set $E \subseteq \mathbb{T}$ as $c(E) = e^{-V(E)}$, where

$$V(E) = \inf_\mu \int_\mathbb{T} \int_\mathbb{T} \log \frac{2}{|\xi - \zeta|} \, d\mu(\xi) \, d\mu(\zeta),$$

in which the infimum is taken over all probability measures μ on E. If $c(E) = 0$, then $m(E) = 0$. However, the Cantor set (in the circle) has Lebesgue measure zero but positive logarithmic capacity. Thus, radial limits of generic functions in the Dirichlet space exist at "more" points than functions in the Hardy space.

This chapter discussed the M_z-invariant subspaces of $\mathcal{D}$. With a notion of logarithmic capacity, one can show that if $E \subseteq \mathbb{T}$ is of positive capacity, then

$$\mathcal{D}_E = \left\{ f \in \mathcal{D} : \lim_{r \to 1^-} f(r\xi) = 0 \text{ for quasi-every } \xi \in E \right\}$$

is an invariant subspace. If u is an inner function, then $uH^2 \cap \mathcal{D}_E$ is also an invariant subspace (which might be the zero subspace). This leads to the question, are all of the invariant subspaces of the form $uH^2 \cap \mathcal{D}_E$? Brown and Shields explored the cyclic vectors for M_z [70]. The Brown–Shields conjecture asserts that the cyclic vectors for the Dirichlet shift are the outer functions in the Dirichlet space whose radial limits are nonzero quasi-everywhere. Some partial results are found in [122, 163, 189].

If $\mathcal{M}_1$ and $\mathcal{M}_2$ are invariant subspaces for the unilateral shift S on H^2, then $S|_{\mathcal{M}_1}$ is unitarily equivalent to $S|_{\mathcal{M}_2}$ (Exercise 5.9.20). Richter [292] explored this property in the Dirichlet space. The problem was fully resolved by Guo and Zhao in [164] where they proved, for invariant subspaces $\mathcal{M}_1, \mathcal{M}_2$ of $\mathcal{D}$, that $M_z|_{\mathcal{M}_1}$ is unitarily equivalent to $M_z|_{\mathcal{M}_2}$ if and only if $\mathcal{M}_1 = \mathcal{M}_2$. To do this, they showed that if $\varphi \in \mathcal{M}_1 \cap (M_z\mathcal{M}_1)^{\perp}$ and $U : \mathcal{M}_1 \to \mathcal{M}_2$ is unitary with $UM_z|_{\mathcal{M}_1} = M_z|_{\mathcal{M}_2}U$, then $\psi = U\varphi \in \mathcal{M}_2 \cap (M_z\mathcal{M}_2)^{\perp}$. Using various properties of the local Dirichlet integral and some results from a paper of Richter and Sundberg [297], they proved that $\varphi = c\psi$ for some constant c. Since φ generates $\mathcal{M}_1$ and ψ generates $\mathcal{M}_2$ (Theorem 9.5.1), it follows that $\mathcal{M}_1 = \mathcal{M}_2$.

Fields medalist J. Douglas [114] derived the important formula

$$\frac{1}{\pi} \int_{\mathbb{D}} |f'|^2 dA = \int_{\mathbb{T}} \int_{\mathbb{T}} \left| \frac{f(\zeta) - f(\xi)}{\zeta - \xi} \right|^2 dm(\zeta)dm(\xi) \quad \text{for all } f \in \mathcal{D},$$

which is used to explore properties of functions in the Dirichlet space (Exercise 9.9.10).

The Dirichlet shift is a 2-isometry (Proposition 9.3.1) and 2-isometries connect to a wide class of operators that are inspired by the spectral theorem for selfadjoint operators. Helton [193] explored the *n-symmetric operators*. These are the $T \in \mathcal{B}(\mathcal{H})$ such that

$$\sum_{k=0}^{n} (-1)^k \binom{n}{k} T^{*(n-k)} T^k = 0.$$

A 1-symmetric operator is selfadjoint and the spectral theorem represents it as M_x on an $L^2(\mu)$ space, where μ is a finite positive Borel measure on $\sigma(T) \subseteq \mathbb{R}$ (Theorem 8.7.1). For other n, these operators are represented by M_x on a space of functions on $\mathbb{R}$ with inner product

$$\langle f, g \rangle = \sum_{i,j=0}^{N} \int f^{(i)} \overline{g^{(j)}} d\mu_{ij},$$

where μ_{ij} are compactly supported finite positive Borel measures on $\mathbb{R}$ and N depends on n. Agler [4], and Agler and Stankus [5, 6, 7], extended these results to the *n-isometric operators*:

$$\sum_{k=0}^{n} (-1)^k \binom{n}{k} T^{*k} T^k = 0.$$

If $n = 1$ this condition becomes $I - T^*T = 0$, that is, T is isometric. When $n = 2$, this condition becomes $I - 2T^*T + T^{*2}T^2 = 0$, that is T is a 2-isometry. There is a version of the representation theorem for certain 2-isometric operators (Theorem 9.3.4) for n-isometric operators that involves M_ξ on the completion of the set of analytic polynomials with respect to the inner product

$$\langle f, g \rangle = \int_{\mathbb{T}} L(f(\xi))\overline{g(\xi)}\, dm(\xi),$$

where L is a certain positive differential operator of order $n - 1$.

9.9 Exercises

Exercise 9.9.1. For an analytic function $f(z) = \sum_{n=0}^{\infty} a_n z^n$ on $\mathbb{D}$, prove that

$$D(f) = \sum_{n=1}^{\infty} n|a_n|^2.$$

Exercise 9.9.2. Prove that $\left(\dfrac{z^n}{\sqrt{n+1}}\right)_{n=0}^{\infty}$ is an orthonormal basis for $\mathcal{D}$.

Exercise 9.9.3.

(a) Prove that the Dirichlet integral of an injective analytic function f on $\mathbb{D}$ is $1/\pi$ times the area of $f(\mathbb{D})$.

(b) What happens if f is analytic but not injective?

Exercise 9.9.4. If $\varphi : \mathbb{D} \to \mathbb{D}$ and $f : \mathbb{D} \to \mathbb{C}$ are analytic, prove that

$$D(f \circ \varphi) = \frac{1}{\pi} \int_{\varphi(\mathbb{D})} |f'(w)|^2 n_\varphi(w)\, dA(w),$$

where $n_\varphi(w)$ is the cardinality of $\{z : \varphi(z) = w\}$.

Exercise 9.9.5. Prove that the inclusion $\mathcal{D} \subseteq H^2$ is proper in two ways.

(a) Use power series.

(b) Use the Dirichlet integral.

Exercise 9.9.6. The containment $\mathcal{D} \subseteq H^2$ follows from (9.1.5). Consider the inclusion operator $i : \mathcal{D} \to H^2$ defined by $i(f) = f$ for $f \in \mathcal{D}$.

(a) Prove that i is bounded, has dense range in H^2, but is not surjective.

(b) Find i^* and prove that it is injective.

(c) Banach's closed range theorem [320] says that $T \in \mathcal{B}(\mathcal{H})$ has closed range if and only if T^* has closed range. Assuming this fact, is i^* surjective?

(d) Does i^* have dense range?

Exercise 9.9.7. Prove that the function

$$k_\lambda(z) = \frac{1}{\bar{\lambda}z} \log\left(\frac{1}{1 - \bar{\lambda}z}\right), \quad \text{where } \lambda, z \in \mathbb{D},$$

is the reproducing kernel for $\mathcal{D}$ as follows.

(a) Prove that $k_\lambda \in \mathcal{D}$ for every $\lambda \in \mathbb{D}$.

(b) Prove that $f(\lambda) = \langle f, k_\lambda \rangle$ for every $\lambda \in \mathbb{D}$ and $f \in \mathcal{D}$.

Exercise 9.9.8. This is a continuation of Exercises 9.9.2 and 9.9.7.

(a) If $e_n(z) = z^n/\sqrt{n+1}$, prove that $k_\lambda(z) = \sum_{n=0}^{\infty} \overline{e_n(\lambda)} e_n(z)$ for all $z, \lambda \in \mathbb{D}$.

(b) Prove that the formula for $k_\lambda(z)$ above is independent of the choice of orthonormal basis $(e_n)_{n=0}^{\infty}$ for $\mathcal{D}$.

Exercise 9.9.9. Prove that

$$\langle f, g \rangle = \lim_{r \to 1^-} \int_{\mathbb{T}} f(r\xi) \overline{(zg)'(r\xi)} \, dm(\xi) \quad \text{for all } f, g \in \mathcal{D}.$$

Exercise 9.9.10. Prove Douglas' [114] formula

$$\frac{1}{\pi} \int_{\mathbb{D}} |f'|^2 dA = \int_{\mathbb{T}} \int_{\mathbb{T}} \left| \frac{f(\zeta) - f(\xi)}{\zeta - \xi} \right|^2 dm(\zeta) dm(\xi) \quad \text{for all } f \in \mathcal{D}$$

as follows (see [121] for more).

(a) Prove that for $n \geq 0$,

$$\int_{\mathbb{T}} \int_{\mathbb{T}} \left| \frac{f(\zeta) - f(\xi)}{\zeta - \xi} \right|^2 dm(\zeta) dm(\xi) = \int_0^{2\pi} \int_0^{2\pi} \left| \frac{f(e^{i(s+t)}) - f(e^{it})}{e^{is} - 1} \right|^2 \frac{dt}{2\pi} \frac{ds}{2\pi}.$$

(b) Use Parseval's theorem to prove that

$$\int_0^{2\pi} |f(e^{i(s+t)}) - f(e^{it})|^2 \frac{dt}{2\pi} = \sum_{n=0}^{\infty} |\hat{f}(n)|^2 |e^{ins} - 1|^2.$$

(c) Prove that

$$\int_0^{2\pi} \left| \frac{e^{ins} - 1}{e^{is} - 1} \right|^2 \frac{ds}{2\pi} = n.$$

(d) Combine the above to obtain Douglas' formula.

Remark: There is a corresponding formula for the harmonically weighted Dirichlet integral (9.3.2):

$$D_\mu(f) = \int_{\mathbb{T}} \int_{\mathbb{T}} \left| \frac{f(\zeta) - f(\xi)}{\zeta - \xi} \right|^2 dm(\zeta) d\mu(\xi).$$

See [295] for details.

Exercise 9.9.11. For an analytic function f on $\mathbb{D}$, prove that $f \in \mathcal{D}$ if and only if $zf \in \mathcal{D}$.

Exercise 9.9.12. For the Dirichlet shift M_z, prove that $\|M_z^k\| = \sqrt{k+1}$ for all $k \geq 0$.

Exercise 9.9.13. Prove that the range of the Dirichlet shift M_z is closed.

Exercise 9.9.14. Prove that the self commutator $M_z^* M_z - M_z M_z^*$ of the Dirichlet shift M_z is compact.

Exercise 9.9.15. For a function $f(z) = \sum_{k=0}^{\infty} a_k z^k \in \mathcal{D}$ and $n \geq 0$, consider the linear transformation $S_n : \mathcal{D} \to \mathcal{D}$ defined by $(S_n f)(z) = \sum_{k=0}^{n} a_k z^k$.

(a) Prove that $S_n \in \mathcal{B}(\mathcal{D})$ and $\|S_n\| = 1$ for all $n \geq 0$.

(b) For each polynomial p prove that $S_n p = p$ for all $n \geq \deg p$.

(c) For each fixed $f \in \mathcal{D}$, prove that $\|S_n f - f\| \to 0$ as $n \to \infty$. In other words, $S_n \to I$ in the strong operator topology.

(d) Does $S_n \to I$ in operator norm?

Exercise 9.9.16. For a function $f(z) = \sum_{k=0}^{\infty} a_k z^k \in \mathcal{D}$ and $0 < r < 1$, consider the linear transformation $A_r : \mathcal{D} \to \mathcal{D}$ defined by $(A_r f)(z) = \sum_{k=0}^{\infty} a_k r^k z^k$.

(a) Prove that $A_r \in \mathcal{B}(\mathcal{D})$ and $\|A_r\| = 1$ for all $0 < r < 1$.

(b) For each polynomial p, prove that $\|A_r p - p\| \to 0$ as $r \to 1^-$.

(c) For each $f \in \mathcal{D}$, prove that $\|A_r f - f\| \to 0$ as $r \to 1^-$. In other words, $A_r \to I$ in the strong operator topology.

(d) Does $A_r \to I$ in operator norm?

Exercise 9.9.17. If $\varphi' \in H^\infty$, prove that φ is a multiplier of $\mathcal{D}$.

Exercise 9.9.18. Let $f(z) = \sum_{n=2}^{\infty} \frac{z^n}{n(\log n)^{3/4}}$.

(a) Prove that $f \in \mathcal{D}$.

(b) Prove that $f^2 \notin \mathcal{D}$.

(c) Conclude that $\mathcal{D}$ is not an algebra.

(d) Prove that the set of multipliers of $\mathcal{D}$ is a proper subset of $\mathcal{D}$.

Exercise 9.9.19. Prove that every 2-isometry $T \in \mathcal{B}(\mathcal{H})$ satisfies $\|T\mathbf{x}\| \geqslant \|\mathbf{x}\|$ for all $\mathbf{x} \in \mathcal{H}$ as follows.

(a) Use the definition of a 2-isometry to prove that

$$\|T^k\mathbf{x}\|^2 - \|T^{k-1}\mathbf{x}\|^2 = \|T\mathbf{x}\|^2 - \|\mathbf{x}\|^2 \quad \text{for all } k \geqslant 1.$$

(b) Prove that $\|T^n\mathbf{x}\|^2 - \|\mathbf{x}\|^2 = n(\|T\mathbf{x}\|^2 - \|\mathbf{x}\|^2)$ for all $n \geqslant 1$.

(c) Deduce that $\|T\mathbf{x}\|^2 \geqslant \|\mathbf{x}\|^2 - \dfrac{1}{n}\|\mathbf{x}\|^2$ for all $n \geqslant 1$.

Exercise 9.9.20. Prove that if p is a polynomial whose zeros lie outside $\mathbb{D}$, then p is a cyclic vector for the Dirichlet shift.

Exercise 9.9.21. Example 9.6.2 showed that $f(z) = 1-z$ is a cyclic vector for the Dirichlet shift. Here is an alternate proof of this fact.

(a) If $g \perp z^n(1 - z)$ for all n, prove that $(n + 1)\hat{g}(n) - (n + 2)\hat{g}(n + 1) = 0$ for all $n \geqslant 0$.

(b) Conclude that $g \equiv 0$.

Remark: See [70] for further examples of cyclic vectors.

Exercise 9.9.22. Here is an operator on $\mathcal{D}$ that is related to the Cesàro operator from Chapter 6. Define

$$(Af)(z) = \frac{1}{z - 1} \int_1^z f(s)\,ds - \int_0^1 f(s)\,ds \quad \text{for } f \in \mathcal{D}.$$

(a) Prove that both integrals in the definition of Af converge for all $f \in \mathcal{D}$.

(b) Consider $\mathcal{D}_0 = \{f \in \mathcal{D} : f(0) = 0\}$ and use the norm $\|f\|^2 = \dfrac{1}{\pi} \displaystyle\int_{\mathbb{D}} |f'|^2 dA$ on $\mathcal{D}_0$ to prove that $(z^n/\sqrt{n})_{n=1}^{\infty}$ is an orthonormal basis for $\mathcal{D}_0$.

(c) Find the matrix representation of A on $\mathcal{D}_0$ with respect to the basis above and use this representation to prove that A is bounded on $\mathcal{D}_0$.

Remark: See [286] for various properties of this operator.

Exercise 9.9.23. This is a continuation of Exercise 9.9.22. Use Exercise 9.9.8 to find a formula for the reproducing kernel $k_\lambda(z)$ for $\mathcal{D}_0$. In other words, find a function $k_\lambda(z)$ such that $k_\lambda \in \mathcal{D}_0$ for all λ and such that $f(\lambda) = \langle f, k_\lambda \rangle$ for all $f \in \mathcal{D}_0$ and $\lambda \in \mathbb{D}$. The inner product on $\mathcal{D}_0$ is the one that comes from the norm given in Exercise 9.9.22.

9.10 Hints for the Exercises

Hint for Ex. 9.9.1: Use polar coordinates and Parseval's theorem.

Hint for Ex. 9.9.3: For (a), consider the Jacobian of f.

Hint for Ex. 9.9.6: For (b), use the formula for inner products based on Taylor coefficients.

Hint for Ex. 9.9.7: Consider

$$\frac{1}{w} \log\left(\frac{1}{1-w}\right) = 1 + \frac{w}{2} + \frac{w^2}{3} + \cdots.$$

Hint for Ex. 9.9.13: Consult Proposition 9.2.3.

Hint for Ex. 9.9.17: If φ' is bounded, show that φ is also bounded.

Hint for Ex. 9.9.22: For (c), use Schur's test.

10

· · • · ·

The Bergman Shift

Key Concepts: Bergman space, Bergman shift (spectrum, adjoint, commutant), invariant subspace, invariant subspace of finite codimension, wandering subspace, index of an invariant subspace.

Outline: The study of the shift $(M_z f)(z) = zf(z)$ on the Bergman space A^2 (the space of analytic functions on $\mathbb{D}$ that are square integrable with respect to area measure) starts out like that of the Hardy and Dirichlet shifts. Many of the same properties hold (the point spectrum is empty, the spectrum is the closed unit disk, the point spectrum of the adjoint is the open unit disk). However, the invariant-subspace structure is dramatically different. With the Hardy and Dirichlet shifts, $\dim(\mathcal{M} \cap (M_z\mathcal{M})^{\perp}) = 1$ for every nonzero invariant subspace $\mathcal{M}$. For the Bergman shift, any dimension is possible. Nevertheless, $\mathcal{M}$ is generated by $\mathcal{M} \cap (M_z\mathcal{M})^{\perp}$ as with the Hardy and Dirichlet shifts.

10.1 The Bergman Space

The *Bergman space A^2* is the set of analytic functions f on $\mathbb{D}$ such that

$$\|f\| = \left(\frac{1}{\pi}\int_{\mathbb{D}}|f|^2 dA\right)^{\frac{1}{2}} \tag{10.1.1}$$

is finite. In the above, dA is two-dimensional planar Lebesgue measure and the $1/\pi$ is a normalizing factor that ensures $\|1\| = 1$. A computation with polar coordinates (Exercise 10.7.1) shows that if $f(z) = \sum_{n=0}^{\infty} a_n z^n$, then

$$\|f\|^2 = \sum_{n=0}^{\infty} \frac{|a_n|^2}{n+1}. \tag{10.1.2}$$

Since

$$\sum_{n=0}^{\infty} \frac{|a_n|^2}{n+1} \leqslant \sum_{n=0}^{\infty} |a_n|^2, \tag{10.1.3}$$

it follows that $H^2 \subseteq A^2$; the containment is strict (Exercise 10.7.4).

Since $A^2 \subseteq L^2(dA)$, and because analyticity is preserved under function addition and scalar multiplication, A^2 is a vector space. In light of (10.1.1), one can define an inner product on A^2 by

$$\langle f, g \rangle = \frac{1}{\pi} \int_{\mathbb{D}} f \bar{g} \, dA.$$

For $f(z) = \sum_{n=0}^{\infty} a_n z^n$ and $g(z) = \sum_{n=0}^{\infty} b_n z^n$ in A^2, this inner product can be written as

$$\langle f, g \rangle = \sum_{n=0}^{\infty} \frac{a_n \overline{b_n}}{n+1}. \tag{10.1.4}$$

Like its cousins, the Hardy and Dirichlet spaces, the Bergman space is a reproducing kernel Hilbert space. We first prove completeness and then discuss the reproducing kernel.

Proposition 10.1.5. *A^2 is a Hilbert space.*

Proof Since A^2 is an inner-product space, it suffices to check that A^2 is complete. First observe that the map

$$\sum_{n=0}^{\infty} a_n z^n \mapsto \left(\frac{a_n}{\sqrt{n+1}} \right)_{n=0}^{\infty}$$

is an isometric and surjective linear transformation from A^2 onto ℓ^2. Since ℓ^2 is complete, Lemma 9.1.8 implies that A^2 is also complete. ∎

The following result (Exercise 10.7.3) completes the verification that A^2 is a reproducing kernel Hilbert space.

Proposition 10.1.6. *Let*

$$k_\lambda(z) = \frac{1}{(1 - \bar{\lambda} z)^2} \quad \text{for } \lambda, z \in \mathbb{D}. \tag{10.1.7}$$

For $f \in A^2$ and $\lambda \in \mathbb{D}$, the following hold.

(a) $k_\lambda \in A^2$ and $\|k_\lambda\|^2 = \dfrac{1}{(1 - |\lambda|^2)^2}$.

(b) $f(\lambda) = \langle f, k_\lambda \rangle$.

(c) $|f(\lambda)| \leqslant \|f\| \dfrac{1}{1 - |\lambda|^2}$.

The Bergman space A^2 has a convenient orthonormal basis (Exercise 10.7.2) that can be used to derive the formula for the reproducing kernel (Exercise 10.7.33).

Proposition 10.1.8. *The sequence $(\sqrt{n+1}\, z^n)_{n=0}^{\infty}$ is an orthonormal basis for A^2. In particular, the polynomials are dense in A^2.*

The Bergman space is a rich class of functions with many mysterious properties that distinguish it from the Hardy and Dirichlet spaces. For example, functions in the Hardy or Dirichlet spaces have radial boundary values almost everywhere on $\mathbb{T}$. Bergman-space functions need not possess radial boundary values at all (see Exercises 10.7.7 and 10.7.8). In addition, the zero sets of A^2 functions may never be completely understood. Since the focus of this book is operator theory, we restrict our study to the Bergman shift. For further function-theoretic properties of A^2, consult [120, 187].

10.2 The Bergman Shift

Since

$$\|M_z f\|^2 = \frac{1}{\pi} \int_{\mathbb{D}} |zf(z)|^2 dA \leqslant \frac{1}{\pi} \Big(\sup_{z \in \mathbb{D}} |z|^2 \Big) \int_{\mathbb{D}} |f|^2 dA = \frac{1}{\pi} \int_{\mathbb{D}} |f|^2 dA = \|f\|^2,$$

for all $f \in A^2$, the *Bergman shift* $M_z : A^2 \to A^2$ defined by

$$(M_z f)(z) = zf(z)$$

is bounded on A^2 and

$$\|M_z\| = \sup_{\|f\|=1} \|M_z f\| \leqslant 1. \tag{10.2.1}$$

The next result establishes equality in (10.2.1).

Proposition 10.2.2. $\|M_z\| = 1$.

Proof We have just seen in (10.2.1) that $\|M_z\| \leqslant 1$. The functions

$$f_n(z) = \sqrt{n+1}\, z^n$$

are unit vectors in A^2 (Proposition 10.1.8) and

$$\|M_z f_n\|^2 = (n+1)\|z^{n+1}\|^2 = \frac{n+1}{n+2} \to 1 \quad \text{as } n \to \infty.$$

Therefore,

$$\|M_z\| \geqslant \sup_{n \geqslant 0} \|M_z f_n\| = 1,$$

and hence $\|M_z\| = 1$. $\blacksquare$

Since $(f_n)_{n=0}^{\infty}$ is an orthonormal basis for A^2 and

$$M_z f_n = \sqrt{\frac{n+1}{n+2}}\, f_{n+1},$$

the matrix representation of M_z with respect to this basis is

$$
\begin{bmatrix}
0 & 0 & 0 & 0 & \cdots \\
\sqrt{\frac{1}{2}} & 0 & 0 & 0 & \cdots \\
0 & \sqrt{\frac{2}{3}} & 0 & 0 & \cdots \\
0 & 0 & \sqrt{\frac{3}{4}} & 0 & \cdots \\
\vdots & \vdots & \vdots & \vdots & \ddots
\end{bmatrix}.
$$

Compare this matrix representation with those for the shifts on the Hardy space (5.1.3) and the Dirichlet space (9.2.2). In particular, the Bergman shift is a weighted shift, in the sense of Exercise 3.6.21.

The shift on the Hardy space is an isometry, that is, $\|M_z f\| = \|f\|$ for all $f \in H^2$, and the Dirichlet shift is expansive, that is, $\|M_z f\| \geqslant \|f\|$ for all $f \in \mathcal{D}$. For the Bergman shift, (10.2.1) shows that M_z is a contraction.

Corollary 10.2.3. $\|M_z f\| \leqslant \|f\|$ for all $f \in A^2$.

The spectral properties of the Bergman shift are similar to those of the Hardy and Dirichlet shifts. We begin our discussion with a lemma that is used several times in this chapter. For fixed $\lambda \in \mathbb{D}$, define the *difference quotient*

$$
(Q_\lambda f)(z) = \frac{f(z) - f(\lambda)}{z - \lambda} \quad \text{for } f \in A^2.
$$

The numerator of $Q_\lambda f$ has a zero of order at least 1 at $z = \lambda$, so $Q_\lambda f$ is analytic on $\mathbb{D}$.

Lemma 10.2.4. *If $\lambda \in \mathbb{D}$ and $f \in A^2$, then $Q_\lambda f \in A^2$. Furthermore, the linear transformation $f \mapsto Q_\lambda f$ is a bounded operator on A^2.*

Proof Let $f \in A^2$ and $\lambda \in \mathbb{D}$. Fix $r > 0$ such that $B_r(\lambda) = \{z : |z - \lambda| < r\}^- \subseteq \mathbb{D}$. Since $f \in A^2$ is analytic on $\mathbb{D}$, the function $Q_\lambda f$ is bounded on $B_r(\lambda)$. Then

$$
\int_{\mathbb{D}} |Q_\lambda f|^2 dA = \int_{B_r(\lambda)} |Q_\lambda f|^2 dA + \int_{\mathbb{D} \backslash B_r(\lambda)} |Q_\lambda f|^2 dA. \tag{10.2.5}
$$

The first term on the right side is bounded above by

$$
\pi r^2 \sup_{z \in B_r(\lambda)^-} |(Q_\lambda f)(z)|^2
$$

and the second term on the right side is bounded above by

$$
\frac{1}{r^2} \int_{\mathbb{D} \backslash B_r(\lambda)} |f(z) - f(\lambda)|^2 dA.
$$

The triangle inequality, along with the fact that the constants belong to A^2, show that both integrals on the right side of (10.2.5) are finite. Thus, $Q_\lambda f \in A^2$. The boundedness of the linear transformation $f \mapsto Q_\lambda f$ on A^2 follows from the closed graph theorem (Theorem 2.2.2). ∎

Proposition 10.2.6. *The Bergman shift M_z satisfies the following.*

(a) $\sigma(M_z) = \mathbb{D}^-$.

(b) $\sigma_p(M_z) = \emptyset$.

(c) $\sigma_{ap}(M_z) = \mathbb{T}$.

Proof (a) For $|\lambda| > 1$,

$$\sup_{z \in \mathbb{D}} \frac{1}{|z - \lambda|} = \frac{1}{|\lambda| - 1},$$

so

$$\frac{1}{\pi} \int_{\mathbb{D}} \left| \frac{1}{z - \lambda} f(z) \right|^2 dA \leqslant \frac{1}{(|\lambda| - 1)^2} \|f\|^2 \quad \text{for all } f \in A^2.$$

It follows that the operator $f \mapsto (z - \lambda)^{-1} f$ is bounded on A^2 and is the inverse of $M_z - \lambda I$. Consequently, $\sigma(M_z) \subseteq \mathbb{D}^-$. For $\lambda \in \mathbb{D}$, $(M_z - \lambda I) A^2 \subseteq \{f \in A^2 : f(\lambda) = 0\}$, which is a proper subset of A^2, and hence $M_z - \lambda I$ is not invertible. This proves that $\mathbb{D} \subseteq \sigma(M_z)$. Since the spectrum of an operator is closed (Theorem 2.4.9), we conclude that $\mathbb{D}^- \subseteq \sigma(M_z)$, so equality follows.

(b) If $\lambda \in \mathbb{C}$ and $f \in A^2$ with $(M_z - \lambda I) f \equiv 0$, then $(z - \lambda) f(z) = 0$ for all $z \in \mathbb{D}$ and hence $f \equiv 0$. Thus, $\sigma_p(M_z) = \emptyset$.

(c) Let $\xi \in \mathbb{T}$ and for each $n \geqslant 0$, define

$$f_n(z) = \frac{1}{c_n}(1 + \bar{\xi}z + \bar{\xi}^2 z^2 + \cdots + \bar{\xi}^{n-1} z^{n-1}),$$

where

$$c_n = \left(1 + \frac{1}{2} + \frac{1}{3} + \cdots + \frac{1}{n}\right)^{\frac{1}{2}}.$$

One can check that each f_n is a unit vector in A^2 that satisfies

$$(M_z - \xi I) f_n = -\frac{1}{c_n}(\xi - \bar{\xi}^{n-1} z^n).$$

Consequently,

$$\|(M_z - \xi I) f_n\| = \frac{1}{c_n}\sqrt{\frac{n + 2}{n + 1}}.$$

Since $c_n \to \infty$, it follows that $\|(M_z - \xi I) f_n\| \to 0$. Thus, $\xi \in \sigma_{ap}(M_z)$ which implies that $\mathbb{T} \subseteq \sigma_{ap}(M_z)$. For each $\lambda \in \mathbb{D}$, Lemma 10.2.4 implies that $Q_\lambda \in \mathcal{B}(A^2)$. In other words,

$$\sup_{\|f\|=1} \left\| \frac{f - f(\lambda)}{z - \lambda} \right\| = \|Q_\lambda\| < \infty.$$

Therefore,

$$\|f\| = \|Q_\lambda(z - \lambda)f\| \leqslant \|Q_\lambda\|\|(z - \lambda)f\| = \|Q_\lambda\|\|(M_z - \lambda I)f\|,$$

and hence

$$\sup_{\|f\|=1} \|(M_z - \lambda I)f\| \geqslant \frac{1}{\|Q_\lambda\|} > 0.$$

Consequently, $\lambda \notin \sigma_{ap}(M_z)$ and hence $\sigma_{ap}(M_z) = \mathbb{T}$. ∎

We begin our study of the adjoint of M_z with the following formula.

Proposition 10.2.7. *If* $g(z) = \sum_{n=0}^{\infty} b_n z^n \in A^2$, *then*

$$(M_z^* g)(z) = \sum_{n=0}^{\infty} \frac{n+1}{n+2} b_{n+1} z^n.$$

Proof If $f = \sum_{n=0}^{\infty} a_n z^n \in A^2$, use (10.1.4) to see that

$$\langle M_z f, g \rangle = \sum_{n=1}^{\infty} \frac{a_{n-1}\overline{b_n}}{n+1}$$

$$= \sum_{n=0}^{\infty} \frac{a_n \overline{b_{n+1}}}{n+2}$$

$$= \sum_{n=0}^{\infty} \frac{1}{n+1} a_n \overline{\left(\frac{n+1}{n+2} b_{n+1}\right)}$$

$$= \langle f, h \rangle,$$

where

$$h(z) = \sum_{n=0}^{\infty} \frac{n+1}{n+2} b_{n+1} z^n.$$

It follows from (10.1.2) that $h \in A^2$ and hence $h = M_z^* g$. ∎

Recall that $k_\lambda(z) = (1 - \bar{\lambda}z)^{-2}$ is the reproducing kernel for A^2 (Proposition 10.1.6). Notice that

$$(M_z^* k_\lambda)(z) = \langle M_z^* k_\lambda, k_z \rangle = \langle k_\lambda, z k_z \rangle = \overline{\bar{\lambda}k_z(\lambda)} = \bar{\lambda}k_\lambda(z) \quad \text{for all } \lambda \in \mathbb{D}.$$

Mimicking the proof of Proposition 5.2.4 leads to the next result.

Proposition 10.2.8. *The adjoint M_z^* of the Bergman shift satisfies the following.*

(a) $\sigma(M_z^*) = \mathbb{D}^-$.

(b) $\sigma_p(M_z^*) = \mathbb{D}$.

(c) $\sigma_{ap}(M_z^*) = \mathbb{D}^-$.

10.3 Invariant Subspaces

The invariant subspaces for the Bergman shift are extremely complicated and a complete description of them is a long-standing open problem. We begin with a description of the more tractable invariant subspaces.

If q is a polynomial, qA^2 is invariant under M_z, as is its closure $(qA^2)^-$. It may turn out that $(qA^2)^- = A^2$. For example, this happens if all the roots of q lie in $\{z : |z| > 1\}$ or if they all lie in $\mathbb{T}$ (see Lemma 10.3.3 below). If all the roots of q lie in $\mathbb{D}$, we will show that qA^2 is closed and the quotient space

$$A^2/qA^2 = \{f + qA^2 : f \in A^2\}$$

has dimension equal to the number of roots of q (counting multiplicity).

Definition 10.3.1. The *codimension* of an M_z-invariant subspace $\mathcal{M}$ of A^2 is $\dim(A^2/\mathcal{M})$.

The following is the main theorem of this section.

Theorem 10.3.2 (Axler–Bourdon [29]). *For an M_z-invariant subspace $\mathcal{M}$ of A^2, the following are equivalent.*

(a) *$\mathcal{M}$ has finite codimension d.*

(b) *There exists a polynomial q of degree d, with all of its zeros in $\mathbb{D}$, such that $\mathcal{M} = qA^2$.*

The proof needs a technical lemma. For $\xi \in \mathbb{T}$, Example 5.7.6 shows that $(z - \xi)H^2$ is dense in H^2. The same holds for A^2.

Lemma 10.3.3. $(z - \xi)A^2$ *is dense in* A^2 *for any* $\xi \in \mathbb{T}$.

Proof Example 5.7.6 provides a sequence $(p_n)_{n=1}^{\infty}$ of polynomials such that $(z-\xi)p_n \to 1$ in H^2 norm. Then (10.1.3) implies that $\|(z - \xi)p_n - 1\|_{A^2} \leqslant \|(z - \xi)p_n - 1\|_{H^2} \to 0$. Thus, $1 \in ((z - \xi)A^2)^-$. The M_z-invariance of $((z - \xi)A^2)^-$, along with the density of the polynomials in A^2 (Proposition 10.1.8), shows that $((z - \xi)A^2)^- = A^2$. ∎

We are now ready for the proof of Theorem 10.3.2.

Proof We follow [29]. First suppose that the roots of q are distinct points $w_1, w_2, ..., w_d$ in $\mathbb{D}$. If q has roots of higher multiplicity, we can apply the argument below with the appropriate number of derivatives (Exercise 10.7.13). We first show that

$$qA^2 = \{f \in A^2 : f(w_j) = 0, \ 1 \leqslant j \leqslant d\}.$$

The $\subseteq$ containment follows by inspection. For the $\supseteq$ containment, use the difference quotient operator Q_w from Lemma 10.2.4 to see that if $f \in A^2$ and $f(w_j) = 0$ for $1 \leqslant j \leqslant d$, then $Q_{w_1} Q_{w_2} Q_{w_3} \cdots Q_{w_d} f \in A^2$. Since

$$Q_{w_1} Q_{w_2} Q_{w_3} \cdots Q_{w_d} f = \frac{c}{q} f$$

for some nonzero constant c, the $\supseteq$ containment follows.

To study $\dim(A^2/qA^2)$, take any $h \in A^2$ and observe that $g = Q_{w_1} Q_{w_2} \cdots Q_{w_d} h \in A^2$. Direct computation shows that

$$g = \frac{h - q_1}{cq},$$

where c is a nonzero constant and q_1 is a polynomial of degree less than $d = \deg q$. Therefore, $h + qA^2 = q_1 + qA^2$, which means that the cosets $z^j + qA^2$ for $0 \leqslant j \leqslant d-1$ span A^2/qA^2. Since $\deg q = d$, these cosets are linearly independent and hence

$$\dim(A^2/qA^2) = d. \tag{10.3.4}$$

We now show that every M_z-invariant subspace $\mathcal{M}$ of codimension d is equal to qA^2 for some d degree polynomial q whose zeros lie in $\mathbb{D}$. Define $T \in \mathcal{B}(A^2/\mathcal{M})$ by

$$T(f + \mathcal{M}) = M_z f + \mathcal{M}.$$

To see that T is well defined, suppose $f_1 + \mathcal{M} = f_2 + \mathcal{M}$. Then $f_1 - f_2 \in \mathcal{M}$ and

$$T(f_1 + \mathcal{M}) = z f_1 + \mathcal{M} = z f_2 + z(f_1 - f_2) + \mathcal{M} = z f_2 + \mathcal{M} = T(f_2 + \mathcal{M}).$$

Note the use of the M_z-invariance of $\mathcal{M}$ and the fact that $f_1 - f_2 \in \mathcal{M}$. For any polynomial p, the M_z-invariance of $\mathcal{M}$ implies that

$$p(T)(f + \mathcal{M}) = pf + \mathcal{M}. \tag{10.3.5}$$

Since $\dim(A^2/\mathcal{M}) = d < \infty$, the cosets $z^j + \mathcal{M}$ for $0 \leqslant j \leqslant d$ are linearly dependent. Thus, there is a polynomial p_0 (not identically zero) such that

$$p_0(z) = \sum_{j=0}^{d} c_j z^j \in \mathcal{M}.$$

Then (10.3.5) implies that $p_0(T)(f + \mathcal{M}) = p_0 f + \mathcal{M} = \mathcal{M}$ for all polynomials f. Consequently, $p_0(T) = 0$ on $A^2/\mathcal{M}$. Factor p_0 as $p_0 = qh$, where q is a polynomial whose roots are in $\mathbb{D}$ and h is a polynomial whose roots are in $\mathbb{C}\backslash\mathbb{D}$. If $|w| > 1$, Proposition 10.2.6a implies that $(z - w)A^2 = A^2$. Lemma 10.3.3 ensures that $(z - \xi)A^2$ is dense in A^2 for every $\xi \in \mathbb{T}$. It follows that hA^2 is dense in A^2 and hence $qA^2 \subseteq (p_0 A^2)^- \subseteq \mathcal{M}$. Use (10.3.4) to obtain

$$\dim(A^2/\mathcal{M}) \leqslant \dim(A^2/qA^2) = \deg q \leqslant \deg p_0 \leqslant d = \dim(A^2/\mathcal{M}).$$

Therefore, $\dim(A^2/\mathcal{M}) = \dim(A^2/qA^2)$ and hence $\mathcal{M} = qA^2$ since $qA^2 \subseteq \mathcal{M}$. ∎

Many of the M_z-invariant subspaces of A^2 demonstrate wild behavior. As a measure of this, consider the *index* $\operatorname{ind} \mathcal{M} := \dim(\mathcal{M}/z\mathcal{M})$. Exercise 10.7.18 shows that

$$\dim(\mathcal{M}/z\mathcal{M}) = \dim(\mathcal{M} \cap (z\mathcal{M})^\perp).$$

In the Hardy and Dirichlet space, the index of every nonzero M_z-invariant subspace is one. Here are some examples of invariant subspaces with index one in the Bergman space. We investigate invariant subspaces of higher index in the next section.

Example 10.3.6. Suppose $f \in A^2 \backslash \{0\}$ and $[f] = \bigvee \{z^n f : n \geqslant 0\}$. Then the index of $[f]$ is 1. Indeed, without loss of generality, suppose that $f(0) = 1$. First we show that $\mathbb{C}f + z[f]$ is closed. Indeed, suppose $(c_n)_{n=1}^{\infty}$ is a sequence of complex numbers and $(g_n)_{n=1}^{\infty}$ is a sequence in $[f]$ such that $(c_n f + z g_n)_{n=1}^{\infty}$ is a Cauchy sequence. Since the linear functional λ on A^2 defined by $\lambda(h) = h(0)$ is continuous, this says that $(c_n)_{n=1}^{\infty}$ is a Cauchy sequence. It follows that $(z g_n)_{n=1}^{\infty}$ is a Cauchy sequence. Lemma 10.2.4 says that Q_0 is a bounded operator on A^2 and thus $g_n = Q_0(z g_n)$ forms a Cauchy sequence. This shows that $\mathbb{C}f + z[f]$ is closed.

Next we show that

$$[f] = \mathbb{C}f + z[f].$$

The $\supseteq$ containment follows from the fact that $[f]$ is an invariant subspace. For the $\subseteq$ containment, let $g \in [f]$ and let $(p_n)_{n=1}^{\infty}$ be a sequence of polynomials such that $p_n f \to g$ in norm. Since $f(0) = 1$, and since norm convergence in A^2 implies pointwise convergence (Proposition 10.1.6), it follows that $p_n(0) \to g(0)$. Define the polynomial

$$q_n(z) = \frac{p_n(z) - p_n(0)}{z}$$

and observe that

$$p_n f = p_n(0)f + z q_n f \in \mathbb{C}f + z[f].$$

Thus, $g \in \mathbb{C}f + z[f]$ from which it follows that $[f] = \mathbb{C}f + z[f]$. Moreover, for any $a \in \mathbb{C}$, the coset $af + z[f]$ is zero precisely when $a = 0$. Thus, the linear transformation $af + z[f] \mapsto a$ is a vector-space isomorphism between

$$[f]/z[f] = (\mathbb{C}f + z[f])/z[f]$$

and the one-dimensional space $\mathbb{C}$. Therefore $\dim([f]/z[f]) = 1$.

Example 10.3.7. Suppose $(z_j)_{j=1}^{\infty}$ is an infinite sequence of distinct points in $\mathbb{D} \backslash \{0\}$ and

$$\mathcal{M} = \{f \in A^2 : f(z_j) = 0 \text{ for all } j \geqslant 1\}.$$

Suppose further that $\mathcal{M} \neq \{0\}$, in other words, there is an $f \in A^2 \backslash \{0\}$ such that $f(z_j) = 0$ for all $j \geqslant 1$. Then $\mathcal{M}$ is an M_z-invariant subspace of A^2 with index 1. To see this, note that $\mathcal{M}$ is M_z-invariant and closed (Proposition 10.1.6). We claim that

$$z\mathcal{M} = \{f \in A^2 : f(0) = 0, f(z_j) = 0 \text{ for all } j \geqslant 1\}.$$

The $\subseteq$ containment follows by definition. For the other containment, let $f \in A^2$ with $f(0) = 0$ and $f(z_j) = 0$ for all $j \geqslant 1$. Recall the difference quotient operator Q_0 from Lemma 10.2.4 and observe that $Q_0 f \in A^2$ and $(Q_0 f)(z_j) = 0$ for all $j \geqslant 1$. In other words, $f/z = Q_0 f \in \mathcal{M}$ and hence $f \in z\mathcal{M}$.

Thus, the coset $f + z\mathcal{M}$ is zero precisely when $f(0) = 0$, and hence the linear transformation $f + z\mathcal{M} \mapsto f(0)$ is a vector-space isomorphism between $\mathcal{M}/z\mathcal{M}$ and $\mathbb{C}$. Therefore, $\dim(\mathcal{M}/z\mathcal{M}) = 1$.

10.4 Invariant Subspaces of Higher Index

Recall that in the Hardy and Dirichlet spaces, a nonzero shift-invariant subspace has index one. The main theorem of this section is the following.

Theorem 10.4.1 (Hedenmalm [186]). *There are M_z-invariant subspaces of A^2 with index two.*

The existence of such invariant subspaces was shown in [19], but the proof was nonconstructive. The example of Hedenmalm, which we present here, is concrete. The proof depends upon several technical results about sampling and interpolation sequences that are beyond the scope of this book. We use the presentation from [187], which has a Banach-space flavor to it and applies to the A^p version of the Bergman spaces as well.

The general idea is to take two zero-based invariant subspaces

$$J(\Lambda_j) = \{f \in A^2 : f|_{\Lambda_j} = 0\} \quad \text{for } j = 1, 2, \tag{10.4.2}$$

where the Λ_j are sequences in $\mathbb{D}$, and then consider their algebraic sum $J(\Lambda_1) + J(\Lambda_2)$. As we show below, one can choose sequences Λ_1 and Λ_2 such that $J(\Lambda_1)$ and $J(\Lambda_2)$ are nonzero and $J(\Lambda_1) + J(\Lambda_2)$ is closed. We know from Example 10.3.7 that each $J(\Lambda_j)$ has index one. It turns out that $J_1(\Lambda_1) + J(\Lambda_2)$ has index two. This is explained with the following lemmas.

Lemma 10.4.3. *Let $\mathcal{M}$ be a nonzero invariant subspace of A^2 such that $\mathcal{M}$ has index one. Then there is a nonzero bounded linear functional $\lambda : \mathcal{M} \to \mathbb{C}$ such that $\ker \lambda = z\mathcal{M}$.*

Proof For the moment, assume that 0 is not a common zero of $\mathcal{M}$. In other words, there is a $g \in \mathcal{M}$ such that $g(0) \neq 0$. Now define $\lambda : \mathcal{M} \to \mathbb{C}$ by $\lambda(f) = f(0)$. Proposition 10.1.6 ensures that λ is bounded on A^2. Thus, λ is bounded on $\mathcal{M}$ and nonzero since $\lambda(g) \neq 0$.

We now show that $\lambda(h) = 0$ if and only if $h \in z\mathcal{M}$. The definition of λ says that $\lambda(z\mathcal{M}) = 0$. For the other direction, since $\mathcal{M}/z\mathcal{M}$ is one dimensional and $g(0) \neq 0$, it follows that $\mathcal{M}/z\mathcal{M} = \mathbb{C}g + z\mathcal{M}$ and that a coset $ag + z\mathcal{M}$ is the zero coset if and only if $a = 0$. If $h \in \mathcal{M}$ and $\lambda(h) = 0$, then $h + z\mathcal{M} = ag + z\mathcal{M}$ for some constant a. Consequently, $0 = h(0) = ag(0)$, which implies $a = 0$. Thus, $h \in z\mathcal{M}$. This shows that λ vanishes precisely on $z\mathcal{M}$.

Suppose $f(0) = 0$ for all $f \in \mathcal{M}$. Let m be the largest integer such that z^m divides every $f \in \mathcal{M}$. Then repeat the argument above with the functional $\lambda(f) = f^{(m)}(0)$. ∎

Lemma 10.4.4. *Suppose that $\mathcal{M}_1$ and $\mathcal{M}_2$ are M_z-invariant subspaces with index one and such that there is a constant $c > 0$ with*

$$\|f_1 + f_2\| \geq c(\|f_1\| + \|f_2\|) \quad \text{for all } f_1 \in \mathcal{M}_1 \text{ and } f_2 \in \mathcal{M}_2. \tag{10.4.5}$$

Then $\mathcal{M} = \mathcal{M}_1 + \mathcal{M}_2$ is an invariant subspace of A^2 with index two.

Proof We first prove that $\mathcal{M} = \mathcal{M}_1 + \mathcal{M}_2$ is closed. Let $(g_n)_{n=1}^{\infty}$ be a sequence in $\mathcal{M}_1$ and $(h_n)_{n=1}^{\infty}$ be a sequence in $\mathcal{M}_2$ such that $(g_n + h_n)_{n=1}^{\infty}$ is a Cauchy sequence in $\mathcal{M}_1 + \mathcal{M}_2$. The inequality in (10.4.5) shows that $(g_n)_{n=1}^{\infty}$ and $(h_n)_{n=1}^{\infty}$ are Cauchy sequences in $\mathcal{M}_1$ and $\mathcal{M}_2$, respectively. Since each $\mathcal{M}_j$ is complete, $g_n \to g \in \mathcal{M}_1$ and $h_n \to h \in \mathcal{M}_2$. Thus, $g_n + h_n \to g + h \in \mathcal{M}_1 + \mathcal{M}_2$ and hence $\mathcal{M}_1 + \mathcal{M}_2$ is closed. Since each $\mathcal{M}_j$ is M_z-invariant, their sum is also M_z-invariant.

We now prove that $\mathcal{M}$ has index two. Lemma 10.4.3 provides nonzero linear functionals $\lambda_j : \mathcal{M}_j \to \mathbb{C}$ whose kernels are precisely $z\mathcal{M}_j$ for $j = 1, 2$. Now define the linear transformation

$$\lambda : \mathcal{M} = \mathcal{M}_1 + \mathcal{M}_2 \to \mathbb{C}^2, \quad \lambda(f_1 + f_2) = (\lambda_1(f_1), \lambda_2(f_2)).$$

By (10.4.5), if $f_1, g_1 \in \mathcal{M}_1$ and $f_2, g_2 \in \mathcal{M}_2$ are such that $f_1 + f_2 = g_1 + g_2$, then $f_1 = g_1$ and $f_2 = g_2$. Thus, λ is well defined.

Each λ_j is surjective with kernel $z\mathcal{M}_j$. Therefore, λ is surjective with kernel $z\mathcal{M}$. By the vector-space isomorphism theorem, the linear transformation

$$[\lambda] : \mathcal{M}/z\mathcal{M} \to \mathbb{C}^2, \quad [\lambda](f_1 + f_2 + z\mathcal{M}) = \lambda(f_1 + f_2),$$

is an isomorphism between $\mathcal{M}/z\mathcal{M}$ and $\mathbb{C}^2$. In particular, $\dim(\mathcal{M}/z\mathcal{M}) = 2$. ∎

The details of the next lemma are beyond the scope of this book and are contained in [186] and [187, p. 177]. For a sequence $\Lambda \subseteq \mathbb{D}$, recall the definition of $J(\Lambda)$ from (10.4.2).

Lemma 10.4.6. *There are disjoint sequences B_1 and B_2 in $\mathbb{D}$ satisfying the following.*

(a) $0 \notin B_1$ and $0 \notin B_2$.

(b) $J(B_j) \neq \{0\}$ for $j = 1, 2$.

(c) *The sequence $B = B_1 \cup B_2 = (z_j)_{j=1}^{\infty}$ has the following property: there are $K_1, K_2 > 0$ such that*

$$K_1 \|f\|^2 \leqslant \sum_{j=1}^{\infty} (1 - |z_j|^2) |f(z_j)|^2 \leqslant K_2 \|f\|^2 \quad \text{for all } f \in A^2. \tag{10.4.7}$$

For convenience, write (10.4.7) as

$$\|f\|^2 \asymp \sum_{j=1}^{\infty} (1 - |z_j|^2) |f(z_j)|^2 \quad \text{for all } f \in A^2.$$

Sequences $(z_j)_{j=1}^{\infty}$ satisfying (10.4.7) are *sampling sequences* and they essentially permit us to discretize the Bergman space norm. The fact that the union of two zero sequences in the Bergman space can be a sampling sequence is one of the remarkable features of the Bergman space. This does not occur in the Hardy and Dirichlet spaces where the union of two zero sets is another zero set.

With all this in place, we are ready to prove Theorem 10.4.1.

Proof It suffices to produce the index-1 invariant subspaces $\mathcal{M}_1$ and $\mathcal{M}_2$ with the properties in Lemma 10.4.4. Let Λ_1 and Λ_2 be disjoint zero sequences for A^2 whose union is a sampling sequence $(z_j)_{j=1}^{\infty}$ for A^2 (Lemma 10.4.6). Each z_j belongs to either Λ_1 or Λ_2, but not both. If $f_1 \in J(\Lambda_1)$ and $f_2 \in J(\Lambda_2)$ then,

$$|f_1(z_j) + f_2(z_j)|^2 = |f_1(z_j)|^2 + |f_2(z_j)|^2$$

since at least one of $f_1(z_j)$ or $f_2(z_j)$ is zero. Hence,

$$\|f_1 + f_2\|^2 \asymp \sum_{j=1}^{\infty}(1 - |z_j|^2)^2|f_1(z_j) + f_2(z_j)|^2$$

$$= \sum_{j=1}^{\infty}(1 - |z_j|^2)^2|f_1(z_j)|^2 + \sum_{j=1}^{\infty}(1 - |z_j|^2)^2|f_2(z_j)|^2$$

$$\asymp \|f_1\|^2 + \|f_2\|^2,$$

which completes the proof. ∎

One can extend this to index $n \in \mathbb{N}$ with some modifications [187, p. 177]. With a different technique, one can create infinite-index invariant subspaces [188]. Despite the fact that the index of $\mathcal{M}$ can be any positive integer (or infinity), the wandering subspace $\mathcal{M} \cap (M_z\mathcal{M})^{\perp}$ generates $\mathcal{M}$, as in the Hardy and Dirichlet spaces.

Theorem 10.4.8 (Aleman–Richter–Sundberg [17]). *If $\mathcal{M}$ is an M_z-invariant subspace of A^2, then $\mathcal{M}$ is the smallest invariant subspace of A^2 containing $\mathcal{M} \cap (z\mathcal{M})^{\perp}$.*

This wandering subspace property holds for other spaces beyond the Hardy, Dirichlet, and Bergman spaces [344].

10.5 Multipliers and Commutant

Proposition 5.5.4 shows that the multiplier algebra of H^2 is H^{∞}. The same is true for the Bergman space.

Proposition 10.5.1. *The multiplier algebra of A^2 is H^{∞}. Moreover, if $\varphi \in H^{\infty}$, the operator $M_{\varphi}f = \varphi f$ is bounded on A^2 and $\|M_{\varphi}\| = \|\varphi\|_{\infty}$.*

Proof For any $\varphi \in H^{\infty}$ and $f \in A^2$,

$$\|M_{\varphi}f\|^2 = \|\varphi f\|^2 = \frac{1}{\pi}\int_{\mathbb{D}}|\varphi|^2|f|^2dA \leqslant \|\varphi\|_{\infty}^2\frac{1}{\pi}\int_{\mathbb{D}}|f|^2dA = \|\varphi\|_{\infty}^2\|f\|^2.$$

Thus, $\varphi f \in A^2$ whenever $f \in A^2$, that is, φ is a multiplier of A^2. Moreover,

$$\|M_{\varphi}\| = \sup_{\|f\|=1}\|\varphi f\| \leqslant \|\varphi\|_{\infty}.$$

If φ is a multiplier of A^2, then the argument used to prove Proposition 5.5.1 shows that φ is a bounded function and the operator M_φ is bounded on A^2. We have already seen that $\|M_\varphi\| \leqslant \|\varphi\|_\infty$. For the reverse inequality, we have

$$\frac{1}{\pi} \int_{\mathbb{D}} |\varphi|^{2n} dA = \|M_\varphi^n 1\|^2 \leqslant \|M_\varphi^n\|^2 \|1\|^2 \leqslant \|M_\varphi\|^{2n} \quad \text{for all } n \geqslant 1.$$

Thus,

$$\left(\frac{1}{\pi} \int_{\mathbb{D}} |\varphi|^{2n} dA\right)^{\frac{1}{2n}} \leqslant \|M_\varphi\| \quad \text{for all } n \geqslant 1.$$

Let $n \to \infty$ and conclude that $\|\varphi\|_\infty \leqslant \|M_\varphi\|$. ∎

Corollary 5.6.2 identifies the commutant of the shift on the Hardy space. The same proof applies to the Bergman space.

Proposition 10.5.2. *For $T \in \mathcal{B}(A^2)$, the following are equivalent.*

(a) $M_z T = T M_z$.

(b) $T = M_\varphi$ *for some* $\varphi \in H^\infty$.

10.6 Notes

Bergman was the first to study square-integrable harmonic functions with respect to area measure in his 1922 paper [45] (see also [46]). Bochner [55] also looked at the reproducing kernel for A^2. There are also A^p versions of the Bergman space (same definition but with 2 replaced by p) where one shows how the zero-set and invariant-subspace structure depend on p [120, 187]. Compare this with the Hardy spaces H^p where the zero sets and the invariant subspaces are the same for each $1 < p < \infty$.

The boundary behavior of Bergman-space functions can be wild in the sense that there are functions in A^2 which have radial limits almost nowhere (see Exercises 10.7.7 and 10.7.8). This is a dramatic change from the Hardy-space setting where radial limits exist almost everywhere (Proposition 5.3.12) and with the Dirichlet space where the radial limits exist quasi-everywhere (see the endnotes for Chapter 9).

Despite the best efforts of excellent mathematicians over the years, our understanding of the Bergman space is relatively thin in comparison with what is known for the Hardy space. For example, the "zero-based" invariant subspaces $J(B)$ discussed in this chapter are not fully understood. A complete description of zero sequences for the Bergman space is not known, although there are many partial results. There is also no complete description of the cyclic vectors for the Bergman shift.

The complexity of the invariant subspaces for the Bergman shift, and the possibility that they may never be completely described, can be measured in many ways. Unlike the nonzero invariant subspaces for the shifts on the Hardy and Dirichlet spaces, which always have index one, in the Bergman space this index can be any positive integer (or even infinity). As another measure, the invariant subspaces of the Bergman space have the property that $M_z|_{\mathcal{M}_1}$ is unitarily equivalent to $M_z|_{\mathcal{M}_2}$ if and only if $\mathcal{M}_1 = \mathcal{M}_2$ [292]. The lack of radial limits, noted earlier, has something to do with the complexity of the invariant subspaces. For example, a result of Aleman, Richter, and Ross says that if $f, g \in A^2$ and f/g has a finite non-tangential limit on a set of positive measure on $\mathbb{T}$, the invariant subspace generated by f and g has index one [16]. This implies that functions in higher index invariant subspaces generally have wild boundary behavior.

The most salient results that display the complexity of the invariant subspaces of the Bergman space are those of Apostol, Bercovici, Foiaş, and Pearcy [19]. They observed that the dimension of $\mathcal{M} \cap (M_z\mathcal{M})^\perp$ could be any $n \geqslant 1$. Theorem 10.4.1 provided a specific example of this for $n = 2$. They also proved that the collection of invariant subspaces is so vast that it contains an isomorphic copy of the lattice of invariant subspaces for any bounded operator on a separable Hilbert space.

10.7 Exercises

Exercise 10.7.1. For any $f(z) = \sum_{n=0}^{\infty} a_n z^n$ analytic on $\mathbb{D}$, prove that

$$\frac{1}{\pi} \int_{\mathbb{D}} |f|^2 \, dA = \sum_{n=0}^{\infty} \frac{|a_n|^2}{n + 1},$$

where $dA = dx\, dy$ is area measure.

Exercise 10.7.2. Prove that $(\sqrt{n + 1}\, z^n)_{n=0}^{\infty}$ is an orthonormal basis for A^2.

Exercise 10.7.3. Prove that

$$k_\lambda(z) = \frac{1}{(1 - \bar{\lambda}z)^2} \quad \text{for } \lambda, z \in \mathbb{D},$$

is the reproducing kernel for A^2 as follows.

(a) Prove that $k_\lambda \in A^2$ for every $\lambda \in \mathbb{D}$.

(b) Prove that $\langle f, k_\lambda \rangle = f(\lambda)$ for every $\lambda \in \mathbb{D}$ and $f \in A^2$.

Exercise 10.7.4. Prove by example that the containment $H^2 \subseteq A^2$ is strict.

Exercise 10.7.5. Another approach to the completeness of A^2 uses *Montel's theorem* [92, p. 153]. This theorem states that if a sequence $(f_n)_{n=1}^{\infty}$ of analytic functions on $\mathbb{D}$ is *locally bounded*, that is, for any compact $K \subseteq \mathbb{D}$ there is a $c_K > 0$ such that $|f_n(z)| \leqslant c_K$ for all $z \in K$ and $n \geqslant 1$, then there is a subsequence $(f_{n_k})_{k=1}^{\infty}$ that converges uniformly on each compact subset $K \subseteq \mathbb{D}$ to an analytic function f on $\mathbb{D}$.

(a) Prove that each Cauchy sequence $(f_n)_{n=1}^{\infty}$ in A^2 is locally bounded.

(b) Since $L^2(\mathbb{D}, dA)$ is complete, $f_n \to f$ for some $f \in L^2(\mathbb{D}, dA)$. Apply Montel's theorem to argue that f is analytic.

Exercise 10.7.6. For $1 < p < \infty$, let A^p denote the space of analytic functions f on $\mathbb{D}$ such that

$$\|f\|_p = \left(\int_{\mathbb{D}} |f|^p dA \right)^{\frac{1}{p}} < \infty.$$

Prove that A^p is a Banach space.

Exercise 10.7.7. A result of Littlewood [118, p. 228] says that if $(a_n)_{n=0}^{\infty}$ is a sequence of complex numbers such that $\lim\limits_{n \to \infty} |a_n|^{\frac{1}{n}} = 1$ and $\sum\limits_{n=0}^{\infty} |a_n|^2 = \infty$, there is a sequence $(\varepsilon_n)_{n=0}^{\infty}$, with $\varepsilon_n = \pm 1$ for each n, such that

$$f(z) = \sum_{n=0}^{\infty} \varepsilon_n a_n z^n$$

is analytic on $\mathbb{D}$ and has radial limits almost nowhere. Use this result to produce an $f \in A^2$ with radial limits almost nowhere.

Exercise 10.7.8. This is an extension of Exercise 10.7.7 and concerns the function

$$f(z) = \sum_{n=1}^{\infty} z^{2^n}.$$

(a) Show that $f \in A^2$.

(b) Show that f has a radius of convergence 1.

Remark: A result of Zygmund [380, Ch. 5] says that f has radial limits almost nowhere.

Exercise 10.7.9. Define the inclusion operator $i : H^2 \to A^2$ by $i(f) = f$; (10.1.3) ensures it is well defined.

(a) Prove that i is bounded, has dense range, but is not surjective.

(b) Compute i^*.

(c) Prove that i^* has dense range.

(d) Prove that the range of i^* is not closed.

Exercise 10.7.10. Repeat the steps in Exercise 5.9.12 for A^2.

Exercise 10.7.11. For each $n \geqslant 0$, let $f_n(z) = z^n$.

(a) Prove that $f_n \to 0$ weakly in H^2.

(b) Prove that $f_n \to 0$ weakly in A^2.

(c) Prove that f_n does not converge weakly in $\mathcal{D}$.

Exercise 10.7.12. Since the Bergman space A^2 is a closed subspace of $L^2(dA)$, there is an orthogonal projection P of $L^2(dA)$ onto A^2 (Proposition 1.7.3).

(a) Prove that $(Pf)(\lambda) = \langle f, k_\lambda \rangle$ for all $f \in L^2(dA)$, where $k_\lambda(z)$ is the reproducing kernel for A^2 from (10.1.7).

(b) Prove that for $m, n \geq 0$, $P(\bar{z}^n z^m) = \begin{cases} \dfrac{m-n+1}{m+1} z^{m-n} & \text{if } m \geq n, \\ 0 & \text{if } m < n. \end{cases}$

Exercise 10.7.13. For distinct $\lambda_1, \lambda_2, ..., \lambda_N$ in $\mathbb{D}$ and positive integers $n_1, n_2, ..., n_N$, consider the polynomial

$$q(z) = \prod_{j=1}^{N} (z - \lambda_j)^{n_j}.$$

Prove that $qA^2 = \{f \in A^2 : f^{(k)}(\lambda_j) = 0, \ 0 \leq k \leq n_j - 1, \ 1 \leq j \leq N\}$.

Exercise 10.7.14. Prove a version of Theorem 10.3.2 for the Hardy space.

Exercise 10.7.15. Prove the following adjoint formula for the Bergman shift M_z:

$$(M_z^* f)(z) = z^{-2}\left(zf(z) - \int_0^z f(w)\, dw\right) \quad \text{for } f \in A^2.$$

Exercise 10.7.16. For the Bergman shift M_z, let $T = M_z^* M_z - M_z M_z^*$. Prove the following formulas.

(a) $(Tf)(z) = z^{-2} \int_0^z (z - w)f(w)\, dw$ for all $f \in A^2$.

(b) $Tz^n = \dfrac{1}{(n+1)(n+2)} z^n$ for all $n \geq 0$.

(c) T is compact and has norm 1.

Exercise 10.7.17. For the Bergman shift M_z, prove that $(M_z^* M_z)^{-1} = 2I - M_z M_z^*$. *Remark:* See [153] for more on this.

Exercise 10.7.18. For an M_z-invariant subspace $\mathcal{M}$ of A^2, prove that

$$\dim(\mathcal{M}/z\mathcal{M}) = \dim(\mathcal{M} \cap (z\mathcal{M})^{\perp}).$$

Exercise 10.7.19. Recall the index of an invariant subspace $\mathcal{M}$ of A^2, denoted by ind $\mathcal{M}$, is $\dim(\mathcal{M}/z\mathcal{M})$. If $\mathcal{M}$ and $\mathcal{N}$ are M_z-invariant subspaces of A^2, prove that $\text{ind}(\mathcal{M} \bigvee \mathcal{N}) \leq$ ind $\mathcal{M} +$ ind $\mathcal{N}$.

Exercise 10.7.20. One can define Bergman spaces for domains other than $\mathbb{D}$ [30, 95, 170]. Let G be a bounded domain in $\mathbb{C}$. By this we mean that G is a nonempty bounded open connected subset of $\mathbb{C}$. Let $A^2(G)$ be the set of analytic functions f on G such that

$$\|f\| = \left(\int_G |f|^2 dA\right)^{\frac{1}{2}} < \infty.$$

One can show that $A^2(G)$ is a vector space and that $\langle f, g \rangle = \int_G f\bar{g}\,dA$ defines an inner product on $A^2(G)$. Prove the following facts about $A^2(G)$.

(a) Suppose that $\lambda \in G$ and $B(\lambda, r)^- = \{z : |z - \lambda| \leqslant r\} \subseteq G$. For a function g that is analytic on a neighborhood of $B(\lambda, r)^-$, prove that

$$g(\lambda) = \frac{1}{2\pi} \int_0^{2\pi} g(\lambda + te^{i\theta})\,d\theta \quad \text{for all } 0 < t < r.$$

(b) Use this to prove that $g(\lambda) = \dfrac{1}{\pi r^2} \displaystyle\int_{B(\lambda,r)} g\,dA$.

(c) Now argue that $|f(\lambda)| \leqslant C_\lambda \|f\|$ for all $f \in A^2(G)$, where $C_\lambda > 0$ depends only on λ.

(d) Prove that $A^2(G)$ is a Hilbert space.

(e) Prove that for each $\lambda \in G$, there is a $k_\lambda \in A^2(G)$ such that $f(\lambda) = \langle f, k_\lambda \rangle$ for every $f \in A^2(G)$.

Exercise 10.7.21. This is a continuation of Exercise 10.7.20. Define the linear transformation M_z on $A^2(G)$ by $M_z f = zf$.

(a) Prove that M_z is bounded on $A^2(G)$.

(b) Compute $\sigma(M_z)$, $\sigma_p(M_z)$, and $\sigma_{ap}(M_z)$.

(c) Show that $G \subseteq \sigma_p(M_z^*)$.

Exercise 10.7.22. Consider $A^2(G)$, the Bergman space of the annulus

$$G = \{z : \tfrac{1}{2} < |z| < 1\}.$$

(a) Prove that there is a constant $c > 0$ such that $|p(0)| \leqslant c\|p\|$ for all polynomials p.

(b) Prove that the polynomials are not dense in $A^2(G)$.

Exercise 10.7.23. If G_1 and G_2 are domains in $\mathbb{C}$ and $\varphi : G_1 \to G_2$ is analytic and bijective, prove that $U : A^2(G_2) \to A^2(G_1)$ defined by $Uf = (f \circ \varphi)\varphi'$ is unitary.

Exercise 10.7.24. Suppose that G is an unbounded domain in $\mathbb{C}$. Can M_z be bounded on $A^2(G)$?

Exercise 10.7.25. An invariant subspace $\mathcal{M}$ for $T \in \mathcal{B}(\mathcal{H})$ is *hyperinvariant* if $A\mathcal{M} \subseteq \mathcal{M}$ for every $A \in \mathcal{B}(\mathcal{H})$ with $AT = TA$.

(a) Prove that if $A = M_\xi$ on $L^2(\mathbb{T})$ (discussed in Chapter 4), then every hyperinvariant subspace $\mathcal{M}$ is of the form $\mathcal{M} = \{f \in L^2(\mathbb{T}) : f|_E = 0$ almost everywhere$\}$ for some measurable $E \subseteq \mathbb{T}$.

(b) Prove that if S is the unilateral shift on H^2, then a subspace $\mathcal{M}$ is S-invariant if and only if $\mathcal{M}$ is hyperinvariant.

Exercise 10.7.26. Use the following steps to prove that a subspace $\mathcal{M} \subseteq A^2$ is M_z-invariant if and only if $\mathcal{M}$ is hyperinvariant.

(a) Prove that $\mathcal{M} \subseteq A^2$ is hyperinvariant if and only if $\varphi\mathcal{M} \subseteq \mathcal{M}$ for every $\varphi \in H^\infty$.

(b) One can prove that if $\varphi \in H^\infty$, then there is a sequence of polynomials $(\varphi_n)_{n=1}^\infty$ such that $\varphi_n(\lambda) \to \varphi(\lambda)$ for each $\lambda \in \mathbb{D}$ and $\sup_{n \geq 1} \|\varphi_n\|_\infty < \infty$ [202, Ch. 3]. Use this to prove that $\langle M_{\varphi_n} f, g \rangle \to \langle M_\varphi f, g \rangle$ for every $f, g \in A^2$.

(c) Prove that if $z\mathcal{M} \subseteq \mathcal{M}$, then $\varphi\mathcal{M} \subseteq \mathcal{M}$ for every $\varphi \in H^\infty$.

(d) If $G = \{\frac{1}{2} < |z| < 1\}$. Prove there is an M_z-invariant subspace of $A^2(G)$ that is not hyperinvariant.

Exercise 10.7.27. For each $\lambda, z \in \mathbb{D}$, let $\widetilde{k}_\lambda(z) = k_\lambda(z)/\|k_\lambda\|$ denote the normalized reproducing kernel for A^2.

(a) If $|\lambda| \to 1^-$, prove that $\widetilde{k}_\lambda \to 0$ weakly in A^2.

(b) If $T \in \mathcal{B}(A^2)$ is compact, prove that the *Berezin transform* [187] $\widetilde{T}(\lambda) = \langle T\widetilde{k}_\lambda, \widetilde{k}_\lambda \rangle$ tends to zero as $|\lambda| \to 1^-$.

Remark: See [44] for one of the first papers on this transform.

Exercise 10.7.28. This is a continuation of Exercise 10.7.27. For any $T \in \mathcal{B}(A^2)$ prove the following facts about the Berezin transform $\widetilde{T}$.

(a) $|\widetilde{T}(z)| \leq \|T\|$ for all $z \in \mathbb{D}$.

(b) $\widetilde{T^*} = \overline{\widetilde{T}}$.

(c) The map $T \mapsto \widetilde{T}$ is linear and injective.

Exercise 10.7.29. This is a continuation of Exercise 10.7.27.

(a) For $f, g \in A^2$, prove that

$$\widetilde{f \otimes g}(z) = \frac{\overline{g(z)}}{\|k_\lambda\|^2} f(z) \quad \text{for all } z \in \mathbb{D}.$$

(b) For $\varphi \in H^\infty$, prove that $\widetilde{M_\varphi}(z) = \varphi(z)$ for all $z \in \mathbb{D}$.

Exercise 10.7.30. This is a continuation of Exercise 10.7.27. For $f \in A^2$, consider $(Tf)(z) = f(-z)$.

(a) Prove that T is a unitary operator on A^2.

(b) Prove that $\widetilde{T}(\lambda) = \dfrac{(1 - |\lambda|^2)^2}{(1 + |\lambda|^2)^2}$ for all $\lambda \in \mathbb{D}$.

(c) Use this to prove that the converse of (b) in Exercise 10.7.27 is not always true.

Remark: Exercise 16.9.33 explores this further.

Exercise 10.7.31.

(a) Prove that the Bergman shift M_z is irreducible.

(b) Prove that M_z^2 is reducible.

Exercise 10.7.32. The Hardy, Dirichlet, and Bergman spaces are examples of reproducing kernel Hilbert spaces. Let $\mathcal{H}$ be a Hilbert space of functions on a set X such that for each $x \in X$, the evaluation functional $\lambda_x(f) = f(x)$ is bounded.

(a) Prove there is a $k(x, y)$ on $X \times X \to \mathbb{C}$ such that $k(x, \cdot) \in \mathcal{H}$ for each $x \in X$ and $f(x) = \langle f, k(x, \cdot) \rangle$ for every $f \in \mathcal{H}$. This function is the *reproducing kernel* for $\mathcal{H}$.

(b) Prove that $k(x, y) = \langle k(x, \cdot), k(y, \cdot) \rangle$ for every $x, y \in X$.

(c) Prove that $k(x, y) = \overline{k(y, x)}$ for all $x, y \in X$.

(d) Prove that for distinct $x_1, x_2, ..., x_n \in X$, the matrix $[k(x_i, x_j)]_{1 \leq i, j \leq n}$ is positive semidefinite.

(e) If $\mathcal{H}$ is separable with orthonormal basis $(f_n)_{n=1}^\infty$, prove that

$$k(x, y) = \sum_{n=1}^\infty f_n(x)\overline{f_n(y)} \quad \text{for all } x, y \in X.$$

Remark: See [264] for more on reproducing kernel Hilbert spaces.

Exercise 10.7.33. Use Exercise 10.7.32, along with the orthonormal basis for A^2 from Proposition 10.1.8, to derive the formula for $k_\lambda(z)$ in Exercise 10.7.3.

Exercise 10.7.34. This is a continuation of Exercise 10.7.32. Consider the Sobolev space W of absolutely continuous functions f on $[0, 1]$ such that $f(0) = f(1) = 0$ and $f' \in L^2[0, 1]$ (see Exercise 1.10.15). Define an inner product on W by $\langle f, g \rangle = \displaystyle\int_0^1 f'(t)\overline{g'(t)}\, dt$.

(a) For each $x \in [0, 1]$, prove that $\lambda_x(f) = f(x)$ is bounded on W.

(b) Prove that

$$k(x, y) = \begin{cases} (1 - y)x & \text{if } x \leq y, \\ (1 - x)y & \text{if } y \leq x, \end{cases}$$

is the reproducing kernel for W.

(c) Prove that $(f_n)_{n \in \mathbb{Z} \setminus \{0\}}$, where $f_n(x) = \dfrac{e^{2\pi inx} - 1}{2\pi n}$, is an orthonormal basis for W.

(d) Verify the formula $k(x, y) = \displaystyle\sum_{n \neq 0} f_n(x)\overline{f_n(y)}$.

10.8 Hints for the Exercises

Hint for Ex. 10.7.1: Use polar coordinates.

Hint for Ex. 10.7.3: Use $\dfrac{1}{(1-w)^2} = 1 + 2w + 3w^2 + 4w^3 + \cdots$ for $w \in \mathbb{D}$.

Hint for Ex. 10.7.6: Make use of the formula

$$f(\lambda) = \frac{1}{\pi} \int_{\mathbb{D}} \frac{f(z)}{(1 - \lambda \overline{z})^2} \, dA$$

for $f \in A^p$ to prove

$$|f(\lambda)| \leqslant c_{p,\lambda} \left(\int_{\mathbb{D}} |f|^p \, dA \right)^{\frac{1}{p}}.$$

Now use Exercise 10.7.5.

Hint for Ex. 10.7.9: For (b), use inner products and Taylor series. For (d), use Exercise 9.9.6.

Hint for Ex. 10.7.13: Examine the proof of Theorem 10.3.2.

Hint for Ex. 10.7.17: First prove that M_z is injective with closed range.

Hint for Ex. 10.7.20: For (d), consider the following: If $(f_n)_{n=1}^{\infty}$ is a Cauchy sequence in $A^2(G)$, then $f_n \to f$ for some $f \in L^2(G, dA)$. Use (c) and Montel's theorem (Exercise 10.7.5) to show that f is analytic.

Hint for Ex. 10.7.23: Use Jacobians.

Hint for Ex. 10.7.25: For (b), consult Theorem 5.4.12 and Theorem 5.6.2.

Hint for Ex. 10.7.31: For (b), use Exercise 5.9.31.

11

$$\cdot \quad \bullet \quad \bullet \quad \bullet \quad \cdot$$

The Fourier Transform

Key Concepts: Fourier transform, convolution, Fourier inversion formula, unitary operator, Plancherel's theorem, spectral properties of the Fourier transform, Hermite functions, Hardy space of the upper half-plane, Paley–Wiener theorem.

Outline: We study the Fourier transform $\mathscr{F} : L^2(\mathbb{R}) \to L^2(\mathbb{R})$ defined by

$$(\mathscr{F}f)(x) = \frac{1}{\sqrt{2\pi}} \int_{-\infty}^{\infty} f(t)e^{-ixt}dt, \tag{11.0.1}$$

show that it is a unitary operator, and compute its spectral decomposition. Along the way, we discuss the Hardy space of the upper half-plane, where the Fourier transform plays a crucial role. Our concern is with the Fourier transform as an operator itself, as opposed to its interactions with other operators or its applications. In particular, a discussion of the Fourier transform's connection to differential operators would require a long digression on unbounded operators, which would draw us too far afield.

11.1 The Fourier Transform on $L^1(\mathbb{R})$

Recall that $L^1(\mathbb{R})$ is the Banach space of Lebesgue-measurable functions on $\mathbb{R}$ such that

$$\|f\|_1 := \int_{-\infty}^{\infty} |f(x)|dx < \infty.$$

We use $\|f\|$, without any subscript, to denote the $L^2(\mathbb{R})$ norm. The proofs in this chapter require some useful dense subsets of $L^p(\mathbb{R})$.

Proposition 11.1.1. *For each $1 \leqslant p < \infty$, the following sets are dense in $L^p(\mathbb{R})$.*

(a) $C_c(\mathbb{R})$, *the set of continuous complex-valued functions on $\mathbb{R}$ with compact support.*

(b) *The set of step functions*

$$f(x) = \sum_{i=1}^{n} c_i \chi_{[a_i,b_i]}(x),$$

where $n \in \mathbb{N}$, $c_j \in \mathbb{C}$, and $[a_j, b_j]$ are closed intervals with disjoint interiors.

We first define the Fourier transform on $L^1(\mathbb{R})$ and then extend it to $L^2(\mathbb{R})$ by a density argument. The discussion of the Fourier transform on both spaces requires some useful harmonic-analysis tools.

Proposition 11.1.2. *For $f \in L^1(\mathbb{R})$, the Fourier transform*

$$(\mathscr{F}f)(x) = \frac{1}{\sqrt{2\pi}} \int_{-\infty}^{\infty} f(t)e^{-ixt}\,dt$$

converges absolutely for all $x \in \mathbb{R}$ and defines a continuous function on $\mathbb{R}$.

Proof For $f \in L^1(\mathbb{R})$ and $x \in \mathbb{R}$,

$$\int_{-\infty}^{\infty} |f(t)e^{-ixt}|\,dt = \int_{-\infty}^{\infty} |f(t)|\,dt = \|f\|_1,$$

so the integral that defines $(\mathscr{F}f)(x)$ converges absolutely for all $x \in \mathbb{R}$. Thus, $(\mathscr{F}f)(x)$ is well defined for every x. For each fixed x, let $(x_n)_{n=1}^{\infty}$ be a sequence in $\mathbb{R}$ such that $x_n \to x$. Then

$$|(\mathscr{F}f)(x_n) - (\mathscr{F}f)(x)| \leqslant \frac{1}{\sqrt{2\pi}} \int_{-\infty}^{\infty} |f(t)||e^{-ix_nt} - e^{-ixt}|\,dt.$$

Since $f \in L^1(\mathbb{R})$ and $|e^{-ix_nt} - e^{-ixt}| \leqslant 2$ for all $n \geqslant 1$ and $t \in \mathbb{R}$, the dominated convergence theorem implies that $(\mathscr{F}f)(x_n) \to (\mathscr{F}f)(x)$. Thus, $\mathscr{F}f$ is continuous on $\mathbb{R}$. ∎

Corollary 11.1.3. *For $f \in L^1(\mathbb{R})$ and $x \in \mathbb{R}$, $|(\mathscr{F}f)(x)| \leqslant \|f\|_1$.*

The next result says that $\mathscr{F}L^1(\mathbb{R}) \subseteq C_0(\mathbb{R})$, the set of continuous functions g on $\mathbb{R}$ that vanish at $\pm\infty$, meaning that

$$g \in C(\mathbb{R}) \quad \text{and} \quad \lim_{|x|\to\infty} g(x) = 0.$$

Proposition 11.1.4 (Riemann–Lebesgue lemma). *If $f \in L^1(\mathbb{R})$, then*

$$\lim_{|x|\to\infty} (\mathscr{F}f)(x) = 0.$$

Proof First observe that

$$(\mathscr{F}\chi_{[a,b]})(x) = \frac{1}{\sqrt{2\pi}} \int_a^b e^{-itx}\,dt = -\frac{1}{\sqrt{2\pi}} \frac{i\left(e^{-iax} - e^{-ibx}\right)}{x} \to 0 \quad \text{as } |x| \to \infty.$$

The linearity of the integral ensures that the same holds for step functions. For any $f \in L^1(\mathbb{R})$ and $\varepsilon > 0$, let g be a step function such that $\|f - g\|_1 < \varepsilon/2$ (Proposition 11.1.1). Now choose $T > 0$ such that $|(\mathscr{F}g)(x)| < \varepsilon/2$ for all $|x| > T$. Then for $|x| > T$, Corollary 11.1.3 implies

$$|(\mathscr{F}f)(x)| \leqslant |(\mathscr{F}f)(x) - (\mathscr{F}g)(x)| + |(\mathscr{F}g)(x)|$$

$$\leqslant \|f - g\|_1 + \frac{\varepsilon}{2} < \frac{\varepsilon}{2} + \frac{\varepsilon}{2} = \varepsilon,$$

which proves the result. ■

See Exercise 4.5.12 for the statement and proof of the $L^1(\mathbb{T})$ version of the Riemann–Lebesgue lemma.

11.2 Convolution and Young's Inequality

The *convolution* of $f, g \in L^1(\mathbb{R})$ is defined formally by

$$(f * g)(t) = \int_{-\infty}^{\infty} f(\tau)g(t - \tau) \, d\tau. \tag{11.2.1}$$

The convergence of this integral comes from the following.

Proposition 11.2.2 (Young's inequality [377]). *If $f, g \in L^1(\mathbb{R})$, then $(f * g)(t)$ is defined for almost every $t \in \mathbb{R}$ and $\|f * g\|_1 \leqslant \|f\|_1 \|g\|_1$.*

Proof Fubini's theorem implies

$$\int_{-\infty}^{\infty} \left(\int_{-\infty}^{\infty} |f(\tau)g(t - \tau)| \, d\tau \right) dt = \int_{-\infty}^{\infty} |f(\tau)| \left(\int_{-\infty}^{\infty} |g(t - \tau)| \, dt \right) d\tau$$

$$= \left(\int_{-\infty}^{\infty} |f(\tau)| \, d\tau \right) \left(\int_{-\infty}^{\infty} |g(t)| \, dt \right)$$

$$= \|f\|_1 \|g\|_1 < \infty.$$

Thus, the integral in (11.2.1) that defines $f * g$ converges almost everywhere and hence $(f * g)(t)$ is well defined for almost all $t \in \mathbb{R}$. Moreover, $f * g \in L^1(\mathbb{R})$ and $\|f * g\|_1 \leqslant \|f\|_1 \|g\|_1$. ■

The previous proposition says that $L^1(\mathbb{R})$ is closed under convolution. What about the other Lebesgue spaces? The answer comes from another inequality of Young. Since it requires no extra effort, we prove the inequality for $L^p(\mathbb{R})$ when $1 \leqslant p < \infty$. Recall that

$$\|f\|_p = \left(\int_{-\infty}^{\infty} |f(x)|^p \, dx \right)^{\frac{1}{p}}$$

is the norm on the Banach space $L^p(\mathbb{R})$. Proving an extension of Young's inequality requires the continuity of translations.

Lemma 11.2.3. *Let $1 \leqslant p < \infty$, $f \in L^p(\mathbb{R})$, and $y \in \mathbb{R}$. If $f_y(x) = f(x - y)$, then*

$$\lim_{y \to 0} \|f_y - f\|_p = 0.$$

Proof If the support of $f \in C_c(\mathbb{R})$ is contained in $[-M, M]$, then

$$\|f - f_y\|_p \leqslant (2M + 2)^{\frac{1}{p}} \|f_y - f\|_\infty \quad \text{for all } |y| \leqslant 1.$$

Since $f_y \to f$ uniformly, it follows that $\|f_y - f\|_p \to 0$ as $y \to 0$. Therefore, the desired result holds for $f \in C_c(\mathbb{R})$. For $f \in L^p(\mathbb{R})$ and $\varepsilon > 0$, use the fact that $C_c(\mathbb{R})$ is dense in $L^p(\mathbb{R})$ (Proposition 11.1.1) to choose $g \in C_c(\mathbb{R})$ such that $\|f - g\|_p < \varepsilon/3$. For this fixed g, let $\delta > 0$ be such that $\|g - g_y\|_p < \varepsilon/3$ whenever $|y| < \delta$. Then

$$\|f - f_y\|_p \leqslant \|f - g\|_p + \|g - g_y\|_p + \|g_y - f_y\|_p \leqslant \frac{\varepsilon}{3} + \frac{\varepsilon}{3} + \frac{\varepsilon}{3} = \varepsilon,$$

which completes the proof. ∎

Lemma 11.2.4 (Young's inequality [377]). *The following hold for $f \in L^p(\mathbb{R})$, with $1 \leqslant p \leqslant \infty$, and $g \in L^1(\mathbb{R})$.*

(a) $(f * g)(t)$ *is well defined for almost every $t \in \mathbb{R}$.*

(b) $f * g \in L^p(\mathbb{R})$.

(c) $\|f * g\|_p \leqslant \|f\|_p \|g\|_1$.

Proof The case $p = 1$ is Proposition 11.2.2. Hölder's inequality (1.8.4) handles the case $p = \infty$ and shows that $(f * g)(t)$ is well defined for every $t \in \mathbb{R}$. For the rest of the proof, assume that $1 < p < \infty$ and let q be the conjugate exponent of p. By Hölder's inequality,

$$\int_{-\infty}^{\infty} |f(x)g(t - x)|\, dx = \int_{-\infty}^{\infty} (|f(x)||g(t - x)|^{\frac{1}{p}})|g(t - x)|^{\frac{1}{q}}\, dx$$

$$\leqslant \left(\int_{-\infty}^{\infty} |f(x)|^p\, |g(t - x)|\, dx \right)^{\frac{1}{p}} \left(\int_{-\infty}^{\infty} |g(t - x)|\, dx \right)^{\frac{1}{q}}$$

$$= \left(\int_{-\infty}^{\infty} |f(x)|^p |g(t - x)|\, dx \right)^{\frac{1}{p}} \|g\|_1^{\frac{1}{q}}. \tag{11.2.5}$$

Hence,

$$\int_{-\infty}^{\infty} \left(\int_{-\infty}^{\infty} |f(x)g(t - x)|dx \right)^p dt$$

$$\leqslant \|g\|_1^{\frac{p}{q}} \int_{-\infty}^{\infty} \left(\int_{-\infty}^{\infty} |f(x)|^p |g(t - x)|\, dx \right) dt \qquad \text{(by (11.2.5))}$$

$$\leqslant \|g\|_1^{\frac{p}{q}} \int_{-\infty}^{\infty} |f(x)|^p \left(\int_{-\infty}^{\infty} |g(t - x)|\, dt \right) dx \qquad \text{(Fubini's theorem)}$$

$$= \|g\|_1^{\frac{p}{q}} \|f\|_p^p \|g\|_1$$

$$= \|f\|_p^p \|g\|_1^p.$$

Therefore,

$$\left(\int_{-\infty}^{\infty}\left(\int_{-\infty}^{\infty}|f(x)g(t-x)|\,dx\right)^{p}dt\right)^{\frac{1}{p}}\leqslant\|f\|_{p}\|g\|_{1}.$$

The inequality above ensures that $(f * g)(t)$ is well defined for almost every $t \in \mathbb{R}$. Furthermore, $f * g \in L^p(\mathbb{R})$ and $\|f * g\|_p \leqslant \|f\|_p \|g\|_1$. ∎

Our final version of Young's inequality is the following.

Lemma 11.2.6 (Young's inequality [377]). *Let* $1 \leqslant p \leqslant \infty$ *and let* q *be its conjugate exponent. The following hold for* $f \in L^p(\mathbb{R})$ *and* $g \in L^q(\mathbb{R})$.

(a) $(f * g)(t)$ *is well defined for all* $t \in \mathbb{R}$.

(b) $f * g$ *is a bounded and uniformly continuous function on* $\mathbb{R}$.

(c) $\|f * g\|_\infty \leqslant \|f\|_p \|g\|_q$.

Proof Suppose that $1 < p < \infty$. Hölder's inequality implies that

$$\int_{-\infty}^{\infty}|f(x)g(t-x)|\,dx\leqslant\left(\int_{-\infty}^{\infty}|f(x)|^{p}\,dx\right)^{\frac{1}{p}}\left(\int_{-\infty}^{\infty}|g(t-x)|^{q}\,dx\right)^{\frac{1}{q}},$$

and hence

$$\int_{-\infty}^{\infty}|f(x)g(t-x)|\,dx\leqslant\|f\|_{p}\|g\|_{q}\quad\text{for all }t\in\mathbb{R}.$$

Thus, $f * g$ is well defined on $\mathbb{R}$ and $\|f * g\|_\infty \leqslant \|f\|_p \|g\|_q$. Next we verify that $f * g$ is uniformly continuous on $\mathbb{R}$. Toward this end, let $\varepsilon > 0$ and $\delta > 0$ be such that $\|g_s - g_t\|_q \leqslant \varepsilon/(1 + \|f\|_p)$ whenever $|s - t| < \delta$ (Lemma 11.2.3). Then,

$$|(f * g)(t) - (f * g)(s)|\leqslant\int_{-\infty}^{\infty}|f(x)|\,|g(t-x)-g(s-x)|\,dx\leqslant\|f\|_{p}\|g_{t}-g_{s}\|_{q}\leqslant\varepsilon.$$

The proof above, subject to some minor changes, also works for $p = 1$ and $p = \infty$. ∎

11.3 Convolution and the Fourier Transform

The Fourier transform behaves well with respect to convolution. Recall from Proposition 11.2.2 that $f * g \in L^1(\mathbb{R})$ for $f, g \in L^1(\mathbb{R})$ and thus $\mathscr{F}f$, $\mathscr{F}g$, and $\mathscr{F}(f * g)$ are well-defined continuous functions on $\mathbb{R}$.

Proposition 11.3.1. *If* $f, g \in L^1(\mathbb{R})$, *then* $\mathscr{F}(f * g) = \sqrt{2\pi}(\mathscr{F}f)(\mathscr{F}g)$.

Proof For $t \in \mathbb{R}$, Fubini's theorem yields

$$(\mathscr{F}(f * g))(t) = \frac{1}{\sqrt{2\pi}}\int_{-\infty}^{\infty}(f * g)(\tau)\,e^{-it\tau}\,d\tau$$

$$
= \frac{1}{\sqrt{2\pi}} \int_{-\infty}^{\infty} \left(\int_{-\infty}^{\infty} f(s)g(\tau - s)\, ds \right) e^{-it\tau}\, d\tau
$$

$$
= \frac{1}{\sqrt{2\pi}} \int_{-\infty}^{\infty} f(s) \left(\int_{-\infty}^{\infty} g(\tau - s)e^{-it\tau}\, d\tau \right) ds
$$

$$
= \frac{1}{\sqrt{2\pi}} \int_{-\infty}^{\infty} f(s) \left(\int_{-\infty}^{\infty} g(\tau)e^{-it(\tau + s)}\, d\tau \right) ds
$$

$$
= \frac{1}{\sqrt{2\pi}} \left(\int_{-\infty}^{\infty} f(s)e^{-its}\, ds \right) \left(\int_{-\infty}^{\infty} g(\tau)e^{-it\tau}\, d\tau \right)
$$

$$
= \sqrt{2\pi}(\mathscr{F}f)(t)(\mathscr{F}g)(t),
$$

which completes the proof. ■

In a similar manner, there is also the following multiplication formula.

Proposition 11.3.2 (Multiplication formula). *If $f, g \in L^1(\mathbb{R})$, then*

$$
\int_{-\infty}^{\infty} (\mathscr{F}f)(t)g(t)\, dt = \int_{-\infty}^{\infty} f(t)(\mathscr{F}g)(t)\, dt.
$$

Proof Fubini's theorem implies

$$
\int_{-\infty}^{\infty} (\mathscr{F}f)(t)g(t)\, dt = \int_{-\infty}^{\infty} \left(\frac{1}{\sqrt{2\pi}} \int_{-\infty}^{\infty} f(x)e^{-itx}\, dx \right) g(t)\, dt
$$

$$
= \int_{-\infty}^{\infty} \left(\frac{1}{\sqrt{2\pi}} \int_{-\infty}^{\infty} g(t)e^{-itx}\, dt \right) f(x)\, dx
$$

$$
= \int_{-\infty}^{\infty} (\mathscr{F}g)(x)f(x)\, dx,
$$

which completes the proof. ■

11.4 The Poisson Kernel

The Poisson kernel appears in Chapter 12 in the study of harmonic functions. Here we use the Poisson kernel as an approximation tool. The following material sets up the Fourier inversion formula that appears in the next section.

The *Poisson kernel* for the upper half-plane $\mathbb{C}_+ := \{z \in \mathbb{C} : \operatorname{Im} z > 0\}$ is

$$
P_y(x) = \frac{1}{\pi} \frac{y}{x^2 + y^2} \quad \text{for } x \in \mathbb{R} \text{ and } y > 0. \tag{11.4.1}
$$

One can verify that

$$
\int_{-\infty}^{\infty} P_y(x)\, dx = 1 \quad \text{for all } y > 0. \tag{11.4.2}
$$

For $x_0 + iy_0 \in \mathbb{C}_+$, the Fourier transform of

$$f(x) = \frac{1}{\pi} \frac{y_0}{(x_0 - x)^2 + y_0^2}$$

is

$$(\mathscr{F}f)(t) = \frac{1}{\sqrt{2\pi}} \int_{-\infty}^{\infty} f(x)e^{-itx} \, dx$$

$$= \frac{1}{\sqrt{2\pi}} \int_{-\infty}^{\infty} P_{y_0}(x_0 - x)e^{-itx} \, dx$$

$$= \frac{1}{\sqrt{2\pi}} \int_{-\infty}^{\infty} P_{y_0}(x)e^{-it(x_0 - x)} \, dx$$

$$= \frac{e^{-ix_0 t}}{\sqrt{2\pi}} \int_{-\infty}^{\infty} P_{y_0}(x)e^{itx} \, dx$$

$$= \frac{e^{-ix_0 t}}{\sqrt{2\pi}} \int_{-\infty}^{\infty} P_{y_0}(x)e^{-itx} \, dx,$$

and hence the last two equations show that

$$(\mathscr{F}f)(t) = \frac{e^{-ix_0 t}}{\sqrt{2\pi}} \int_{-\infty}^{\infty} P_{y_0}(x)e^{i|t|x} \, dx.$$

For $R > y_0$, let Γ_R denote the positively oriented curve created by the interval $[-R, R]$ and the semicircle $\{Re^{i\theta} : 0 \leqslant \theta \leqslant \pi\}$; see Figure 11.4.1. Then,

$$(\mathscr{F}f)(t) = \frac{e^{-ix_0 t}}{\sqrt{2\pi}} \lim_{R \to \infty} \int_{-R}^{R} \frac{y_0}{\pi(x^2 + y_0^2)} e^{i|t|x} \, dx$$

$$= \frac{e^{-ix_0 t}}{\sqrt{2\pi}} \lim_{R \to \infty} \int_{\Gamma_R} \frac{y_0}{\pi(w^2 + y_0^2)} e^{i|t|w} \, dw.$$

The integrand

$$w \mapsto \frac{y_0}{\pi(w^2 + y_0^2)} e^{i|t|w}$$

has one pole inside Γ_R, namely iy_0, with residue $\frac{1}{2\pi i} e^{-y_0|t|}$. Therefore,

$$(\mathscr{F}f)(t) = \frac{1}{\sqrt{2\pi}} e^{-ix_0 t - y_0|t|}.$$

This next identity follows from Proposition 11.3.2 and Exercise 11.10.3.

Corollary 11.4.3. *Let $u \in L^1(\mathbb{R})$. Then for all $z = x + iy \in \mathbb{C}_+$,*

$$\int_{-\infty}^{\infty} P_y(x - t)u(t) \, dt = \frac{1}{\sqrt{2\pi}} \int_{-\infty}^{\infty} e^{-y|t|}(\mathscr{F}u)(t)e^{ixt} \, dt.$$

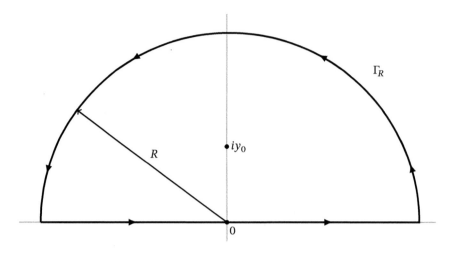

Figure 11.4.1 The contour Γ_R.

The left side of the equation above is the *Poisson integral* of u. It can be also be written as $(P_y * u)(x)$.

Lemma 11.4.4. *If $f \in L^\infty(\mathbb{R})$ is continuous at $x = 0$, then*

$$\lim_{y \to 0^+} \int_{-\infty}^{\infty} P_y(t) f(t)\, dt = f(0).$$

Proof Since f is continuous at $t = 0$, given $\varepsilon > 0$, there is a $\delta > 0$ such that $|f(t) - f(0)| < \varepsilon$ for all $|t| < \delta$. By (11.4.2),

$$\left| \int_{-\infty}^{\infty} P_y(t) f(t)\, dt - f(0) \right| = \left| \int_{-\infty}^{\infty} P_y(t)(f(t) - f(0))\, dt \right| \leq \int_{-\infty}^{\infty} P_y(t)|f(t) - f(0)|\, dt.$$

The integral on the right equals

$$\int_{|t| < \delta} P_y(t)|f(t) - f(0)|\, dt + \int_{|t| \geq \delta} P_y(t)|f(t) - f(0)|\, dt. \tag{11.4.5}$$

For the first integral in (11.4.5), use (11.4.2) to observe that

$$\int_{|t| < \delta} P_y(t)|f(t) - f(0)|\, dt \leq \varepsilon \int_{|t| < \delta} P_y(t)\, dt \leq \varepsilon.$$

For the second integral in (11.4.5), we have

$$\int_{|t| \geq \delta} P_y(t)|f(t) - f(0)|\, dt \leq 2\|f\|_\infty \frac{1}{\pi} \int_{|t| \geq \delta} \frac{y}{t^2 + y^2}\, dt = 4\|f\|_\infty \frac{1}{\pi}\left(\frac{\pi}{2} - \tan^{-1}\left(\frac{\delta}{y}\right)\right),$$

which tends to zero as $y \to 0^+$. Put this all together and obtain

$$\limsup_{y \to 0^+} \left| \int_{-\infty}^{\infty} P_y(t) f(t) dt - f(0) \right| \leq \varepsilon,$$

which completes the proof. ∎

Lemma 11.4.6. *For $1 \leq p < \infty$ and $f \in L^p(\mathbb{R})$,*

$$\lim_{y \to 0^+} \|P_y * f - f\|_p = 0.$$

Proof From (11.4.2) it follows that

$$(P_y * f)(x) - f(x) = \int_{-\infty}^{\infty} P_y(t)(f(x-t) - f(t)) \, dt$$

and hence

$$|(P_y * f)(x) - f(x)| \leq \int_{-\infty}^{\infty} P_y(t)|f(x-t) - f(t)| \, dt.$$

Jensen's inequality [319, p. 63] applied to the probability measure $P_y(t) \, dt$ (recall (11.4.2)) and the convex function $w \mapsto w^p$ on $[0, \infty)$ yields

$$|(P_y * f)(x) - f(x)|^p \leq \int_{-\infty}^{\infty} P_y(t)|f(x-t) - f(t)|^p \, dt.$$

Integrate both sides and use Fubini's theorem to get

$$\|P_y * f - f\|_p^p \leq \int_{-\infty}^{\infty} P_y(t)\|f_t - f\|_p^p \, dt.$$

Lemma 11.2.3 implies that the function $t \mapsto \|f_t - f\|_p^p$ is bounded and continuous at $t = 0$. An application of Lemma 11.4.4 finishes the proof. ∎

11.5 The Fourier Inversion Formula

In this section, we invert the Fourier transform.

Theorem 11.5.1 (Inversion formula). *Suppose that f and $\mathscr{F}f$ belong to $L^1(\mathbb{R})$. Then for almost every $t \in \mathbb{R}$,*

$$f(t) = \frac{1}{\sqrt{2\pi}} \int_{-\infty}^{\infty} (\mathscr{F}f)(\tau) e^{it\tau} \, d\tau.$$

Proof Let

$$F_y(x) = \frac{1}{\pi} \int_{-\infty}^{\infty} \frac{y}{(x-t)^2 + y^2} f(t) \, dt \quad \text{for } x + iy \in \mathbb{C}_+.$$

Proposition 11.2.2 says that $F_y \in L^1(\mathbb{R})$ and Lemma 11.4.6 says that $\|F_y - f\|_1 \to 0$ as $y \to 0^+$. By [319, p. 70], a convergent sequence in $L^1(\mathbb{R})$ has a subsequence that converges almost everywhere. Consequently, there is a sequence $(y_n)_{n=1}^\infty$ of positive real numbers such that $y_n \to 0$ and

$$\lim_{n \to \infty} F_{y_n}(x) = f(x) \tag{11.5.2}$$

for almost all $x \in \mathbb{R}$. On the other hand, Corollary 11.4.3 implies that

$$F_y(x) = \frac{1}{\sqrt{2\pi}} \int_{-\infty}^\infty e^{-y|t|} (\mathscr{F}f)(t) e^{ixt} \, dt \quad \text{for } x \in \mathbb{R} \text{ and } y > 0.$$

The assumption $\mathscr{F}f \in L^1(\mathbb{R})$ and the dominated convergence theorem imply that

$$\lim_{y \to 0^+} F_y(x) = \frac{1}{\sqrt{2\pi}} \int_{-\infty}^\infty (\mathscr{F}f)(t) e^{ixt} \, dt \quad \text{for all } x \in \mathbb{R}. \tag{11.5.3}$$

Combine (11.5.2) and (11.5.3) to deduce the desired inversion formula. ∎

Rewrite the inversion formula as

$$f(t) = \frac{1}{\sqrt{2\pi}} \int_{-\infty}^\infty (\mathscr{F}f)(\tau) e^{-i(-t)\tau} \, d\tau$$

and observe that

$$f(t) = (\mathscr{F}(\mathscr{F}f))(-t) \quad \text{for almost all } t \in \mathbb{R}, \tag{11.5.4}$$

provided that f and $\mathscr{F}f$ belong to $L^1(\mathbb{R})$.

Theorem 11.5.1 and the uniqueness theorem imply that the Fourier transform maps $L^1(\mathbb{R})$ to $C_0(\mathbb{R})$ injectively. The following result is an important step in extending the definition of the Fourier transform to $L^2(\mathbb{R})$.

Corollary 11.5.5. *Let $f \in L^1(\mathbb{R})$ be continuous at $x = 0$ and $\mathscr{F}f \geqslant 0$. Then $\mathscr{F}f \in L^1(\mathbb{R})$ and*

$$f(t) = \frac{1}{\sqrt{2\pi}} \int_{-\infty}^\infty (\mathscr{F}f)(\tau) e^{it\tau} \, d\tau$$

for almost all $t \in \mathbb{R}$. Furthermore,

$$f(0) = \frac{1}{\sqrt{2\pi}} \int_{-\infty}^\infty (\mathscr{F}f)(\tau) \, d\tau.$$

Proof For $y > 0$ and $x \in \mathbb{R}$, Corollary 11.4.3 says that

$$\frac{1}{\pi} \int_{-\infty}^\infty \frac{y}{(x-t)^2 + y^2} f(t) \, dt = \frac{1}{\sqrt{2\pi}} \int_{-\infty}^\infty e^{-y|t|} (\mathscr{F}f)(t) e^{ixt} \, dt. \tag{11.5.6}$$

If $x = 0$ and $y > 0$, then

$$\frac{1}{\pi} \int_{-\infty}^{\infty} \frac{y}{t^2 + y^2} f(t)\, dt = \frac{1}{\sqrt{2\pi}} \int_{-\infty}^{\infty} e^{-y|t|} (\mathscr{F} f)(t)\, dt.$$

The monotone convergence theorem ensures that

$$\lim_{y \to 0^+} \frac{1}{\sqrt{2\pi}} \int_{-\infty}^{\infty} e^{-y|t|} (\mathscr{F} f)(t)\, dt = \frac{1}{\sqrt{2\pi}} \int_{-\infty}^{\infty} (\mathscr{F} f)(t)\, dt.$$

By the approximation properties of the Poisson kernel (Lemma 11.4.4),

$$f(0) = \lim_{y \to 0^+} \frac{1}{\pi} \int_{-\infty}^{\infty} \frac{y}{t^2 + y^2} f(t)\, dt.$$

Thus,

$$f(0) = \frac{1}{\sqrt{2\pi}} \int_{-\infty}^{\infty} (\mathscr{F} f)(t)\, dt,$$

which also shows that $\mathscr{F} f \in L^1(\mathbb{R})$ (since we are assuming that $\mathscr{F} f \geqslant 0$). Since $\mathscr{F} f \in L^1(\mathbb{R})$, the inversion formula (Theorem 11.5.1) applies and ensures that

$$f(t) = \frac{1}{\sqrt{2\pi}} \int_{-\infty}^{\infty} (\mathscr{F} f)(\tau) e^{it\tau}\, d\tau$$

for almost all $t \in \mathbb{R}$. ∎

11.6 The Fourier–Plancherel Transform

Corollary 11.5.5 is the main ingredient needed to establish that the Fourier transform maps $L^1(\mathbb{R}) \cap L^2(\mathbb{R})$ into $L^2(\mathbb{R})$. This is the first step in extending the definition of the Fourier transform to $L^2(\mathbb{R})$.

Theorem 11.6.1 (Plancherel). *If $f \in L^1(\mathbb{R}) \cap L^2(\mathbb{R})$, then $\mathscr{F} f \in L^2(\mathbb{R})$ and $\|\mathscr{F} f\| = \|f\|$.*

Proof Let $g(x) = \overline{f(-x)}$ and define $h = f * g$. By Young's inequality (Lemma 11.2.6), $h \in L^1(\mathbb{R}) \cap C(\mathbb{R})$ and, by Proposition 11.3.1,

$$\mathscr{F} h = \sqrt{2\pi}(\mathscr{F} f)(\mathscr{F} g) = \sqrt{2\pi}|\mathscr{F} f|^2 \geqslant 0. \qquad (11.6.2)$$

Note the use of $\mathscr{F} g = \overline{\mathscr{F} f}$ above. Corollary 11.5.5 ensures that $\mathscr{F} h \in L^1(\mathbb{R})$, which is equivalent to $\mathscr{F} f \in L^2(\mathbb{R})$. Moreover,

$$h(0) = \frac{1}{\sqrt{2\pi}} \int_{-\infty}^{\infty} (\mathscr{F} h)(t)\, dt.$$

Let us look closely at both sides of the identity above. First observe that (11.6.2) yields

$$\frac{1}{\sqrt{2\pi}} \int_{-\infty}^{\infty} (\mathscr{F}h)(t)\,dt = \int_{-\infty}^{\infty} |(\mathscr{F}f)(t)|^2\,dt = \|\mathscr{F}f\|^2.$$

Moreover,

$$h(0) = \int_{-\infty}^{\infty} f(t)g(0-t)\,dt = \int_{-\infty}^{\infty} f(t)\overline{f(t)}\,dt = \|f\|^2,$$

which shows that $\|\mathscr{F}f\| = \|f\|$. ∎

Let $f \in L^2(\mathbb{R})$ and choose a sequence $(f_n)_{n=1}^{\infty}$ in $L^1(\mathbb{R}) \cap L^2(\mathbb{R})$ such that $\|f_n - f\| \to 0$. Then $(f_n)_{n=1}^{\infty}$ is a Cauchy sequence in $L^2(\mathbb{R})$ and, by Theorem 11.6.1,

$$\|\mathscr{F}f_n - \mathscr{F}f_m\| = \|\mathscr{F}(f_n - f_m)\| = \|f_n - f_m\|.$$

Thus, $(\mathscr{F}f_n)_{n=1}^{\infty}$ is a Cauchy sequence in $L^2(\mathbb{R})$. Since $L^2(\mathbb{R})$ is complete, $\mathscr{F}f_n$ converges in $L^2(\mathbb{R})$. If $(g_n)_{n=1}^{\infty}$ is another sequence in $L^1(\mathbb{R}) \cap L^2(\mathbb{R})$ approximating f in $L^2(\mathbb{R})$, then $\mathscr{F}g_n$ also converges in $L^2(\mathbb{R})$. However, Theorem 11.6.1 implies

$$\|\mathscr{F}f_n - \mathscr{F}g_n\| = \|\mathscr{F}(f_n - g_n)\| = \|f_n - g_n\| \leqslant \|f_n - f\| + \|g_n - f\| \to 0,$$

and hence

$$\lim_{n\to\infty} \mathscr{F}g_n = \lim_{n\to\infty} \mathscr{F}f_n,$$

where convergence is in the $L^2(\mathbb{R})$ norm. Therefore, the limit of $\mathscr{F}f_n$ is independent of the choice of $f_n \in L^1(\mathbb{R}) \cap L^2(\mathbb{R})$ as long as $f_n \to f$ in $L^2(\mathbb{R})$. This enables us to make the following definition.

Definition 11.6.3. The *Fourier–Plancherel transform* of $f \in L^2(\mathbb{R})$ is

$$\mathscr{F}f = \lim_{n\to\infty} \mathscr{F}f_n,$$

where $(f_n)_{n=1}^{\infty}$ is any sequence in $L^1(\mathbb{R}) \cap L^2(\mathbb{R})$ such that $\|f_n - f\| \to 0$. One standard choice of approximating sequence is

$$f_n(t) = \begin{cases} f(t) & \text{if } |t| \leqslant n, \\ 0 & \text{if } |t| > n. \end{cases}$$

Therefore,

$$(\mathscr{F}f)(t) = \lim_{n\to\infty} \frac{1}{\sqrt{2\pi}} \int_{-n}^{n} f(\tau)e^{-it\tau}\,d\tau. \tag{11.6.4}$$

Here the limit is taken in $L^2(\mathbb{R})$, and (11.6.4) is referred to as a *limit in the mean*. Some textbooks write

$$(\mathscr{F}f)(t) = \text{l.i.m.} \frac{1}{\sqrt{2\pi}} \int_{-\infty}^{\infty} f(\tau)e^{-it\tau}\,d\tau.$$

The main result of this section is the next theorem.

Theorem 11.6.5. *The Fourier–Plancherel transform* $\mathscr{F} : L^2(\mathbb{R}) \to L^2(\mathbb{R})$ *is unitary.*

Proof Let $f \in L^2(\mathbb{R})$ and let $(f_n)_{n=1}^{\infty}$ be a sequence in $L^1(\mathbb{R}) \cap L^2(\mathbb{R})$ such that $f_n \to f$ in $L^2(\mathbb{R})$. Theorem 11.6.1 implies that

$$\|\mathscr{F} f\| = \lim_{n\to\infty} \|\mathscr{F} f_n\| = \lim_{n\to\infty} \|f_n\| = \|f\|, \tag{11.6.6}$$

and hence $\mathscr{F}$ is isometric. We next show that $\mathscr{F}$ is surjective. An important step involves the identity

$$\int_{-\infty}^{\infty} f(t)(\mathscr{F} g)(t)\,dt = \int_{-\infty}^{\infty} (\mathscr{F} f)(t)g(t)\,dt \quad \text{for all } f, g \in L^2(\mathbb{R}). \tag{11.6.7}$$

To verify this identity, observe that

$$\lim_{n\to\infty} \|\mathscr{F} f_n - \mathscr{F} f\| = \lim_{n\to\infty} \|\mathscr{F} g_n - \mathscr{F} g\| = 0,$$

where $(g_n)_{n=1}^{\infty}$ is any sequence in $L^1(\mathbb{R}) \cap L^2(\mathbb{R})$ such that $g_n \to g$ in $L^2(\mathbb{R})$. The Cauchy–Schwarz inequality yields

$$\|g_n \mathscr{F} f_n - g\mathscr{F} f\|_1 \leqslant \|g_n(\mathscr{F} f_n - \mathscr{F} f)\|_1 + \|(g_n - g)\mathscr{F} f\|_1$$
$$\leqslant \|g_n\|\|\mathscr{F} f_n - \mathscr{F} f\| + \|g_n - g\|\|\mathscr{F} f\|,$$

and hence

$$\lim_{n\to\infty} \|g_n \mathscr{F} f_n - g\mathscr{F} f\|_1 = 0.$$

In a similar way

$$\lim_{n\to\infty} \|f_n \mathscr{F} g_n - f\mathscr{F} g\|_1 = 0.$$

Since f_n and g_n belong to $L^1(\mathbb{R})$, Proposition 11.3.2 implies that

$$\int_{-\infty}^{\infty} f(t)(\mathscr{F} g)(t)\,dt = \lim_{n\to\infty} \int_{-\infty}^{\infty} f_n(t)(\mathscr{F} g_n)(t)\,dt$$
$$= \lim_{n\to\infty} \int_{-\infty}^{\infty} (\mathscr{F} f_n)(t)g_n(t)\,dt$$
$$= \int_{-\infty}^{\infty} (\mathscr{F} f)(t)g(t)\,dt,$$

which verifies (11.6.7). We are now in a position to show that the Fourier–Plancherel transform $\mathscr{F}$ is surjective. Assume that $g \in (\operatorname{ran} \mathscr{F})^{\perp}$, that is

$$\int_{-\infty}^{\infty} (\mathscr{F} f)(t)\overline{g(t)}\,dt = 0 \quad \text{for all } f \in L^2(\mathbb{R}).$$

Then (11.6.7) yields

$$\int_{-\infty}^{\infty} f(t)(\mathscr{F}\overline{g})(t)\,dt = 0,$$

which implies $\mathscr{F}\overline{g} = 0$ and hence $\overline{g} = 0$ by (11.6.6). In other words, the Fourier–Plancherel transform $\mathscr{F}$ is surjective. ∎

Corollary 11.6.8. $\langle f,g \rangle = \langle \mathscr{F}f, \mathscr{F}g \rangle$ *for all* $f,g \in L^2(\mathbb{R})$.

Proof Use the polarization identity (Exercise 1.10.17) on (11.6.6). ∎

11.7 Eigenvalues and Hermite Functions

Let $U : L^2(\mathbb{R}) \to L^2(\mathbb{R})$ be defined by $(Uf)(x) = f(-x)$, and note that U is unitary, $U^* = U$, and hence $U^2 = I$. One can check that $U\mathscr{F}$ is unitary (the product of two unitary operators is unitary) and that $U\mathscr{F} = \mathscr{F}U$. Moreover, $U\mathscr{F}^2 = I$ holds on $C_c(\mathbb{R})$ by (11.5.4), which is a dense subset of $L^2(\mathbb{R})$. Therefore, $U\mathscr{F}^2 = I$ holds on $L^2(\mathbb{R})$ and hence $\mathscr{F}^* = U\mathscr{F}$. Also note that $\mathscr{F}^4 = U^{*2} = I$. The identity $\mathscr{F}^4 = I$ permits us to compute the spectrum of the Fourier–Plancherel transform.

Theorem 11.7.1. $\sigma(\mathscr{F}) = \sigma_p(\mathscr{F}) = \{\pm 1, \pm i\}$.

Proof Since $\mathscr{F}^4 = I$, one can verify directly (Exercise 11.10.15) that

$$(\mathscr{F} - zI)^{-1} = \frac{1}{1 - z^4}(\mathscr{F}^3 + z\mathscr{F}^2 + z^2\mathscr{F} + z^3 I) \quad \text{for all } z \notin \{\pm 1, \pm i\}$$

and hence $\sigma(\mathscr{F}) \subseteq \{\pm 1, \pm i\}$. To prove the reverse containment, we need to discuss the Hermite functions $(h_n)_{n=0}^{\infty}$. We follow [263]. The exponential generating function for the Hermite functions is

$$e^{\frac{-x^2}{2} + 2xt - t^2} = \sum_{n=0}^{\infty} h_n(x)\frac{t^n}{n!} \quad \text{for } x, t \in \mathbb{R}. \tag{11.7.2}$$

The first few are

$$h_0(x) = e^{-\frac{x^2}{2}},$$

$$h_1(x) = e^{-\frac{x^2}{2}} 2x,$$

$$h_2(x) = e^{-\frac{x^2}{2}} (4x^2 - 2),$$

$$h_3(x) = e^{-\frac{x^2}{2}} (8x^3 - 12x),$$

$$h_4(x) = e^{-\frac{x^2}{2}} (16x^4 - 48x^2 + 12x).$$

Since

$$e^{-\frac{x^2}{2}+2xt-t^2} = e^{-\frac{(x-2t)^2}{2}} e^{t^2},$$

it follows that

$$(\mathscr{F}e^{-\frac{u^2}{2}+2ut-t^2})(x) = e^{t^2}(\mathscr{F}e^{-\frac{(u-2t)^2}{2}})(x).$$

Use

$$(\mathscr{F}e^{-\frac{s^2}{2}})(u) = e^{-\frac{u^2}{2}} \quad \text{and} \quad (\mathscr{F}f_y)(x) = e^{-iyx}(\mathscr{F}f)(x)$$

(see Exercises 11.10.1 and 11.10.2) to deduce that

$$e^{t^2}(\mathscr{F}e^{-\frac{(u-2t)^2}{2}})(x) = e^{t^2}e^{-2itx}(\mathscr{F}e^{-u^2/2})(x) = e^{t^2}\,e^{-2itx}e^{-\frac{x^2}{2}} = e^{(-(-it)^2-2ixt-\frac{x^2}{2})}.$$

The above is the generating function (11.7.2) for the Hermite functions at $-it$. Thus,

$$\sum_{n=0}^{\infty} \mathscr{F}(h_n(u))(x)\frac{t^n}{n!} = \mathscr{F}\Big(\sum_{n=0}^{\infty} h_n(u)\frac{t^n}{n!}\Big)(x)$$

$$= (\mathscr{F}e^{-\frac{u^2}{2}+2ut-t^2})(x)$$

$$= e^{(-(-it)^2-2ixt-\frac{x^2}{2})}$$

$$= \sum_{n=0}^{\infty} h_n(x)\frac{(-it)^n}{n!}.$$

The identities above are justified since, for fixed t, the series in (11.7.2) converges in $L^2(\mathbb{R})$. Compare the corresponding power-series coefficients in the variable t to get

$$\mathscr{F}h_n = (-i)^n h_n. \tag{11.7.3}$$

Thus, $\sigma(\mathscr{F}) = \sigma_p(\mathscr{F}) = \{\pm 1, \pm i\}$. This also yields a set of eigenvectors for $\mathscr{F}$. ∎

The Hermite functions $(h_n)_{n=0}^{\infty}$ form an orthogonal basis for $L^2(\mathbb{R})$ (Exercise 11.10.12) and $\|h_n\| = \sqrt{2^n n!\sqrt{\pi}}$. Thus, with respect to the orthonormal basis

$$\Big(\frac{h_n}{\sqrt{2^n n!\sqrt{\pi}}}\Big)_{n=0}^{\infty},$$

the Fourier–Plancherel transform has the matrix representation

$$
\begin{bmatrix}
1 & 0 & 0 & 0 & 0 & 0 & 0 & 0 & \cdots \\
0 & -i & 0 & 0 & 0 & 0 & 0 & 0 & \cdots \\
0 & 0 & -1 & 0 & 0 & 0 & 0 & 0 & \cdots \\
0 & 0 & 0 & i & 0 & 0 & 0 & 0 & \cdots \\
0 & 0 & 0 & 0 & 1 & 0 & 0 & 0 & \cdots \\
0 & 0 & 0 & 0 & 0 & -i & 0 & 0 & \cdots \\
0 & 0 & 0 & 0 & 0 & 0 & -1 & 0 & \cdots \\
0 & 0 & 0 & 0 & 0 & 0 & 0 & i & \cdots \\
\vdots & \vdots & \vdots & \vdots & \vdots & \vdots & \vdots & \vdots & \ddots
\end{bmatrix}.
\tag{11.7.4}
$$

11.8 The Hardy Space of the Upper Half-Plane

In this section we develop the Hardy space of the upper half-plane $\mathbb{C}_+$. As with the Hardy space of $\mathbb{D}$ (Chapter 5), we only outline the main ideas and refer the reader to [149] for the details. For $f \in L^2(\mathbb{R})$, define

$$
F(z) = \frac{1}{\sqrt{2\pi}} \int_{-\infty}^{\infty} f(t) e^{izt} \, dt,
$$

(whenever this integral exists) where $z = x + iy$ is a complex variable. This definition can be interpreted, formally at least, as the inverse Fourier transform evaluated at z. For $y > 0$, $e^{izt} = e^{itx-ty}$, which diverges as $t \to -\infty$ and converges to zero as $t \to \infty$. Thus, if the support of f is contained in $[0, \infty)$, Morera's theorem confirms that

$$
F(z) = \frac{1}{\sqrt{2\pi}} \int_{0}^{\infty} f(t) e^{izt} \, dt
$$

is analytic on $\mathbb{C}_+$. Plancherel's formula (Theorem 11.6.5) yields

$$
\int_{-\infty}^{\infty} |F(x+iy)|^2 dx = \int_{0}^{\infty} e^{-2yt} |f(t)|^2 dt \leq \int_{0}^{\infty} |f(t)|^2 dt,
$$

and hence

$$
\sup_{y>0} \int_{-\infty}^{\infty} |F(x+iy)|^2 dx < \infty.
\tag{11.8.1}
$$

The upper half-plane analogue of Proposition 5.3.12 says that

$$
\lim_{y\to 0^+} F(x+iy) = F(x) = \frac{1}{\sqrt{2\pi}} \int_{0}^{\infty} f(t) e^{ixt} \, dt
$$

for almost every $x \in \mathbb{R}$ and

$$
\int_{-\infty}^{\infty} |F(x)|^2 dx = \int_{0}^{\infty} |f(t)|^2 dt.
$$

The class of analytic functions F on $\mathbb{C}_+$ which satisfy (11.8.1) is the *Hardy space of the upper half-plane* and is denoted by $H^2(\mathbb{C}_+)$. Note that

$$H^2(\mathbb{C}_+) = \left\{ \frac{1}{\sqrt{2\pi}} \int_0^\infty f(t) e^{izt} dt \ : \ f \in L^2(0, \infty) \right\} \tag{11.8.2}$$

and

$$H^2(\mathbb{R}) := \{ g \in L^2(\mathbb{R}) \ : \ (\mathscr{F}g)(x) = 0 \text{ for almost every } x < 0 \}$$

is the set of corresponding boundary functions. The description above of $H^2(\mathbb{C}_+)$ in (11.8.2) is the *Paley–Wiener theorem*.

The map

$$(Uf)(x) = \frac{1}{\sqrt{\pi}} \frac{1}{x+i} f\left(\frac{x-i}{x+i}\right)$$

defines a unitary operator from $L^2(\mathbb{T})$ onto $L^2(\mathbb{R})$ and a unitary operator from $H^2(\mathbb{T})$ onto $H^2(\mathbb{R})$ (Exercise 11.10.7). It can be used to prove (Exercise 11.10.11) a version of the Cauchy integral formula from Proposition 5.3.13, namely

$$g(z) = \frac{1}{2\pi i} \int_{-\infty}^\infty \frac{g(t)}{t-z} dt \quad \text{for all } g \in H^2(\mathbb{C}_+).$$

11.9 Notes

A rudimentary version of the Fourier inversion formula (Theorem 11.5.1) was discovered by Fourier in 1822 [131] in the less familiar form

$$f(x) = \frac{1}{2\pi} \int_a^b d\alpha \, f(\alpha) \int_{-\infty}^\infty dp \, \cos(px - p\alpha).$$

The term "Fourier transform" was used in 1924 by Titchmarsh [361, 362] who stated the inversion formula in more familiar terms as Fourier cosine transforms

$$f(x) = \frac{2}{\pi} \int_0^\infty \cos(xu) \, du \int_0^\infty \cos(xt) f(t) \, dt.$$

In other words

$$f(x) = \sqrt{\frac{2}{\pi}} \int_0^\infty \cos(xu) F(u) \, du \quad \text{and} \quad F(x) = \sqrt{\frac{2}{\pi}} \int_0^\infty \cos(xu) f(u) \, du.$$

In 1910, Plancherel [271] proved that $\|f\| = \|\mathscr{F}f\|$ for all $f \in L^2(\mathbb{R})$. All of this work was made precise over the years in treatments by several others and finally given a modern presentation by Wiener in 1930 [375, 376].

The eigenvalues and eigenvectors of the Fourier transform can be found in Titchmarsh's book [362], but they have a long history dating back to Gauss and Sylvester in a different form. In 1867, Sylvester, as part of his efforts to generalize the quaternions, explored the $N \times N$ discrete Fourier transform matrix

$$F = \frac{1}{\sqrt{N}} [\omega_N^{jk}]_{j,k=0}^{N-1},$$

where $\omega_N = e^{2\pi i/N}$. This matrix is unitary and appears in many places in applied mathematics. As with the Fourier transform on $L^2(\mathbb{R})$, there is the identity $F^4 = I$, so the eigenvalues of F are contained in $\{\pm 1, \pm i\}$ for $N < 4$ and equal to this set for $N \geqslant 4$. In 1972, McClellan and Parks [240] worked out the multiplicities for the eigenvalues of F. In 1982, Dickenson and Steiglitz [111] (see also [156]) showed this is equivalent to Gauss' evaluation of the quadratic Gauss sum. One can also compute the eigenvectors for F by various means [240], although they seem to lack a convenient closed-form expression.

The Paley–Wiener theorem was developed by Paley and Wiener in 1934 [258] (see also [319, Ch. 19]) and appears in two main forms. The first says that if $f \in H^2(\mathbb{C}_+)$, then

$$f(z) = \int_0^\infty g(t)e^{itz} \, dt$$

for some $g \in L^2(0, \infty)$. The other statement deals with entire functions and says that if f is entire with

$$|f(z)| \leqslant Ce^{A|z|} \quad \text{for all } z \in \mathbb{C}$$

and $f|_\mathbb{R} \in L^2(\mathbb{R})$, then there exists a function $g \in L^2(-A, A)$ such that

$$f(z) = \int_{-A}^A g(t)e^{itz} \, dt.$$

11.10 Exercises

Exercise 11.10.1. Prove that $(\mathscr{F}e^{-\frac{t^2}{2}})(x) = e^{-\frac{x^2}{2}}$.

Exercise 11.10.2. Prove that $(\mathscr{F}f_y)(x) = e^{-iyx}(\mathscr{F}f)(x)$, where $f_y(x) = f(x + iy)$ and $f \in L^2(\mathbb{R})$.

Exercise 11.10.3. Let $z = x + iy \in \mathbb{C}_+$ and let $f(t) = \frac{1}{\sqrt{2\pi}}e^{ixt-y|t|}$. Prove that

$$(\mathscr{F}f)(t) = \frac{1}{\pi}\frac{y}{(x-t)^2 + y^2}.$$

Exercise 11.10.4. Let $z = x + iy \in \mathbb{C}_+$ and let $f(t) = \frac{-i}{\sqrt{2\pi}}\operatorname{sgn}(t)e^{ixt-y|t|}$. Prove that

$$(\mathscr{F}f)(t) = \frac{1}{\pi}\frac{x-t}{(x-t)^2 + y^2}.$$

Exercise 11.10.5. Let $z = x + iy \in \mathbb{C}_+$ and let

$$f(t) = \begin{cases} \dfrac{1}{\sqrt{2\pi}} e^{izt} & \text{if } t > 0, \\ 0 & \text{if } t < 0. \end{cases}$$

Prove that $(\mathscr{F}f)(t) = \dfrac{1}{2\pi i} \dfrac{1}{t - z}$.

Exercise 11.10.6. Let $z = x + iy \in \mathbb{C}_+$ and let

$$f(t) = \begin{cases} 0 & \text{if } t > 0, \\ \dfrac{1}{\sqrt{2\pi}} e^{i\bar{z}t} & \text{if } t < 0. \end{cases}$$

Prove that $(\mathscr{F}f)(t) = -\dfrac{1}{2\pi i} \dfrac{1}{t - \bar{z}}$.

Exercise 11.10.7.

(a) Prove that

$$(Uf)(x) = \frac{1}{\sqrt{\pi}(x + i)} f\left(\frac{x - i}{x + i}\right) \quad \text{for } x \in \mathbb{R},$$

is a unitary operator from $L^2(\mathbb{T})$ onto $L^2(\mathbb{R})$.

(b) Prove that

$$(U^*g)(\xi) = \frac{2\sqrt{\pi i}}{1 - \xi} g\left(i\frac{1 + \xi}{1 - \xi}\right) \quad \text{for } \xi \in \mathbb{T}.$$

(c) Prove that $UH^2(\mathbb{T}) = H^2(\mathbb{R})$.

Exercise 11.10.8. This is a continuation of Exercise 11.10.7. Let $\mathcal{M}$ be a subspace of $L^2(\mathbb{R})$. Prove the following are equivalent.

(a) $e^{i\lambda x}\mathcal{M} \subseteq \mathcal{M}$ for all $\lambda \geqslant 0$.

(b) $w_t U^*\mathcal{M} \subseteq U^*\mathcal{M}$ for all $t \geqslant 0$, where

$$w_t(\xi) = \exp\left(-t\frac{1 + \xi}{1 - \xi}\right) \quad \text{for } \xi \in \mathbb{T}\backslash\{1\}.$$

(c) $M_\xi U^*\mathcal{M} \subseteq U^*\mathcal{M}$.

Remark: See [253] for more.

Exercise 11.10.9. Use Exercise 11.10.8 to prove that if $e^{i\lambda x}\mathcal{M} \subseteq \mathcal{M}$ for all $\lambda \geqslant 0$, then $\mathcal{M} = \chi_E L^2(\mathbb{R})$ for some Lebesgue measurable set $E \subseteq \mathbb{R}$ or $\mathcal{M} = wH^2(\mathbb{R})$ for some measurable unimodular function w on $\mathbb{R}$.

Exercise 11.10.10.

(a) Prove that the functions

$$
f_n(x) = \begin{cases} \dfrac{1}{\sqrt{\pi}} \dfrac{(x-i)^n}{(x+i)^{n+1}} & \text{if } n \geqslant 0, \\[3mm] \dfrac{1}{\sqrt{\pi}} \dfrac{(x+i)^{-n-1}}{(x-i)^{-n}} & \text{if } n \leqslant -1, \end{cases}
$$

form an orthonormal basis for $L^2(\mathbb{R})$.

(b) Prove that $(f_n)_{n=0}^{\infty}$ is an orthonormal basis for $H^2(\mathbb{R})$.

Exercise 11.10.11.

(a) Prove the following Cauchy integral formula for $H^2(\mathbb{C}_+)$:

$$
g(z) = \frac{1}{2\pi i} \int_{-\infty}^{\infty} \frac{g(t)}{t-z} \, dt \quad \text{for all } g \in H^2(\mathbb{C}_+) \text{ and } z \in \mathbb{C}_+.
$$

(b) Use (a) to obtain the estimate $|g(x+iy)| \leqslant \dfrac{1}{2\sqrt{\pi y}} \|g\|$ for all $x+iy \in \mathbb{C}_+$.

Exercise 11.10.12. Prove that the normalized Hermite functions $(h_n/\|h_n\|)_{n=0}^{\infty}$ from (11.7.2) form an orthonormal basis for $L^2(\mathbb{R})$.

Exercise 11.10.13. Consider $N : L^2(\mathbb{R}) \to L^2(\mathbb{R})$ defined by $(Nf)(t) = f(t+1)$.

(a) Prove that N is bounded and normal.

(b) Determine $\sigma(N)$.

(c) Prove that N is unitarily equivalent to a multiplication operator M_ψ on some $L^2(\mu)$ space and identify ψ and μ.

Exercise 11.10.14. For $\varphi \in L^1(\mathbb{R})$, consider the convolution operator $X_\varphi : L^2(\mathbb{R}) \to L^2(\mathbb{R})$ defined by $X_\varphi = f * \varphi$.

(a) Prove that X_φ is bounded on $L^2(\mathbb{R})$.

(b) Compute $\|X_\varphi\|$.

(c) Compute X_φ^*.

(d) Prove that X_φ is normal.

(e) Determine $\sigma(X_\varphi)$.

(f) Prove that X_φ is unitarily equivalent to a multiplication operator M_ψ on some $L^2(\mu)$ space and identify ψ and μ.

Exercise 11.10.15. For the Fourier transform $\mathscr{F}$ on $L^2(\mathbb{R})$, prove that

$$(\mathscr{F} - zI)^{-1} = \frac{1}{1 - z^4}(\mathscr{F}^3 + z\mathscr{F}^2 + z^2\mathscr{F} + z^3 I) \quad \text{for all } z \notin \{\pm 1, \pm i\}.$$

Exercise 11.10.16. For an integer $N \geqslant 1$, let $\omega_N = e^{2\pi i/N}$. Prove that the *discrete Fourier transform* matrix $F_N = \frac{1}{\sqrt{N}}[\omega_N^{jk}]_{j,k=0}^{N-1}$ is unitary and satisfies $F_N^4 = I$.

Exercise 11.10.17. This is a continuation of Exercise 11.10.16.

(a) Prove that the set $\mathcal{A}$ of $N \times N$ matrices A such that $F_N A F_N^*$ is diagonal is an N-dimensional commutative complex algebra of normal matrices.

(b) Find an orthonormal basis for $\mathcal{A}$ with respect to the Hilbert–Schmidt norm on $\mathcal{M}_N$.

Exercise 11.10.18. These next several exercises follow [119] and develop a Paley–Wiener theorem for the Bergman space of the upper half-plane. They also relate the Bergman shift from Chapter 10 to a Volterra-type operator.

(a) Define a norm on $\mathscr{L} = L^2\left((0, \infty), \dfrac{dx}{x}\right)$ by

$$\|f\|_{\mathscr{L}} = \left(\pi \int_0^\infty \frac{|f(x)|^2}{x}\,dx\right)^{\frac{1}{2}}.$$

Prove that the linear transformation $(Tf)(x) = e^{-x} \displaystyle\int_0^x e^t f(t)\,dt$ is bounded on $\mathscr{L}$.

(b) Prove that $Tf = h * f$, where $h(t) = \begin{cases} e^{-t} & \text{if } t > 0, \\ 0 & \text{if } t < 0. \end{cases}$

Exercise 11.10.19. Let $A^2(\mathbb{C}_+)$ denote the Bergman space of analytic functions F on $\mathbb{C}_+$ with norm

$$\|F\|_{A^2(\mathbb{C}_+)} = \left(\int_0^\infty \int_{-\infty}^\infty |F(x + iy)|^2\,dx\,dy\right)^{\frac{1}{2}}.$$

(a) Recall from Chapter 10 that A^2 is the Bergman space on $\mathbb{D}$ with norm

$$\|g\|_{A^2} = \left(\int_{\mathbb{D}} |g(x + iy)|^2\,dx\,dy\right)^{\frac{1}{2}}.$$

Prove that $G(z) = \dfrac{2}{(z + i)^2} g\left(\dfrac{z - i}{z + i}\right)$ belongs to $A^2(\mathbb{C}_+)$ and $\|g\|_{A^2} = \|G\|_{A^2(\mathbb{C}_+)}$.

(b) Prove that the bounded operator $g \mapsto G$ from (a) is surjective and therefore unitary.

(c) For $n \geqslant 0$, define

$$G_n(z) = \frac{2}{(z + i)^2}\left(\frac{z - i}{z + i}\right)^n.$$

Prove that $(G_n)_{n=0}^\infty$ is an orthogonal basis for $A^2(\mathbb{C}_+)$.

(d) Prove that the shift operator M_z on A^2 is unitarily equivalent to the multiplication operator $(MF)(z) = \dfrac{z - i}{z + i} F(z)$ on $A^2(\mathbb{C}_+)$.

Exercise 11.10.20. Continue with the notation from Exercises 11.10.18 and 11.10.19.

(a) For $f \in \mathscr{L}$, define $F(z) = \displaystyle\int_0^\infty f(t)e^{itz}\, dt$. Prove that $F \in A^2(\mathbb{C}_+)$ and $\|f\|_{\mathscr{L}} = \|F\|_{A^2(\mathbb{C}_+)}$.

(b) Define $\widehat{G_n}(t) = \dfrac{1}{2\pi} \displaystyle\int_{-\infty}^\infty e^{-ixt} G_n(t)\, dt$ (a slightly different form of the Fourier transform) and prove that

$$\widehat{G_n}(t) = \begin{cases} c_n te^{-t} L_n(2t) & \text{if } t \geqslant 0, \\ 0 & \text{if } t < 0, \end{cases}$$

where $c_n \in \mathbb{C}$ and L_n is a polynomial of degree n.

Remark: L_n is the *Laguerre polynomial* of degree n.

(c) Use this to prove that the operator $f \mapsto F$ is surjective. This yields a Paley–Wiener theorem for the Bergman space: every $F \in A^2(\mathbb{C}_+)$ can be written as $F(z) = \displaystyle\int_0^\infty f(t)e^{itz}\, dt$ for some $f \in \mathscr{L}$.

Exercise 11.10.21. Continue with the notation from Exercises 11.10.18 and 11.10.20.

(a) Prove that $MG_n = G_{n+1}$ and that $\widehat{MG_n}(t) = \widehat{G_n}(t) - 2(T\widehat{G_n})(t)$.

(b) Prove that M on $A^2(\mathbb{C}_+)$ is unitarily equivalent to $I - 2T$ on $\mathscr{L}$.

(c) Prove that M_z on A^2 is unitarily equivalent to $I - 2T$ on $\mathscr{L}$. Then conclude that the M_z-invariant subspaces of A^2 are in bijective and order-preserving correspondence with the T-invariant subspaces of $\mathscr{L}$.

(d) For $a > 0$, the subspace of functions in $\mathscr{L}$ that vanish almost everywhere on $[0, a]$ is T-invariant. Prove that the corresponding M_z-invariant subspace of A^2 is

$$\exp\left(-a\frac{1 + z}{1 - z}\right) A^2.$$

Exercise 11.10.22. The formula $\mathscr{F}h_n = (-i)^n h_n$ from (11.7.3), where h_n is the nth Hermite function, can be used to define a square root of the Fourier transform.

(a) Let $Qh_n = e^{-\frac{in\pi}{4}} h_n$ and prove that Q extends to a unitary operator on $L^2(\mathbb{R})$.

(b) Prove that $Q^2 = \mathscr{F}$.

(c) Prove that there are infinitely many unitary square roots of $\mathscr{F}$.

(d) Are there any non-unitary square roots of $\mathscr{F}$?

Remark: A formula of Mehler [242] (see also [125]) shows that

$$(Qf)(x) = \sqrt{\frac{1-i}{2\pi}} e^{i\frac{x^2}{2}} \int_{-\infty}^{\infty} e^{-i(\sqrt{2}xt - \frac{t^2}{2})} f(t)\, dt.$$

Exercise 11.10.23. Let $\mu = \delta_1 + \delta_{-1} + \delta_i + \delta_{-i}$ and define the Hilbert space

$$(L^2(\mu))^{(\infty)} = \Big\{ (f_i)_{i=1}^{\infty} : f_i \in L^2(\mu), \sum_{i=1}^{\infty} \|f_i\|^2 < \infty \Big\}.$$

These types of infinite direct sums are discussed in Chapter 14. Define $M : L^2(\mu) \to L^2(\mu)$ by $(Mf)(\xi) = \xi f(\xi)$.

(a) Prove that M is unitary.

(b) Prove that $M^{(\infty)}$ on $(L^2(\mu))^{(\infty)}$ defined by $M^{(\infty)}(f_i)_{i=1}^{\infty} = (Mf_i)_{i=1}^{\infty}$ is unitary.

(c) Use the matrix representation of the Fourier transform in (11.7.4) to prove that the Fourier transform is unitarily equivalent to $M^{(\infty)}$.

11.11 Hints for the Exercises

Hint for Ex. 11.10.1: Complete the square in the integral and notice that it is a Gaussian integral (or integrate over a well-chosen contour).

Hint for Ex. 11.10.7: Prove that $\varphi(z) = \dfrac{z-i}{z+i}$ is a conformal map from $\mathbb{C}_+$ onto $\mathbb{D}$.

Hint for Ex. 11.10.8: Consult Exercise 11.10.7. For the proof that (b) implies (c), consider the function $\psi_t(\xi) = \dfrac{w_t(\xi) - 1 + t}{w_t(\xi) - 1 - t}$ and show that $|\psi_t(\xi)| \leqslant 1$ for all $\xi \in \mathbb{T}\setminus\{1\}$ and that $\psi_t(\xi) = \xi + o(1)$ as $t \to 0$.

Hint for Ex. 11.10.10: Consult Exercise 11.10.7.

Hint for Ex. 11.10.11: For (a), start with Proposition 5.3.13 and use Exercise 11.10.7.

Hint for Ex. 11.10.12: To show completeness, suppose that $f \in L^2(\mathbb{R})$ is orthogonal to h_n for every $n \geqslant 0$. Now use the fact that the Hermite polynomials are dense in a certain weighted L^2 space (see Exercise 1.10.44).

Hint for Ex. 11.10.19: For (c), consult Proposition 10.1.8 and use (a).

Hint for Ex. 11.10.20: For (b), make use of the identities

$$(1-u)^{-k-2} = \sum_{j=0}^{\infty} \binom{j+k+1}{k+1} u^j \quad \text{and} \quad \int_{-\infty}^{\infty} t^j e^{-t}\, dt = j!.$$

Hint for Ex. 11.10.21: For (b), use the Fourier transform.

12

· ▪ ● ▪ ·

The Hilbert Transform

Key Concepts: Harmonic conjugate, Poisson integral, Fatou's theorem, Hilbert transform on the circle, partial isometry, Hilbert transform on the real line, spectral properties of Hilbert transforms.

Outline: This chapter concerns two versions of the *Hilbert transform*. The first version is $\mathscr{H} : L^2(\mathbb{R}) \to L^2(\mathbb{R})$ defined by

$$(\mathscr{H}f)(x) = \frac{1}{\pi} \, \mathrm{PV} \int_{-\infty}^{\infty} \frac{f(t)}{x - t} dt,$$

and the second is $\mathscr{D} : L^2(\mathbb{T}) \to L^2(\mathbb{T})$ defined by

$$(\mathscr{D}f)(e^{i\theta}) = \frac{1}{2\pi} \, \mathrm{PV} \int_{-\pi}^{\pi} \cot\left(\frac{\theta - t}{2}\right) f(e^{it}) dt,$$

where PV denotes the principal value of an integral. Although these operators are defined on the Hilbert spaces $L^2(\mathbb{R})$ and $L^2(\mathbb{T})$, respectively, they also act on various Banach spaces of functions. This chapter emphasizes the Hilbert-space properties of these operators. In particular, we study the boundedness, norm, adjoint, and spectral properties of Hilbert transforms and how these properties relate to harmonic conjugation and Riemann–Hilbert problems.

12.1 The Poisson Integral on the Circle

If Ω is a simply connected planar region and $u : \Omega \to \mathbb{R}$, then u is *harmonic* if its second-order partial derivatives are continuous and satisfy the *Laplace equation*

$$\partial_x^2 u + \partial_y^2 u = 0 \tag{12.1.1}$$

on Ω. The Cauchy–Riemann equations imply that $u = \mathrm{Re}\, f$ is harmonic when $f : \Omega \to \mathbb{C}$ is analytic. Since Ω is simply connected, any harmonic function u on Ω is the real part of an analytic function on Ω. That is, there exists a harmonic function $v : \Omega \to \mathbb{R}$ such that

$u + iv$ is analytic on Ω [92, p. 252]. The function v is a *harmonic conjugate* of u and it is unique up to an additive constant.

The first type of harmonic functions explored in this chapter are defined on the unit disk $\mathbb{D}$ and are of the form

$$(Pf)(re^{i\theta}) = \int_{-\pi}^{\pi} P_r(\theta - t)f(e^{it})\frac{dt}{2\pi}. \tag{12.1.2}$$

In the above, $f \in L^2(\mathbb{T})$ and

$$P_r(t) = \mathrm{Re}\left(\frac{1+z}{1-z}\right), \quad z = re^{it}, \quad 0 \leqslant r < 1, \quad \theta \in [-\pi, \pi], \tag{12.1.3}$$

denotes the *Poisson kernel* of the unit disk. Since

$$z \mapsto \frac{1+z}{1-z}$$

is analytic on $\mathbb{D}$, it follows that

$$z \mapsto \mathrm{Re}\left(\frac{1+z}{1-z}\right)$$

is harmonic. Differentiating under the integral sign reveals that $(Pf)(re^{i\theta})$ is harmonic on $\mathbb{D}$. A power series calculation (Exercise 12.5.1) shows that the Poisson kernel $P_r(t)$ satisfies

$$P_r(t) = \sum_{n=-\infty}^{\infty} r^{|n|}e^{int} = \frac{1 - r^2}{1 - 2r\cos t + r^2} = \frac{1 - r^2}{|1 - re^{it}|^2}. \tag{12.1.4}$$

The right side of (12.1.4) shows that $P_r(t) > 0$ on $\mathbb{D}$. Moreover,

$$\int_{-\pi}^{\pi} P_r(t)\frac{dt}{2\pi} = 1 \quad \text{for all } 0 \leqslant r < 1; \tag{12.1.5}$$

see Exercise 12.5.2. The graph of $P_r(t)$ peaks sharply at the origin as $r \to 1^-$ (Figure 12.1.1). This suggests that $(Pf)(re^{i\theta}) \to f(e^{i\theta})$ as $r \to 1^-$, which was confirmed in 1906 by Fatou.

Theorem 12.1.6 (Fatou [127]). *If $f \in L^2(\mathbb{T})$, then*

$$(Pf)(re^{i\theta}) = \int_{-\pi}^{\pi} P_r(\theta - t)f(e^{it})\frac{dt}{2\pi}$$

is a harmonic function on $\mathbb{D}$ such that

$$\lim_{r \to 1^-} (Pf)(re^{i\theta}) = f(e^{i\theta})$$

for almost every $\theta \in [-\pi, \pi]$.

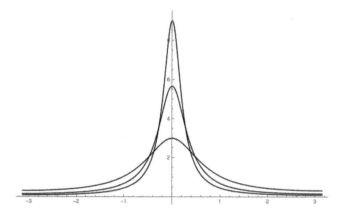

Figure 12.1.1 The graphs of $P_r(\theta)$ on $[-\pi, \pi]$ for $r = 0.5, 0.7, 0.8$. The area under each curve is 2π, but the graphs peak higher as $r \to 1^-$.

Proof The discussion above proves that Pf is harmonic. The next step is to prove that

$$\lim_{r \to 1^-} (Pf)(re^{i\theta}) = f(e^{i\theta})$$

when f is continuous at $e^{i\theta}$. We refer the reader to [202, p. 34] for a proof in the general case. Given $\varepsilon > 0$, let $I_\delta = (\theta - \delta, \theta + \delta)$ be an open interval containing θ such that $|f(e^{it}) - f(e^{i\theta})| < \varepsilon$ for all $t \in I_\delta$. Observe that

$$\left| \int_{-\pi}^{\pi} P_r(\theta - t) f(e^{it}) \frac{dt}{2\pi} - f(e^{i\theta}) \right|$$

$$= \left| \int_{-\pi}^{\pi} P_r(\theta - t)(f(e^{it}) - f(e^{i\theta})) \frac{dt}{2\pi} \right| \qquad \text{(by (12.1.5))}$$

$$\leqslant \int_{-\pi}^{\pi} P_r(\theta - t)|f(e^{it}) - f(e^{i\theta})| \frac{dt}{2\pi}$$

$$= \int_{I_\delta} P_r(\theta - t)|f(e^{it}) - f(e^{i\theta})| \frac{dt}{2\pi}$$

$$+ \int_{[-\pi,\pi]\backslash I_\delta} P_r(\theta - t)|f(e^{it}) - f(e^{i\theta})| \frac{dt}{2\pi}$$

$$\leqslant \varepsilon + \int_{[-\pi,\pi]\backslash I_\delta} P_r(\theta - t)|f(e^{it}) - f(e^{i\theta})| \frac{dt}{2\pi}.$$

From (12.1.4), there is a $c > 0$ such that

$$P_r(\theta - t) = \frac{1 - r^2}{1 - 2r\cos(\theta - t) + r^2} \leqslant c(1 - r) \quad \text{for } t \in [-\pi, \pi]\backslash I_\delta \text{ and } 0 \leqslant r < 1.$$

This ensures that ensures the last integral above tends to zero as $r \to 1^-$. The result now follows. ∎

One says that Pf is the *harmonic extension* of f to $\mathbb{D}$. Moreover, if $f, g \in L^2(\mathbb{T})$ and $Pf = Pg$ on $\mathbb{D}$, then $f = g$ almost everywhere on $\mathbb{T}$ [202, Ch. 3]. We now relate the harmonic extension of f to

$$\sum_{n=-\infty}^{\infty} \widehat{f}(n) e^{in\theta}, \quad \text{where } \widehat{f}(n) = \int_{-\pi}^{\pi} f(e^{it}) e^{-int} \frac{dt}{2\pi},$$

the Fourier series of f.

Term-by-term integration in (12.1.2), permissible by uniform convergence, yields

$$(Pf)(re^{i\theta}) = \int_{-\pi}^{\pi} P_r(\theta - t) f(e^{it}) \frac{dt}{2\pi} \qquad \text{(by (12.1.2))}$$

$$= \int_{-\pi}^{\pi} \left(\sum_{n=-\infty}^{\infty} r^{|n|} e^{in\theta} e^{-int} \right) f(e^{it}) \frac{dt}{2\pi} \qquad \text{(by (12.1.4))}$$

$$= \sum_{n=-\infty}^{\infty} r^{|n|} e^{in\theta} \left(\int_{-\pi}^{\pi} f(e^{it}) e^{-int} \frac{dt}{2\pi} \right)$$

$$= \sum_{n=-\infty}^{\infty} \widehat{f}(n) r^{|n|} e^{in\theta}. \qquad (12.1.7)$$

Fatou's theorem (Theorem 12.1.6) says that $(Pf)(re^{i\theta}) \to f(e^{i\theta})$ pointwise almost everywhere. This next result shows convergence in the $L^2(\mathbb{T})$ norm.

Proposition 12.1.8. *If $f \in L^2(\mathbb{T})$, then*

$$\lim_{r \to 1^-} \int_{-\pi}^{\pi} |(Pf)(re^{i\theta}) - f(e^{i\theta})|^2 \frac{d\theta}{2\pi} = 0.$$

Proof Using (12.1.7) and Parseval's theorem,

$$\int_{-\pi}^{\pi} |(Pf)(re^{i\theta}) - f(e^{i\theta})|^2 \frac{d\theta}{2\pi} = \sum_{n=-\infty}^{\infty} |\widehat{f}(n)|^2 (1 - r^{|n|})^2,$$

which tends to zero as $r \to 1^-$ by the dominated convergence theorem. ∎

12.2 The Hilbert Transform on the Circle

The previous section suggests a way to find the harmonic conjugate of a function of the form Pf, where $f \in L^2(\mathbb{T})$ is real valued. Define the *conjugate Poisson kernel*

$$Q_r(t) = \text{Im}\left(\frac{1+z}{1-z}\right), \qquad (12.2.1)$$

where $z = re^{it}$ with $0 \leqslant r < 1$ and $-\pi \leqslant t \leqslant \pi$. The reader can verify that (Exercise 12.5.3)

$$Q_r(t) = \frac{2r \sin(t)}{1 - 2r \cos(t) + r^2} = \sum_{n=-\infty}^{\infty} -i \, \text{sgn}(n) r^{|n|} e^{int}. \qquad (12.2.2)$$

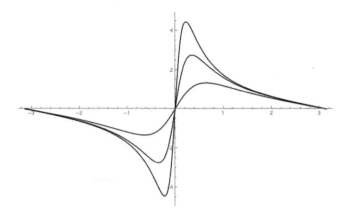

Figure 12.2.1 The graphs of $Q_r(\theta)$ on $[-\pi, \pi]$ for $r = 0.5, 0.7, 0.8$. The graphs gravitate towards the imaginary axis as r increases.

In the above,

$$\text{sgn}(n) = \begin{cases} 1 & \text{if } n > 0, \\ -1 & \text{if } n < 0, \\ 0 & \text{if } n = 0. \end{cases}$$

See Figure 12.2.1 for the graph of $Q_r(t)$ for several values of r. From (12.2.1), the function Qf defined by

$$(Qf)(re^{i\theta}) = \frac{1}{2\pi} \int_{-\pi}^{\pi} Q_r(\theta - t) f(e^{it}) \, dt$$

is harmonic on $\mathbb{D}$ and

$$(Qf)(re^{i\theta}) = -i \sum_{n=-\infty}^{\infty} \hat{f}(n) \, \text{sgn}(n) r^{|n|} e^{in\theta}. \tag{12.2.3}$$

Moreover,

$$(Pf)(z) + i(Qf)(z) = \frac{1}{2\pi} \int_{-\pi}^{\pi} \frac{e^{it} + z}{e^{it} - z} f(e^{it}) \, dt$$

is analytic on $\mathbb{D}$. The integral on the right side of the previous line is the *Herglotz integral* of f. If f is sufficiently smooth, which causes $\hat{f}(n)$ to decay quickly, the limit in (12.2.3) as $r \to 1^-$ exists for all $\theta \in [-\pi, \pi]$. For general $f \in L^2(\mathbb{T})$, there is the following.

Theorem 12.2.4. *For $f \in L^2(\mathbb{T})$, the function $\mathcal{Q}f$ defined by*

$$(\mathcal{Q}f)(e^{i\theta}) := \lim_{r \to 1^-} (Qf)(re^{i\theta})$$

exists for almost every θ *and defines an* $L^2(\mathbb{T})$ *function with Fourier series*

$$-i \sum_{n=-\infty}^{\infty} \widehat{f}(n)\,\mathrm{sgn}(n)e^{in\theta}. \tag{12.2.5}$$

We now focus on the operator-theoretic properties of $f \mapsto \mathscr{Q}f$, called the *Hilbert transform* on the circle. Below, we use the notation 1 to denote the constant function and $L^2(\mathbb{T}) \ominus \mathrm{span}\{1\}$ for $(\mathrm{span}\{1\})^{\perp}$.

Proposition 12.2.6. $\mathscr{Q}$ *is a bounded operator on* $L^2(\mathbb{T})$ *which satisfies the following.*

(a) $\ker \mathscr{Q} = \mathrm{span}\{1\}$.

(b) $\mathrm{ran}\, \mathscr{Q} = L^2(\mathbb{T}) \ominus \mathrm{span}\{1\}$.

(c) $\mathscr{Q}|_{L^2(\mathbb{T})\ominus\mathrm{span}\{1\}}$ *is an isometry.*

Proof From (12.2.5) and Parseval's theorem,

$$\|\mathscr{Q}f\|^2 = \sum_{n\in\mathbb{Z}\setminus\{0\}} |\widehat{f}(n)|^2 = \|f\|^2 - |\widehat{f}(0)|^2 \leqslant \|f\|^2 \quad \text{for all } f \in L^2(\mathbb{T}). \tag{12.2.7}$$

Thus, $\mathscr{Q}$ is bounded on $L^2(\mathbb{T})$.

(a) If $\|\mathscr{Q}f\| = 0$, (12.2.5) shows that

$$0 = \|f\|^2 - |\widehat{f}(0)|^2 = \sum_{n\in\mathbb{Z}\setminus\{0\}} |\widehat{f}(n)|^2.$$

This implies that $\widehat{f}(n) = 0$ for all $n \neq 0$, hence $f = \widehat{f}(0)$ is constant. This proves that $\ker \mathscr{Q} \subseteq \mathrm{span}\{1\}$. To see that $\mathrm{span}\{1\} \subseteq \ker \mathscr{Q}$, simply reverse the argument above.

(b) By (12.2.5), $\mathrm{ran}\, \mathscr{Q} \subseteq L^2(\mathbb{T}) \ominus \mathrm{span}\{1\}$. If $g \in L^2(\mathbb{T}) \ominus \mathrm{span}\{1\}$ with Fourier series

$$g = \sum_{n\in\mathbb{Z}\setminus\{0\}} \widehat{g}(n)e^{in\theta},$$

let

$$f = i \sum_{n=-\infty}^{\infty} \widehat{g}(n)\,\mathrm{sgn}(n)e^{in\theta}.$$

Then $f \in L^2(\mathbb{T})$ and $\mathscr{Q}f = g$.

(c) By (12.2.7), $\mathscr{Q}$ is isometric on $L^2(\mathbb{T}) \ominus \mathrm{span}\{1\}$. ∎

Operators that are isometric on the orthogonal complement of their kernels are called *partial isometries* and are explored further in Chapter 14.

If $r = 1$ in (12.2.2), then

$$Q_1(\theta - t) = \cot\left(\frac{\theta - t}{2}\right)$$

and the integral in the formula

$$(\mathscr{Q}f)(e^{i\theta}) = \int_{-\pi}^{\pi} \cot\left(\frac{\theta - t}{2}\right) f(e^{it}) \frac{dt}{2\pi} \tag{12.2.8}$$

becomes problematic due to the singularity at $t = \theta$. For general $f \in L^2(\mathbb{T})$, the integral in (12.2.8) must be understood as a *principal-value* integral

$$\text{PV} \int_{-\pi}^{\pi} \cot\left(\frac{\theta - t}{2}\right) f(e^{it}) \frac{dt}{2\pi} = \lim_{\varepsilon \to 0^+} \int_{|\theta - t| > \varepsilon} \cot\left(\frac{\theta - t}{2}\right) f(e^{it}) \frac{dt}{2\pi}. \tag{12.2.9}$$

This principal-value integral equals $\mathscr{Q}f$ almost everywhere [149, Ch. 3]. Thus, (12.2.9) is another formula for the Hilbert transform. One can also compute the adjoint of the Hilbert transform $\mathscr{Q}$.

Proposition 12.2.10. $\mathscr{Q}^* = -\mathscr{Q}$.

Proof For $f, g \in L^2(\mathbb{T})$, Parseval's theorem yields

$$\langle \mathscr{Q}f, g \rangle = \sum_{n=-\infty}^{\infty} -i\,\text{sgn}(n)\widehat{f}(n)\overline{\widehat{g}(n)} = \sum_{n=-\infty}^{\infty} \widehat{f}(n)\overline{i\,\text{sgn}(n)\widehat{g}(n)} = \langle f, -\mathscr{Q}g \rangle.$$

This shows that $\mathscr{Q}^* = -\mathscr{Q}$. ∎

We end this section with the observation that the Fourier-series formula (12.2.5) shows that, with respect to the orthonormal basis $(e^{in\theta})_{n=-\infty}^{\infty}$ of $L^2(\mathbb{T})$, the matrix representation of $\mathscr{Q}$ is the diagonal matrix

$$
\begin{bmatrix}
\ddots & \vdots & \vdots & \vdots & \vdots & \vdots & \iddots \\
\cdots & i & 0 & 0 & 0 & 0 & \cdots \\
\cdots & 0 & i & 0 & 0 & 0 & \cdots \\
\cdots & 0 & 0 & \boxed{0} & 0 & 0 & \cdots \\
\cdots & 0 & 0 & 0 & -i & 0 & \cdots \\
\cdots & 0 & 0 & 0 & 0 & -i & \cdots \\
\iddots & \vdots & \vdots & \vdots & \vdots & \vdots & \ddots
\end{bmatrix}.
$$

The box denotes the $(0, 0)$ entry. The matrix above illustrates the identity $\mathscr{Q}^* = -\mathscr{Q}$, as well as the following.

Corollary 12.2.11. $\sigma(\mathscr{Q}) = \sigma_p(\mathscr{Q}) = \{-i, 0, i\}$.

Proof Apply Theorem 2.4.7. ∎

These two results imply that $\mathscr{Q}$ is a normal operator with the following eigenspaces:

$$\ker(\mathscr{Q}) = \text{span}\{1\},$$
$$\ker(\mathscr{Q} - iI) = \bigvee\{e^{in\theta} : n < 0\},$$
$$\ker(\mathscr{Q} + iI) = \bigvee\{e^{in\theta} : n > 0\}.$$

Finally, observe that $\mathcal{Q}$ is a rank-one perturbation of any of the unitary operators $\mathcal{Q} + \alpha(1 \otimes 1)$, for $\alpha \in \mathbb{T}$. These unitary operators have the matrix representations

$$\begin{bmatrix} \ddots & \vdots & \vdots & \vdots & \vdots & \vdots & \reflectbox{$\ddots$} \\ \cdots & i & 0 & 0 & 0 & 0 & \cdots \\ \cdots & 0 & i & 0 & 0 & 0 & \cdots \\ \cdots & 0 & 0 & \boxed{\alpha} & 0 & 0 & \cdots \\ \cdots & 0 & 0 & 0 & -i & 0 & \cdots \\ \cdots & 0 & 0 & 0 & 0 & -i & \cdots \\ \reflectbox{$\ddots$} & \vdots & \vdots & \vdots & \vdots & \vdots & \ddots \end{bmatrix}.$$

12.3 The Hilbert Transform on the Real Line

We now consider the harmonic conjugation problem for harmonic functions of the form

$$(Ug)(z) = \int_{-\infty}^{\infty} \mathcal{P}_z(t)g(t)\,dt.$$

In the above, $g \in L^2(\mathbb{R})$ is real valued and

$$\mathcal{P}_z(t) = \mathrm{Re}\left(\frac{1}{i\pi}\frac{1}{t-z}\right), \quad \text{where } t \in \mathbb{R} \text{ and } z = x + iy \in \mathbb{C}_+, \tag{12.3.1}$$

is the *Poisson kernel* of the upper-half plane. A computation (Exercise 12.5.5) reveals that

$$\mathcal{P}_z(t) = \frac{1}{\pi}\frac{y}{(x-t)^2 + y^2}.$$

A similar argument as in the previous section shows that Ug is harmonic on $\mathbb{C}_+$. Moreover, there is the corresponding version of Fatou's theorem (Theorem 12.1.6; see also Lemma 11.4.4).

Theorem 12.3.2. *For $g \in L^2(\mathbb{R})$,*

$$g(x) = \lim_{y \to 0^+} (Ug)(x+iy)$$

for almost every $x \in \mathbb{R}$.

As shown in Lemma 11.4.6,

$$\lim_{y \to 0^+} \int_{-\infty}^{\infty} |(Uf)(x+iy) - f(x)|^2 dx = 0.$$

Consider the *conjugate Poisson kernel*

$$\mathcal{Q}_z(t) = \mathrm{Im}\left(\frac{1}{i\pi}\frac{1}{t-z}\right), \tag{12.3.3}$$

where $t \in \mathbb{R}$ and $z = x + iy \in \mathbb{C}_+$. A computation (Exercise 12.5.6) reveals that

$$Q_z(t) = \frac{1}{\pi} \frac{x - t}{(x - t)^2 + y^2}.$$

By (12.3.3) the function

$$(Vg)(z) = \int_{-\infty}^{\infty} Q_z(t) g(t) \, dt$$

is harmonic on $\mathbb{C}_+$ and by (12.3.1)

$$(Ug)(z) + i(Vg)(z) = \frac{1}{i\pi} \int_{-\infty}^{\infty} \frac{g(t)}{t - z} \, dt$$

is analytic on $\mathbb{C}_+$. The right side of the previous equation is the *Cauchy integral* of g. When g is real valued, Vg is a harmonic conjugate of Ug. The next theorem is the analogue of Theorem 12.2.4.

Theorem 12.3.4. *For $g \in L^2(\mathbb{R})$,*

$$(\mathscr{H}g)(x) := \lim_{y \to 0^+} (Vg)(x + iy)$$

exists for almost every $x \in \mathbb{R}$.

The function $\mathscr{H}g$ defined above is the *Hilbert transform* of g. As in the circle case (see (12.2.9)), the Hilbert transform can be written as a principal-value integral in the sense that for almost every $x \in \mathbb{R}$,

$$(\mathscr{H}g)(x) = \frac{1}{\pi} \operatorname{PV} \int_{-\infty}^{\infty} \frac{g(t)}{x - t} \, dt = \lim_{\varepsilon \to 0^+} \int_{|x - t| > \varepsilon} \frac{g(t)}{x - t} \, dt.$$

In the circle case, the corresponding Hilbert transform $\mathscr{Q}$ is a rank-one perturbation of a unitary operator. For the Hilbert transform on $\mathbb{R}$, something different occurs.

Theorem 12.3.5. *The Hilbert transform $\mathscr{H}$ is a unitary operator on $L^2(\mathbb{R})$.*

Proof The first step is to prove that

$$(\mathscr{F}\mathscr{H}g)(s) = -i \operatorname{sgn}(s)(\mathscr{F}g)(s) \quad \text{for all } s \in \mathbb{R}, \tag{12.3.6}$$

where

$$(\mathscr{F}h)(s) = \frac{1}{\sqrt{2\pi}} \int_{-\infty}^{\infty} h(x) e^{-ixs} \, dx$$

is the Fourier transform of $h \in L^2(\mathbb{R})$ (Chapter 11). To avoid getting sidetracked by technical details, we proceed formally. The precise details are found in [161, Ch. 4]. Observe that

$$(\mathscr{F}\mathscr{H}g)(s) = \frac{1}{\sqrt{2\pi}} \int_{-\infty}^{\infty} e^{-ixs} \left(\frac{1}{\pi} \int_{-\infty}^{\infty} \frac{g(x - t)}{t} \, dt \right) dx$$

$$= \frac{1}{\sqrt{2\pi}} \int_{-\infty}^{\infty} \frac{1}{\pi t} \Big(\int_{-\infty}^{\infty} g(x-t) e^{-ixs} dx \Big) dt$$

$$= \int_{-\infty}^{\infty} \frac{1}{\pi t} e^{-ist} (\mathscr{F}g)(s) \, dt$$

$$= (\mathscr{F}g)(s) \int_{-\infty}^{\infty} \frac{1}{\pi t} e^{-ist} dt$$

$$= -i(\mathscr{F}g)(s) \operatorname{sgn}(s).$$

The last formula is understood in the distributional sense [161]. This verifies (12.3.6). Plancherel's theorem and (12.3.6) reveal that the Hilbert transform is isometric on $L^2(\mathbb{R})$. To complete the proof that $\mathscr{H}$ is unitary, it suffices to show that $\mathscr{H}$ is surjective. This is done by proving that $\mathscr{H}^2 = -I$. Indeed, define $W : L^2(\mathbb{R}) \to L^2(\mathbb{R})$ by $(Wf)(s) = -i \operatorname{sgn}(s) f(s)$. Then (12.3.6) shows that

$$\mathscr{F} \mathscr{H} \mathscr{F}^{-1} = W, \tag{12.3.7}$$

and hence

$$\mathscr{H}^2 = \mathscr{F}^{-1} W \mathscr{F} \mathscr{F}^{-1} W \mathscr{F} = \mathscr{F}^{-1} W^2 \mathscr{F} = \mathscr{F}^{-1}(-I)\mathscr{F} = -I.$$

Since $\mathscr{H}$ is isometric and surjective, it is unitary. ∎

Corollary 12.3.8. $\mathscr{H}^* = -\mathscr{H}$.

What is the spectrum of the Hilbert transform on $\mathbb{R}$?

Corollary 12.3.9. $\sigma(\mathscr{H}) = \sigma_p(\mathscr{H}) = \{-i, i\}$.

Proof By (12.3.7), $\mathscr{H}$ is unitarily equivalent to the multiplication operator $(Wf)(s) = -i \operatorname{sgn}(s) f(s)$ on $L^2(\mathbb{R})$. Notice that W is a multiplication operator whose symbol is a step function with essential range equal to $\{-i, i\}$. Therefore, $\sigma(W) = \sigma_p(W) = \{-i, i\}$ (Proposition 8.1.12) and hence $\sigma(\mathscr{H}) = \sigma_p(\mathscr{H}) = \{-i, i\}$ since unitary equivalence preserves spectra. ∎

The following matrix representation provides the full spectral decomposition of $\mathscr{H}$. To see this, observe that the *Cayley transform*

$$c(z) = \frac{z-i}{z+i}$$

is a conformal map from $\mathbb{C}_+$ onto $\mathbb{D}$ that maps the real line onto $\mathbb{T} \setminus \{1\}$. A change of variables (Exercise 11.10.7) shows that

$$(Uf)(x) = \frac{1}{\sqrt{\pi}} \frac{1}{x+i} f(c(x))$$

defines a unitary operator from $L^2(\mathbb{T})$ onto $L^2(\mathbb{R})$. Since $(\xi^n)_{n=0}^{\infty}$ is an orthonormal sequence in $L^2(\mathbb{T})$,

$$U(\xi^n) = \frac{1}{\sqrt{\pi}} \frac{(x-i)^n}{(x+i)^{n+1}} \quad \text{for } n \geqslant 0$$

is an orthonormal sequence in $L^2(\mathbb{R})$. A computation (Exercise 12.5.18) shows that

$$\mathcal{H}\left(\frac{1}{\sqrt{\pi}} \frac{(x-i)^n}{(x+i)^{n+1}} \right) = -i \frac{1}{\sqrt{\pi}} \frac{(x-i)^n}{(x+i)^{n+1}}.$$

Similarly,

$$U(\overline{\xi}^n) = \frac{1}{\sqrt{\pi}} \frac{(x+i)^{n-1}}{(x-i)^n} \quad \text{for } n \geqslant 1$$

is an orthonormal sequence in $L^2(\mathbb{R})$ and

$$\mathcal{H}\left(\frac{1}{\sqrt{\pi}} \frac{(x+i)^{n-1}}{(x-i)^n} \right) = i \frac{1}{\sqrt{\pi}} \frac{(x+i)^{n-1}}{(x-i)^n}.$$

Therefore,

$$\ker(\mathcal{H} + iI) = \bigvee \left\{ \frac{1}{\sqrt{\pi}} \frac{(x-i)^n}{(x+i)^{n+1}} : n \geqslant 0 \right\}$$

and

$$\ker(\mathcal{H} - iI) = \bigvee \left\{ \frac{1}{\sqrt{\pi}} \frac{(x+i)^{n-1}}{(x-i)^n} : n \geqslant 1 \right\}.$$

Since all of these eigenvectors together form an orthonormal basis for $L^2(\mathbb{R})$ (they are images of an orthonormal basis under a unitary operator), the matrix representation of $\mathcal{H}$ is the diagonal matrix

$$\begin{bmatrix} \ddots & \vdots & \vdots & \vdots & \vdots & \vdots & \iddots \\ \cdots & i & 0 & 0 & 0 & 0 & \cdots \\ \cdots & 0 & i & 0 & 0 & 0 & \cdots \\ \cdots & 0 & 0 & \boxed{-i} & 0 & 0 & \cdots \\ \cdots & 0 & 0 & 0 & -i & 0 & \cdots \\ \cdots & 0 & 0 & 0 & 0 & -i & \cdots \\ \iddots & \vdots & \vdots & \vdots & \vdots & \vdots & \ddots \end{bmatrix}.$$

The box denotes the $(0, 0)$ entry. Notice how this is a rank-one unitary perturbation of the matrix representation of $\mathcal{Q}$.

12.4 Notes

A two-volume treatment of Hilbert transforms is [216, 217]. The identity $\mathscr{H}^2 = -I$ sometimes appears in terms of Hilbert transform pairs, in the sense that if

$$g(x) = \frac{1}{\pi} \text{PV} \int_{-\infty}^{\infty} \frac{f(t)}{x-t}\, dt,$$

then

$$f(x) = -\frac{1}{\pi} \text{PV} \int_{-\infty}^{\infty} \frac{g(t)}{x-t}\, dt.$$

Hilbert [195] understood these pairs as

$$u(\sigma) = \frac{1}{\pi} \text{PV} \int_{-\pi}^{\pi} \frac{dv}{ds} \log\left(2\sin\left|\frac{s-\sigma}{2}\right|\right) ds$$

and

$$v(\sigma) = -\frac{1}{\pi} \text{PV} \int_{-\pi}^{\pi} \frac{du}{ds} \log\left(2\sin\left|\frac{s-\sigma}{2}\right|\right) ds.$$

Hardy [176, 177] understood these pairs as

$$g(x) = \frac{1}{\pi}\frac{d}{dx} \int_{-\infty}^{\infty} f(t) \log\left|1 - \frac{x}{t}\right| dt$$

and

$$f(x) = -\frac{1}{\pi}\frac{d}{dx} \int_{-\infty}^{\infty} g(t) \log\left|1 - \frac{x}{t}\right| dt.$$

The mapping properties of the Hilbert transform were first investigated by Privalov [275], who proved that the Hilbert transform preserves certain Lipschitz classes, and by M. Riesz, [308] who proved that the Hilbert transform preserves the $L^p(\mathbb{R})$ classes for $1 < p < \infty$. Further work of Pichorides [269] computed the norm of the Hilbert transform $\mathscr{H}$ on $L^p(\mathbb{T})$ as $\tan(\pi/2p)$ for $1 < p \leqslant 2$ and $\cot(\pi/2p)$ for $2 < p < \infty$. The classes $L^1(\mathbb{R})$ and $L^\infty(\mathbb{R})$ are not preserved by the Hilbert transform [149] (Exercise 12.5.4). There is a large body of work on certain operators related to the Hilbert transform called Calderón–Zygmund operators [350]. This family of operators has mapping properties analogous to those of the Hilbert transform.

The Hilbert transform was considered by Hilbert in 1905 to study a boundary-value problem Riemann considered as part of his doctoral thesis. In a version of this problem, one is given an $f : \mathbb{R} \to \mathbb{R}$ and asked to find analytic functions F_+ and F_- on the upper and lower half planes, respectively, such that $F_+(x) - F_-(x) = f(x)$ on $\mathbb{R}$. As it turns out, $(\mathscr{H}f)(x) = -i(F_+(x) + F_-(x))$. See [259] for a survey of Riemann–Hilbert problems.

12.5 Exercises

Exercise 12.5.1. Let $0 \leqslant r < 1$ and $-\pi \leqslant t \leqslant \pi$.

(a) Prove that the Poisson kernel from (12.1.3) satisfies

$$P_r(t) = \frac{1 - r^2}{1 - 2r \cos t + r^2} = \mathrm{Re}\left(\frac{1 + re^{it}}{1 - re^{it}}\right) = \frac{1 - r^2}{|1 - re^{it}|^2}.$$

(b) Prove that $P_r(t) = \displaystyle\sum_{n=-\infty}^{\infty} r^{|n|} e^{int}$.

Exercise 12.5.2. Prove that $\displaystyle\int_{-\pi}^{\pi} P_r(t) \frac{dt}{2\pi} = 1$ for all $0 \leqslant r < 1$.

Exercise 12.5.3. Let $0 \leqslant r < 1$ and let $-\pi \leqslant t \leqslant \pi$.

(a) Prove that the conjugate Poisson kernel from (12.2.1) satisfies

$$Q_r(t) = \mathrm{Im}\left(\frac{1 + re^{it}}{1 - re^{it}}\right) = \frac{2r \sin t}{1 - 2r \cos t + r^2}.$$

(b) Prove that $Q_r(t) = \displaystyle\sum_{n=-\infty}^{\infty} -i\,\mathrm{sgn}(n) r^{|n|} e^{int}$.

Exercise 12.5.4. Let $h(t) = \displaystyle\sum_{n=2}^{\infty} \frac{\cos(nt)}{n \log n}$.

(a) Prove that $h \in L^2[-\pi, \pi]$.

(b) Prove that h is continuous on $[-\pi, \pi] \backslash \{0\}$.

(c) Prove that $\displaystyle\lim_{t \to 0} h(t) = +\infty$.

(d) Prove that $(\mathcal{Q}h)(t) = -\displaystyle\sum_{n=2}^{\infty} \frac{\sin(nt)}{n \log n}$ and defines a continuous function on $[-\pi, \pi]$.

Exercise 12.5.5. For $t \in \mathbb{R}$ and $z = x + iy \in \mathbb{C}_+$, prove that

$$\mathrm{Re}\left(\frac{1}{i\pi} \frac{1}{t - z}\right) = \frac{1}{\pi} \frac{y}{(x - t)^2 + y^2}.$$

Exercise 12.5.6. For $t \in \mathbb{R}$ and $z = x + iy \in \mathbb{C}_+$, prove that

$$\mathrm{Im}\left(\frac{1}{i\pi} \frac{1}{t - z}\right) = \frac{1}{\pi} \frac{x - t}{(x - t)^2 + y^2}.$$

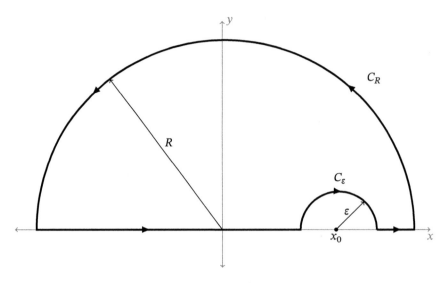

Figure 12.5.1 The contour for Exercise 12.5.8.

Exercise 12.5.7. An important aspect of the Hilbert transform is the principal-value integral. For example, the integral

$$\int_\alpha^\beta \frac{dt}{t-x}$$

is undefined for all $x \in (\alpha, \beta)$. However, for each $x \in (\alpha, \beta)$, prove that

$$\mathrm{PV} \int_\alpha^\beta \frac{dt}{t-x} = \lim_{\varepsilon \to 0} \int_{|x-t|>\varepsilon} \frac{dt}{t-x} = \log\left|\frac{x-\alpha}{\beta-x}\right|.$$

Exercise 12.5.8. Let $f = p/q$, where p and q are polynomials with $\deg q \geqslant 2 + \deg p$ and the roots of q lie in the lower half plane. Do the computations below from [216].

(a) Pick $x_0 \in \mathbb{R}, \varepsilon > 0$, and $R > 0$. Consider the contour C in Figure 12.5.1. Write

$$\int_C \frac{f(z)}{z-x_0} \, dz = \int_{-R}^{x_0-\varepsilon} + \int_{C_\varepsilon} + \int_{x_0+\varepsilon}^{R} + \int_{C_R}$$

and prove that as $R \to \infty$ and $\varepsilon \to 0$ the sum of the first and third integrals on the right side tend to

$$\mathrm{PV} \int_{-\infty}^{\infty} \frac{f(x)}{x-x_0} \, dx.$$

(b) Prove that the fourth integral tends to zero as $R \to \infty$.

(c) Prove that the second integral tends to $-i\pi f(x_0)$ as $\varepsilon \to 0^+$.

(d) Prove that the integral on the left side equals zero.

(e) Combine the steps above to obtain $f(x_0) = \dfrac{1}{i\pi}\,\mathrm{PV}\displaystyle\int_{-\infty}^{\infty}\frac{f(x)}{x - x_0}\,dx.$

(f) Let $f = u + iv$, where $u = \operatorname{Re} f$ and $v = \operatorname{Im} f$, and deduce that

$$u(x_0) = -\frac{1}{\pi}\,\mathrm{PV}\int_{-\infty}^{\infty}\frac{v(x)}{x - x_0}\,dx \quad\text{and}\quad v(x_0) = \frac{1}{\pi}\,\mathrm{PV}\int_{-\infty}^{\infty}\frac{u(x)}{x - x_0}\,dx.$$

Remark: Compare this pair with the formulas of Hilbert and Hardy mentioned in the endnotes for this chapter.

Exercise 12.5.9. Here is the circle version of Exercise 12.5.8. Let $f \in L^2[-\pi, \pi]$ have the Fourier series

$$f(x) = \sum_{n=1}^{\infty}(a_n \cos nx + b_n \sin nx)$$

and corresponding conjugate series

$$g(x) = \sum_{n=1}^{\infty}(-a_n \sin nx + b_n \cos nx).$$

Show that the Hilbert-transform pair becomes

$$g(x) = -\,\mathrm{PV}\int_{-\pi}^{\pi} f(t)\cot\left(\frac{x - t}{2}\right)\frac{dt}{2\pi} \quad\text{and}\quad f(x) = \mathrm{PV}\int_{-\pi}^{\pi} g(t)\cot\left(\frac{x - t}{2}\right)\frac{dt}{2\pi}.$$

Exercise 12.5.10. Let $a > 0$ and

$$f(x) = \frac{a}{a^2 + x^2}.$$

For $x \in \mathbb{R}$, prove that $(\mathscr{H} f)(x) = \dfrac{x}{a^2 + x^2}.$

Exercise 12.5.11. Let $a > 0$ and

$$f(x) = \frac{x}{a^2 + x^2}.$$

For $x \in \mathbb{R}$, prove that $(\mathscr{H} f)(x) = \dfrac{-a}{a^2 + x^2}.$

Exercise 12.5.12. Let $a > 0$ and $f(x) = \sin(ax)$.

(a) For $x \in \mathbb{R}$, prove that $(\mathscr{H} f)(x) = -\cos(ax)$.

(b) What is the Hilbert transform of $\cos(ax)$?

Remark: Technically, $\sin(ax)$ does not belong to $L^2(\mathbb{R})$, but one can take its Hilbert transform as a principal-value integral.

Exercise 12.5.13. For $a > 0$, evaluate $\mathcal{H} f$ for

$$f(x) = \begin{cases} e^{-ax} & \text{if } x \geq 0, \\ 0 & \text{if } x < 0. \end{cases}$$

Exercise 12.5.14. Evaluate $\mathcal{H} f$ for

$$f(x) = \begin{cases} 1 & \text{if } |x| \leq 1, \\ 0 & \text{if } |x| > 1. \end{cases}$$

Exercise 12.5.15. If

$$g(x) = \frac{1}{\pi} \, \text{PV} \int_{-\infty}^{\infty} \frac{f(t)}{x-t} \, dt,$$

prove that $g(x) = \dfrac{1}{\pi} \displaystyle\int_{-\infty}^{\infty} f'(t) \log\left|1 - \frac{x}{t}\right| dt.$

Exercise 12.5.16.

(a) Prove that if $f \in L^2(\mathbb{T})$ is real valued, then $\langle \mathcal{2} f, f \rangle = 0$.

(b) Prove that if $f \in L^2(\mathbb{R})$ is real valued, then $\langle \mathcal{H} f, f \rangle = 0$.

Exercise 12.5.17. Let $f \in L^2(\mathbb{R})$.

(a) If f is even, prove that $(\mathcal{H} f)(x) = \dfrac{2x}{\pi} \, \text{PV} \displaystyle\int_0^{\infty} \frac{f(t)}{x^2 - t^2} \, dt.$

(b) If f is odd, prove that $(\mathcal{H} f)(x) = \dfrac{2}{\pi} \, \text{PV} \displaystyle\int_0^{\infty} \frac{t f(t)}{x^2 - t^2} \, dt.$

Exercise 12.5.18. Verify the identities

$$\mathcal{H}\left(\frac{1}{\sqrt{\pi}} \frac{(x-i)^n}{(x+i)^{n+1}} \right) = -i \frac{1}{\sqrt{\pi}} \frac{(x-i)^n}{(x+i)^{n+1}} \quad \text{for } n \geq 0,$$

and

$$\mathcal{H}\left(\frac{1}{\sqrt{\pi}} \frac{(x+i)^{n-1}}{(x-i)^n} \right) = i \frac{1}{\sqrt{\pi}} \frac{(x+i)^{n-1}}{(x-i)^n} \quad \text{for } n \geq 1.$$

Exercise 12.5.19. Recall convolution of $f, g \in L^2(\mathbb{R})$ from (11.2.1). Prove that $\mathcal{H}(f * g) = (\mathcal{H} f) * g = f * (\mathcal{H} g)$.

Exercise 12.5.20. For the operators $H_+ = \frac{1}{2}(I + i\mathcal{H})$ and $H_- = \frac{1}{2}(I - i\mathcal{H})$, prove the following.

(a) H_+ and H_- are idempotent.

(b) $H_+^* = H_-$.

(c) $H_-H_+ = H_+H_- = 0$.

(d) $\ker H_- = \operatorname{ran} H_+ = H^2(\mathbb{R})$.

(e) $\operatorname{ran} H_- = \ker H_+ = \overline{H^2(\mathbb{R})}$.

Exercise 12.5.21. Let $A = \frac{1}{\sqrt{2}}(I + \mathcal{H})$. Prove that $A^2 = \mathcal{H}$. Thus, A is a square root of the Hilbert transform $\mathcal{H}$ on $L^2(\mathbb{R})$.

Exercise 12.5.22. Let $A = \frac{1}{\sqrt{2}}(I + \mathcal{Q} - 1 \otimes 1)$. Prove that $A^2 = \mathcal{Q}$. Thus, A is a square root of the Hilbert transform $\mathcal{Q}$ on $L^2(\mathbb{T})$.

Exercise 12.5.23. For $a \in \mathbb{R}$ and $\rho > 0$, define the translation operator T_a on $L^2(\mathbb{R})$ by $(T_a f)(x) = f(x - a)$ and the dilation operator D_ρ on $L^2(\mathbb{R})$ by $(D_\rho f)(x) = f(\rho x)$. Prove that for any $\alpha, \beta \in \mathbb{C}$, the operator $\alpha I + \beta \mathcal{H}$ commutes with D_ρ for all $\rho > 0$ and T_a for all $a \in \mathbb{R}$.
Remark: The converse is true but more difficult to prove [216].

Exercise 12.5.24. Let $f, g \in L^2(\mathbb{R})$ such that $\mathscr{F}f$ vanishes on $|t| > a$ and $\mathscr{F}g$ vanishes on $|t| < a$. Prove Bedrosian's Hilbert transform product theorem [36]: $\mathcal{H}(fg) = f\mathcal{H}g$.
Remark: See [216] for more on this.

Exercise 12.5.25. Contrary to some of the other commutative diagrams mentioned in this book, the following diagram

$$
\begin{array}{ccc}
L^2(\mathbb{T}) & \xrightarrow{\;\mathcal{Q}\;} & L^2(\mathbb{T}) \\
\big\downarrow{\scriptstyle U} & & \big\uparrow{\scriptstyle U^*} \\
L^2(\mathbb{R}) & \xrightarrow[\;\mathcal{H}\;]{} & L^2(\mathbb{R})
\end{array}
$$

does not quite hold. In the above, U is from Exercise 11.10.7. Prove that $\mathcal{Q}\xi^n = U^*\mathcal{H}U\xi^n$ holds for all $n \neq 0$ but does not hold when $n = 0$.

Exercise 12.5.26. Suppose f and $(T_a f)(x) = (x - a)f(x)$ belong to $L^2(\mathbb{R})$ for some $a \in \mathbb{C}$.

(a) Prove that $\mathcal{H}(T_a f) = T_a \mathcal{H}f - \frac{1}{\pi}\int_{-\infty}^{\infty} f(t)\, dt$.

(b) Use (a) to prove the following formula of Akhiezer [2]:

$$
(x - i)\mathcal{H}\!\left(\frac{f(t)}{t - i}\right) = \mathcal{H}f + \frac{1}{\pi}\int_{-\infty}^{\infty} \frac{f(t)}{t - i}\, dt.
$$

Exercise 12.5.27. Show that the set

$$
\{(f + \mathcal{H}f)|_{[0,1]} : f \in C^\infty(\mathbb{R}),\ f \text{ is compactly supported in } (0, \infty)\}
$$

is dense in $L^2[0, 1]$ as follows.

(a) Prove that if $g \in L^2[0,1]$ is orthogonal to every such $(f + \mathcal{H}f)|_{[0,1]}$ above, then $g - \mathcal{H}g = 0$ on $(0, \infty)$.

(b) Prove that $\mathcal{H}g = g$ on $\mathbb{R}$.

(c) Prove $g = 0$.

Exercise 12.5.28. The *discrete Hilbert transform* $H : \ell^2(\mathbb{Z}) \to \ell^2(\mathbb{Z})$ is defined by

$$(Hx)_n = \sum_{\substack{m \in \mathbb{Z} \\ m \neq n}} \frac{x_m}{n - m} \quad \text{for } x = (x_n)_{n=-\infty}^{\infty} \in \ell^2(\mathbb{Z}).$$

Show that H is bounded on $\ell^2(\mathbb{Z})$ using the following steps.

(a) For $n \in \mathbb{Z}$, define $I_n = [n - \frac{1}{4}, n + \frac{1}{4}]$. For $x = (x_n)_{n=-\infty}^{\infty} \in \ell^2(\mathbb{Z})$, define

$$f(x) = 2 \sum_{n=-\infty}^{\infty} x_n \chi_{I_n}(x).$$

Prove that $f \in L^2(\mathbb{R})$ and $\|f\| = \|x\|$.

(b) For a fixed $m \in \mathbb{Z}$ and $x \in I_m$, prove that

$$\sum_{n \neq m} x_n \int_{I_n} \frac{dt}{x - t} = \frac{1}{2}(\pi \mathcal{H} f)(x) - x_m(\pi \mathcal{H} \chi_{I_m})(x).$$

(c) Use the identity

$$\sum_{n \neq m} \frac{x_n}{n - m} = \sum_{n \neq m} \left(\frac{1}{n - m} - (\pi \mathcal{H} \chi_{I_n})(x) + (\pi \mathcal{H} \chi_{I_n})(x) \right) x_n,$$

valid for $x \in I_m$, to prove that

$$|(Hx)_m| \leq \sum_{n \neq m} |x_n| \int_{I_n} \left| \frac{1}{n - m} - \frac{1}{x - t} \right| dt + \sum_{n \neq m} |x_n| |(\pi \mathcal{H} \chi_{I_n})(x)|.$$

(d) For $x \in I_n$ and $y \in I_m$, prove that $\left| \frac{1}{x - y} - \frac{1}{n - m} \right| \leq \frac{2}{(n - m)^2}$.

(e) Prove that

$$|(Hx)_m| \leq \sum_{n \neq m} \frac{|x_n|}{(n - m)^2} + \frac{1}{2}|(\pi \mathcal{H} f)(x)| + |x_n| |(\pi \mathcal{H} \chi_{I_m})(x)|.$$

(f) Integrate each term over I_m, square, sum over m, and use the fact that $\mathcal{H}$ is unitary on $L^2(\mathbb{R})$ to prove that H is bounded on $\ell^2(\mathbb{Z})$.

(g) Find the matrix representation of H with respect to the basis $(e_n)_{n=-\infty}^{\infty}$ for $\ell^2(\mathbb{Z})$.

Remark: See [241] for more on this.

Exercise 12.5.29. Another version of the discrete Hilbert transform is the operator H_D : $\ell^2(\mathbb{Z}) \to \ell^2(\mathbb{Z})$ defined by

$$(H_D\mathbf{x})_n = \frac{1}{\pi} \sum_{\substack{m \in \mathbb{Z} \\ m \neq n}} \frac{x_m(1 - (-1)^{n-m})}{n - m} \quad \text{for } \mathbf{x} = (x_n)_{n=-\infty}^{\infty} \in \ell^2(\mathbb{Z}).$$

Use the identities

$$\sum_{\substack{k \in \mathbb{Z} \\ k \neq j}} \frac{(1 - (-1)^{k-j})^2}{(k - j)^2} = 8 \sum_{k=0}^{\infty} \frac{1}{(2k + 1)^2} = \pi^2$$

and

$$\sum_{\substack{k \in \mathbb{Z} \\ j \neq m}} \frac{(1 - (-1)^{k-j})(1 - (-1)^{k-m})}{(k - j)(k - m)} = 0$$

to prove that

(a) $\|H_D\mathbf{x}\| = \|\mathbf{x}\|$ for all $\mathbf{x} \in \ell^2(\mathbb{Z})$.

(b) $H_D^2 = -I$.

(c) Find the matrix representation of this operator with respect to the standard basis $(\mathbf{e}_n)_{n=-\infty}^{\infty}$ for $\ell^2(\mathbb{Z})$.

Remark: See [216] for more on this.

12.6 Hints for the Exercises

Hint for Ex. 12.5.1: Write $\dfrac{1 + u}{1 - u}$ for $|u| < 1$ as a power series.

Hint for Ex. 12.5.10: Consider $F(z) = \dfrac{1}{a - iz} \in H^2(\mathbb{C}_+)$.

Hint for Ex. 12.5.11: Exercise 12.5.10 and $\mathscr{H}^2 = -I$.

Hint for Ex. 12.5.24: Use inverse Fourier transforms and the relationship between the Hilbert transform and the Fourier transform.

Hint for Ex. 12.5.27: For (b), recall that the Hilbert transform is unitary on $L^2(\mathbb{R})$.

13

. . • . .

Bishop Operators

Key Concepts: The invariant subspace problem, reducing subspace, universal operator, Caradus' theorem, Bishop operator (norm, spectrum, invariant subspace, reducing subspace).

Outline: This chapter covers *Bishop operators* $(T_\alpha f)(x) = xf(\{x + \alpha\})$ on $L^2[0,1]$. In the above, $\alpha \in [0,1)$ and $\{t\}$ denotes the fractional part of $t \in \mathbb{R}$. Interest in these operators comes from attempts to provide counterexamples to the invariant subspace problem. We begin with a short survey on the invariant subspace problem and universal operators.

13.1 The Invariant Subspace Problem

At various points in this book, we describe all of the invariant subspaces for particular operators, such as certain multiplication operators, the Volterra operator, the unilateral shift on H^2, and so forth. This suggests a natural question: does every bounded operator on a Hilbert space have an invariant subspace? We should be careful to phrase this problem precisely.

First of all, $\{0\}$ and $\mathcal{H}$ are invariant for any $A \in \mathcal{B}(\mathcal{H})$. Consequently, we restrict our attention to proper nonzero subspaces; note that $\dim \mathcal{H} \geqslant 2$ is necessary for the problem to be interesting. If $\mathcal{H}$ is finite dimensional and $\dim \mathcal{H} \geqslant 2$, then A has an eigenvalue and hence the span of a single eigenvector is a proper nonzero A-invariant subspace. Thus, the finite-dimensional case is settled in the affirmative. At the other extreme, if $\mathcal{H}$ is nonseparable (see Exercise 1.10.37 for an example), then for any nonzero $\mathbf{x} \in \mathcal{H}$ and any nonzero $A \in \mathcal{B}(\mathcal{H})$, the cyclic subspace $\bigvee\{A^n\mathbf{x} : n \geqslant 0\}$ is a proper nonzero invariant subspace.

The previous discussion brings us to the most famous open problem in operator theory. The *invariant subspace problem* asks whether every bounded operator on a separable Hilbert space of dimension at least two has a proper nonzero invariant subspace. Many operators (for example compact operators and operators that commute with a nonzero compact operator) have invariant subspaces. The text [80] surveys the latest developments on the invariant subspace problem. For reference, let us formally state several important definitions.

Definition 13.1.1. Let $\mathcal{M}$ be a subspace of complex Hilbert space $\mathcal{H}$ and let $A \in \mathcal{B}(\mathcal{H})$.

(a) $\mathcal{M}$ is *invariant* for A if $A\mathcal{M} \subseteq \mathcal{M}$.

(b) $\mathcal{M}$ is *hyperinvariant* for A if $\mathcal{M}$ is invariant for every $T \in \mathcal{B}(\mathcal{H})$ such that $AT = TA$.

(c) $\mathcal{M}$ is *reducing* for A if $A\mathcal{M} \subseteq \mathcal{M}$ and $A^*\mathcal{M} \subseteq \mathcal{M}$.

(d) $\mathcal{M}$ is *irreducible* for A if it is invariant, $\mathcal{M} \neq \mathcal{H}$, but $\mathcal{M}$ contains no nonzero reducing subspace.

The following theorem allows us to focus on the reducing and irreducible subspaces separately [356, p. 8].

Theorem 13.1.2. *Every invariant subspace for a bounded Hilbert space operator can be written as an orthogonal direct sum of a reducing subspace and an irreducible subspace.*

E. Bishop suggested that for irrational $\alpha \in [0, 1)$, the transformation $x \mapsto \{x + \alpha\}$ on $L^2[0, 1]$ is sufficiently "ergodic" so that the operator T_α has no proper nonzero invariant subspaces. Davie [106] showed this is not always the case. However, not all irrational α are included in Davie's paper and it remains an intriguing open problem whether all Bishop operators have proper nonzero invariant subspaces. Our survey of Bishop operators begins in Section 13.4.

13.2 Lomonosov's Theorem

In an unpublished work, von Neumann proved that every compact operator on a complex Hilbert space of dimension at least two has a proper nonzero invariant subspace. Aronszajn and Smith generalized the result to Banach-space operators.

Theorem 13.2.1 (von Neumann, Aronszajn–Smith [23]). *If $\mathcal{H}$ is a complex Hilbert space with $\dim \mathcal{H} \geqslant 2$ and $A \in \mathcal{B}(\mathcal{H})$ is compact, then A has a proper nonzero invariant subspace.*

In 1973, Lomonosov proved a sweeping generalization of this result [230]. Not only is his result much stronger but its proof is simpler than that of Aronszajn and Smith. Traditionally, "Lomonosov's theorem" refers to Corollary 13.2.4 below. It is a consequence of Theorem 13.2.2, which is a hyperinvariant-subspace version of the result. The proof below is from A. J. Michaels' exposition of Hilden's proof [243] (see also [280]). Michaels claims that "the ideas are all Lomonosov's and Hilden's. Even the exposition of the proof is derivative; it is largely based on Wallen's 1973 Wabash conference lecture."

Theorem 13.2.2. *Every nonzero compact operator on a complex Hilbert space has a proper nonzero hyperinvariant subspace.*

Proof Let $K \in \mathcal{B}(\mathcal{H})$ be a nonzero compact operator. Riesz's theorem (Theorem 2.6.9) says that $\sigma(K)$ is the union of $\{0\}$ and a (possibly empty) discrete set of points that accumulate only at 0. Moreover, every nonzero element of $\sigma(K)$ is an eigenvalue of finite multiplicity. There are two possibilities: $\sigma(K) \neq \{0\}$ or $\sigma(K) = \{0\}$.

If $\sigma(K) \neq \{0\}$, then K has a nonzero eigenvalue λ. Consequently, $\mathcal{M} = \ker(K - \lambda I)$, the eigenspace for λ, is K-invariant; it is nonzero by definition. Since K is compact, $K \neq \lambda I$ and hence $\mathcal{M} \neq \mathcal{H}$. If $A \in \mathcal{B}(\mathcal{H})$ and $AK = KA$, then for each $\mathbf{x} \in \mathcal{M}$, it follows that $K(A\mathbf{x}) = A(K\mathbf{x}) = A(\lambda \mathbf{x}) = \lambda(A\mathbf{x})$, and hence $A\mathbf{x} \in \mathcal{M}$. In other words, $\mathcal{M}$ is an invariant subspace for A. Thus, $\mathcal{M}$ is a proper nonzero hyperinvariant subspace for K.

Suppose that $\sigma(K) = \{0\}$. Then the spectral radius formula (Theorem 8.4.4) ensures that

$$\lim_{n \to \infty} \|(\alpha K)^n\|^{\frac{1}{n}} = \lim_{n \to \infty} |\alpha| \|K^n\|^{\frac{1}{n}} = 0 \quad \text{for all } \alpha \in \mathbb{C}. \qquad (13.2.3)$$

Since $K \neq 0$, assume that $\|K\| = 1$. Indeed, multiply by $1/\|K\|$ which does not change the invariant-subspace structure for K. Then there is an $\mathbf{x}_0 \in \mathcal{H}$ such that $\|K\mathbf{x}_0\| = 1 + \delta$ for some $\delta > 0$. In particular, this implies that $\|\mathbf{x}_0\| > 1$ since $\|K\| = 1$. Let

$$\mathfrak{B} = \{\mathbf{x} \in \mathcal{H} : \|\mathbf{x} - \mathbf{x}_0\| \leqslant 1\}$$

and observe that $\mathbf{0} \notin \mathfrak{B}$ and $\mathbf{0} \notin (K\mathfrak{B})^-$ since for $\mathbf{x} \in \mathfrak{B}$,

$$\|K\mathbf{x}\| \geqslant \|K\mathbf{x}_0\| - \|K(\mathbf{x} - \mathbf{x}_0)\| > \delta > 0.$$

For each $\mathbf{v} \in \mathcal{H}$, let

$$\mathcal{M}_{\mathbf{v}} = \{A\mathbf{v} : A \in \mathcal{B}(\mathcal{H}), \ AK = KA\}.$$

Each $\mathcal{M}_{\mathbf{v}}$ is a vector space that is invariant under any operator that commutes with K (hence invariant under K itself) since $A(A\mathbf{v}) = A^2\mathbf{v}$ and $A^2K = KA^2$. Since each $\mathcal{M}_{\mathbf{v}}^-$ is a hyperinvariant subspace for K, it suffices to prove that there is a $\mathbf{v} \in \mathcal{H}$ such that $\mathcal{M}_{\mathbf{v}}^-$ is proper and nonzero. If $\mathbf{v} \neq \mathbf{0}$, then $\mathcal{M}_{\mathbf{v}} \neq \{\mathbf{0}\}$ since $\mathbf{v} = I\mathbf{v}$ and $IK = KI$. For $\mathbf{v} \in \mathcal{H}\backslash\{\mathbf{0}\}$, it may be that $\mathcal{M}_{\mathbf{v}}^- = \mathcal{H}$. It turns out that this cannot occur for all $\mathbf{v} \neq \mathbf{0}$. Suppose toward a contradiction that $\mathcal{M}_{\mathbf{v}}$ is dense in $\mathcal{H}$ for all $\mathbf{v} \in \mathcal{H}\backslash\{\mathbf{0}\}$. Then for each $\mathbf{v} \neq \mathbf{0}$, there is an $A \in \mathcal{B}(\mathcal{H})$ such that $AK = KA$ and $\|A\mathbf{v} - \mathbf{x}_0\| < 1$. Let

$$\mathcal{U}(A) = \{\mathbf{v} : \|A\mathbf{v} - \mathbf{x}_0\| < 1\},$$

and observe that

$$\bigcup_{\substack{A \in \mathcal{B}(\mathcal{H}) \\ AK=KA}} \mathcal{U}(A) = \mathcal{H}\backslash\{\mathbf{0}\}.$$

Since K is compact, $(K\mathfrak{B})^-$ is a compact subset of $\mathcal{H}\backslash\{\mathbf{0}\}$. Each $\mathcal{U}(A)$ is open, so there are $A_1, A_2, \ldots, A_n \in \mathcal{B}(\mathcal{H})$ that commute with K such that

$$K\mathfrak{B} \subseteq \bigcup_{i=1}^{n} \mathcal{U}(A_i).$$

Since $K\mathbf{x}_0 \in K\mathfrak{B}$ (since $\mathbf{x}_0 \in \mathfrak{B}$), there is an $i_1 \in \{1, 2, \ldots, n\}$ such that $K\mathbf{x}_0 \in \mathcal{U}(A_{i_1})$. In other words, $A_{i_1}K\mathbf{x}_0 \in \mathfrak{B}$, and hence $KA_{i_1}K\mathbf{x}_0 \in K\mathfrak{B}$. In a similar way, there is an $i_2 \in \{1, 2,..., n\}$ such that $KA_{i_1}K\mathbf{x}_0 \in \mathcal{U}(A_{i_2})$, and hence $A_{i_2}KA_{i_1}K\mathbf{x}_0 \in \mathcal{B}$. This process can be continued indefinitely to produce $i_1, i_2,...,$ such that

$$A_{i_m}KA_{i_{m-1}}K \cdots A_{i_1}KA_{i_1}K\mathbf{x}_0 \in \mathcal{B}$$

for each $m \geqslant 1$. Let $c = \max\{\|A_i\| : 1 \leqslant i \leqslant m\}$ and recall that each A_i commutes with K. Therefore,

$$(c^{-1}A_{i_m})(c^{-1}A_{i_{m-1}}) \cdots (c^{-1}A_{i_1})(cK)^m\mathbf{x}_0 \in \mathcal{B}.$$

Since (13.2.3) ensures that $\|(cK)^m\| \to 0$ and each $\|c^{-1}A_{i_j}\| \leqslant 1$, it follows that $\mathbf{0} \in \mathcal{B}$ because $\mathcal{B}$ is closed. This contradicts the fact that $\mathbf{0} \notin \mathcal{B}$, so there exists a $\mathbf{v} \neq \mathbf{0}$ such that $\mathcal{M}_\mathbf{v}^- \neq \mathcal{H}$. This yields a hyperinvariant subspace for K. ∎

Corollary 13.2.4 (Lomonosov [230]). *If $A \in \mathcal{B}(\mathcal{H})$ commutes with a nonzero compact operator K, then A has a proper nonzero invariant subspace.*

Proof If $A \in \mathcal{B}(\mathcal{H})$ commutes with a nonzero compact operator K, then Theorem 13.2.2 ensures that K has a proper nonzero hyperinvariant subspace $\mathcal{M}$. Since A commutes with K, the definition of hyperinvariance ensures that $\mathcal{M}$ is A-invariant. ∎

13.3 Universal Operators

One approach to the invariant subspace problem is to study the invariant subspaces of a special type of operator which boasts a sufficiently complicated lattice of invariant subspaces. Recall that $A \in \mathcal{B}(\mathcal{H})$ and $B \in \mathcal{B}(\mathcal{K})$ are *similar* if there is an invertible $S \in \mathcal{B}(\mathcal{H}, \mathcal{K})$ such that $S^{-1}BS = A$. This is represented by the following commutative diagram:

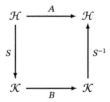

Note that similarity preserves invariant subspaces: if $\mathcal{M} \subseteq \mathcal{H}$ is an invariant subspace for A, then $S\mathcal{M} \subseteq \mathcal{K}$ is an invariant subspace for B.

Definition 13.3.1. $U \in \mathcal{B}(\mathcal{H})$ is *universal* for $\mathcal{H}$ if for any nonzero $A \in \mathcal{B}(\mathcal{H})$, there is an invariant subspace $\mathcal{M}$ for U and a nonzero constant λ such that A is similar to $\lambda U|_\mathcal{M}$.

Notice that $U|_\mathcal{M}$ and $\lambda U|_\mathcal{M}$ have the same invariant subspaces. Thus, the collection of invariant subspaces of $U|_\mathcal{M}$ is in bijective and order-preserving correspondence with the collection of invariant subspaces of A. The invariant subspaces of a universal operator are so plentiful that they can model the invariant subspace structure of any operator on $\mathcal{H}$. Although the existence of universal operators seems impossible, in 1960 Rota produced an example of one [316]. Since this early example, they have been shown to exist in abundance. One method for producing them is due to Caradus.

Theorem 13.3.2 (Caradus [75]). *If $\mathcal{H}$ is an infinite-dimensional separable Hilbert space and $U \in \mathcal{B}(\mathcal{H})$ satisfies*

(a) $\dim \ker U = \infty$ *and*

(b) $\operatorname{ran} U = \mathcal{H}$,

then U is universal for $\mathcal{H}$.

Proof Let $\mathcal{K} = \ker U$ and consider the orthogonal decomposition $\mathcal{H} = \mathcal{K} \oplus \mathcal{K}^\perp$. Since $\operatorname{ran} U = \mathcal{H}$, it follows that $\tilde{U} = U|_{\mathcal{K}^\perp} : \mathcal{K}^\perp \to \mathcal{H}$ is a bijection. Consequently, $V = \tilde{U}^{-1}$ is a bijection from $\mathcal{H}$ to $\mathcal{K}^\perp$ such that

$$UV = I \quad \text{and} \quad \operatorname{ran} V = \mathcal{K}^\perp.$$

Since $\dim \mathcal{K} = \infty$, there is a unitary $W : \mathcal{H} \to \mathcal{K}$. By construction,

$$UW = 0, \quad \ker W = \{\mathbf{0}\}, \quad \text{and} \quad \operatorname{ran} W = \mathcal{K}.$$

Suppose that $T \in \mathcal{B}(\mathcal{H})$ and let $\lambda \neq 0$ satisfy $|\lambda| \, \|T\| \, \|V\| < 1$, so that the series defining

$$Q = \sum_{n=0}^{\infty} \lambda^n V^n W T^n$$

converges in operator norm (apply Theorem 1.8.11 to the Banach space $\mathcal{B}(\mathcal{H})$). Then

$$Q = W + \lambda VQT, \tag{13.3.3}$$

and hence

$$UQ = \lambda QT \tag{13.3.4}$$

since $UV = I$ and $UW = 0$.

If we can show that $\mathcal{M} = \operatorname{ran} Q$ is closed and $Q : \mathcal{H} \to \mathcal{M}$ is injective, then the bounded inverse theorem will ensure that Q is invertible. In light of (13.3.4), this will show that λT is similar to the restriction of U to $\mathcal{M}$. From (13.3.4) it follows that $\mathcal{M}$ is U-invariant.

If $Q\mathbf{x} = \mathbf{0}$, we know that $W\mathbf{x} + \lambda VQT\mathbf{x} = \mathbf{0}$. Now use the fact that $\operatorname{ran} W \perp \operatorname{ran} V$ to conclude that $W\mathbf{x} = \mathbf{0}$. But $\ker W = \{\mathbf{0}\}$, which implies that $\mathbf{x} = \mathbf{0}$ and hence Q is injective.

To show that Q has closed range, suppose that $Q\mathbf{x}_n \to \mathbf{y}$. From (13.3.3), it follows that $W\mathbf{x}_n + \lambda VQT\mathbf{x}_n \to \mathbf{y}$. Since $\operatorname{ran} W = \mathcal{K}$ and $\operatorname{ran} V = \mathcal{K}^\perp$, one sees that $W\mathbf{x}_n \to P_\mathcal{K}\mathbf{y}$ (the orthogonal projection of $\mathbf{y}$ onto $\mathcal{K}$). From here recall that $W : \mathcal{H} \to \mathcal{K}$ is unitary and so $\mathbf{x}_n \to \mathbf{x} = W^* P_\mathcal{K}\mathbf{y}$. The boundedness of Q implies that $Q\mathbf{x}_n \to Q\mathbf{x}$, and hence $\mathcal{M} = \operatorname{ran} Q$ is closed. ∎

Example 13.3.5. Caradus' theorem produces many universal operators beyond Rota's initial example of the backward shift of infinite multiplicity (see Chapter 14 for a formal treatment of infinite direct sums of Hilbert spaces). Specifically, we mean the operator U on the Hilbert space

$$(H^2)^{(\infty)} = \left\{ (f_j)_{j=0}^{\infty} : f_j \in H^2, \| (f_j)_{j=0}^{\infty} \|^2 = \sum_{j=0}^{\infty} \| f_j \|_{H^2}^2 < \infty \right\}$$

defined by

$$U(f_0, f_1, f_2, f_3, \dots) = (f_1, f_2, f_3, \dots). \tag{13.3.6}$$

Observe that $\ker U = \{ (f, 0, 0, \dots) : f \in H^2 \}$, which is infinite dimensional, and $\operatorname{ran} U = (H^2)^{(\infty)}$. Thus, the conditions of Caradus' theorem are satisfied and hence U is universal. Of course, Rota did not have the luxury of Caradus' theorem and so his proof was more complicated.

Example 13.3.7. For each $a > 0$, the operator $U_a : L^2(\mathbb{R}_+) \to L^2(\mathbb{R}_+)$ defined by

$$(U_a f)(x) = f(x + a)$$

is universal since U_a is similar to the backward shift of infinite multiplicity discussed above (Exercise 13.9.9). One can also see this by applying Caradus' theorem.

Later chapters of this book contain examples of universal operators among the translation, Toeplitz, and composition operators. Two good surveys of universal operators are [101] and [80, Ch. 8].

13.4 Properties of Bishop Operators

A Bishop operator $(T_\alpha f)(x) = x f(\{x + \alpha\})$ on $L^2[0, 1]$ can be written as

$$T_\alpha = M_x U_\alpha,$$

where $M_x : L^2[0, 1] \to L^2[0, 1]$, defined by

$$(M_x f)(x) = x f(x),$$

is the multiplication operator covered earlier in Chapter 4, and $U_\alpha : L^2[0, 1] \to L^2[0, 1]$ is defined by

$$U_\alpha f = f(u_\alpha),$$

where $u_\alpha(x) = \{x + \alpha\}$, is a composition operator. One can check that u_α is a (Lebesgue) measure-preserving transformation of $[0, 1)$ to itself and hence the composition operator U_α is unitary. Note that

$$u_\alpha^{\circ j}(x) = u_{j\alpha}(x) \quad \text{for all } x \in [0, 1],$$

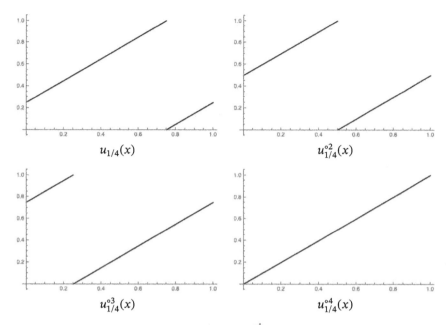

Figure 13.4.1 The graphs of $u_{1/4}^{\circ j}(x)$ for $j = 1, 2, 3, 4$.

where $u_\alpha^{\circ j}$ denotes the function u_α composed with itself j times. If $p, q \in \mathbb{N}$ are relatively prime, $p < q$, and $\alpha = p/q$, then $u_\alpha^{\circ q}(x) = x$ for $x \in [0, 1]$; see Figures 13.4.1 and 13.4.2. Furthermore, q is the smallest positive integer for which this holds.

If $E = [0, 1/q)$, then the sets

$$u_\alpha^{\circ j}(E) \quad \text{for } 0 \leqslant j \leqslant q - 1,$$

are pairwise disjoint and

$$\bigcup_{j=0}^{q-1} u_\alpha^{\circ j}(E) = [0, 1). \tag{13.4.1}$$

These facts translate into operator identities. For example,

$$U_\alpha^n f = f \circ u_{n\alpha} \quad \text{for all } n \in \mathbb{Z} \text{ and } f \in L^2[0, 1],$$

and, since U_α is unitary,

$$U_\alpha^* = U_{-\alpha}. \tag{13.4.2}$$

If $p, q \in \mathbb{N}$ are relatively prime, $p < q$, and $\alpha = p/q$, then $U_\alpha^q = I$.

The iterates of T_α are important in what follows. Starting with

$$T_\alpha f = M_x U_\alpha f = M_x (f \circ u_\alpha),$$

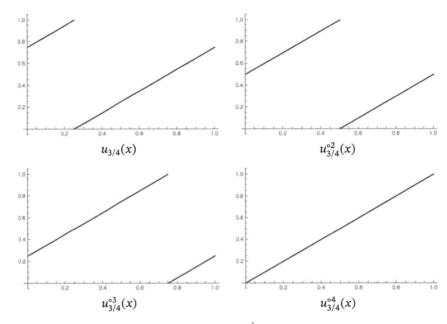

Figure 13.4.2 Graphs of $u_{3/4}^{\circ j}(x)$ for $j = 1, 2, 3, 4$.

we have

$$T_\alpha^2 f = T_\alpha(M_x f(u_\alpha)) = xu_\alpha(x)f(u_\alpha \circ u_\alpha),$$

and hence $(T_\alpha^2 f)(x) = x\{x + \alpha\}f(\{x + 2\alpha\})$. In general,

$$(T_\alpha^n f)(x) = g_{\alpha,n}(x)f(\{x + n\alpha\}), \qquad (13.4.3)$$

where

$$g_{\alpha,n}(x) = x\{x + \alpha\}\{x + 2\alpha\}\cdots\{x + (n-1)\alpha\}.$$

When $\alpha = p/q$ as above, observe that

$$(T_\alpha^q f)(x) = g_{\alpha,q}(x)f(x). \qquad (13.4.4)$$

In other words, T_α^q is the multiplication operator with symbol $g_{\alpha,q}$. Let us now compute the norm of a Bishop operator.

Proposition 13.4.5. $\|T_\alpha\| = 1$ *for every* $\alpha \in [0, 1)$.

Proof Proposition 4.1.2 says that $\|M_x\| = 1$. The operator U_α is unitary, and hence $\|U_\alpha\| = 1$. From Proposition 2.1.12,

$$\|T_\alpha\| = \|M_x U_\alpha\| \leqslant \|M_x\|\|U_\alpha\| = 1$$

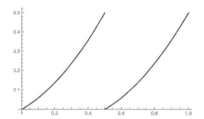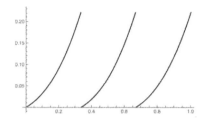

Figure 13.5.1 The graphs of $g_{1/2,2}(x)$ (left) and $g_{2/3,3}(x)$ (right).

and hence $\|T_\alpha\| \leqslant 1$. The functions $h_n(x) = \sqrt{2n+1}\, x^n$ are unit vectors in $L^2[0,1]$, and, since U_α is unitary (see (13.4.2)), the $f_n = U_{-\alpha} h_n$ are also unit vectors. Moreover, since $U_\alpha U_{-\alpha} = I$, it follows that

$$T_\alpha f_n = M_x U_\alpha f_n = M_x U_\alpha U_{-\alpha} h_n = M_x h_n = \sqrt{2n+1}\, x^{n+1} = \frac{\sqrt{2n+1}}{\sqrt{2n+3}} h_{n+1}.$$

Thus,

$$\|T_\alpha f_n\| = \frac{\sqrt{2n+1}}{\sqrt{2n+3}} \to 1 \quad \text{as } n \to \infty,$$

and hence

$$1 \geqslant \|T_\alpha\| = \sup\{\|T_\alpha f\| : \|f\| = 1\} \geqslant \lim_{n\to\infty} \|T_\alpha f_n\| = 1.$$

Consequently, $\|T_\alpha\| = 1$. ∎

13.5 Rational Case: Spectrum

We begin our discussion of the spectral properties of the Bishop operators with the following. If $p, q \in \mathbb{N}$ are relatively prime, let $\alpha = p/q$ and define

$$g_{\alpha,q}(x) = x\left\{x + \frac{p}{q}\right\}\left\{x + 2\frac{p}{q}\right\}\left\{x + 3\frac{p}{q}\right\}\cdots\left\{x + (q-1)\frac{p}{q}\right\}.$$

These functions are piecewise continuous with only a finite number of zeros; see Figure 13.5.1.

Proposition 13.5.1. $\sigma_p(T_\alpha) = \varnothing$ *for any rational* α.

Proof Suppose $\alpha = p/q$, with p, q relatively prime, and suppose that $T_\alpha f = \lambda f$ for some $\lambda \in \mathbb{C}$ and $f \in L^2[0,1]$. Then $T_\alpha^q f = \lambda^q f$. On the other hand, (13.4.4) yields $T_\alpha^q f = g_{\alpha,q} f$. Together, these say that $(g_{\alpha,q} - \lambda^q)f = 0$ almost everywhere on $[0,1]$. Since the equation $g_{\alpha,q}(x) = \lambda^q$ has only a finite number of solutions, f is zero almost everywhere and hence $\sigma_p(T_\alpha) = \varnothing$. ∎

We need a new technique to characterize the eigenvalues of T_α if α is irrational, which we take up later in this chapter. We now present a spectral result from the 1965 doctoral thesis of Parrott.

Theorem 13.5.2 (Parrott [260]). *If $\alpha = p/q$, where $p, q \in \mathbb{N}$ are relatively prime and $p < q$, then $\sigma(T_\alpha) = \{\lambda \in \mathbb{C} : \lambda^q \in g_{\alpha,q}([0,1])\}$.*

The proof of this theorem requires the following symmetry lemma.

Lemma 13.5.3. *Let $\alpha = p/q$ where p and q are relatively prime. If $\xi \in \mathbb{T}$ and $\xi^q = 1$, then T_α is similar to ξT_α.*

Proof From (13.4.1), the sets $u_\alpha^{\circ j}(E)$, where $E = [0, 1/q)$ and $0 \leqslant j \leqslant q - 1$, are pairwise disjoint and

$$\bigcup_{j=0}^{q-1} u_\alpha^{\circ j}(E) = [0, 1).$$

Define $g : [0,1] \to \mathbb{T}$ by $g(x) = \xi^j$ on $u_\alpha^{\circ j}(E) = u_{j\alpha}(E)$ and notice that $g \circ u_\alpha = \xi g$. Proposition 8.1.12 implies that M_g (multiplication by g on $L^2[0,1]$) is invertible with inverse $M_{1/g}$. Moreover, for each $f \in L^2[0,1]$,

$$
\begin{aligned}
M_g^{-1} T_\alpha M_g f &= M_{1/g} T_\alpha M_g f \\
&= M_{1/g} M_x U_\alpha M_g f \\
&= M_{1/g} M_x g(u_\alpha) f(u_\alpha) \\
&= M_{1/g} M_x (\xi g) f(u_\alpha) \\
&= \xi x f(u_\alpha) \\
&= \xi T_\alpha f.
\end{aligned}
$$

Thus, T_α is similar to ξT_α. ∎

We are now ready to prove Theorem 13.5.2.

Proof Equation (13.4.4) yields

$$T_\alpha^q = M_{g_{\alpha,q}}, \tag{13.5.4}$$

while Proposition 8.1.12 implies that

$$\sigma(M_{g_{\alpha,q}}) = g_{\alpha,q}([0,1]). \tag{13.5.5}$$

By the spectral mapping theorem,

$$\sigma(T_\alpha^q) = \{\lambda^q : \lambda \in \sigma(T_\alpha)\} \tag{13.5.6}$$

and

$$\sigma(\xi T_\alpha) = \xi \sigma(T_\alpha). \tag{13.5.7}$$

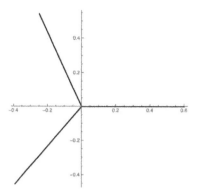

Figure 13.5.2 The spectrum of $T_{2/3}$.

Equation (13.5.6), along with (13.5.4) and (13.5.5), show that at least one qth root of every number in $g_{\alpha,q}([0,1])$ belongs to $\sigma(T_\alpha)$. Equation (13.5.7) shows that every qth root belongs to $\sigma(T_\alpha)$. ∎

Here are a few examples of $\sigma(T_{p/q})$.

Example 13.5.8. If $\frac{p}{q} = \frac{1}{2}$, then $g_{1/2,2}(x) = x\{x + \frac{1}{2}\}$. From Figure 13.5.1, one can see that $g_{1/2,2}([0,1]) = [0, \frac{1}{2}]$. Thus,

$$\sigma(T_{1/2}) = \left\{\lambda \in \mathbb{C} : \lambda^2 \in g_{1/2,2}([0,1]) = \left[0, \tfrac{1}{2}\right]\right\} = \left[-\tfrac{1}{\sqrt{2}}, \tfrac{1}{\sqrt{2}}\right].$$

Example 13.5.9. If $\frac{p}{q} = \frac{2}{3}$, then $g_{2/3,3}(x) = x\{x + \frac{2}{3}\}\{x + \frac{4}{3}\}$. The graph of $g_{2/3,3}(x)$ (see Figure 13.5.1) says that $g_{2/3,3}([0,1]) = [0, \frac{2}{9}]$. Thus,

$$\sigma(T_{2/3}) = \left\{\lambda \in \mathbb{C} : \lambda^3 \in \left[0, \tfrac{2}{9}\right]\right\}$$

consists of three equally spaced line segments from the origin; see Figure 13.5.2.

13.6 Rational Case: Invariant Subspaces

The Bishop operator T_α has plenty of invariant subspaces if $\alpha \in \mathbb{Q} \cap [0,1)$. In order to make this more readable, let us focus on the representative case $T_{1/2}$. Consult [80, Ch. 5] for a survey of what happens for general $\alpha = p/q$. Here the function $u_{1/2}$ plays an important role (see Figure 13.6.1).

Example 13.6.1. Let

$$E = \left[0, \tfrac{1}{8}\right] \cup \left[\tfrac{3}{8}, \tfrac{5}{8}\right] \cup \left[\tfrac{7}{8}, 1\right]$$

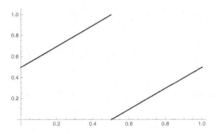

Figure 13.6.1 Plot of $u_{1/2}(x)$.

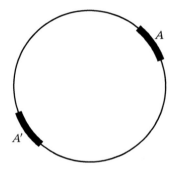

Figure 13.6.2 The set A rotated by $1/2$ (mod 1) to form A'. The set $A \cup A'$ is invariant under rotation by $1/2$ (mod 1).

and note that $u_{1/2}E = E$ and $u_{1/2}E^c = E^c$. Thus, if f vanishes almost everywhere on E^c (see Figure 13.6.1), then $(T_{1/2}f)(x) = xf(\{x + \frac{1}{2}\})$ also vanishes for almost every $x \in E^c$. Thus, $\chi_E L^2[0,1]$ is a $T_{1/2}$-invariant subspace.

One can extend the example above to include invariant subspaces of the form $\chi_F L^2[0,1]$, where F is a measurable subset of $[0,1]$ such that $u_{1/2}F = F$. Such sets can be created by taking any measurable subset $A \subseteq [0,1]$ and shifting it by $1/2$ (mod 1) to form the set A'. Then $F = A \cup A'$ is invariant modulo a shift by $1/2$ (mod 1). It often helps to equate $[0,1]$ with a circle (see Figure 13.6.2), where rotation modulo one is easier to visualize. The subspaces $\chi_F L^2[0,1]$, for which $u_{1/2}F = F$, are not only invariant for $T_{1/2}$, but they are also reducing.

Theorem 13.6.2 (Parrott [260]). *If $F \subseteq [0,1]$ is measurable and $u_{1/2}F = F$, then $\chi_F L^2[0,1]$ reduces $T_{1/2}$.*

Proof If $f \in \chi_F L^2[0,1]$, then $f = 0$ almost everywhere on F^c. But then $(T_{1/2}f)(x) = xf(\{x + \frac{1}{2}\})$ is zero on F^c since $u_{1/2}F^c = F^c$. Also observe that $(\chi_F L^2[0,1])^{\perp} = \chi_{F^c}L^2[0,1]$. Now apply the previous argument, using the fact that $u_{1/2}F^c = F^c$, to show that $\chi_{F^c}L^2[0,1]$ is also $T_{1/2}$-invariant. ∎

The converse of Theorem 13.6.2 requires the following fact.

Proposition 13.6.3. *If $\mathcal{M} \subseteq L^2[0,1]$ is reducing for $T_{1/2}$, then $\mathcal{M}$ is reducing for both M_x and $U_{1/2}$.*

Proof Compute $T_{1/2}T_{1/2}^*$ as follows:

$$T_{1/2}T_{1/2}^* = M_x U_{1/2} U_{1/2}^* M_x = M_x I M_x = M_x^2.$$

Thus, $|T_{1/2}^*| = M_x$ (see Definition 14.9.1). As a consequence of the spectral theorem for selfadjoint operators (Theorem 8.7.1), there is a sequence of polynomials $(p_n)_{n=1}^\infty$ such that $p_n(T_{1/2}T_{1/2}^*) \to |T_{1/2}^*| = M_x$ in operator norm. Thus, if $\mathcal{M}$ reduces $T_{1/2}$, then $\mathcal{M}$ is invariant for each of the operators $p_n(T_{1/2}T_{1/2}^*)$, so $\mathcal{M}$ is invariant for M_x. Since M_x is selfadjoint, $\mathcal{M}$ is reducing for M_x.

We now show that $U_{1/2}\mathcal{M} \subseteq \mathcal{M}$. Suppose $f \in \mathcal{M}$ and $U_{1/2}f = h + k$ with $h \in \mathcal{M}$ and $k \in \mathcal{M}^\perp$. Then $T_{1/2}f \in \mathcal{M}$ and $T_{1/2}f = M_x U_{1/2}f = M_x h + M_x k$. We already proved that $\mathcal{M}$ is reducing for M_x, so $M_x h \in \mathcal{M}$ and $M_x k \in \mathcal{M}^\perp$. Thus, $M_x k = 0$ almost everywhere, which makes $k = 0$ almost everywhere. It follows that $U_{1/2}f = h \in \mathcal{M}$, and hence $\mathcal{M}$ is an invariant subspace for $U_{1/2}$.

We finish by showing that $U_{1/2}^*\mathcal{M} \subseteq \mathcal{M}$. We first argue that $M_x\mathcal{M}$ is dense in $\mathcal{M}$. This follows because $M_x L^2[0,1]$ is dense in $L^2[0,1]$ and both $\mathcal{M}$ and $\mathcal{M}^\perp$ are invariant for M_x. To complete the proof that $\mathcal{M}$ is invariant for $U_{1/2}^*$, let $f \in \mathcal{M}$ and choose a sequence $(f_n)_{n=1}^\infty$ of functions in $\mathcal{M}$ such that $xf_n \to f$ in $L^2[0,1]$. Then $T_{1/2}^* f_n \in \mathcal{M}$ and $T_{1/2}^* f_n = U_{1/2}^* M_x f_n = U_{1/2}^*(xf_n)$. It follows that $U_{1/2}^*(xf_n) \in \mathcal{M}$ and $U_{1/2}^*(xf_n) \to U_{1/2}^* f$. Hence, $U_{1/2}^* f \in \mathcal{M}$. $\blacksquare$

The converse of Theorem 13.6.2 yields the reducing subspaces for $T_{1/2}$.

Theorem 13.6.4 (Parrott [260])**.** *If $\mathcal{M} \subseteq L^2[0,1]$ is a reducing subspace for $T_{1/2}$, then there is a measurable subset $F \subseteq [0,1]$ such that $u_{1/2}F = F$ and $\mathcal{M} = \chi_F L^2[0,1]$.*

Proof By Proposition 13.6.3, $\mathcal{M}$ is reducing for both M_x and $U_{1/2}$. By Theorem 4.1.7, $\mathcal{M} = \chi_F L^2[0,1]$ for some measurable subset $F \subseteq [0,1]$. Since $\chi_F L^2[0,1]$ is reducing for $U_{1/2}$, it follows that $f \in \chi_F L^2[0,1]$ if and only if $U_{1/2}f \in \chi_F L^2[0,1]$. This says that f vanishes almost everywhere on F^c if and only if $U_{1/2}f$ vanishes almost everywhere on F^c. Applying this to $f = \chi_F$, it must be the case that $u_{1/2}F^c = F^c$, and hence $u_{1/2}F = F$. $\blacksquare$

Parrott also characterized the irreducible invariant subspaces for $T_{1/2}$. However, we do not go into the details here.

Theorem 13.6.5 (Parrott [260])**.** *Suppose $w \in L^\infty[0,1]$, $w \circ u_{1/2} = w$, and $\mathcal{M}$ is the set of $f \in L^2[0,1]$ such that $T_{1/2}f = w \circ f$ and $w^{-1}(\{0\}) \subseteq f^{-1}(\{0\})$ almost everywhere. Then $\mathcal{M}$ is an irreducible invariant subspace for $T_{1/2}$ and every irreducible invariant subspace for $T_{1/2}$ is of this form.*

Combining this theorem with Theorem 13.6.2 yields all of the invariant subspaces of $T_{1/2}$. The text [80, Ch. 5] gives an equivalent description of the invariant subspaces for $T_{1/2}$ and also those for $T_{p/q}$, in which $p, q \in \mathbb{N}$.

13.7 Irrational Case

Although the details of this go beyond the scope of this book, we mention some of the properties of T_α when $\alpha \in \mathbb{Q}^c \cap [0, 1]$. The main difference between the rational and irrational case comes from the fact that unlike the $U_{1/2}$ case, where there were plenty of sets E such that $u_{1/2}E = E$ and plenty of functions w such that $w \circ u_{1/2} = w$, when α is irrational there are essentially no such E or w. Thus, there are no obvious candidates for invariant subspaces or obvious elements of the spectrum. Unlike the case when α is rational, where the spectrum depends on α (Theorem 13.5.2), the spectrum of T_α for irrational α does not depend on α and it is much richer.

Theorem 13.7.1 (Parrott [260]). $\sigma(T_\alpha) = \{z \in \mathbb{C} : |z| \leqslant e^{-1}\}$ *for all irrational* $\alpha \in [0, 1]$.

Proposition 13.5.1 showed that $\sigma_p(T_\alpha) = \varnothing$ when α is rational. A more involved proof using results from number theory shows that the same holds when α is irrational.

Theorem 13.7.2 (Davie [106]). $\sigma_p(T_\alpha) = \varnothing$ *for all irrational* $\alpha \in [0, 1]$.

As mentioned earlier, there are no obvious examples of proper nonzero invariant subspaces for T_α when α is irrational. The reason Bishop suggested studying these operators was to explore the possibility that they have no nonzero proper invariant subspaces. It turns out that many of them do.

Theorem 13.7.3 (Davie [106]). *For almost every* $\alpha \in [0, 1]$, *the operator* T_α *has proper nonzero invariant subspaces.*

We do not get into the details here but the class of possible exceptional α is a small set that Davie describes exactly. It is also the case that "invariant subspace" can be replaced with "hyperinvariant subspace" in Davie's theorem.

13.8 Notes

Many operators on Hilbert spaces have proper nonzero invariant subspaces (such as normal operators, compact operators, operators commuting with compact operators). This is all summarized in a recent survey [80]. Another good text concerning invariant subspaces is [280]. Although the invariant subspace problem for Hilbert spaces is still open, the invariant subspace problem for certain Banach spaces was resolved in the negative by Enflo [124]. Read provided a bounded operator on ℓ^1 with no proper nontrivial invariant subspaces [281, 282].

The invariant subspace problem is connected to the invariant subspaces of the Bergman shift M_z. As discussed in Chapter 10, the M_z-invariant subspaces of A^2 are complicated.

A remarkable result from [19] says that if it is true that for all M_z-invariant subspaces $\mathcal{M}$ and $\mathcal{N}$ of A^2 with $\mathcal{M} \subseteq \mathcal{N}$ and $\dim(\mathcal{N}/\mathcal{M}) > 1$ there exists an invariant subspace $\mathcal{L}$ such that $\mathcal{M} \subsetneq \mathcal{L} \subsetneq \mathcal{N}$, then the invariant subspace problem is true. This remarkable result says that if we can understand the structure of one operator, the Bergman shift, then we can solve the invariant subspace problem. Of course, as we have seen in Chapter 10, this one operator has a particularly complicated collection of invariant subspaces.

Various properties of Bishop-type operators $f(x) \mapsto \varphi(x)(f \circ T)(x)$, where $\varphi \in L^\infty[0, 1]$ and T is a measure-preserving transformation on $[0, 1]$, have been studied in [54, 236, 260]. In particular, [260] contains a description of the numerical ranges of these operators. More detailed properties of the classical (and generalized) Bishop operators are covered in [81].

13.9 Exercises

Exercise 13.9.1. Prove that every selfadjoint operator on a complex Hilbert space $\mathcal{H}$ with $\dim \mathcal{H} \geqslant 2$ has a proper nonzero invariant subspace.

Exercise 13.9.2. Prove that if $A \in \mathcal{B}(\mathcal{H})$ and $p(A) = 0$ for some nonzero polynomial, then A has a proper nonzero invariant subspace.

Exercise 13.9.3. This is an extension of Exercise 13.9.2. An operator $T \in \mathcal{B}(\mathcal{H})$ is *polynomially compact* if $p(T)$ is compact for some polynomial p. Use Lomonosov's theorem (Corollary 13.2.4) to prove that every polynomially compact operator on a Hilbert space of dimension at least two has a proper nonzero invariant subspace.
Remark: This result was first proved by Bernstein and Robinson in 1966 with nonstandard analysis [47]. Halmos produced a standard proof [167].

Exercise 13.9.4. Let $A \in \mathcal{B}(\mathcal{H})$ and let $\mathcal{M} \subseteq \mathcal{H}$ be a linear manifold (not necessarily topologically closed). Prove that if $A\mathcal{M} \subseteq \mathcal{M}$, then $\mathcal{M}^-$ is an invariant subspace of A.

Exercise 13.9.5. This exercise from [280] shows the prevalence of cyclic vectors. Recall that $\mathbf{x}_0 \in \mathcal{H}$ is a *cyclic vector* for $A \in \mathcal{B}(\mathcal{H})$ if $\bigvee\{A^n\mathbf{x}_0 : n \geqslant 0\} = \mathcal{H}$. Define

$$\mathbf{x}_n = \left(I - \frac{1}{2\|A\|}A\right)^n \mathbf{x}_0 \quad \text{for } n \geqslant 0$$

and prove the following.

(a) $\mathbf{x}_n$ is a cyclic vector for A for every $n \geqslant 0$.

(b) $\bigvee\{\mathbf{x}_n : n \geqslant 0\} = \mathcal{H}$.

Thus, if A has a cyclic vector, it has a set of cyclic vectors with dense linear span.

Exercise 13.9.6. Prove the following result from [329] (see also [168]). If T is a linear transformation on an infinite-dimensional vector space V, then there is a linear manifold (closed under vector addition and scalar multiplication) $\mathcal{M}$ such that $\{0\} \subsetneq \mathcal{M} \subsetneq V$ and $T\mathcal{M} \subseteq \mathcal{M}$.

(a) Let $\mathbf{x} \in V \setminus \{\mathbf{0}\}$ and consider $\beta = \{\mathbf{x}, T\mathbf{x}, T^2\mathbf{x}, \ldots\}$. If β is linearly dependent, explain why the proof is complete.

(b) If β is linearly independent, every vector in the span of β can be written uniquely as $\sum_{n=0}^{\infty} a_n T^n \mathbf{x}$, in which only finitely many coefficients are nonzero. Consider the linear functional

$$\varphi\left(\sum_{n=0}^{\infty} a_n T^n \mathbf{x} \right) = \sum_{n=0}^{\infty} a_n$$

and prove that $\ker \varphi$ is a T-invariant linear manifold.

(c) Prove that $\{\mathbf{0}\} \subsetneq \ker \varphi \subsetneq V$.

Remark: The point of this exercise is that nonzero proper invariant subspaces always exist if the requirement that they be topologically closed is dropped.

Exercise 13.9.7. Suppose A is a linear transformation on a vector space V such that $A\mathcal{M} \subseteq \mathcal{M}$ for every linear manifold $\mathcal{M}$ in V. Prove that $A = \lambda I$, where $\lambda \in \mathbb{C}$.

Exercise 13.9.8. For the invariant subspace problem, one needs $\mathcal{H}$ to be a Hilbert space and not just an inner product space. Consider the inner product space $\mathbb{C}[x]$ of polynomials in the real variable x endowed with the $L^2[0, 1]$ inner product. The operator $M_x f = xf$ is bounded on $\mathbb{C}[x]$. Use the following steps to prove that there are no proper nonzero topologically closed subspaces $\mathcal{M}$ which are invariant for M_x.

(a) Suppose $\mathcal{M} \subseteq \mathbb{C}[x]$ is closed and M_x-invariant. Prove that $\mathcal{M} = \{f \in \mathbb{C}[x] : f|_E = 0\}$ for some finite set E.

(b) If $E \neq \varnothing$, prove that $\mathcal{M}$ is not closed in $\mathbb{C}[x]$.

Remark: See [280] for related examples.

Exercise 13.9.9. Define W on $L^2(0, \infty)$ by $(Wg)(x) = g(x + 1)$. Use the following steps to prove that W is similar to the backward shift of infinite multiplicity B on $(H^2)^{(\infty)}$ defined in (13.3.6) by $B(f_0, f_1, f_2, \ldots) = (f_1, f_2, f_3, \ldots)$.

(a) For $g \in L^2(0, \infty)$ and $n \geqslant 0$, define $g_n = \chi_{[n,n+1]}g$. Prove that $Wg_0 = 0$ and $Wg_n = g_{n-1}$ for $n \geqslant 1$.

(b) If $(u_j)_{j=0}^{\infty}$ is an orthonormal basis for $L^2[0, 1]$ and $n \geqslant 0$, prove that $(u_j(x + n))_{j=0}^{\infty}$ is an orthonormal basis for $L^2[n, n + 1]$.

(c) Use (b) to construct an invertible operator $V : L^2(0, \infty) \to (H^2)^{(\infty)}$ such that $VWV^{-1} = B$.

Exercise 13.9.10. If $\alpha = p/q$, where $p, q \in \mathbb{N}$ are relatively prime, consider the unitary portion $(U_\alpha f)(x) = f(\{x + \alpha\})$ of a Bishop operator.

(a) Use the ideas of the proof of Theorem 13.5.2 to prove that $\sigma(U_\alpha) = \sigma_p(U_\alpha) = \{\xi \in \mathbb{T} : \xi^q = 1\}$.

(b) For each $\xi \in \sigma_p(U_\alpha)$, find an orthogonal basis for $\ker(U_\alpha - \xi I)$.

Exercise 13.9.11. If α is irrational, consider the unitary portion $(U_\alpha f)(x) = f(\{x + \alpha\})$ of a Bishop operator. Prove that $\sigma(U_\alpha) = \mathbb{T}$.

Exercise 13.9.12. For the unitary portion $(U_\alpha f)(x) = f(\{x + \alpha\})$ of a Bishop operator, prove that U_α is cyclic if and only if α is irrational.

Exercise 13.9.13. Compute $\sigma(T_{4/5})$.

Exercise 13.9.14. In his doctoral thesis [260], Parrott proved a more general version of Theorem 13.5.2. If $\varphi \in L^\infty[0,1]$ and $\alpha \in [0,1)$, define the Bishop-type operator $(T_{\varphi,\alpha} f)(x) = \varphi(x) f(\{x + \alpha\})$. Use the techniques of the proof of Theorem 13.5.2 to prove that if $\alpha = p/q \in [0,1)$, where $p, q \in \mathbb{N}$ are relatively prime, then $\sigma(T_{\varphi,\alpha}) = \{\lambda \in \mathbb{C} : \lambda^q \in \mathscr{R}_\psi\}$, where $\psi(x) = \varphi(x)\varphi(\{x + \alpha\})\varphi(\{x + 2\alpha\}) \cdots \varphi(\{x + (q-1)\alpha\})$ and $\mathscr{R}_\psi$ is the essential range of ψ (Definition 8.1.11).

Exercise 13.9.15. Use the notation from Exercise 13.9.14 to compute $\sigma(T_{x^n, 1/n})$ for $n \geqslant 1$.

Exercise 13.9.16. For each α, show that the self commutator $T_\alpha^* T_\alpha - T_\alpha T_\alpha^*$ of the Bishop operator T_α is the multiplication operator M_φ on $L^2[0,1]$ for some piecewise-linear function φ on $[0,1]$.

Exercise 13.9.17. This exercise outlines a proof that the constant function 1 on $[0,1]$ is a cyclic vector for the Bishop operator $(T_{1/2} f)(x) = x f(\{x + \frac{1}{2}\})$.

(a) For $f \in L^2[0,1]$, find two $\frac{1}{2}$-periodic functions $f_1, f_2 \in L^2[0,1]$ such that $f(x) = f_1(x) + x f_2(x)$.

(b) Let $w(x) = x\{x + 1/2\}$; see Figure 13.9.1. Prove that $\{p \circ w : p \in \mathbb{C}[x]\}$ is dense in $L^2[0, 1/2]$.

(c) Prove that $(T_{1/2}^{2n} 1)(x) = w(x)^n$ and $(T_{1/2}^{2n+1} 1)(x) = x w(x)^n$ for all $n \geqslant 0$.

(d) Prove that $\bigvee \{T_{1/2}^k 1 : k \geqslant 0\} \supseteq \{p(w) + x q(w) : p, q \in \mathbb{C}[x]\}$.

(e) Use the previous parts to prove that $\bigvee \{T_{1/2}^k 1 : k \geqslant 0\} = L^2[0,1]$.

Remark: See [80, Ch. 5] for more on the cyclic vectors for $T_{p/q}$.

Exercise 13.9.18. Prove that if $f = \chi_E$, where E is a Lebesgue-measurable subset of $[0,1]$ such that $\{x \in [0,1] : x \in E \text{ or } \{x + 1/2\} \in E\}$ has measure one, then f is a cyclic vector for $T_{1/2}$.
Remark: See [80, Ch. 5] for more.

Exercise 13.9.19. Is the constant function 1 on $[0,1]$ a cyclic vector for $T_{1/3}$?

Exercise 13.9.20. For $\varphi \in L^\infty(\mathbb{T})$, consider the Bishop-type operator B_φ on $L^2(\mathbb{T})$ defined by $(B_\varphi f)(\xi) = \varphi(\xi) f(\bar{\xi})$ for $f \in L^2(\mathbb{T})$. Compute the matrix representation of B_φ with respect to the orthonormal basis $(\xi^n)_{n=-\infty}^\infty$ for $L^2(\mathbb{T})$.

Exercise 13.9.21. This is a continuation of Exercise 13.9.20. Show that $\sigma(B_\varphi) = \{\xi \in \mathbb{T} : \xi^2 \in \mathscr{R}_{\varphi(w)\varphi(\bar{w})}\}$.
Remark: See [260] for more on these generalizations of Bishop-type operators.

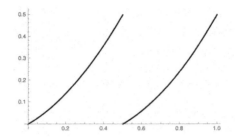

Figure 13.9.1 The function $w(x) = x\{x + 1/2\}$.

13.10 Hints for the Exercises

Hint for Ex. 13.9.1: Consult Theorem 8.7.1.

Hint for Ex. 13.9.7: The assumption says that for each $\mathbf{x} \in \mathcal{V}$, there is a $\lambda(\mathbf{x}) \in \mathbb{C}$ such that $A\mathbf{x} = \lambda(\mathbf{x})\mathbf{x}$.

Hint for Ex. 13.9.8: For (a), use Theorem 4.1.7.

Hint for Ex. 13.9.10: Consider the orthonormal basis $(e^{2\pi i n x})_{n=-\infty}^{\infty}$.

Hint for Ex. 13.9.17: For (b), use the Stone–Weierstrass theorem.

14

· · ● · ·

Operator Matrices

Key Concepts: Direct sum of Hilbert spaces, block operator, invariant subspace, reducing subspace, spectrum of a block matrix, idempotent operator, Parrott's theorem, Douglas' factorization theorem, polar decomposition.

Outline: This chapter concerns operators represented by matrices of operators. We examine the norm, spectrum, and invariant subspaces of such operators in terms of their entries.

14.1 Direct Sums of Hilbert Spaces

For Hilbert spaces $\mathcal{H}_1$ and $\mathcal{H}_2$, form $\mathcal{H}_1 \oplus \mathcal{H}_2$, the *direct sum*, as follows. First, consider the set of all ordered pairs $(\mathbf{x}_1, \mathbf{x}_2)$, where $\mathbf{x}_1 \in \mathcal{H}_1$ and $\mathbf{x}_2 \in \mathcal{H}_2$. For convenience, we use column-vector notation:

$$\mathcal{H}_1 \oplus \mathcal{H}_2 = \left\{ \begin{bmatrix} \mathbf{x}_1 \\ \mathbf{x}_2 \end{bmatrix} : \mathbf{x}_1 \in \mathcal{H}_1, \mathbf{x}_2 \in \mathcal{H}_2 \right\}.$$

Then make $\mathcal{H}_1 \oplus \mathcal{H}_2$ into a vector space with the operations of addition

$$\begin{bmatrix} \mathbf{x}_1 \\ \mathbf{x}_2 \end{bmatrix} + \begin{bmatrix} \mathbf{y}_1 \\ \mathbf{y}_2 \end{bmatrix} = \begin{bmatrix} \mathbf{x}_1 + \mathbf{y}_1 \\ \mathbf{x}_2 + \mathbf{y}_2 \end{bmatrix}$$

and scalar multiplication

$$\lambda \begin{bmatrix} \mathbf{x}_1 \\ \mathbf{x}_2 \end{bmatrix} = \begin{bmatrix} \lambda \mathbf{x}_1 \\ \lambda \mathbf{x}_2 \end{bmatrix}.$$

Finally, endow $\mathcal{H}_1 \oplus \mathcal{H}_2$ with the inner product

$$\left\langle \begin{bmatrix} \mathbf{x}_1 \\ \mathbf{x}_2 \end{bmatrix}, \begin{bmatrix} \mathbf{y}_1 \\ \mathbf{y}_2 \end{bmatrix} \right\rangle := \langle \mathbf{x}_1, \mathbf{y}_1 \rangle_{\mathcal{H}_1} + \langle \mathbf{x}_2, \mathbf{y}_2 \rangle_{\mathcal{H}_2}. \tag{14.1.1}$$

The corresponding norm satisfies

$$\left\|\begin{bmatrix} \mathbf{x}_1 \\ \mathbf{x}_2 \end{bmatrix}\right\|^2 = \|\mathbf{x}_1\|^2_{\mathcal{H}_1} + \|\mathbf{x}_2\|^2_{\mathcal{H}_2}.$$

Since $\mathcal{H}_1$ and $\mathcal{H}_2$ are complete, one can show that $\mathcal{H}_1 \oplus \mathcal{H}_2$ is also complete, and hence a Hilbert space (Exercise 14.11.1).

This notation is compatible with the orthogonal decomposition notation for a subspace $\mathcal{M}$ of $\mathcal{H}$. Indeed, recall from Definition 3.1.1 that

$$\mathcal{M}^\perp = \{\mathbf{x} \in \mathcal{H} : \langle \mathbf{x}, \mathbf{y} \rangle = 0 \text{ for all } \mathbf{y} \in \mathcal{M}\}$$

is the orthogonal complement of $\mathcal{M}$. Then $\mathcal{H} = \mathcal{M} \oplus \mathcal{M}^\perp$ and every $\mathbf{x} \in \mathcal{H}$ can be written uniquely as $\mathbf{x} = \mathbf{x}_1 + \mathbf{x}_2$, where $\mathbf{x}_1 \in \mathcal{M}$ and $\mathbf{x}_2 \in \mathcal{M}^\perp$. Furthermore, $\|\mathbf{x}\|^2 = \|\mathbf{x}_1\|^2 + \|\mathbf{x}_2\|^2$, and hence we can equate $\mathcal{M} \oplus \mathcal{M}^\perp$ with the set of column vectors

$$\left\{ \begin{bmatrix} \mathbf{x}_1 \\ \mathbf{x}_2 \end{bmatrix} : \mathbf{x}_1 \in \mathcal{M}, \ \mathbf{x}_2 \in \mathcal{M}^\perp \right\}$$

and regard the orthogonal decomposition of $\mathcal{M}$ and $\mathcal{M}^\perp$ as the direct sum of the Hilbert spaces $\mathcal{M}$ and $\mathcal{M}^\perp$. Thus, the use of the same notation $\oplus$ in two seemingly different contexts is justified. Some authors use the phrase *external direct sum* for $\mathcal{H}_1 \oplus \mathcal{H}_2$ and *internal direct sum* for $\mathcal{M} \oplus \mathcal{M}^\perp$, although we make no such distinction here.

One can consider the direct sum of a finite number of Hilbert spaces in a similar way. For example, if $\mathcal{H}$ is a Hilbert space and $n \in \mathbb{N}$, define

$$\mathcal{H}^{(n)} = \{(\mathbf{x}_1, \mathbf{x}_2, \ldots, \mathbf{x}_n) : \mathbf{x}_1, \mathbf{x}_2, \ldots, \mathbf{x}_n \in \mathcal{H}\}.$$

When endowed with the norm

$$\|(\mathbf{x}_i)_{i=1}^n\| = \left(\sum_{i=1}^n \|\mathbf{x}_i\|^2 \right)^{\frac{1}{2}}$$

and its corresponding inner product, $\mathcal{H}^{(n)}$ is a Hilbert space.

One can extend the construction above and define $\mathcal{H}^{(\infty)}$. As expected, there are some convergence issues to address. For a Hilbert space $\mathcal{H}$, consider

$$\mathcal{H}^{(\infty)} := \left\{ (\mathbf{x}_i)_{i=1}^\infty : \mathbf{x}_i \in \mathcal{H}, \sum_{i=1}^\infty \|\mathbf{x}_i\|^2 < \infty \right\} \tag{14.1.2}$$

with corresponding norm

$$\|(\mathbf{x}_i)_{i=1}^\infty\| = \left(\sum_{i=1}^\infty \|\mathbf{x}_i\|^2 \right)^{\frac{1}{2}}.$$

One can modify the proof of Proposition 1.2.5 (the completeness of ℓ^2) to prove that $\mathcal{H}^{(\infty)}$ is a Hilbert space (Exercise 14.11.2).

14.2 Block Operators

Let $T \in \mathcal{B}(\mathcal{H})$ and let $\mathcal{M}$ be a subspace of $\mathcal{H}$. The present goal is to represent T in a manner that isolates how T interacts with the components of the orthogonal decomposition

$$\mathcal{H} = \mathcal{M} \oplus \mathcal{M}^{\perp}. \tag{14.2.1}$$

To do this, let P denote the orthogonal projection onto $\mathcal{M}$. Then P is selfadjoint and idempotent, and $I - P$ is the orthogonal projection onto $\mathcal{M}^{\perp}$ (Proposition 3.1.2). A computation using the identity $P + (I - P) = I$ confirms that

$$T = A + B + C + D, \tag{14.2.2}$$

where

$$A = PTP, \quad B = PT(I - P), \quad C = (I - P)TP, \quad \text{and} \quad D = (I - P)T(I - P).$$

Since

$$\operatorname{ran} P = \ker(I - P) = \mathcal{M} \quad \text{and} \quad \operatorname{ran}(I - P) = \ker P = \mathcal{M}^{\perp}, \tag{14.2.3}$$

we restrict the domain and codomains of A, B, C, and D, initially defined on $\mathcal{H}$, and regard these operators as maps between $\mathcal{M}$ and $\mathcal{M}^{\perp}$ as follows:

$$
\left.
\begin{aligned}
A &= PTP \in \mathcal{B}(\mathcal{M}), \\
B &= PT(I - P) \in \mathcal{B}(\mathcal{M}^{\perp}, \mathcal{M}), \\
C &= (I - P)TP \in \mathcal{B}(\mathcal{M}, \mathcal{M}^{\perp}), \\
D &= (I - P)T(I - P) \in \mathcal{B}(\mathcal{M}^{\perp}).
\end{aligned}
\right\} \tag{14.2.4}
$$

This information is compactly represented in the block-operator notation

$$T = \begin{bmatrix} PTP & PT(I - P) \\ (I - P)TP & (I - P)T(I - P) \end{bmatrix}, \tag{14.2.5}$$

which is more conveniently displayed as

$$T = \begin{bmatrix} A & B \\ C & D \end{bmatrix}. \tag{14.2.6}$$

An expression of the form (14.2.6) implicitly comes equipped with an orthogonal decomposition of the underlying Hilbert space, with respect to which A, B, C, and D are defined as in (14.2.4). Two block operators of the form (14.2.6), with respect to the same orthogonal decomposition of the underlying Hilbert space, are equal if and only if their corresponding entries are equal.

We equate $\mathbf{x}_1 + \mathbf{x}_2 \in \mathcal{M} \oplus \mathcal{M}^{\perp}$, where $\mathbf{x}_1 \in \mathcal{M}$ and $\mathbf{x}_2 \in \mathcal{M}^{\perp}$, with the column vector

$$\begin{bmatrix} \mathbf{x}_1 \\ \mathbf{x}_2 \end{bmatrix}$$

and use (14.2.2) to see that T acts as matrix multiplication

$$\begin{bmatrix} A & B \\ C & D \end{bmatrix} \begin{bmatrix} \mathbf{x}_1 \\ \mathbf{x}_2 \end{bmatrix} = \begin{bmatrix} A\mathbf{x}_1 + B\mathbf{x}_2 \\ C\mathbf{x}_1 + D\mathbf{x}_2 \end{bmatrix}.$$

This dissection of the operator T with respect to the decomposition (14.2.1) respects the adjoint and operator composition. Since T was arbitrary, replace T with T^* in (14.2.5) and deduce

$$\begin{aligned} T^* &= \begin{bmatrix} PT^*P & PT^*(I-P) \\ (I-P)T^*P & (I-P)T^*(I-P) \end{bmatrix} \\ &= \begin{bmatrix} (PTP)^* & [(I-P)TP]^* \\ [PT(I-P)]^* & [(I-P)T(I-P)]^* \end{bmatrix} \\ &= \begin{bmatrix} A^* & C^* \\ B^* & D^* \end{bmatrix}. \end{aligned} \tag{14.2.7}$$

Consequently, the block-operator representation of T^* with respect to the orthogonal decomposition (14.2.1) is the formal adjoint of the 2×2 block-operator matrix (14.2.6):

$$\begin{bmatrix} A & B \\ C & D \end{bmatrix}^* = \begin{bmatrix} A^* & C^* \\ B^* & D^* \end{bmatrix}.$$

Block-operator representations are also compatible with operator composition. Suppose that $S, T \in \mathcal{B}(\mathcal{H})$ and observe that

$$ST = \begin{bmatrix} PSTP & PST(I-P) \\ (I-P)STP & (I-P)ST(I-P) \end{bmatrix} \tag{14.2.8}$$

by (14.2.5). This is also what one obtains by formally multiplying the block-operator representations of S and T and using the idempotence of P:

$$ST = \begin{bmatrix} PSP & PS(I-P) \\ (I-P)SP & (I-P)S(I-P) \end{bmatrix} \begin{bmatrix} PTP & PT(I-P) \\ (I-P)TP & (I-P)T(I-P) \end{bmatrix}. \tag{14.2.9}$$

We suppress the details here, although the reader is invited to carry them out (Exercise 14.11.3).

Back in Theorem 8.3.1, we presented an efficient proof of Fuglede's theorem. Here is a generalization, due to Putnam, whose proof uses block-operator techniques.

Theorem 14.2.10 (Fuglede–Putnam). *Let $T, M, N \in B(\mathcal{H})$. If M, N are normal and $MT = TN$, then $M^*T = TN^*$.*

Proof Consider the block operators on $\mathcal{H} \oplus \mathcal{H}$ defined by

$$T' = \begin{bmatrix} 0 & 0 \\ T & 0 \end{bmatrix} \quad \text{and} \quad N' = \begin{bmatrix} N & 0 \\ 0 & M \end{bmatrix}.$$

Note that N' is normal and T' commutes with N'. By Fuglede's theorem (Theorem 8.3.1), $T'(N')^* = (N')^*T'$, that is,

$$\begin{bmatrix} 0 & 0 \\ T & 0 \end{bmatrix}\begin{bmatrix} N^* & 0 \\ 0 & M^* \end{bmatrix} = \begin{bmatrix} N^* & 0 \\ 0 & M^* \end{bmatrix}\begin{bmatrix} 0 & 0 \\ T & 0 \end{bmatrix}.$$

Matrix multiplication yields

$$\begin{bmatrix} 0 & 0 \\ TN^* & 0 \end{bmatrix} = \begin{bmatrix} 0 & 0 \\ M^*T & 0 \end{bmatrix},$$

and hence $TN^* = M^*T$. ∎

More generally, we may consider $n \times n$ block-operator decompositions. Let $P_1, P_2, ..., P_n \in \mathcal{B}(\mathcal{H})$ be orthogonal projections onto proper nonzero subspaces $\mathcal{H}_1, \mathcal{H}_2, ..., \mathcal{H}_n \subseteq \mathcal{H}$, respectively. If $P_1 + P_2 + \cdots + P_n = I$, then $\mathcal{H}_1, \mathcal{H}_2, ..., \mathcal{H}_n$ are pairwise orthogonal (Exercise 14.11.4). For $T \in \mathcal{B}(\mathcal{H})$, write

$$T = \begin{bmatrix} T_{11} & T_{12} & \cdots & T_{1n} \\ T_{21} & T_{22} & \cdots & T_{2n} \\ \vdots & \vdots & \ddots & \vdots \\ T_{n1} & T_{n2} & \cdots & T_{nn} \end{bmatrix}, \tag{14.2.11}$$

in which $T_{ij} = P_i T P_j$ is regarded as an element of $\mathcal{B}(\mathcal{H}_j, \mathcal{H}_i)$ by restricting the domain and codomain of $P_i T P_j$ to $\mathcal{H}_j$ and $\mathcal{H}_i$, respectively. One can verify that operator composition and the adjoint respect such decompositions as in the 2×2 case (Exercise 14.11.5).

14.3 Invariant Subspaces

If we apply (14.2.5) to the orthogonal projection P of $\mathcal{H}$ onto a proper nonzero subspace $\mathcal{M}$ of $\mathcal{H}$, then (14.2.3) ensures that

$$P = \begin{bmatrix} I_{\mathcal{M}} & 0 \\ 0 & 0_{\mathcal{M}^\perp} \end{bmatrix} \quad \text{and} \quad I_{\mathcal{H}} - P = \begin{bmatrix} 0_{\mathcal{M}} & 0 \\ 0 & I_{\mathcal{M}^\perp} \end{bmatrix}. \tag{14.3.1}$$

We henceforth suppress the subscripts on zero and identity operators since the spaces upon which they operate can be deduced from context. Observe that the selfadjointness and idempotence of P and $I - P$, along with the identities $P(I - P) = (I - P)P = 0$ and $P + (I - P) = I$, are reflected in the block-operator representations (14.3.1).

Now suppose that $T\mathcal{M} \subseteq \mathcal{M}$; that is, $\mathcal{M}$ is T-invariant. Then

$$C = (I - P)TP = 0 \tag{14.3.2}$$

since $\operatorname{ran} P = \mathcal{M} = \ker(I - P)$, and hence T has the block-operator representation

$$T = \begin{bmatrix} A & B \\ 0 & D \end{bmatrix}, \tag{14.3.3}$$

in which $A = T|_{\mathcal{M}}$ and $D = T|_{\mathcal{M}^\perp}$. Conversely, if $T \in \mathcal{B}(\mathcal{H})$ is represented by a block-operator of this form, then $\mathcal{M}$ is T-invariant. Since $(I - P)TP = 0$ if and only if $PTP = TP$, we obtain the following result.

Theorem 14.3.4. *For a proper nonzero subspace $\mathcal{M}$ of $\mathcal{H}$, the following are equivalent.*

(a) *$\mathcal{M}$ is T-invariant.*

(b) $T = \begin{bmatrix} A & B \\ 0 & D \end{bmatrix}$ *with respect to the decomposition $\mathcal{H} = \mathcal{M} \oplus \mathcal{M}^\perp$.*

(c) *$PTP = TP$, in which P is the orthogonal projection onto $\mathcal{M}$.*

Now suppose that $\mathcal{M}$ is invariant for both T and T^*; that is, suppose that $\mathcal{M}$ is a reducing subspace. In addition to (14.3.2), we also have $B = PT(I - P) = 0$ since $\operatorname{ran}(I - P) = \mathcal{M}^\perp = \ker P$. Therefore, the block-operator representation of T is block diagonal:

$$T = \begin{bmatrix} A & 0 \\ 0 & D \end{bmatrix},$$

which we write as $T = A \oplus D$. In particular, T commutes with P (and with $I - P$), which can be seen from the block-operator representation of P given in (14.3.1). One can also deduce this algebraically as follows:

$$
\begin{aligned}
TP &= PTP && (\mathcal{M} \text{ is } T\text{-invariant}) \\
&= (PT^*P)^* && (P = P^*) \\
&= (T^*P)^* && (\mathcal{M} \text{ is } T^*\text{-invariant}) \\
&= PT && (P = P^*).
\end{aligned}
$$

We record this in the following theorem.

Theorem 14.3.5. *For a proper nonzero subspace $\mathcal{M}$ of $\mathcal{H}$, the following are equivalent.*

(a) *$\mathcal{M}$ is a reducing subspace for T.*

(b) *$T = A \oplus D$ with respect to the decomposition $\mathcal{H} = \mathcal{M} \oplus \mathcal{M}^\perp$.*

(c) *$PT = TP$, in which P is the orthogonal projection onto $\mathcal{M}$.*

14.4 Inverses and Spectra

Inverses are another domain in which block operators can be manipulated like matrices, so long as one remembers that the entries need not commute. However, there are a few unexpected surprises.

If $\mathcal{H} = \mathcal{H}_1 \oplus \mathcal{H}_2$ is an orthogonal decomposition, consider a 2×2 block operator

$$\begin{bmatrix} A & B \\ 0 & D \end{bmatrix}, \tag{14.4.1}$$

in which $A \in \mathcal{B}(\mathcal{H}_1)$ and $D \in \mathcal{B}(\mathcal{H}_2)$ are invertible and $B \in \mathcal{B}(\mathcal{H}_2, \mathcal{H}_1)$. To avoid trivialities, we assume that $\mathcal{H}_1$ and $\mathcal{H}_2$ are both proper and nonzero. A computation (Exercise 14.11.6) confirms that

$$\begin{bmatrix} A & B \\ 0 & D \end{bmatrix}^{-1} = \begin{bmatrix} A^{-1} & -A^{-1}BD^{-1} \\ 0 & D^{-1} \end{bmatrix}. \tag{14.4.2}$$

Hence, a block upper-triangular operator matrix is invertible if the blocks on its diagonal are invertible (the same holds true for lower-triangular block operators). Surprisingly, the converse is false [169, Pr. 71].

Example 14.4.3. A 2×2 upper-triangular block operator may be invertible while both of the blocks on the diagonal are not invertible. For example, let S denote the unilateral shift on ℓ^2 (Chapter 5). Then neither S nor S^*, the forward and backward shifts, respectively, are invertible (Proposition 5.1.4 and Proposition 5.2.4). Nevertheless, a computation (Exercise 14.11.7) confirms that

$$\begin{bmatrix} S & I - SS^* \\ 0 & S^* \end{bmatrix} \quad \text{is invertible with inverse} \quad \begin{bmatrix} S^* & 0 \\ I - SS^* & S \end{bmatrix}. \tag{14.4.4}$$

In particular, the operators above are block triangular and unitary. In contrast, if the spaces involved are finite dimensional, this cannot occur unless the matrix is block diagonal with unitary operators on the diagonal (Exercise 14.11.8).

Suppose that (14.4.1) is invertible. In light of the previous example, we cannot conclude that A or D is invertible. However, if one of them is invertible, then so is the other (Exercise 14.11.12). We use this observation in the second part of the following theorem.

Theorem 14.4.5. *If*

$$T = \begin{bmatrix} A & X \\ 0 & B \end{bmatrix},$$

then

$$\sigma(A) \, \Delta \, \sigma(B) \subseteq \sigma(T) \subseteq \sigma(A) \cup \sigma(B). \tag{14.4.6}$$

Here Δ denotes the symmetric difference of two sets. In particular, if $\sigma(A) \cap \sigma(B) = \varnothing$, then $\sigma(T) = \sigma(A) \cup \sigma(B)$.

Proof If $\lambda \in \sigma(A) \, \Delta \, \sigma(B)$, then exactly one of $A - \lambda I$ and $B - \lambda I$ is invertible. Therefore,

$$T - \lambda I = \begin{bmatrix} A - \lambda I & X \\ 0 & B - \lambda I \end{bmatrix}$$

is not invertible (Exercise 14.11.12), and hence $\lambda \in \sigma(T)$.

If $\lambda \notin \sigma(A) \cup \sigma(B)$, then $A - \lambda I$ and $B - \lambda I$ are invertible and hence $T - \lambda I$ is invertible by (14.4.2). Thus, $\lambda \notin \sigma(T)$. ■

Example 14.4.7. The 2×2 block operator T from Example 14.4.3 is unitary, and hence $\sigma(T) \subseteq \mathbb{T}$. However, the spectra of its diagonal blocks are $\sigma(S) = \sigma(S^*) = \mathbb{D}^-$ (Proposition 5.1.4), and hence both containments in (14.4.6) can be strict. On the other hand, if the spaces that A and B operate on are finite dimensional, then $\sigma(T) = \sigma(A) \cup \sigma(B)$ (Exercise 14.11.13).

Block-operator techniques can sometimes reveal things that are unexpected. Although AB and BA are generally unequal, their spectra are closely related. See Exercise 14.11.18 for an extension of the next result.

Theorem 14.4.8. *If $A, B \in \mathcal{B}(\mathcal{H})$, then $\sigma(AB) \cup \{0\} = \sigma(BA) \cup \{0\}$.*

Proof Let $\lambda \neq 0$ and observe that

$$\begin{bmatrix} I & 0 \\ B & I \end{bmatrix} \begin{bmatrix} AB - \lambda I & A \\ 0 & -\lambda I \end{bmatrix} = \begin{bmatrix} -\lambda I & A \\ 0 & BA - \lambda I \end{bmatrix} \begin{bmatrix} I & 0 \\ B & I \end{bmatrix}.$$

Let

$$S = \begin{bmatrix} AB - \lambda I & A \\ 0 & -\lambda I \end{bmatrix} \quad \text{and} \quad T = \begin{bmatrix} -\lambda I & A \\ 0 & BA - \lambda I \end{bmatrix}.$$

Since

$AB - \lambda I$ is invertible $\iff$ S is invertible	(by (14.4.2))
$\iff$ T is invertible	(S and T are similar)
$\iff$ $BA - \lambda I$ is invertible	(by Exercise 14.11.12),

it follows that $\sigma(AB)$ and $\sigma(BA)$ have the same nonzero elements. ∎

Example 14.4.9. The restriction $\lambda \neq 0$ is essential in the previous proof. If S denotes the unilateral shift on ℓ^2, then $\sigma(S^*S) = \{1\}$ and $\sigma(SS^*) = \{0, 1\}$ since $S^*S = I$ and $SS^* = I - \mathbf{e}_0 \otimes \mathbf{e}_0$ is the orthogonal projection onto $\{\mathbf{e}_0\}^\perp = \bigvee\{\mathbf{e}_n : n \geq 1\}$.

14.5 Idempotents

Recall that $A \in \mathcal{B}(\mathcal{H})$ is *idempotent* if $A^2 = A$. Examples of idempotents include orthogonal projections and the 2×2 matrices

$$\begin{bmatrix} 1 & 1 \\ 0 & 0 \end{bmatrix} \quad \text{and} \quad \begin{bmatrix} \frac{1}{2} & \frac{1}{2} \\ \frac{1}{2} & \frac{1}{2} \end{bmatrix}.$$

From the first matrix above, one sees that an idempotent operator need not be selfadjoint.

Theorem 14.5.1. *If $A \in \mathcal{B}(\mathcal{H})$ is idempotent and $A \notin \{0, I\}$, then $\sigma(A) = \sigma_p(A) = \{0, 1\}$.*

Proof Suppose that $\lambda \notin \{0, 1\}$ and let

$$B = \frac{1}{1 - \lambda}A - \frac{1}{\lambda}(I - A).$$

Then

$$(A - \lambda I)B = (A - \lambda I)\left(\frac{1}{1 - \lambda}A - \frac{1}{\lambda}(I - A)\right)$$

$$= \frac{1}{1 - \lambda}(A^2 - \lambda A) - \frac{1}{\lambda}(A - A^2 - \lambda I + \lambda A)$$

$$= \frac{A}{1 - \lambda}(1 - \lambda) + (I - A)$$

$$= I.$$

Since B is a polynomial in A, it follows that $B(A - \lambda I) = I$, and hence $A - \lambda I$ has an inverse in $\mathcal{B}(\mathcal{H})$. Thus, $\lambda \notin \sigma(A)$, and hence $\sigma(A) \subseteq \{0, 1\}$.

Since $I - A \neq 0$, there is an $\mathbf{x}$ such that $\mathbf{y} = (I - A)\mathbf{x} \neq \mathbf{0}$. Then $A\mathbf{y} = A(I - A)\mathbf{x} = \mathbf{0}$, so $0 \in \sigma_p(A)$. Similarly, $A \neq 0$, and hence there is an $\mathbf{x} \in \mathcal{H}$ such that $\mathbf{y} = A\mathbf{x} \neq \mathbf{0}$. Then, $(A - I)\mathbf{y} = (A - I)A\mathbf{x} = A^2\mathbf{x} - A\mathbf{x} = 0$ and so $A\mathbf{y} = \mathbf{y}$. Thus, $1 \in \sigma_p(A)$. This shows that $\{0, 1\} \subseteq \sigma_p(A) \subseteq \sigma(A)$. ∎

If $A \in \mathcal{B}(\mathcal{H})$ is idempotent, the previous theorem ensures that $A\mathbf{y} = \mathbf{y}$ if and only if $\mathbf{y} \in \operatorname{ran} A$. In other words, $\operatorname{ran} A$ is the eigenspace of A corresponding to the eigenvalue 1. Indeed, the condition $A\mathbf{y} = \mathbf{y}$ ensures that $\mathbf{y} \in \operatorname{ran} A$. On the other hand, if $\mathbf{y} \in \operatorname{ran} A$, then $\mathbf{y} = A\mathbf{x}$ for some $\mathbf{x} \in \mathcal{H}$ and hence $\mathbf{y} = A\mathbf{x} = A^2\mathbf{x} = A(A\mathbf{x}) = A\mathbf{y}$.

Proposition 14.5.2. *If A is idempotent, then $\operatorname{ran} A$ is closed.*

Proof Let $(\mathbf{y}_n)_{n=1}^\infty$ be a sequence in $\operatorname{ran} A$ such that $\mathbf{y}_n \to \mathbf{y}$. Then

$$\mathbf{y} = \lim_{n\to\infty} \mathbf{y}_n = \lim_{n\to\infty} A\mathbf{y}_n = A\mathbf{y}$$

and hence $\mathbf{y} \in \operatorname{ran} A$. ∎

Since $I = A + (I - A)$, it follows that $\mathbf{x} = A\mathbf{x} + (I - A)\mathbf{x}$. Observe that $A\mathbf{x} \in \operatorname{ran} A$ and $(I - A)\mathbf{x} \in \ker A$ since $A(I - A)\mathbf{x} = (A - A^2)\mathbf{x} = 0\mathbf{x} = \mathbf{0}$. Thus, if $A \in \mathcal{B}(\mathcal{H})$ is idempotent, then $\mathcal{H} = \ker A + \operatorname{ran} A$, in which the subspaces involved are closed, although not necessarily orthogonal, and $\ker A \cap \operatorname{ran} A = \{\mathbf{0}\}$. Recall that we only use the symbol $\oplus$ to signify an orthogonal direct sum. Since A fixes each element of its range, we deduce the following result.

Proposition 14.5.3. *If $A \in \mathcal{B}(\mathcal{H})$ is idempotent, then*

$$A = \begin{bmatrix} I & B \\ 0 & 0 \end{bmatrix}$$

with respect to the direct sum $\mathcal{H} = \operatorname{ran} A \oplus (\operatorname{ran} A)^\perp$. Moreover, any block operator of the form above is idempotent.

Exercise 14.11.14 yields the following theorem.

Theorem 14.5.4. *For $A \in \mathcal{B}(\mathcal{H})$, the following are equivalent.*

(a) *A is an orthogonal projection.*

(b) *$A = A^*$ and $\sigma(A) \subseteq \{0, 1\}$.*

(c) *A is selfadjoint and idempotent.*

14.6 The Douglas Factorization Theorem

Let $A, B \in \mathcal{B}(\mathcal{H})$. When is $\operatorname{ran} A \subseteq \operatorname{ran} B$? Note that this question concerns operator ranges, which are vector spaces but need not be topologically closed (and hence not subspaces). A corollary to the following theorem of Douglas provides the answer.

Theorem 14.6.1 (Douglas [115]). *Let $\mathcal{K}_1, \mathcal{K}_2$, and $\mathcal{H}$ be Hilbert spaces. For $A \in \mathcal{B}(\mathcal{K}_1, \mathcal{H})$ and $B \in \mathcal{B}(\mathcal{K}_2, \mathcal{H})$, the following are equivalent.*

(a) *There is a contraction $C \in \mathcal{B}(\mathcal{K}_1, \mathcal{K}_2)$ such that $A = BC$.*

(b) *$AA^* \leqslant BB^*$.*

Proof (a) $\Rightarrow$ (b) Suppose that C is a contraction and $A = BC$. Since $I - CC^* \geqslant 0$ (Exercise 14.11.20), it follows that $BB^* - AA^* = BB^* - BCC^*B^* = B(I - CC^*)B^* \geqslant 0$. Thus, $AA^* \leqslant BB^*$.

(b) $\Rightarrow$ (a) Assume that $AA^* \leqslant BB^*$. Then

$$\|A^*\mathbf{x}\| \leqslant \|B^*\mathbf{x}\| \quad \text{for all } \mathbf{x} \in \mathcal{H}, \tag{14.6.2}$$

and hence we can define a linear transformation

$$D : \operatorname{ran} B^* \to \operatorname{ran} A^*, \quad D(B^*\mathbf{x}) = A^*\mathbf{x}.$$

Note that the vector spaces $\operatorname{ran} B^*$ and $\operatorname{ran} A^*$ are not necessarily topologically closed. By (14.6.2), D is well defined and $\|D(B^*\mathbf{x})\| \leqslant \|B^*\mathbf{x}\|$ for all $\mathbf{x} \in \mathcal{H}$. Therefore, we can extend D to a contraction from $(\operatorname{ran} B^*)^-$ into $\mathcal{K}_1$. Finally, let $D\mathbf{z} = 0$ for $\mathbf{z} \in (\operatorname{ran} B^*)^\perp$ so that D extends by linearity to a bounded operator on all of $\mathcal{K}_2$. Then D is a contraction that satisfies $DB^* = A^*$. Take adjoints to obtain $A = BD^*$, and thus $C = D^*$ is the desired contraction. ∎

The construction above yields the following two sets of equalities:

$$\ker C = \ker D^* = (\operatorname{ran} D)^\perp = (\operatorname{ran} A^*)^\perp = \ker A \tag{14.6.3}$$

and

$$(\operatorname{ran} C)^- = (\operatorname{ran} D^*)^- = (\ker D)^\perp = (\operatorname{ran} B^*)^- = (\ker B)^\perp. \tag{14.6.4}$$

Corollary 14.6.5. *Let $A : \mathcal{H} \to \mathcal{K}_1$ and $B : \mathcal{H} \to \mathcal{K}_2$ be contractions. Then*

$$\begin{bmatrix} A \\ B \end{bmatrix} : \mathcal{H} \to \mathcal{K}_1 \oplus \mathcal{K}_2, \quad \mathbf{x} \mapsto \begin{bmatrix} A\mathbf{x} \\ B\mathbf{x} \end{bmatrix}$$

is a contraction if and only if there is a contraction $C : \mathcal{H} \to \mathcal{K}_1$ such that $A = C(I - B^*B)^{\frac{1}{2}}$.

Proof From Exercise 14.11.20,

$$T = \begin{bmatrix} A \\ B \end{bmatrix}$$

is a contraction if and only if $T^*T \leqslant I$. Since $T^* = [A^* \ B^*]$, the inequality $T^*T \leqslant I$ is equivalent to $T^*T = A^*A + B^*B \leqslant I$. Write this as $A^*A \leqslant I - B^*B$, and then apply Douglas' factorization theorem (Theorem 14.6.1). ∎

Theorem 14.6.1 is often used as the following range-inclusion theorem.

Corollary 14.6.6. *For $A, B \in \mathcal{B}(\mathcal{H})$, the following are equivalent.*

(a) $\operatorname{ran} A \subseteq \operatorname{ran} B$.

(b) $AA^* \leqslant \lambda BB^*$ *for some $\lambda > 0$.*

(c) *There exists a $C \in \mathcal{B}(\mathcal{H})$ such that $A = BC$.*

Proof (a) $\Rightarrow$ (c) If $\mathbf{x} \in \mathcal{H}$, then $A\mathbf{x} \in \operatorname{ran} A \subseteq \operatorname{ran} B$. Since $\mathcal{H} = \ker B \oplus (\ker B)^\perp$, there is a unique $\mathbf{z} \in (\ker B)^\perp$ such that $B\mathbf{z} = A\mathbf{x}$. Define a linear transformation on $\mathcal{H}$ by $C\mathbf{x} = \mathbf{z}$. Use the closed graph theorem (Theorem 2.2.2) to show that C is bounded as follows. Suppose $C\mathbf{x}_n = \mathbf{z}_n$ where $(\mathbf{x}_n)_{n=1}^\infty$ is a sequence in $\mathcal{H}$ and $(\mathbf{z}_n)_{n=1}^\infty$ is a sequence in $(\ker B)^\perp$ such that $\mathbf{x}_n \to \mathbf{x}$ and $\mathbf{z}_n \to \mathbf{z}$. Then $A\mathbf{x}_n \to A\mathbf{x}$ and $B\mathbf{z}_n \to B\mathbf{z}$. The assumption $B\mathbf{z}_n = A\mathbf{x}_n$ implies $B\mathbf{z} = A\mathbf{x}$. Moreover, since $(\ker B)^\perp$ is closed, it follows that $\mathbf{z} \in (\ker B)^\perp$. Thus, the uniqueness of $\mathbf{z}$ ensures that $C\mathbf{x} = \mathbf{z}$, so C is bounded. Finally, observe that $A\mathbf{x} = B\mathbf{z} = BC\mathbf{x}$ for all $\mathbf{x}$, and hence $A = BC$.

(c) $\Rightarrow$ (a) This follows from $A\mathcal{H} = BC\mathcal{H} \subseteq B\mathcal{H}$.

(b) $\Rightarrow$ (c) This is a minor alteration of the proof of Theorem 14.6.1.

(c) $\Rightarrow$ (b) If $A = BC$, then

$$AA^* = BCC^*B^* = \|C\|^2 BB^* - B(\|C\|^2 I - CC^*)B^* \leqslant \|C\|^2 BB^*,$$

which completes the proof. ∎

14.7 The Julia Operator of a Contraction

In this section, we examine Möbius transformations of operators, following the presentation [134, Ch. 7]. This material is needed in our treatment of Parrott's theorem which plays a crucial role in our analysis of Hankel operators (see Chapter 17).

Suppose that

$$U = \begin{bmatrix} A & B \\ C & D \end{bmatrix}$$

is a unitary operator on $\mathcal{H} = \mathcal{H}_1 \oplus \mathcal{H}_2$. Let P denote the orthogonal projection of $\mathcal{H}$ onto $\mathcal{H}_1$ and observe that

$$\|C\| = \|(I - P)UP\| \leqslant \|U\| = 1.$$

The operator-valued function

$$f_U(X) = B - AX(I + CX)^{-1}D$$

is well defined on

$$\mathscr{D}(U) = \{X \in \mathcal{B}(\mathcal{H}_2, \mathcal{H}_1) : I + CX \text{ is invertible in } \mathcal{B}(\mathcal{H}_2)\}.$$

By Proposition 2.3.9, notice that $\mathscr{D}(U)$ contains every $X \in \mathcal{B}(\mathcal{H}_2, \mathcal{H}_1)$ such that $\|X\| < 1$ and that f_U maps $\mathscr{D}(U)$ into $\mathcal{B}(\mathcal{H}_2, \mathcal{H}_1)$.

Proposition 14.7.1. *If* $X \in \mathscr{D}(U)$ *and* $E = (I + CX)^{-1}D$, *then*

$$I - f_U(X)^* f_U(X) = E^*(I - X^*X)E.$$

Proof Since $U^*U = I$, it follows that

$$\begin{bmatrix} I & 0 \\ 0 & I \end{bmatrix} = \begin{bmatrix} A^* & C^* \\ B^* & D^* \end{bmatrix} \begin{bmatrix} A & B \\ C & D \end{bmatrix} = \begin{bmatrix} A^*A + C^*C & A^*B + C^*D \\ B^*A + D^*C & B^*B + D^*D \end{bmatrix}.$$

Comparing entries in the matrices above yields the operator identities

$$A^*A + C^*C = I,$$
$$A^*B + C^*D = 0,$$
$$B^*A + D^*C = 0,$$
$$B^*B + D^*D = I,$$

from which we deduce the desired identity. ∎

Corollary 14.7.2. *If* $X \in \mathscr{D}(U)$ *is a contraction, then so is* $f_U(X)$.

Proof Exercise 14.11.20 says that $I - X^*X \geqslant 0$ and Proposition 14.7.1 implies that

$$I - f_U(X)^* f_U(X) \geqslant 0.$$

Exercise 14.11.20 implies that $f_U(X)$ is a contraction. ∎

Let $B : \mathcal{H}_2 \to \mathcal{H}_1$ be a contraction. The corresponding *Julia operator* on $\mathcal{H}_1 \oplus \mathcal{H}_2$ is

$$J(B) = \begin{bmatrix} (I - BB^*)^{\frac{1}{2}} & B \\ -B^* & (I - B^*B)^{\frac{1}{2}} \end{bmatrix}.$$

Lemma 14.7.3 (Julia [207]). *$J(B)$ is a unitary operator on $\mathcal{H}_1 \oplus \mathcal{H}_2$.*

Proof We need to check that $J(B)J(B)^* = J(B)^*J(B) = I$. Observe that $J(B)^*J(B)$ equals

$$
\begin{bmatrix}
I & (I - BB^*)^{\frac{1}{2}}B - B(I - B^*B)^{\frac{1}{2}} \\
B^*(I - BB^*)^{\frac{1}{2}} - (I - B^*B)^{\frac{1}{2}}B^* & I
\end{bmatrix}.
$$

Moreover (Exercise 8.10.27),

$$
(I - BB^*)^{\frac{1}{2}}B = B(I - B^*B)^{\frac{1}{2}} \quad \text{and} \quad B^*(I - BB^*)^{\frac{1}{2}} = (I - B^*B)^{\frac{1}{2}}B^*.
$$

Thus, $J(B)^*J(B) = I$. The proof that $J(B)J(B)^* = I$ is similar. ∎

Since $J(B)$ is a unitary operator on $\mathcal{H}_1 \oplus \mathcal{H}_2$, Corollary 14.7.2 gives the following result.

Corollary 14.7.4. *Let $B, X \in \mathcal{B}(\mathcal{H}_2, \mathcal{H}_1)$ be contractions such that $I - B^*X$ is invertible in $\mathcal{B}(\mathcal{H}_2)$. Then*

$$
f_{J(B)}(X) = B - (I - BB^*)^{\frac{1}{2}}X(I - B^*X)^{-1}(I - B^*B)^{\frac{1}{2}} \tag{14.7.5}
$$

is a contraction in $\mathcal{B}(\mathcal{H}_2, \mathcal{H}_1)$.

14.8 Parrott's Theorem

This next result of Parrott plays an important role in our presentation of Hankel operators (Chapter 17). If

$$
\begin{bmatrix} A & B \\ C & D \end{bmatrix} : \mathcal{H}_1 \oplus \mathcal{H}_2 \to \mathcal{H}_3 \oplus \mathcal{H}_4
$$

is a contraction, then the restrictions

$$
\begin{bmatrix} B \\ D \end{bmatrix} : \mathcal{H}_2 \to \mathcal{H}_3 \oplus \mathcal{H}_4 \quad \text{and} \quad \begin{bmatrix} C & D \end{bmatrix} : \mathcal{H}_1 \oplus \mathcal{H}_2 \to \mathcal{H}_4
$$

are contractions as well. In the light of the conditions above, suppose that B, C, D are given such that

$$
\begin{bmatrix} B \\ D \end{bmatrix} \in \mathcal{B}(\mathcal{H}_2, \mathcal{H}_3 \oplus \mathcal{H}_4) \quad \text{and} \quad \begin{bmatrix} C & D \end{bmatrix} \in \mathcal{B}(\mathcal{H}_1 \oplus \mathcal{H}_2, \mathcal{H}_4)
$$

are contractions. Is there an $A \in \mathcal{B}(\mathcal{H}_1, \mathcal{H}_3)$ such that

$$
\begin{bmatrix} A & B \\ C & D \end{bmatrix} \in \mathcal{B}(\mathcal{H}_1 \oplus \mathcal{H}_2, \mathcal{H}_3 \oplus \mathcal{H}_4)
$$

is a contraction? The affirmative answer is known as Parrott's theorem.

Theorem 14.8.1 (Parrott [261]). *Let*

$$\begin{bmatrix} B \\ D \end{bmatrix} \in \mathcal{B}(\mathcal{H}_2, \mathcal{H}_3 \oplus \mathcal{H}_4) \quad \text{and} \quad \begin{bmatrix} C & D \end{bmatrix} \in \mathcal{B}(\mathcal{H}_1 \oplus \mathcal{H}_2, \mathcal{H}_4)$$

be contractions. Then there is an $A \in \mathcal{B}(\mathcal{H}_1, \mathcal{H}_3)$ such that

$$\begin{bmatrix} A & B \\ C & D \end{bmatrix} \in \mathcal{B}(\mathcal{H}_1 \oplus \mathcal{H}_2, \mathcal{H}_3 \oplus \mathcal{H}_4)$$

is a contraction.

Proof Corollary 14.6.5 provides contractions $E \in \mathcal{B}(\mathcal{H}_2, \mathcal{H}_3)$ and $F \in \mathcal{B}(\mathcal{H}_1, \mathcal{H}_4)$ such that

$$B = E(I - D^*D)^{\frac{1}{2}} \quad \text{and} \quad C = (I - DD^*)^{\frac{1}{2}}F.$$

Define

$$X = \begin{bmatrix} 0 & E \\ F & 0 \end{bmatrix} \quad \text{and} \quad Y = \begin{bmatrix} 0 & 0 \\ 0 & -D \end{bmatrix}.$$

A computation shows that

$$I - X^*X = \begin{bmatrix} I - F^*F & 0 \\ 0 & I - E^*E \end{bmatrix} \tag{14.8.2}$$

and

$$I - Y^*Y = \begin{bmatrix} I & 0 \\ 0 & I - D^*D \end{bmatrix}. \tag{14.8.3}$$

Since E, F, and D are contractions, the operators on the diagonals in (14.8.2) and (14.8.3) are positive and hence X and Y are contractions. Moreover,

$$I - Y^*X = I - \begin{bmatrix} 0 & 0 \\ 0 & -D \end{bmatrix}^* \begin{bmatrix} 0 & E \\ F & 0 \end{bmatrix} = \begin{bmatrix} I & 0 \\ D^*F & I \end{bmatrix}.$$

Therefore, $I - Y^*X$ is invertible and

$$(I - Y^*X)^{-1} = \begin{bmatrix} I & 0 \\ -D^*F & I \end{bmatrix}.$$

Use Corollary 14.7.4 to define the contraction $f(X) \in \mathcal{B}(\mathcal{H}_1 \oplus \mathcal{H}_2, \mathcal{H}_3 \oplus \mathcal{H}_4)$, that is $f(X) = f_{J(Y)}(X)$. To explicitly calculate $f(X)$, observe that

$$(I - YY^*)^{\frac{1}{2}} = \begin{bmatrix} I & 0 \\ 0 & (I - DD^*)^{\frac{1}{2}} \end{bmatrix}$$

and

$$J(Y) = \begin{bmatrix} (I - YY^*)^{\frac{1}{2}} & Y \\ -Y^* & (I - Y^*Y)^{\frac{1}{2}} \end{bmatrix}$$

$$= \left[\begin{array}{cc|cc} I & 0 & 0 & 0 \\ 0 & (I - DD^*)^{\frac{1}{2}} & 0 & -D \\ \hline 0 & 0 & I & 0 \\ 0 & D^* & 0 & (I - D^*D)^{\frac{1}{2}} \end{array} \right].$$

Thus, calculating according to (14.7.5) by using the above $J(Y)$,

$$f(X) = Y - (I - YY^*)^{\frac{1}{2}} X (I - Y^*X)^{-1} (I - Y^*Y)^{\frac{1}{2}}$$

$$= \begin{bmatrix} 0 & 0 \\ 0 & -D \end{bmatrix} - \begin{bmatrix} I & 0 \\ 0 & (I - DD^*)^{\frac{1}{2}} \end{bmatrix} \begin{bmatrix} 0 & E \\ F & 0 \end{bmatrix} \begin{bmatrix} I & 0 \\ -D^*F & I \end{bmatrix} \begin{bmatrix} I & 0 \\ 0 & (I - D^*D)^{\frac{1}{2}} \end{bmatrix}$$

$$= \begin{bmatrix} -ED^*F & -B \\ -C & -D \end{bmatrix}.$$

Thus, $A = ED^*F$ solves Parrott's problem. ∎

14.9 Polar Decomposition

If z is a nonzero complex number, then z can be written in polar form as $z = u|z|$, where $u \in \mathbb{T}$ and $|z| = \sqrt{\bar{z}z}$. Furthermore, this representation is unique. There is an analogue of this factorization for Hilbert space operators. For $A \in \mathcal{B}(\mathcal{H})$, note that A^*A is a positive operator, and hence it has a unique positive square root (Theorem 8.6.4).

Definition 14.9.1. For $A \in \mathcal{B}(\mathcal{H})$, the *modulus* of A is $|A| = \sqrt{A^*A}$.

Example 14.9.2. For a diagonal operator D_Λ with eigenvalues $\Lambda = (\lambda_n)_{n=0}^{\infty}$, recall from Chapter 2 that $D_\Lambda^* = D_{\overline{\Lambda}}$, where $\overline{\Lambda} = (\overline{\lambda_n})_{n=0}^{\infty}$. Therefore,

$$|D_\Lambda| = \begin{bmatrix} |\lambda_0| & 0 & 0 & 0 & \cdots \\ 0 & |\lambda_1| & 0 & 0 & \cdots \\ 0 & 0 & |\lambda_2| & 0 & \cdots \\ 0 & 0 & 0 & |\lambda_3| & \cdots \\ \vdots & \vdots & \vdots & \vdots & \ddots \end{bmatrix}.$$

More generally, consider a compact selfadjoint operator $A \in \mathcal{B}(\mathcal{H})$. By the spectral theorem for compact selfadjoint operators (Theorem 2.6.7), $A = \sum_{n=0}^{\infty} \lambda_n (\mathbf{x}_n \otimes \mathbf{x}_n)$, where $(\lambda_n)_{n=0}^{\infty}$ is the sequence of eigenvalues of A, which are real and tend to zero, and $(\mathbf{x}_n)_{n=0}^{\infty}$ is the corresponding sequence of orthonormal eigenvectors. Then

$$|A| = \sum_{n=0}^{\infty} |\lambda_n|(\mathbf{x}_n \otimes \mathbf{x}_n). \tag{14.9.3}$$

Example 14.9.4. For the unilateral shift S on ℓ^2, one sees that $S^*S = I$, and hence $|S| = I$. Since $SS^* = I - e_0 \otimes e_0$,

$$|S^*| = \begin{bmatrix} 0 & 0 & 0 & 0 & \cdots \\ 0 & 1 & 0 & 0 & \cdots \\ 0 & 0 & 1 & 0 & \cdots \\ 0 & 0 & 0 & 1 & \cdots \\ \vdots & \vdots & \vdots & \vdots & \ddots \end{bmatrix}.$$

Example 14.9.5. For a multiplication operator M_φ on $L^2(\mu)$ (see Chapter 8), $M_\varphi^* = M_{\bar\varphi}$ so

$$|M_\varphi| = M_{|\varphi|}.$$

Example 14.9.6. From Chapter 7, the Volterra operator

$$(Vf)(x) = \int_0^x f(t)\,dt$$

on $L^2[0,1]$ has an orthonormal basis of eigenvectors for V^*V

$$f_n(x) = \sqrt{2}\cos\left(\frac{2n+1}{2}\pi x\right) \quad \text{for } n \geq 0,$$

with corresponding eigenvalues

$$\lambda_n = \frac{4}{(2n+1)^2\pi^2} \quad \text{for } n \geq 0$$

(Proposition 7.2.1). Moreover, V^*V is compact and $V^*V = \sum_{n=0}^{\infty} \lambda_n(f_n \otimes f_n)$. Thus, as in (14.9.3),

$$|V| = \sum_{n=0}^{\infty} \sqrt{\lambda_n}(f_n \otimes f_n).$$

Example 14.9.7. The Bishop operator

$$(T_\alpha f)(x) = xf(\{x + \alpha\})$$

on $L^2[0,1]$ (see Chapter 13) factors as $T_\alpha = M_x U_\alpha$, where $(M_x f)(x) = xf(x)$ and $U_\alpha f = f(\{x + \alpha\})$ is unitary. Then $|T_\alpha^*| = M_x$. Moreover, by Exercise 14.11.26, $|T_\alpha| = U_{-\alpha}M_x U_\alpha$.

A complex number is invertible if and only if it is nonzero. For an operator, there are issues with its kernel that need to be taken into account.

Definition 14.9.8. $A \in \mathcal{B}(\mathcal{H})$ is a *partial isometry* if $\|Ax\| = \|x\|$ for all $x \in (\ker A)^\perp$. The *initial space* of A is $(\ker A)^\perp$ and the *final space* of A is $\operatorname{ran} A$.

A partial isometry has closed range since it is isometric on $(\ker A)^\perp$. Indeed, if $\mathcal{H} = \ker A \oplus (\ker A)^\perp$, then $\|Ax\| = \|AP_{(\ker A)^\perp}x\| = \|P_{(\ker A)^\perp}x\|$. This says that if $(Ax_n)_{n=1}^{\infty}$ is a Cauchy sequence in $\operatorname{ran} A$, then $(P_{(\ker A)^\perp}x_n)_{n=1}^{\infty}$ is a Cauchy sequence in $(\ker A)^\perp$. Since $(\ker A)^\perp$ is closed, there is an $x \in \mathcal{H}$ such that $P_{(\ker A)^\perp}x_n \to P_{(\ker A)^\perp}x$. Then $Ax_n = AP_{(\ker A)^\perp}x_n \to AP_{(\ker A)^\perp}x = Ax$ which shows that $\operatorname{ran} A$ is closed.

Example 14.9.9. A unitary operator is a partial isometry with initial and final space equal to $\mathcal{H}$. In particular, a partial isometry is invertible if and only if it is unitary.

Example 14.9.10. An isometry $A \in \mathcal{B}(\mathcal{H})$ is a partial isometry with initial space $\mathcal{H}$ and final space ran A. In particular, the unilateral shift S on ℓ^2 is a partial isometry with initial space ℓ^2 and final space $\{e_0\}^\perp = \bigvee\{e_n : n \geqslant 1\}$.

Example 14.9.11. The backward shift S^* on ℓ^2 is a partial isometry with initial space $\bigvee\{e_n : n \geqslant 1\}$ and final space ℓ^2.

Example 14.9.12. If $\Lambda = (\lambda_n)_{n=0}^\infty$ with $|\lambda_j| \in \{0,1\}$, then D_Λ is a partial isometry with initial space and final space $\bigvee\{e_n : \lambda_n \neq 0\}$.

Example 14.9.13. If $\varphi \in L^\infty(\mu)$ and $|\varphi(z)| \in \{0,1\}$ μ-almost everywhere, then M_φ is a partial isometry on $L^2(\mu)$ with initial space and final space $\varphi L^2(\mu) = \chi_E L^2(\mu)$, where $E = \{|\varphi| = 1\}$.

Proposition 14.9.14. *For $A \in \mathcal{B}(\mathcal{H})$, the following are equivalent.*

(a) *A is a partial isometry.*

(b) *A^* is a partial isometry.*

(c) *$A = AA^*A$.*

(d) *$A^* = A^*AA^*$.*

(e) *A^*A is an orthogonal projection.*

(f) *AA^* is an orthogonal projection.*

Moreover, if A is a partial isometry, then A^*A is the orthogonal projection of $\mathcal{H}$ onto $(\ker A)^\perp$ and AA^* is the orthogonal projection of $\mathcal{H}$ onto ran A.

Proof (a) $\Rightarrow$ (c) For a selfadjoint operator $B \in \mathcal{B}(\mathcal{H})$, (2.6.6) says that $\langle Bx, x \rangle = 0$ for all $x \in \mathcal{H}$ if and only if $B = 0$. Apply this to $B = (I - A^*A)|_{(\ker A)^\perp}$ to see that if A is isometric on $(\ker A)^\perp$, then $A^*A = I$ on $(\ker A)^\perp$. Thus, $A = AA^*A$ on $\mathcal{H}$.

(c) $\Rightarrow$ (e) If $AA^*A = A$, then $(A^*A)^2 = A^*A$. Since A^*A is selfadjoint and idempotent, it is an orthogonal projection (Theorem 14.5.4).

(c) $\Rightarrow$ (a) Since $\ker A^*A = \ker A$ (Exercise 14.11.22), it follows that A^*A is the orthogonal projection onto $(\ker A)^\perp$. Indeed, if $y \in \ker A$, then $\langle A^*Ax, y \rangle = \langle Ax, Ay \rangle = 0$ for all $x \in \mathcal{H}$. Thus, $A^*A\mathcal{H} = (\ker A)^\perp$. If $x \in (\ker A)^\perp$, then $A^*Ax = x$, and hence

$$\|Ax\|^2 = \langle Ax, Ax \rangle = \langle A^*Ax, x \rangle = \langle x, x \rangle = \|x\|^2.$$

Thus, A is a partial isometry.

(e) $\Rightarrow$ (c) If A^*A is an orthogonal projection, then, as we have seen earlier, it is the orthogonal projection onto $(\ker A)^\perp$. For $x \in \ker A$, we have $Ax = 0 = AA^*Ax$. If $x \in (\ker A)^\perp$, then $x = A^*Ax$, and hence $Ax = A(A^*Ax) = (AA^*A)x$. Thus, $A = AA^*A$.

(c) $\Leftrightarrow$ (d) Take the adjoint of one of the equations to obtain the other.

(b) $\Leftrightarrow$ (d) Replace A with A^* in the proof of (a) $\Leftrightarrow$ (c)

(d) $\Leftrightarrow$ (f) Replace A with A^* in the proof of (c) $\Leftrightarrow$ (e) ■

With the notions of the analogues of the "modulus" and "argument" of an operator in hand, we are now ready to prove the polar decomposition theorem.

Theorem 14.9.15 (Polar Decomposition). *If $A \in \mathcal{B}(\mathcal{H})$, then there is a partial isometry $U \in \mathcal{B}(\mathcal{H})$ with initial space $(\ker A)^{\perp}$ and final space $(\operatorname{ran} A)^-$ such that $A = U|A|$. Furthermore, this decomposition is unique in the sense that if $A = WQ$, where $Q \geqslant 0$ and W is a partial isometry with $\ker W = \ker Q$, then $W = U$ and $Q = |A|$.*

Proof For any $\mathbf{x} \in \mathcal{H}$,

$$
\begin{aligned}
\|A\mathbf{x}\|^2 &= \langle A\mathbf{x}, A\mathbf{x} \rangle \\
&= \langle A^*A\mathbf{x}, \mathbf{x} \rangle \\
&= \langle |A|^2\mathbf{x}, \mathbf{x} \rangle && \text{(definition of $|A|$)} \\
&= \langle |A|\mathbf{x}, |A|\mathbf{x} \rangle && \text{($|A|$ is selfadjoint)} \\
&= \| |A|\mathbf{x} \|^2. && (14.9.16)
\end{aligned}
$$

Since $A^*A = |A||A|$, the Douglas factorization theorem (Theorem 14.6.1) produces a $C \in \mathcal{B}(\mathcal{H})$ such that $A^* = |A|C$. By (14.6.3) and (14.6.4),

$$
\ker C = \ker A^* \quad \text{and} \quad (\operatorname{ran} C)^- = (\ker |A|)^{\perp}. \qquad (14.9.17)
$$

Taking adjoints yields $A = C^*|A|$ with

$$
\begin{aligned}
\ker C^* &= (\operatorname{ran} C)^{\perp} && \text{(Proposition 3.1.7)} \\
&= \ker |A| && \text{(by (14.9.17))} \\
&= \ker A && \text{(by (14.9.16))}
\end{aligned}
$$

and, in a similar way,

$$
(\operatorname{ran} C^*)^- = (\ker C)^{\perp} = (\ker A^*)^{\perp} = (\operatorname{ran} A)^-.
$$

Furthermore, $A = C^*|A|$ together with (14.9.16), implies that $\|C^*|A|\mathbf{x}\| = \|A\mathbf{x}\| = \||A|\mathbf{x}\|$ and hence C^* is a partial isometry on

$$
(\operatorname{ran} |A|)^- = (\ker |A|)^{\perp} = (\ker A)^{\perp} = (\ker C^*)^{\perp}.
$$

Also note that $\operatorname{ran} C^*$ is closed and $\operatorname{ran} C^* = (\operatorname{ran} C^*)^- = (\operatorname{ran} A)^-$. Letting $U = C^*$ proves the existence of the polar decomposition.

If $A = WQ$, then $A^*A = QW^*WQ$. Proposition 14.9.14 implies that W^*W is the orthogonal projection onto $(\ker W)^{\perp} = (\ker Q)^{\perp} = (\operatorname{ran} Q)^-$. Thus, $A^*A = Q^2$. By the uniqueness of the positive square root (Theorem 8.6.4), $Q = |A|$. Since $U|A|\mathbf{x} = A\mathbf{x} = W|A|\mathbf{x}$ for all $\mathbf{x} \in \mathcal{H}$, one sees that W and U agree on $\operatorname{ran} |A|$. Therefore, $W = U$. ■

Example 14.9.18. If $\Lambda = (\lambda_n)_{n=0}^{\infty}$ is a bounded sequence, the polar decomposition of the diagonal operator D_Λ is

$$
D_\Lambda = \begin{bmatrix}
u_0 & 0 & 0 & 0 & \cdots \\
0 & u_1 & 0 & 0 & \cdots \\
0 & 0 & u_2 & 0 & \cdots \\
0 & 0 & 0 & u_3 & \cdots \\
\vdots & \vdots & \vdots & \vdots & \ddots
\end{bmatrix}
\begin{bmatrix}
|\lambda_0| & 0 & 0 & 0 & \cdots \\
0 & |\lambda_1| & 0 & 0 & \cdots \\
0 & 0 & |\lambda_2| & 0 & \cdots \\
0 & 0 & 0 & |\lambda_3| & \cdots \\
\vdots & \vdots & \vdots & \vdots & \ddots
\end{bmatrix},
$$

where

$$
u_j = \begin{cases}
e^{i \arg \lambda_j} & \text{if } \lambda_j \neq 0, \\
0 & \text{if } \lambda_j = 0.
\end{cases}
$$

Example 14.9.19. If $\varphi \in L^\infty(\mu)$, then the polar decomposition of M_φ on $L^2(\mu)$ is $M_{e^{i \arg \varphi}} M_{|\varphi|}$, where we interpret (as above with a diagonal operator) $e^{i \arg \varphi}$ to be zero when φ is zero.

Example 14.9.20. For the Volterra operator V, work with the representation of V with respect to the orthonormal basis $(f_n)_{n=0}^{\infty}$ from Example 14.9.6 to see that

$$
|V| = \begin{bmatrix}
\sqrt{\lambda_0} & 0 & 0 & 0 & 0 & \cdots \\
0 & \sqrt{\lambda_1} & 0 & 0 & 0 & \cdots \\
0 & 0 & \sqrt{\lambda_2} & 0 & 0 & \cdots \\
0 & 0 & 0 & \sqrt{\lambda_3} & 0 & \cdots \\
0 & 0 & 0 & 0 & \sqrt{\lambda_4} & \cdots \\
\vdots & \vdots & \vdots & \vdots & \vdots & \ddots
\end{bmatrix}.
$$

To obtain the partial isometric factor U, observe that $V f_n = U|V| f_n = \sqrt{\lambda_n} U f_n$ and hence

$$
U f_n = \frac{1}{\sqrt{\lambda_n}} V f_n = \sqrt{2} \sin\left(\frac{2n+1}{2}\pi x\right).
$$

The (m, n) entry of the matrix with respect to the orthonormal basis $(f_n)_{n=0}^{\infty}$ is

$$
\langle U f_n, f_m \rangle = \int_0^1 (U f_n)(x) \overline{f_m(x)}\, dx.
$$

Thus,

$$
U = \frac{1}{\pi}
\begin{bmatrix}
2 & 2 & \frac{2}{3} & \frac{2}{3} & \frac{2}{5} & \frac{2}{5} & \cdots \\
-2 & \frac{2}{3} & 2 & \frac{2}{5} & \frac{2}{3} & \frac{2}{7} & \cdots \\
\frac{2}{3} & -2 & \frac{2}{5} & 2 & \frac{2}{7} & \frac{2}{3} & \cdots \\
-\frac{2}{3} & \frac{2}{5} & -2 & \frac{2}{7} & 2 & \frac{2}{9} & \cdots \\
\frac{2}{5} & -\frac{2}{3} & \frac{2}{7} & -2 & \frac{2}{9} & 2 & \cdots \\
-\frac{2}{5} & \frac{2}{7} & -\frac{2}{3} & \frac{2}{9} & -2 & \frac{2}{11} & \cdots \\
\vdots & \vdots & \vdots & \vdots & \vdots & \vdots & \ddots
\end{bmatrix}.
$$

14.10 Notes

Corollary 14.6.6 says a bit more. Namely, $\operatorname{ran} A \subseteq \operatorname{ran} B$ if and only if there is a C such that $A = BC$. Moreover, C is the unique operator that satisfies the following:

(a) $\|C\|^2 = \inf\{\lambda > 0 : AA^* \leqslant \lambda BB^*\}$,

(b) $\ker A = \ker C$, and

(c) $(\operatorname{ran} C)^- \subseteq (\operatorname{ran} B^*)^-$.

This theorem plays an important role in defining de Branges–Rovnyak spaces, which are used as model spaces for certain contractions [134, 135].

The Julia lemma (Lemma 14.7.3) relates a contraction to a unitary operator on a larger space. This result was significantly extended by Sz.-Nagy [355] in a result known as the Sz.-Nagy dilation theorem. This result says that if B is a contraction on a Hilbert space $\mathcal{H}$, there is a Hilbert space $\mathcal{K}$ containing $\mathcal{H}$ and a unitary operator U on $\mathcal{K}$ such that $P_{\mathcal{H}} U^n |_{\mathcal{H}} = B^n$ for all $n \geqslant 1$.

Partial isometries on finite-dimensional spaces, often called partial isometric matrices, are well studied and much is known about them. In fact, they have a complete description: $A \in M_n$ with rank r is a partial isometric matrix if and only if $A = U(I_r \oplus 0_{n-r})V$ for some unitary matrices $U, V \in M_n$. Furthermore, A is a partial isometric matrix if and only if $A = WP$, where $P \in M_n$ is an orthogonal projection and $W \in M_n$ is unitary. There is also a unitary invariant for partial isometric matrices and a beautiful theory of the numerical range of a partial isometric matrix. See [145] for a survey of all of this.

14.11 Exercises

Exercise 14.11.1. If $\mathcal{H}_1$ and $\mathcal{H}_2$ are Hilbert spaces, prove that the inner product space $\mathcal{H}_1 \oplus \mathcal{H}_2$ defined in (14.1.1) is complete, and hence a Hilbert space.

Exercise 14.11.2. If $\mathcal{H}$ is a Hilbert space, prove that $\mathcal{H}^{(\infty)}$, as defined in (14.1.2), is a Hilbert space.

Exercise 14.11.3. Expand the right side of (14.2.9) and prove that it yields the right side of (14.2.8).

Exercise 14.11.4. Suppose that $P_1, P_2, \ldots, P_r \in \mathcal{B}(\mathcal{H})$ are orthogonal projections such that $P_1 + P_2 + \cdots + P_r = I$. Prove that $P_i P_j = 0$ if $i \neq j$ and conclude that the ranges of $P_1, P_2, \ldots, P_r$ are pairwise orthogonal.

Exercise 14.11.5. Prove that $n \times n$ block-operator notation (14.2.11) respects adjoints and operator composition.

Exercise 14.11.6. Prove the formula (14.4.2) for the inverse of a 2×2 block operator whose diagonal blocks are invertible.

Exercise 14.11.7. Prove (14.4.4) and explain how it relates to Lemma 14.7.3.

Exercise 14.11.8. Let $\mathcal{H} = \mathcal{H}_1 \oplus \mathcal{H}_2 \oplus \cdots \oplus \mathcal{H}_n$, in which $\mathcal{H}_1, \mathcal{H}_2,..., \mathcal{H}_n$ are finite dimensional. Prove that if $T = [T_{ij}]_{i,j=1}^n \in \mathcal{B}(\mathcal{H})$ is an upper-triangular block operator that is unitary, then T is block diagonal.
Remark: Example 14.4.3 shows that this can fail if the spaces involved are infinite dimensional.

Exercise 14.11.9. Let

$$T = \begin{bmatrix} A & B \\ C & D \end{bmatrix}.$$

Prove that $\|T\| \geqslant \max\{\|A\|, \|B\|, \|C\|, \|D\|\}$.

Exercise 14.11.10. Let

$$T = \begin{bmatrix} A & B \\ C & D \end{bmatrix}.$$

Prove that

$$\|T\| \leqslant \left\| \begin{bmatrix} \|A\| & \|B\| \\ \|C\| & \|D\| \end{bmatrix} \right\|.$$

Exercise 14.11.11. Let

$$T = \begin{bmatrix} A & 0 \\ 0 & 0 \end{bmatrix}.$$

Prove that $\|T\| = \|A\|$.

Exercise 14.11.12. Suppose that

$$T = \begin{bmatrix} A & B \\ 0 & D \end{bmatrix}$$

is invertible. Prove that A is invertible if and only if D is invertible.
Remark: Example 14.4.3 shows that in the infinite-dimensional setting T may be invertible while neither A nor D is invertible.

Exercise 14.11.13. Prove that if A and B are operators on finite-dimensional Hilbert spaces, then $\sigma(T) = \sigma(A) \cup \sigma(B)$ holds in Theorem 14.4.5.

Exercise 14.11.14. Prove that for $A \in \mathcal{B}(\mathcal{H})$, the following are equivalent.

(a) A is an orthogonal projection.

(b) $A = A^*$ and $\sigma(A) \subseteq \{0, 1\}$.

(c) A is selfadjoint and idempotent.

Exercise 14.11.15. Prove that if $A \in \mathcal{B}(\mathcal{H})$ is idempotent and $\|Ax\| \leqslant \|x\|$ for all $x \in \mathcal{H}$, then A is an orthogonal projection.

Exercise 14.11.16. Let $A, B \in \mathcal{B}(\mathcal{H})$ be idempotent.

(a) Prove that $A + B$ is idempotent if and only if $AB = BA = 0$.

(b) Prove that $A - B$ is idempotent if and only if $AB = BA = B$.

Exercise 14.11.17. Let $A, B \in \mathcal{B}(\mathcal{H})$ be idempotent and $AB = BA$.

(a) Prove that $C = AB$ is idempotent.

(b) Prove that $\ker C = \ker A + \ker B$.

(c) Prove that $\operatorname{ran} C = \operatorname{ran} A \cap \operatorname{ran} B$.

Exercise 14.11.18. If $A \in \mathcal{B}(\mathcal{H}, \mathcal{K})$ and $B \in \mathcal{B}(\mathcal{K}, \mathcal{H})$, prove that

$$\sigma(AB) \cup \{0\} = \sigma(BA) \cup \{0\}.$$

Exercise 14.11.19. Let $\dim \mathcal{H} \geqslant 2$. Prove that

$$\{\|T\| : T \in \mathcal{B}(\mathcal{H}) \text{ is idempotent}\} = \{0\} \cup [1, \infty).$$

Exercise 14.11.20. Prove that $A \in \mathcal{B}(\mathcal{H})$ is a contraction if and only if $A^*A \leqslant I$.

Exercise 14.11.21. Prove that if $A \in \mathcal{B}(\mathcal{H})$ is a contraction, then there is a Hilbert space $\mathcal{K}$ containing $\mathcal{H}$ and a unitary operator U on $\mathcal{K}$ such that $P_{\mathcal{H}} U|_{\mathcal{H}} = A$.

Exercise 14.11.22. For $A \in \mathcal{B}(\mathcal{H})$, prove that $\ker A = \ker A^*A$.

Exercise 14.11.23. Let $A \in \mathcal{B}(\mathcal{H})$ be a contraction.

(a) Prove that the operator

$$M(A) = \begin{bmatrix} A & (I - AA^*)^{\frac{1}{2}} \\ 0 & 0 \end{bmatrix}$$

on $\mathcal{H} \oplus \mathcal{H}$ is a partial isometry.

(b) Identify the initial and final spaces of $M(A)$.

(c) Prove that if A is unitarily equivalent to B, then $M(A)$ is unitarily equivalent to $M(B)$.

Remark: See [171] for more on this.

Exercise 14.11.24. Find the polar decomposition of the backward shift S^*.

Exercise 14.11.25. For $A \in \mathcal{B}(\mathcal{H})$, let $A = U_A|A|$ and $A^* = U_{A^*}|A^*|$ be the polar decompositions of A and A^*. Prove the following.

(a) $|A| = U_A^* A$.

(b) $A = |A^*| U_A$.

(c) $U_{A^*} = U_A^*$.

Exercise 14.11.26. Let $T_\alpha = M_x U_\alpha$ be the Bishop operator from Example 14.9.7.

(a) Prove that $|T_\alpha| = U_\alpha^* M_x U_\alpha$.

(b) What is the polar decomposition of T_α?

Exercise 14.11.27. Use the polar decomposition to prove that if $A \in \mathcal{B}(\mathcal{H})$ and $\ker A = \ker A^* = \{0\}$, then A^*A is unitarily equivalent to AA^*.

Exercise 14.11.28. For the Volterra operator V from Chapter 7, compute the polar decomposition of VV^*.

Exercise 14.11.29. Let P be a positive invertible operator with $\|P\| \leqslant 1$.

(a) Prove that $P \pm i(I - P^2)^{\frac{1}{2}}$ is unitary.

(b) Prove that P is the average of two unitary operators.

(c) Prove that each invertible contraction can be written as the average of two unitary operators.

Exercise 14.11.30.

(a) If $T \in \mathcal{B}(\mathcal{H})$, prove that $\sum_{n=1}^{\infty} \langle |T|\mathbf{x}_n, \mathbf{x}_n \rangle$ is independent of the choice of orthonormal basis $(\mathbf{x}_n)_{n=1}^{\infty}$ for $\mathcal{H}$.

(b) $T \in \mathcal{B}(\mathcal{H})$ is a *trace-class operator* if $\|T\|_{\mathrm{tr}} := \sum_{n=1}^{\infty} \langle |T|\mathbf{x}_n, \mathbf{x}_n \rangle$ is finite. Prove that the set of trace-class operators is a normed vector space with respect to $\| \cdot \|_{\mathrm{tr}}$.

(c) When is a diagonal operator a trace-class operator?

(d) Prove that $\|T\| \leqslant \|T\|_{\mathrm{tr}}$.

(e) Recall the Hilbert–Schmidt operators from Exercise 3.6.31. Prove that T is a trace-class operator if and only if $|T|^{\frac{1}{2}}$ is a Hilbert–Schmidt operator.

Exercise 14.11.31. Show that every trace-class operator is compact but not every compact operator is trace-class.

Exercise 14.11.32. This is a continuation of Exercise 14.11.30. Let $A \in \mathcal{B}(\mathcal{H})$ and let $T \in \mathcal{B}(\mathcal{H})$ be a trace-class operator. Prove the following.

(a) $\|T\|_{\mathrm{tr}} = \|T^*\|_{\mathrm{tr}}$.

(b) $\|AT\|_{\mathrm{tr}} \leqslant \|T\|_{\mathrm{tr}} \|A\|$.

(c) $\|TA\|_{\mathrm{tr}} \leqslant \|T\|_{\mathrm{tr}} \|A\|$.

(d) The trace-class operators form a two-sided ideal of compact operators in $\mathcal{B}(\mathcal{H})$.

Exercise 14.11.33. Let A, X, Y be $n \times n$ matrices. Prove that

$$\|XAX^* + YAY^*\| \leqslant \|A\| \, \|XX^* + YY^*\|.$$

Remark: Direct verification of this inequality is feasible. However, to appreciate the advantage of considering matrices of operators, try to derive the inequality above from the inequality $\|ZBZ^*\| \leqslant \|Z\| \, \|B\| \|Z^*\| = \|B\| \, \|Z\|^2 = \|B\| \, \|ZZ^*\|$, with appropriate choices of B and Z.

Exercise 14.11.34. Let $\mathcal{H}$ be an infinite-dimensional Hilbert space and let $T \in \mathcal{B}(\mathcal{H})$.

(a) If T is compact, normal, and $T^2 = 0$, prove that $T = 0$.

(b) Let $A \in \mathcal{B}(\mathcal{H})$ be a noncompact operator and define $T \in \mathcal{B}(\mathcal{H} \oplus \mathcal{H})$ by

$$T = \begin{bmatrix} 0 & A \\ 0 & 0 \end{bmatrix}.$$

Verify that $T^2 = 0$.

Exercise 14.11.35. Let $T_1, T_2, ..., T_n$ belong to $\mathcal{B}(\mathcal{H})$ such that $T_i T_j = 0$ for $i \neq j$. Prove that $\sigma\left(\sum_{i=1}^{n} T_i\right) \subseteq \bigcup_{i=1}^{n} \sigma(T_i)$.

Exercise 14.11.36. This is a continuation of Exercise 14.11.35. Let $T_1, T_2, ..., T_n$ belong to $B(\mathcal{H})$ such that $T_i T_j = 0$ for $i \neq j$. Prove that $\sigma\left(\sum_{i=1}^{n} T_i\right) \backslash \{0\} = \left(\bigcup_{i=1}^{n} \sigma(T_i)\right) \backslash \{0\}$.

14.12 Hints for the Exercises

Hint for Ex. 14.11.3: P and $I - P$ are idempotent.

Hint for Ex. 14.11.4: Show that $P_i P_j P_i$ is a positive operator and use the equation $P_i = P_i I P_i$.

Hint for Ex. 14.11.7: $I - SS^*$ is an orthogonal projection.

Hint for Ex. 14.11.12: Write a potential inverse of T as a 2×2 block operator and examine its entries.

Hint for Ex. 14.11.14: If A is an idempotent, make use of the formula

$$(\lambda I - A)^{-1} = \frac{1}{\lambda} I + \frac{1}{\lambda(\lambda - 1)} A \quad \text{for all } \lambda \neq 0, 1.$$

Hint for Ex. 14.11.15: Start with the direct-sum decomposition $\mathcal{H} = \ker P + \operatorname{ran} P$ and show that $\ker P \subseteq (\operatorname{ran} P)^{\perp}$. Use the fact that $\mathbf{v} \perp \mathbf{w}$ if and only if $\|\mathbf{w}\| \leq \|c\mathbf{v} + \mathbf{w}\|$ for all $c \in \mathbb{C}$.

Hint for Ex. 14.11.18: Mimic the proof of Theorem 14.4.8.

Hint for Ex. 14.11.19: Consider 2×2 matrices.

Hint for Ex. 14.11.21: Consider the Julia operator for $-B^*$ and an operator permutation matrix.

Hint for Ex. 14.11.28: Use Exercise 7.7.16.

Hint for Ex. 14.11.29: For (c), suppose that $(T^*T)^{\frac{1}{2}} = P$.

Hint for Ex. 14.11.30: For (a), prove that for any two orthonormal bases $(\mathbf{u}_n)_{n=1}^{\infty}$ and $(\mathbf{v}_n)_{n=1}^{\infty}$ for $\mathcal{H}$,

$$\sum_{n=1}^{\infty} \|T\mathbf{u}_n\|^2 = \sum_{n=1}^{\infty} \|T^*\mathbf{v}_n\|^2 = \sum_{n=1}^{\infty} \sum_{m=1}^{\infty} |\langle T\mathbf{u}_n, \mathbf{v}_m\rangle|^2.$$

Hint for Ex. 14.11.31: Consult Exercise 14.11.30 and Theorem 2.5.1.

Hint for Ex. 14.11.32: For (a), consult Exercise 14.11.25. For (b), use the polar decomposition and the spectral theorem for positive compact operators.

Hint for Ex. 14.11.33: Consider

$$B = \begin{bmatrix} A & 0 \\ 0 & A \end{bmatrix} \quad \text{and} \quad Z = \begin{bmatrix} X & Y \\ 0 & 0 \end{bmatrix}.$$

Hint for Ex. 14.11.34: For part (a), consider the spectral theorem.

Constructions with the Shift Operator

Key Concepts: von Neumann–Wold decomposition of an isometry, spectral representation of $S + S^*$, properties of $S \oplus S^*$, tensor product of operators, properties of $S \otimes S^*$.

Outline: This chapter explores three operators created from the unilateral shift S and its adjoint S^*: the sum $S + S^*$, the direct sum $S \oplus S^*$, and the tensor product $S \otimes S^*$. We give Hilbert's spectral representation of the selfadjoint operator $S + S^*$. The operator $S \oplus S^*$ on $H^2 \oplus H^2$ is complex symmetric. We examine its invariant subspaces and spectral properties. The operator $S \otimes S^*$ on $H^2 \otimes H^2$ is also complex symmetric. We discuss its reducing subspaces and spectral properties.

15.1 The von Neumann–Wold Decomposition

When studying an invariant subspace $\mathcal{M}$ of the unilateral shift S on H^2, Beurling examined the *wandering subspace* $\mathcal{M} \cap (S\mathcal{M})^\perp$. Wandering subspaces play an important role in studying isometries on general Hilbert spaces $\mathcal{H}$. We follow the presentation from [253].

Theorem 15.1.1 (von Neumann–Wold Decomposition). *Let $T \in \mathcal{B}(\mathcal{H})$ be an isometry and $W = (T\mathcal{H})^\perp$.*

(a) $T^m W \perp T^n W$ *for all $m, n \geqslant 0$ with $m \neq n$.*

(b) *If $W_\infty = \bigcap_{n \geqslant 0} T^n \mathcal{H}$, then $TW_\infty = W_\infty$ and $T|_{W_\infty}$ is unitary.*

(c) *If $W_0 = \bigoplus_{n \geqslant 0} T^n W$, then $TW_0 \subseteq W_0$ and there is no nonzero invariant subspace of W_0 upon which the restriction of T is unitary.*

(d) $\mathcal{H} = W_0 \oplus W_\infty$.

Proof (a) Since T is an isometry, $T^*T = I$. Iterate this identity and obtain

$$T^{*m}T^n = T^{n-m} \quad \text{for all } n \geqslant m \geqslant 0. \qquad (15.1.2)$$

Then for any $\mathbf{x}, \mathbf{y} \in W$ and $n > m \geqslant 0$,

$$\langle T^n\mathbf{x}, T^m\mathbf{y}\rangle = \langle T^{*m}T^n\mathbf{x}, \mathbf{y}\rangle = \langle T^{n-m}\mathbf{x}, \mathbf{y}\rangle,$$

which equals zero since $T^{n-m}\mathbf{x} \in T\mathcal{H}$ and $\mathbf{y} \in W = (T\mathcal{H})^\perp$.

(b) Let $\mathbf{x} \in W_\infty$. The definition of W_∞ ensures that for each $n \geqslant 0$, there is an $\mathbf{x}_n \in \mathcal{H}$ such that $\mathbf{x} = T^n\mathbf{x}_n$. Thus, $T\mathbf{x} = T^{n+1}\mathbf{x}_n$, so $T\mathbf{x} \in \bigcap_{n \geqslant 0} T^n\mathcal{H} = W_\infty$. It follows that $TW_\infty \subseteq W_\infty$. The next step is to verify the reverse inclusion. The identity $\mathbf{x} = T^n\mathbf{x}_n$ says that $T^n\mathbf{x}_n = T^{n+k}\mathbf{x}_{n+k}$ for all $k, n \geqslant 0$. Now use (15.1.2) and conclude $\mathbf{x}_n = T^k\mathbf{x}_{n+k}$ for all $k \geqslant 0$, and hence $\mathbf{x}_n \in W_\infty$. Therefore, $\mathbf{x} = T^n\mathbf{x}_n$, so $\mathbf{x} \in TW_\infty$. This implies $W_\infty \subseteq TW_\infty$ and hence, $TW_\infty = W_\infty$. Since T maps W_∞ isometrically onto W_∞, it follows that $T|_{W_\infty}$ is unitary.

(c) Observe that

$$W_0 = \left\{ \mathbf{x} = \sum_{n=0}^\infty T^n\mathbf{x}_n \; : \; \mathbf{x}_n \in W, \; \sum_{n=0}^\infty \|T^n\mathbf{x}_n\|^2 = \sum_{n=0}^\infty \|\mathbf{x}_n\|^2 < \infty \right\}.$$

It follows that $TW_0 \subseteq W_0$ and $\bigcap_{n=0}^\infty T^n W_0 = \{\mathbf{0}\}$. If $W_0' \subseteq W_0$ is an invariant subspace for T such that $T|_{W_0'}$ is unitary, then

$$W_0' = \bigcap_{n=0}^\infty T^n W_0' \subseteq \bigcap_{n=0}^\infty T^n W_0 = \{\mathbf{0}\}.$$

(d) If $\mathbf{y}_\infty \in W_\infty$ and $\mathbf{y}_0 \in W_0$, then for each $n \geqslant 1$ there exists a $\mathbf{z}_n \in W$ such that $\mathbf{y}_\infty = T^n\mathbf{z}_n$. Moreover, there are $\mathbf{x}_n \in W$ such that $\mathbf{y}_0 = \sum_{n=0}^\infty T^n\mathbf{x}_n$. Thus,

$$\langle \mathbf{y}_\infty, \mathbf{y}_0 \rangle = \left\langle \mathbf{y}_\infty, \sum_{n=0}^\infty T^n\mathbf{x}_n \right\rangle = \sum_{n=0}^\infty \langle \mathbf{y}_\infty, T^n\mathbf{x}_n \rangle = \sum_{n=0}^\infty \langle T^{n+1}\mathbf{z}_{n+1}, T^n\mathbf{x}_n \rangle = 0$$

since $T^m W \perp T^n W$ for all $m \neq n$. So far, we know that $W_0 \perp W_\infty$ and $W_0 \oplus W_\infty \subseteq \mathcal{H}$. To show the reverse inclusion, suppose that $\mathbf{x} \in W_0^\perp$. Then

$$\mathbf{x} \perp \bigoplus_{n=0}^\infty T^n W,$$

hence $\mathbf{x} \perp T^n W$ for all $n \geqslant 0$. Since $W = (T\mathcal{H})^\perp$, it follows from Exercise 15.6.1 that $\mathbf{x} \perp T^n\mathcal{H} \cap (T^{n+1}\mathcal{H})^\perp$ for all $n \geqslant 0$, so $\mathbf{x} \in W_\infty$. A similar argument shows that if $\mathbf{x} \in W_\infty^\perp$, then $\mathbf{x} \in W_0$. ∎

Example 15.1.3. Let S denote the unilateral shift on H^2. As discussed in Chapter 5, S is an isometry. Moreover, $H^2 = \bigvee\{z^n : n \geqslant 0\}$ and hence

$$(SH^2)^\perp = \{f \in H^2 : \langle f, Sz^n \rangle = 0 \text{ for all } n \geqslant 0\}$$
$$= \{f \in H^2 : \langle f, z^n \rangle = 0 \text{ for all } n \geqslant 1\}$$
$$= \{f \in H^2 : \hat{f}(n) = 0 \text{ for all } n \geqslant 1\}$$
$$= \text{span}\{1\},$$

where 1 denotes the constant function. Thus, $W = (SH^2)^\perp = \text{span}\{1\}$. One also notes that

$$W_\infty = \bigcap_{n=0}^{\infty} S^n H^2 = \{0\},$$

since any $f \in W_\infty$ is an analytic function on $\mathbb{D}$ with a zero of infinite order at $z = 0$, and hence is identically equal to zero. Finally, recall the definition of an infinite direct sum of Hilbert spaces from (14.1.2) to see that

$$
\begin{aligned}
W_0 &= \text{span}\{1\} \oplus S\,\text{span}\{1\} \oplus S^2\,\text{span}\{1\} \oplus \cdots \\
&= \text{span}\{1\} \oplus \text{span}\{z\} \oplus \text{span}\{z^2\} \oplus \cdots \\
&= H^2.
\end{aligned}
$$

Example 15.1.4. Suppose that q is an inner function and $\mathcal{H} = qH^2$, one of the so-called Beurling subspaces of H^2 discussed in Chapter 5. Since $\mathcal{H}$ is an invariant subspace of S, it follows that $T = S|_{\mathcal{H}} \in \mathcal{B}(\mathcal{H})$ is an isometry. Observe that

$$W = qH^2 \ominus TqH^2 = \{qf : f \in H^2, \langle qf, z^n q \rangle = 0 \text{ for all } n \geq 1\}.$$

The fact that q is inner implies $q\bar{q} = 1$ almost everywhere on $\mathbb{T}$ and hence

$$\langle qf, qz^n \rangle = \int_{\mathbb{T}} q(\xi) f(\xi) \overline{q(\xi)\xi^n} dm(\xi) = \int_{\mathbb{T}} f(\xi) \bar{\xi}^n \, dm(\xi) = \hat{f}(n).$$

Thus, $W = \{qf : f \in H^2, \hat{f}(n) = 0 \text{ for all } n \geq 1\} = \text{span}\{q\}$. For the same reason as in Example 15.1.3, $W_\infty = \bigcap_{n=0}^{\infty} T^n qH^2 = \{0\}$. Finally,

$$
\begin{aligned}
W_0 &= \text{span}\{q\} \oplus \text{span}\{Tq\} \oplus \text{span}\{T^2 q\} \oplus \text{span}\{T^3 q\} \oplus \cdots \\
&= \text{span}\{q\} \oplus z\,\text{span}\{q\} \oplus z^2\,\text{span}\{q\} \oplus z^3\,\text{span}\{q\} \oplus \cdots \\
&= qH^2.
\end{aligned}
$$

Example 15.1.5. Consider the isometry $T = S^2$ on H^2. Following the approach of Example 15.1.3, we deduce that

$$
\begin{aligned}
W &= H^2 \ominus TH^2 \\
&= \{f \in H^2 : \langle f, S^2 z^n \rangle = 0 \text{ for all } n \geq 0\} \\
&= \{f \in H^2 : \langle f, z^n \rangle = 0 \text{ for all } n \geq 2\} \\
&= \{f \in H^2 : \hat{f}(n) = 0 \text{ for all } n \geq 2\} \\
&= \text{span}\{1, z\}.
\end{aligned}
$$

Observe that $W_\infty = \{0\}$ and

$$W_0 = \text{span}\{1, z\} \oplus \text{span}\{z^2, z^3\} \oplus \text{span}\{z^4, z^5\} \oplus \cdots = H^2.$$

Example 15.1.6. Define $T \in \mathcal{B}(H^2)$ by

$$(Tf)(z) = \widehat{f}(0) + z(f - \widehat{f}(0)).$$

Since $f(0) \perp z(f - f(0))$, it follows that

$$\begin{aligned}
\|Tf\|^2 &= |\widehat{f}(0)|^2 + \|z(f - \widehat{f}(0))\|^2 \\
&= |\widehat{f}(0)|^2 + \|f - \widehat{f}(0)\|^2 \\
&= |\widehat{f}(0)|^2 + \sum_{n=1}^{\infty} |\widehat{f}(n)|^2 \\
&= \|f\|^2.
\end{aligned}$$

Thus, T is an isometry. Observe that $W_\infty = \bigcap_{n=0}^{\infty} T^n H^2 = \operatorname{span}\{1\}$ and that $T|_{W_0}$ is unitary. Moreover,

$$W = (TH^2)^\perp = \left(\bigvee\{1, z^2, z^3, z^4, \ldots\}\right)^\perp = \operatorname{span}\{z\},$$

$T^n W = \operatorname{span}\{z^{n+1}\}$, and $W_0 = \bigoplus_{n=1}^{\infty} \operatorname{span}\{z^n\}$. Finally, $W_0 \oplus W_\infty = H^2$.

Example 15.1.7. For $\zeta \in \mathbb{T}$, define $T_\zeta \in \mathcal{B}(H^2)$ by

$$(T_\zeta f)(z) = \frac{1}{2}z^2(f(z) + f(-z)) + \frac{1}{2}(f(\zeta z) - f(-\zeta z)).$$

Since $T_\zeta z^{2n} = z^{2n+2}$ and $T_\zeta z^{2n+1} = \zeta^{2n+1} z^{2n+1}$ for $n \geqslant 0$, it follows that T_ζ is an isometry. Moreover,

$$W_\infty = \bigcap_{n=0}^{\infty} T_\zeta^n H^2 = \bigvee\{z^{2n+1} : n \geqslant 0\},$$

the subspace of odd functions in H^2. Observe that $T_\zeta|_{W_\infty}$ is unitary, since it is the composition operator $g(z) \mapsto g(\zeta z)$ on the subspace of odd functions in H^2. One can also see that

$$W = (T_\zeta H^2)^\perp = \left(\bigvee\{z^n : n \geqslant 1\}\right)^\perp = \operatorname{span}\{1\}$$

and $W_0 = \bigoplus_{n=0}^{\infty} \operatorname{span}\{z^{2n}\}$, the subspace of even functions in H^2. This yields the orthogonal decomposition $W_0 \oplus W_\infty = H^2$.

Definition 15.1.8. An isometry T is *pure* if $W_\infty = \{0\}$.

The following corollary of the von Neumann–Wold theorem shows that for pure isometries, every T-invariant subspace has the *wandering subspace property*.

Corollary 15.1.9. *If $T \in \mathcal{B}(\mathcal{H})$ is a pure isometry, then any T-invariant subspace $\mathcal{M}$ satisfies*

$$\mathcal{M} = \bigvee\{T^n \mathbf{x} : \mathbf{x} \in \mathcal{M} \cap (T\mathcal{M})^\perp\}.$$

The corollary above is reminiscent of Beurling's theorem (Theorem 5.4.12) and it says that any invariant subspace $\mathcal{M}$ of T is generated by its wandering subspace $\mathcal{M} \cap (T\mathcal{M})^\perp$.

15.2 The Sum of S and S^*

For historical reasons, we consider

$$T = \frac{1}{2}(S + S^*)$$

instead of $S + S^*$. From (5.1.3) and (5.2.3), the matrix representation of T with respect to the orthonormal basis $(z^n)_{n=0}^{\infty}$ for H^2 is the selfadjoint Toeplitz matrix

$$
\begin{bmatrix}
0 & \frac{1}{2} & 0 & 0 & 0 & \cdots \\
\frac{1}{2} & 0 & \frac{1}{2} & 0 & 0 & \cdots \\
0 & \frac{1}{2} & 0 & \frac{1}{2} & 0 & \cdots \\
0 & 0 & \frac{1}{2} & 0 & \frac{1}{2} & \cdots \\
0 & 0 & 0 & \frac{1}{2} & 0 & \cdots \\
\vdots & \vdots & \vdots & \vdots & \vdots & \ddots
\end{bmatrix}.
$$

This is the Toeplitz operator with symbol $\cos\theta$. Toeplitz operators appear in Chapter 16.

Proposition 15.2.1. $\|T\| = 1$.

Proof Since $\|S\| = \|S^*\| = 1$ (by (5.1.2)), it follows that $\|S + S^*\| \leqslant 2$ and hence $\|T\| \leqslant 1$. To prove the reverse inequality, consider the unit vector

$$f_n(z) = \frac{1}{\sqrt{n}} \sum_{j=1}^{n} z^j$$

in H^2. Then

$$(S + S^*)f_n = \frac{1}{\sqrt{n}} \sum_{j=1}^{n} z^{j+1} + \frac{1}{\sqrt{n}} \sum_{j=1}^{n} z^{j-1} = \frac{1}{\sqrt{n}}\left(1 + z + 2\left(\sum_{j=2}^{n-1} z^j\right) + z^n + z^{n+1}\right),$$

and so

$$\|(S + S^*)f_n\|^2 = \frac{1}{n}\left(1^2 + 1^2 + 2^2(n-2) + 1^2 + 1^2\right) = \frac{4 + 4(n-2)}{n} \to 4.$$

Thus,

$$\|S + S^*\| = \sup_{\|f\|=1} \|(S + S^*)f\| \geqslant \lim_{n\to\infty} \|(S + S^*)f_n\| = 2$$

and hence $\|T\| = 1$. ∎

An exercise (Exercise 15.6.3) shows that T is cyclic. Thus, the spectral theorem for selfadjoint operators (Theorem 8.7.1) ensures that T is unitarily equivalent to M_x on $L^2(\mu)$ for some finite positive compactly supported Borel measure on $\mathbb{R}$. A 1912 result of Hilbert explicitly computes this measure and diagonalizes T [197]. We follow the treatment from [313, Ch. 3].

Theorem 15.2.2. *For the operator $T = \frac{1}{2}(S + S^*)$, the following hold.*

(a) $\sigma(T) = [-1, 1]$.

(b) $\sigma_p(T) = \emptyset$.

(c) *Let $\rho(x) = \sqrt{1 - x^2}$ for $-1 \leqslant x \leqslant 1$, and let $L^2(\rho)$ denote the space $L^2(\rho(x)\,dx)$. Then the operator $V : L^2(\rho) \to H^2$ defined by*

$$(Vf)(z) = \sqrt{\frac{2}{\pi}} \int_{-1}^{1} \frac{f(x)}{1 - 2xz + z^2} \rho(x)\,dx \quad \text{for } z \in \mathbb{D},$$

is unitary and

$$V^*TV = M_x, \tag{15.2.3}$$

where $(M_x f)(x) = xf(x)$ on $L^2(\rho)$.

One can visualize this theorem with the following commutative diagram:

$$
\begin{array}{ccc}
L^2(\rho) & \xrightarrow{\ M_x\ } & L^2(\rho) \\
\Big\downarrow{\scriptstyle V} & & \Big\uparrow{\scriptstyle V^*} \\
H^2 & \xrightarrow[\ T\ ]{} & H^2
\end{array}
$$

Proof We first show that V is a unitary operator from $L^2(\rho)$ onto H^2. Use the identity

$$\frac{1}{1 - 2x\bar{\lambda} + \bar{\lambda}^2} = \sum_{n=0}^{\infty} u_n(x)\bar{\lambda}^{-n}, \tag{15.2.4}$$

where $u_n(x)$ is the nth *Chebyshev polynomial* of the second kind [357]. The first few of these polynomials are

$$u_0(x) = 1, \quad u_1(x) = 2x, \quad u_2(x) = 4x^2 - 1, \quad u_3(x) = 8x^3 - 4x,$$

and they satisfy the recurrence

$$u_{n+1}(x) = 2xu_n(x) - u_{n-1}(x) \quad \text{for all } n \geqslant 1.$$

An induction argument with this recurrence implies that

$$\deg u_n = n \quad \text{for all } n \geqslant 0. \tag{15.2.5}$$

It is also known that

$$u_n(\cos\theta) = \frac{\sin((n+1)\theta)}{\sin\theta},$$

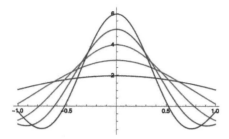

Figure 15.2.1 The graphs of $\sin((n+1)x)/\sin x$ for $n = 1, 2, 3, 4, 5$.

which yields

$$|u_n(x)| \leqslant n+1 \quad \text{for all } x \in [-1, 1]$$

(see Figure 15.2.1). For fixed $\lambda \in \mathbb{D}$, the series (15.2.4) converges uniformly for $x \in [-1, 1]$. The polynomials u_n are eigenfunctions for a certain selfadjoint differential operator and they satisfy the orthogonality conditions

$$\langle u_n, u_m \rangle_{L^2(\rho)} = \int_{-1}^{1} u_n(x)\overline{u_m(x)}\rho(x)\,dx = \begin{cases} 0 & \text{if } n \neq m, \\ \dfrac{\pi}{2} & \text{if } n = m. \end{cases} \tag{15.2.6}$$

By (15.2.5), the linear span of $\{u_n(x) : n \geqslant 1\}$ contains every polynomial, so it is dense in $L^2(\rho)$, and hence

$$\left(\sqrt{\frac{2}{\pi}} u_n \right)_{n=0}^{\infty}$$

is an orthonormal basis for $L^2(\rho)$. For $f \in L^2(\rho)$, use Parseval's theorem (Theorem 1.4.9) to see that

$$\frac{\pi}{2} \int_{-1}^{1} |f(x)|^2 \rho(x)\,dx = \frac{\pi}{2}\|f\|_{L^2(\rho)}^2$$

$$= \sum_{n=0}^{\infty} |\langle f, u_n \rangle_{L^2(\rho)}|^2$$

$$= \sum_{n=0}^{\infty} \left| \int_{-1}^{1} f(x)u_n(x)\rho(x)\,dx \right|^2. \tag{15.2.7}$$

Now consider the functions

$$h_\lambda(x) = \frac{1}{1 - 2x\overline{\lambda} + \overline{\lambda}^2} \quad \text{for } \lambda \in \mathbb{D}$$

from (15.2.4). Each of these rational functions has a pole at $x_0 = w(\overline{\lambda})$, where

$$w(z) = \frac{1}{2}\left(z + \frac{1}{z}\right).$$

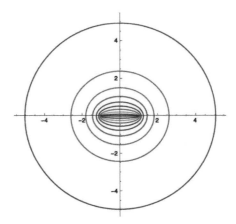

Figure 15.2.2 Images of $w(r\mathbb{T})$ for $r = 0.1, 0.2, 0.3, ..., 0.9$.

For each $r \in [-1, 1]$, the quadratic formula shows that $w(z) = r$ has no solution $z \in \mathbb{D}$ (see Figure 15.2.2). Therefore, $w(\mathbb{D}) \cap [-1, 1] = \varnothing$, and hence $h_\lambda \in L^2(\rho)$ for each fixed $\lambda \in \mathbb{D}$.

Define

$$(Vf)(z) = \sqrt{\frac{2}{\pi}} \langle f, h_z \rangle_{L^2(\rho)} \quad \text{for } f \in L^2(\rho) \text{ and } z \in \mathbb{D}.$$

From the expansion (15.2.4),

$$(Vf)(z) = \sqrt{\frac{2}{\pi}} \int_{-1}^{1} \frac{f(x)}{1 - 2xz + z^2} \rho(x)\, dx = \sqrt{\frac{2}{\pi}} \sum_{n=0}^{\infty} z^n \int_{-1}^{1} f(x) u_n(x) \rho(x)\, dx.$$

Thus, the nth Taylor-series coefficient of Vf is

$$\sqrt{\frac{2}{\pi}} \int_{-1}^{1} f(x) u_n(x) \rho(x)\, dx,$$

and hence, by (15.2.7) and the definition of the norm in H^2, we have

$$\|Vf\|_{H^2}^2 = \sum_{n=0}^{\infty} \left| \sqrt{\frac{2}{\pi}} \int_{-1}^{1} f(x) u_n(x) \rho(x)\, dx \right|^2$$

$$= \frac{2}{\pi} \sum_{n=0}^{\infty} \left| \int_{-1}^{1} f(x) u_n(x) \rho(x)\, dx \right|^2$$

$$= \frac{2}{\pi} \cdot \frac{\pi}{2} \int_{-1}^{1} |f(x)|^2 \rho(x)\, dx$$

$$= \|f\|_{L^2(\rho)}^2.$$

Thus, V maps $L^2(\rho)$ to H^2 isometrically.

To show that V is surjective, observe that for each $n \geqslant 0$,

$$
\begin{aligned}
(Vu_n)(z) &= \sqrt{\frac{2}{\pi}} \sum_{m=0}^{\infty} z^m \int_{-1}^{1} u_n(x)u_m(x)\rho(x)\,dx \\
&= \sqrt{\frac{2}{\pi}} \frac{\pi}{2} z^n \qquad\qquad\text{(by (15.2.6))} \\
&= \sqrt{\frac{\pi}{2}} z^n,
\end{aligned}
$$

so the range of V contains every (analytic) polynomial and is therefore dense in H^2. Since V is an isometry, its range is closed (Exercise 2.8.9). Therefore, $VL^2(\rho) = H^2$. Recall that

$$
k_\lambda(z) = \frac{1}{1 - \bar{\lambda}z} \quad \text{for } \lambda, z \in \mathbb{D},
$$

is the reproducing kernel for H^2. That is, $k_\lambda \in H^2$ and $g(\lambda) = \langle g, k_\lambda \rangle_{H^2}$ for all $g \in H^2$ and $\lambda \in \mathbb{D}$. Furthermore (Exercise 5.9.8),

$$
\bigvee\{k_\lambda : \lambda \in \mathbb{D}\} = H^2. \tag{15.2.8}
$$

For fixed $\lambda, z \in \mathbb{D}$,

$$
\begin{aligned}
(Vh_\lambda)(z) &= \sqrt{\frac{2}{\pi}} \sum_{n=0}^{\infty} z^n \int_{-1}^{1} h_\lambda(x)u_n(x)\rho(x)\,dx \\
&= \sqrt{\frac{2}{\pi}} \sum_{n=0}^{\infty} z^n \sum_{m=0}^{\infty} \bar{\lambda}^{m} \int_{-1}^{1} u_m(x)u_n(x)\rho(x)\,dx \\
&= \sqrt{\frac{2}{\pi}} \frac{\pi}{2} \sum_{n=0}^{\infty} \bar{\lambda}^{n} z^n \\
&= \sqrt{\frac{\pi}{2}} \frac{1}{1 - \bar{\lambda}z} \\
&= \sqrt{\frac{\pi}{2}} k_\lambda(z).
\end{aligned}
$$

The next step is to verify that

$$
V^*TV = M_x. \tag{15.2.9}
$$

Since $\bigvee\{k_\lambda : \lambda \in \mathbb{D}\} = H^2$, V is unitary, and $h_\lambda = \sqrt{\frac{\pi}{2}} V^* k_\lambda$, it follows that

$$
\bigvee\{h_\lambda : \lambda \in \mathbb{D}\} = L^2(\rho).
$$

To prove (15.2.9), it suffices to verify that

$$
(V^*TVh_\lambda)(x) = xh_\lambda(x) \quad \text{for all } \lambda \in \mathbb{D}.
$$

Observe that

$$Sk_\lambda = -\frac{1}{\bar{\lambda}} + \frac{1}{\bar{\lambda}}k_\lambda \quad \text{and} \quad S^*k_\lambda = \bar{\lambda}k_\lambda.$$

This yields

$$
\begin{aligned}
(V^*TVh_\lambda)(x) &= \frac{1}{2}\sqrt{\frac{\pi}{2}}(V^*(S + S^*)k_\lambda)(x) \\
&= \frac{1}{2}\sqrt{\frac{\pi}{2}}\left(V^*\left(-\frac{1}{\bar{\lambda}} + \frac{1}{\bar{\lambda}}k_\lambda + \bar{\lambda}k_\lambda\right)\right)(x) \\
&= \frac{1}{2}\sqrt{\frac{\pi}{2}}\left(V^*\left(-\frac{1}{\bar{\lambda}}k_0 + \frac{1}{\bar{\lambda}}k_\lambda + \bar{\lambda}k_\lambda\right)\right)(x) \\
&= \frac{1}{2}\sqrt{\frac{\pi}{2}}\sqrt{\frac{2}{\pi}}\left(-\frac{1}{\bar{\lambda}}h_0 + \frac{1}{\bar{\lambda}}h_\lambda + \bar{\lambda}h_\lambda\right)(x) \\
&= \frac{1}{2}\left(-\frac{1}{\bar{\lambda}}\cdot 1 + (\frac{1}{\bar{\lambda}} + \bar{\lambda})h_\lambda\right)(x) \\
&= \frac{1}{2}h_\lambda(x)\left(-\frac{1}{\bar{\lambda}}(1 - 2x\bar{\lambda} + \bar{\lambda}^2) + \frac{1}{\bar{\lambda}} + \bar{\lambda}\right) \\
&= \frac{1}{2}h_\lambda(x)\,2x \\
&= x\,h_\lambda(x).
\end{aligned}
$$

This proves (c). Parts (a) and (b) follow from (c) and Proposition 8.1.12. ∎

The previous theorem and Corollary 8.8.7 yield the following.

Corollary 15.2.10. *The invariant subspaces of T are $V\chi_E L^2(\rho)$, where E is a Lebesgue-measurable subset of $[-1, 1]$.*

15.3 The Direct Sum of S and S^*

The operator $S \oplus S^*$, the direct sum of S and S^*, is defined on

$$H^2 \oplus H^2 = \left\{ \begin{bmatrix} f \\ g \end{bmatrix} : f, g \in H^2 \right\}$$

by

$$(S \oplus S^*)(f \oplus g) = Sf \oplus S^*g.$$

This can be written in matrix form as

$$\begin{bmatrix} S & 0 \\ 0 & S^* \end{bmatrix},$$

in the sense that

$$\begin{bmatrix} S & 0 \\ 0 & S^* \end{bmatrix}\begin{bmatrix} f \\ g \end{bmatrix} = \begin{bmatrix} Sf \\ S^*g \end{bmatrix} = (S \oplus S^*)(f \oplus g).$$

Proposition 15.3.1. $S \oplus S^*$ *satisfies the following.*

(a) $\|S \oplus S^*\| = 1$.

(b) $\sigma(S \oplus S^*) = \mathbb{D}^-$.

(c) $\sigma_p(S \oplus S^*) = \mathbb{D}$.

(d) $(S \oplus S^*)^* = S^* \oplus S$.

Proof (a) For any $f \oplus g \in H^2 \oplus H^2$, use the facts that $\|Sf\| = \|f\|$ and $\|S^*g\| \leqslant \|g\|$ for all $f, g \in H^2$ to conclude that

$$\|(S \oplus S^*)(f \oplus g)\|^2 = \|Sf \oplus S^*g\|^2 = \|Sf\|^2 + \|S^*g\|^2 \leqslant \|f\|^2 + \|g\|^2 = \|f \oplus g\|^2.$$

Thus, $\|S \oplus S^*\| \leqslant 1$. Equality follows since $\|(S \oplus S^*)(1 \oplus 0)\| = \|z \oplus 0\| = 1$.
(b) For any $\lambda \in \mathbb{D}$,

$$(S \oplus S^*)(0 \oplus k_\lambda) = 0 \oplus \overline{\lambda}k_\lambda = \overline{\lambda}(0 \oplus k_\lambda)$$

and hence

$$\mathbb{D} \subseteq \sigma_p(S \oplus S^*) \subseteq \sigma(S \oplus S^*).$$

Thus, $\mathbb{D}^- \subseteq \sigma(S \oplus S^*)$. Since $\|S \oplus S^*\| = 1$, it follows that $\sigma(S \oplus S^*) \subseteq \mathbb{D}^-$ (Theorem 2.4.9) and thus $\sigma(S \oplus S^*) = \mathbb{D}^-$.
(c) In light of (b), it remains to show that $\sigma_p(S \oplus S^*) \cap \mathbb{T} = \emptyset$. If $\xi \in \mathbb{T}$ and $(S \oplus S^*)(f \oplus g) = \xi(f \oplus g)$, then $Sf = \xi f$ and $S^*g = \xi g$, which only hold when f and g are zero (Propositions 5.1.4 and 5.2.4).
(d) This follows from the matrix form of $S \oplus S^*$ and (14.2.7). ■

The operator $S \oplus S^*$ is a complex symmetric operator. To see this, define the conjugation $J : H^2 \to H^2$ by

$$(Jf)(z) = \overline{f(\overline{z})}$$

and observe that

$$J\left(\sum_{n=0}^{\infty} a_n z^n\right) = \sum_{n=0}^{\infty} \overline{a_n} z^n.$$

The above map J is isometric, involutive, and conjugate linear. Furthermore,

$$JS = SJ \quad \text{and} \quad JS^* = S^*J. \tag{15.3.2}$$

Exercise 15.6.8 shows that the mapping C on $H^2 \oplus H^2$ defined by

$$C = \begin{bmatrix} 0 & J \\ J & 0 \end{bmatrix} \tag{15.3.3}$$

is a conjugation.

Proposition 15.3.4. *If $T = S \oplus S^*$, then $T = CT^*C$.*

Proof Using the block-matrix representations for C and T and (15.3.2), it follows that

$$
CT^*C = \begin{bmatrix} 0 & J \\ J & 0 \end{bmatrix} \begin{bmatrix} S^* & 0 \\ 0 & S \end{bmatrix} \begin{bmatrix} 0 & J \\ J & 0 \end{bmatrix}
$$

$$
= \begin{bmatrix} 0 & JS \\ JS^* & 0 \end{bmatrix} \begin{bmatrix} 0 & J \\ J & 0 \end{bmatrix}
$$

$$
= \begin{bmatrix} JSJ & 0 \\ 0 & JS^*J \end{bmatrix}
$$

$$
= \begin{bmatrix} J^2 S & 0 \\ 0 & J^2 S^* \end{bmatrix}
$$

$$
= \begin{bmatrix} S & 0 \\ 0 & S^* \end{bmatrix}
$$

$$
= T,
$$

which proves the result. ∎

Beurling's theorem (Theorem 5.4.12) says that the invariant subspaces for S are $\{0\}$ or uH^2, where u is an inner function. Thus, the invariant subspaces for S^* are either H^2 itself or of the form $(uH^2)^\perp$. The subspaces $uH^2 \oplus (vH^2)^\perp$, where u and v are inner functions, along with $\{0\} \oplus (vH^2)^\perp$, $uH^2 \oplus \{0\}$, and $uH^2 \oplus H^2$, are invariant for $S \oplus S^*$ and comprise the *splitting* invariant subspaces. As it turns out, there are many other invariant subspaces. The following example is from [74].

Example 15.3.5. For $\lambda \in \mathbb{D}$, consider the $S \oplus S^*$-invariant subspace generated by $k_\lambda \oplus k_\lambda$, that is,

$$
\bigvee \{(S \oplus S^*)^n (k_\lambda \oplus k_\lambda) : n \geqslant 0\}.
$$

For any polynomial p, we have $p(S \oplus S^*)(k_\lambda \oplus k_\lambda) = p k_\lambda \oplus p(\bar{\lambda})k_\lambda$. Suppose that $(p_n)_{n=1}^\infty$ is a sequence of polynomials such that

$$
p_n(S \oplus S^*)(k_\lambda \oplus k_\lambda) \to f \oplus g
$$

in $H^2 \oplus H^2$. Thus, $p_n k_\lambda \to f$ and $p_n(\bar{\lambda})k_\lambda \to g$ in H^2 norm. Since $p_n k_\lambda \to f$, it follows that $p_n \to (1 - \bar{\lambda}z)f$ in norm. Proposition 5.3.8 yields

$$
p_n(z) \to (1 - \bar{\lambda}z)f(z) \quad \text{for all } z \in \mathbb{D},
$$

and, in particular, $p_n(\bar{\lambda}) \to (1 - \bar{\lambda}^2)f(\bar{\lambda})$. In other words, $g = f(\bar{\lambda})(1 - \bar{\lambda}^2)k_\lambda$. Thus,

$$
\bigvee \{(S \oplus S^*)^n (k_\lambda \oplus k_\lambda) : n \geqslant 0\} \subseteq \{f \oplus f(\bar{\lambda})(1 - \bar{\lambda}^2)k_\lambda : f \in H^2\}.
$$

If $f \in H^2$ and $\lambda \in \mathbb{D}$, then $(1 - \bar{\lambda}z)f \in H^2$ and there are polynomials p_n such that $p_n \to (1 - \bar{\lambda}z)f$ in norm. The discussion above implies that

$$p_n(S \oplus S^*)(k_\lambda \oplus k_\lambda) \to f \oplus f(\bar{\lambda})(1 - \bar{\lambda}^2)k_\lambda,$$

and hence

$$\bigvee \left\{ (S \oplus S^*)^n(k_\lambda \oplus k_\lambda) : n \geqslant 0 \right\} = \{ f \oplus f(\bar{\lambda})(1 - \bar{\lambda}^2)k_\lambda : f \in H^2 \}.$$

This subspace is contained in $H^2 \oplus (uH^2)^\perp$, where

$$u(z) = \frac{z - \lambda}{1 - \bar{\lambda}z},$$

but the containment is proper. Indeed, $1 \oplus k_\lambda \in H^2 \oplus (uH^2)^\perp$. However,

$$1 \oplus k_\lambda \notin \{ f \oplus f(\bar{\lambda})(1 - \bar{\lambda}^2)k_\lambda : f \in H^2 \}.$$

The complete description of the $S \oplus S^*$-invariant subspaces is contained in a paper of Timotin [360] and requires a diversion into dilation theory, which would take us far afield. Timotin's paper also contains the following.

Theorem 15.3.6. *The only nonzero reducing subspaces of $S \oplus S^*$ are $H^2 \oplus \{0\}$ and $\{0\} \oplus H^2$.*

15.4 The Tensor Product of S and S^*

Although there is a general and abstract approach to tensor products of Hilbert spaces and operators (see the end notes for the references), we take a more heuristic and explicit approach. Recall that the tensor product of vectors $\mathbf{x}$ and $\mathbf{y}$ in a Hilbert space $\mathcal{H}$ is identified with the rank-one operator

$$\mathbf{x} \otimes \mathbf{y} : \mathcal{H} \to \mathcal{H}, \quad (\mathbf{x} \otimes \mathbf{y})\mathbf{z} = \langle \mathbf{z}, \mathbf{y} \rangle \mathbf{x} \quad \text{for } \mathbf{z} \in \mathcal{H}.$$

The operator $\mathbf{x} \otimes \mathbf{y}$ is a *simple tensor*.

Proposition 15.4.1. $\|\mathbf{x} \otimes \mathbf{y}\| = \|\mathbf{x}\| \|\mathbf{y}\|$.

Proof By the Cauchy–Schwarz inequality,

$$\|(\mathbf{x} \otimes \mathbf{y})\mathbf{z}\| = \|\langle \mathbf{z}, \mathbf{y} \rangle \mathbf{x}\| = |\langle \mathbf{z}, \mathbf{y} \rangle| \|\mathbf{x}\| \leqslant \|\mathbf{z}\| \|\mathbf{y}\| \|\mathbf{x}\|.$$

Equality is attained for $\mathbf{z} = \mathbf{y}/\|\mathbf{y}\|$. ∎

Consider the vector space $\mathcal{H} \odot \mathcal{H}$ of all finite linear combinations of simple tensors, that is, expressions of the form

$$\sum_{j=1}^{n} c_j(\mathbf{x}_j \otimes \mathbf{y}_j), \quad \text{where } n \in \mathbb{N}, c_j \in \mathbb{C}, \text{ and } \mathbf{x}_j, \mathbf{y}_j \in \mathcal{H}.$$

We define an inner product on $\mathcal{H} \odot \mathcal{H}$ first on simple tensors $\mathbf{x} \otimes \mathbf{y}$ and $\mathbf{z} \otimes \mathbf{w}$ by

$$\langle \mathbf{x} \otimes \mathbf{y}, \mathbf{z} \otimes \mathbf{w} \rangle = \langle \mathbf{x}, \mathbf{z} \rangle \langle \mathbf{y}, \mathbf{w} \rangle.$$

Extend this to $\mathcal{H} \odot \mathcal{H}$ by

$$\left\langle \sum_{j=1}^{n} a_j (\mathbf{x}_j \otimes \mathbf{y}_j), \sum_{k=1}^{n} b_k (\mathbf{z}_k \otimes \mathbf{w}_k) \right\rangle = \sum_{j,k=1}^{n} a_j \overline{b_k} \langle \mathbf{x}_j, \mathbf{z}_k \rangle \langle \mathbf{y}_j, \mathbf{w}_k \rangle \tag{15.4.2}$$

and verify it satisfies the required properties of an inner product (Definition 1.4.1). The corresponding norm satisfies

$$\left\| \sum_{j=1}^{N} a_j (\mathbf{x}_j \otimes \mathbf{y}_j) \right\|^2 = \sum_{j,k=1}^{N} a_j \overline{a_k} \langle \mathbf{x}_j, \mathbf{x}_k \rangle \langle \mathbf{y}_j, \mathbf{y}_k \rangle. \tag{15.4.3}$$

This makes $\mathcal{H} \odot \mathcal{H}$ an inner product space.

Definition 15.4.4. The *tensor product* $\mathcal{H} \otimes \mathcal{H}$ is the completion of $\mathcal{H} \odot \mathcal{H}$ with respect to the norm (15.4.3).

Proposition 15.4.5. *If $(\mathbf{u}_n)_{n=1}^{\infty}$ is an orthonormal basis for $\mathcal{H}$, then $(\mathbf{u}_m \otimes \mathbf{u}_n)_{m,n=1}^{\infty}$ is an orthonormal basis for $\mathcal{H} \otimes \mathcal{H}$.*

Proof The definition of the inner product on $\mathcal{H} \otimes \mathcal{H}$ from (15.4.2) implies that the sequence $(\mathbf{u}_m \otimes \mathbf{u}_n)_{m,n=1}^{\infty}$ is orthonormal. Let

$$\mathcal{M} = \bigvee \{ \mathbf{u}_m \otimes \mathbf{u}_n : m, n \geqslant 1 \}$$

in the norm of $\mathcal{H} \otimes \mathcal{H}$. The next step is to show that $\mathcal{M} = \mathcal{H} \otimes \mathcal{H}$.

Let $c_{mn} \in \mathbb{C}$ for $m, n \geqslant 1$. Then for each $N \in \mathbb{N}$, the definition (15.4.3) implies that

$$\left\| \sum_{m,n=1}^{N} c_{mn} (\mathbf{u}_m \otimes \mathbf{u}_n) \right\|^2 = \sum_{\substack{m,n=1 \\ m',n'=1}}^{N} c_{mn} \overline{c_{m'n'}} \langle \mathbf{u}_m, \mathbf{u}_{m'} \rangle \langle \mathbf{u}_n, \mathbf{u}_{n'} \rangle = \sum_{m,n=1}^{N} |c_{mn}|^2.$$

Consequently,

$$\sum_{m,n=1}^{\infty} c_{mn} (\mathbf{u}_m \otimes \mathbf{u}_n) \in \mathcal{M} \quad \text{if and only if} \quad \sum_{m,n=1}^{\infty} |c_{mn}|^2 < \infty.$$

If $\mathbf{x}, \mathbf{y} \in \mathcal{H}$, then

$$\mathbf{x} = \sum_{m=1}^{\infty} a_m \mathbf{u}_m \quad \text{and} \quad \mathbf{y} = \sum_{n=1}^{\infty} b_n \mathbf{u}_n$$

where $\sum_{m=1}^{\infty} |a_m|^2$ and $\sum_{n=1}^{\infty} |b_n|^2$ are finite. Since

$$\sum_{m,n=1}^{\infty} |a_m b_n|^2 = \left(\sum_{m=1}^{\infty} |a_m|^2 \right) \left(\sum_{n=1}^{\infty} |b_n|^2 \right) < \infty,$$

it follows that

$$\sum_{m,n=1}^{\infty} a_m \overline{b_n}(\mathbf{u}_m \otimes \mathbf{u}_n) \in \mathcal{M}.$$

For any $\mathbf{z} \in \mathcal{H}$,

$$(\mathbf{x} \otimes \mathbf{y})(\mathbf{z}) = \langle \mathbf{z}, \mathbf{y} \rangle \mathbf{x}$$

$$= \Big(\sum_{n=1}^{\infty} \overline{b_n} \langle \mathbf{z}, \mathbf{u}_n \rangle \Big) \mathbf{x}$$

$$= \Big(\sum_{n=1}^{\infty} \overline{b_n} \langle \mathbf{z}, \mathbf{u}_n \rangle \Big) \Big(\sum_{m=1}^{\infty} a_m \mathbf{u}_m \Big)$$

$$= \sum_{m,n=1}^{\infty} a_m \overline{b_n} \langle \mathbf{z}, \mathbf{u}_n \rangle \mathbf{u}_m$$

$$= \Big(\sum_{m,n=1}^{\infty} a_m \overline{b_n} (\mathbf{u}_m \otimes \mathbf{u}_n) \Big)(\mathbf{z}).$$

Therefore,

$$\mathbf{x} \otimes \mathbf{y} = \sum_{m,n=1}^{\infty} a_m \overline{b_n}(\mathbf{u}_m \otimes \mathbf{u}_n) \in \mathcal{M},$$

and hence $\mathcal{M}$ contains $\{\mathbf{x} \otimes \mathbf{y} : \mathbf{x}, \mathbf{y} \in \mathcal{H}\}$, a set whose closed linear span is $\mathcal{H} \otimes \mathcal{H}$. This proves that $\mathcal{M} = \mathcal{H} \otimes \mathcal{H}$ and thus $(\mathbf{u}_m \otimes \mathbf{u}_n)_{m,n=1}^{\infty}$ is an orthonormal basis for $\mathcal{H} \otimes \mathcal{H}$. ∎

For $A, B \in \mathcal{B}(\mathcal{H})$, define the *tensor product*

$$A \otimes B : \mathcal{H} \otimes \mathcal{H} \to \mathcal{H} \otimes \mathcal{H}$$

first on simple tensors $\mathbf{x} \otimes \mathbf{y}$ by

$$(A \otimes B)(\mathbf{x} \otimes \mathbf{y}) = (A\mathbf{x}) \otimes (B\mathbf{y}),$$

and extend the tensor product by linearity to $\mathcal{H} \otimes \mathcal{H}$.

Let us focus on the tensor product $S \otimes S^*$ on $H^2 \otimes H^2$. The orthonormal basis $(z^n)_{n=0}^{\infty}$ for H^2 is a natural one to choose, so

$$H^2 \otimes H^2 = \Big\{ \sum_{m,n=0}^{\infty} c_{mn}(z^m \otimes z^n) : \sum_{m,n=0}^{\infty} |c_{mn}|^2 < \infty \Big\}.$$

Moreover,

$$(S \otimes S^*) \Big(\sum_{m,n=0}^{\infty} c_{mn}(z^m \otimes z^n) \Big) = \sum_{m,n=0}^{\infty} c_{mn}(Sz^m \otimes S^* z^n).$$

Note that

$$(Sz^m \otimes S^* z^n) = z^{m+1} \otimes z^{n-1},$$

unless $n = 0$, for which $S^* 1 = 0$ and the tensor above is zero.

Proposition 15.4.6. $S \otimes S^*$ *satisfies the following.*

(a) $\|S \otimes S^*\| = 1$.

(b) $\sigma(S \otimes S^*) = \mathbb{D}^-$.

(c) $(S \otimes S^*)^* = S^* \otimes S$.

Proof (a) Observe that

$$\left\| (S \otimes S^*)\left(\sum_{m,n=0}^{\infty} c_{mn}(z^m \otimes z^n) \right) \right\|^2 = \left\| \sum_{\substack{m=0 \\ n=1}}^{\infty} c_{mn}(z^{m+1} \otimes z^{n-1}) \right\|^2$$

$$= \sum_{\substack{m=0 \\ n=1}}^{\infty} |c_{mn}|^2$$

$$\leqslant \sum_{\substack{m=0 \\ n=0}}^{\infty} |c_{mn}|^2$$

$$= \left\| \sum_{\substack{m=0 \\ n=0}}^{\infty} c_{mn}(z^m \otimes z^n) \right\|^2.$$

Thus, $\|S \otimes S^*\| \leqslant 1$. Equality follows from $\|(S \otimes S^*)(1 \otimes z)\| = \|z \otimes 1\| = 1$ (Proposition 15.4.1).

(b) Since $\|S \otimes S^*\| = 1$, one concludes that $\sigma(S \otimes S^*) \subseteq \mathbb{D}^-$. If $\lambda \in \mathbb{D}$, then $(S^* - \bar{\lambda}I)k_\lambda = 0$ so $1 \otimes k_\lambda$ belongs to the kernel of $S \otimes S^* - \bar{\lambda}(I \otimes I)$. Thus, $\mathbb{D}^- \subseteq \sigma(S \otimes S^*)$.

(c) It suffices to prove

$$\langle (S \otimes S^*)(z^m \otimes z^n), z^j \otimes z^k \rangle = \langle z^m \otimes z^n, (S^* \otimes S)(z^j \otimes z^k) \rangle.$$

The left side equals

$$\langle z^{m+1} \otimes z^{n-1}, z^j \otimes z^k \rangle = \langle z^{m+1}, z^j \rangle + \langle z^{n-1}, z^k \rangle = \delta_{m+1,j} + \delta_{n-1,k},$$

while the right side equals

$$\langle z^m \otimes z^n, z^{j-1} \otimes z^{k+1} \rangle = \langle z^m, z^{j-1} \rangle + \langle z^n, z^{k+1} \rangle = \delta_{m,j-1} + \delta_{n,k+1}.$$

Since $\delta_{m,n} = \delta_{m+j,n+j}$, the result follows. $\blacksquare$

Another perspective furthers our understanding of $S \otimes S^*$. The tensor-product space $H^2 \otimes H^2$ is understood as

$$\sum_{m,n=0}^{\infty} c_{mn}(z^m \otimes z^n),$$

where

$$\sum_{m,n=0}^{\infty} |c_{mn}|^2 < \infty, \tag{15.4.7}$$

and (15.4.7) is the square of the norm on $H^2 \otimes H^2$. Associate the tensor $z^m \otimes z^n$ with the two-variable monomial $z^m w^n$ and consider the *Hardy space* $H^2(\mathbb{D}^2)$ of the *bidisk* $\mathbb{D}^2 = \{(z, w) : z, w \in \mathbb{D}\}$, where

$$H^2(\mathbb{D}^2) := \Big\{ f(z, w) = \sum_{m,n=0}^{\infty} c_{mn} z^m w^n : \|f\|_{H^2(\mathbb{D}^2)}^2 = \sum_{m,n=0}^{\infty} |c_{mn}|^2 < \infty \Big\}. \tag{15.4.8}$$

There is the natural unitary operator

$$U : H^2 \otimes H^2 \to H^2(\mathbb{D}^2), \quad U\Big(\sum_{m,n=0}^{\infty} c_{mn}(z^m \otimes w^n) \Big) = \sum_{m,n=0}^{\infty} c_{mn} z^m w^n.$$

Theorem 15.4.9. $U(S \otimes S^*)U^* = T$, *where*

$$(Tf)(z, w) = z\Big(\frac{f(z, w) - f(z, 0)}{w} \Big).$$

Proof Verify this identity on $z^m w^n$ and extend by linearity. ∎

This setting reveals additional structure. Fix $N \geqslant 0$ and consider

$$\mathcal{P}_N = \bigvee \{z^k w^{N-k} : 0 \leqslant k \leqslant N\},$$

the set of homogeneous polynomials of degree N together with the zero polynomial. Then

$$T(z^k w^{N-k}) = \begin{cases} z^{k+1} w^{N-k-1} & \text{if } 0 \leqslant k \leqslant N-1, \\ 0 & \text{if } k = N, \end{cases}$$

so $\mathcal{P}_N$ is a reducing subspace for T (Exercise 15.6.28). Furthermore, with respect to the inner product on $H^2(\mathbb{D}^2)$ from (15.4.8), the spaces $\mathcal{P}_N$ and $\mathcal{P}_M$ are orthogonal for $N \neq M$, and

$$H^2(\mathbb{D}^2) = \bigoplus_{N=0}^{\infty} \mathcal{P}_N.$$

For $\sum_{k=0}^{N} a_k z^k w^{N-k} \in \mathcal{P}_N$, note that

$$T\Big(\sum_{k=0}^{N} a_k z^k w^{N-k} \Big) = \sum_{k=0}^{N-1} a_k z^{k+1} w^{N-k-1}.$$

In other words, T sends $(a_0, a_1, a_2,..., a_N)$ to $(0, a_0, a_1, a_2,..., a_{N-1})$. From the matrix identity

$$
\begin{bmatrix}
0 & 0 & 0 & 0 & \cdots & 0 \\
1 & 0 & 0 & 0 & \cdots & 0 \\
0 & 1 & 0 & 0 & \cdots & 0 \\
0 & 0 & 1 & 0 & \cdots & 0 \\
0 & 0 & 0 & \ddots & \ddots & 0 \\
0 & 0 & 0 & \cdots & 1 & 0
\end{bmatrix}
\begin{bmatrix}
a_0 \\ a_1 \\ a_2 \\ a_3 \\ \vdots \\ a_N
\end{bmatrix}
=
\begin{bmatrix}
0 \\ a_0 \\ a_1 \\ a_2 \\ \vdots \\ a_{N-1}
\end{bmatrix},
$$

it follows that $T|_{\mathcal{P}_N} \cong J_{N+1}(0)^*$, where $J_N(0)$ is the $N \times N$ nilpotent Jordan block. Since $J_N(0)^*$ is unitarily equivalent to $J_N(0)$ (Exercise 15.6.18), we obtain the following.

Theorem 15.4.10. $S \otimes S^* \cong \displaystyle\bigoplus_{N=1}^{\infty} J_N(0)$.

The conjugation C_n on $\mathbb{C}^n$ given by $C_n(z_1, z_2, z_3,..., z_n) = (\overline{z_n}, \overline{z_{n-1}}, \overline{z_{n-2}},..., \overline{z_1})$ satisfies $J_n(0) = C_n J_n(0)^* C_n$. Hence, each $J_n(0)$ is unitarily equivalent to a complex symmetric matrix (Exercise 15.6.18). Combine this theorem with Exercise 15.6.19 to obtain the following.

Corollary 15.4.11. $S \otimes S^*$ *is a complex symmetric operator.*

If $\mathcal{M}$ reduces T, then, with respect to the orthogonal decomposition $\mathcal{H} = \mathcal{M} \oplus \mathcal{M}^{\perp}$, we may write T as

$$
T = \begin{bmatrix} T_1 & 0 \\ 0 & T_2 \end{bmatrix}.
$$

Moreover, $\mathcal{M}$ reduces T if and only if $PT = TP$, where P is the orthogonal projection of $\mathcal{H}$ onto $\mathcal{M}$ (Theorem 14.3.5). This yields a correspondence between reducing subspaces for T and the orthogonal projections that commute with T.

Let us characterize the reducing subspaces for $T = S \otimes S^*$. Since T is unitarily equivalent to

$$
J = \bigoplus_{N=1}^{\infty} J_N(0),
$$

it suffices to describe the reducing subspaces of J. To this end, we identify the orthogonal projections P on

$$
\mathcal{H} = \bigoplus_{N=1}^{\infty} \mathbb{C}^N
$$

that commute with J. Write

$$
J = \begin{bmatrix}
J_1(0) & 0 & 0 & \cdots \\
0 & J_2(0) & 0 & \cdots \\
0 & 0 & J_3(0) & \cdots \\
\vdots & \vdots & \vdots & \ddots
\end{bmatrix}
\quad \text{and} \quad
P = \begin{bmatrix}
P_{11} & P_{12} & P_{13} & \cdots \\
P_{21} & P_{22} & P_{23} & \cdots \\
P_{31} & P_{32} & P_{33} & \cdots \\
\vdots & \vdots & \vdots & \ddots
\end{bmatrix},
$$

in which each P_{jk} is a $j \times k$ matrix and $P_{jk} = P_{kj}^*$ since $P = P^*$. The equation $PJ = JP$ yields matrix equations of the form

$$P_{mn}J_n = J_m P_{mn} \quad \text{for } m, n \geqslant 1.$$

A computation confirms that each P_{mn} is of the form

$$P_{mn} = \begin{cases} T_n & \text{if } m = n, \\ \begin{bmatrix} T_n \\ 0 \end{bmatrix} & \text{if } m > n, \\ [0 \; T_m] & \text{if } m < n, \end{cases}$$

in which T_m and T_n are arbitrary upper-triangular Toeplitz matrices of size $m \times m$ and $n \times n$, respectively [155]. For example, the upper-left 10×10 principal submatrix of P is

$$\begin{bmatrix} p_{11} & 0 & p_{13} & 0 & 0 & p_{16} & 0 & 0 & 0 & p_{1,10} \\ p_{21} & p_{22} & p_{23} & 0 & p_{25} & p_{26} & 0 & 0 & p_{29} & p_{2,10} \\ 0 & 0 & p_{22} & 0 & 0 & p_{25} & 0 & 0 & 0 & p_{29} \\ p_{41} & p_{42} & p_{43} & p_{55} & p_{45} & p_{46} & 0 & p_{59} & p_{49} & p_{4,10} \\ 0 & 0 & p_{42} & 0 & p_{55} & p_{45} & 0 & 0 & p_{59} & p_{49} \\ 0 & 0 & 0 & 0 & 0 & p_{55} & 0 & 0 & 0 & p_{59} \\ p_{71} & p_{72} & p_{73} & p_{85} & p_{75} & p_{76} & p_{99} & p_{89} & p_{79} & p_{7,10} \\ 0 & 0 & p_{72} & 0 & p_{85} & p_{75} & 0 & p_{99} & p_{89} & p_{79} \\ 0 & 0 & 0 & 0 & 0 & p_{85} & 0 & 0 & p_{99} & p_{89} \\ 0 & 0 & 0 & 0 & 0 & 0 & 0 & 0 & 0 & p_{99} \end{bmatrix}. \quad (15.4.12)$$

Since the selfadjointness of P ensures that $P_{mn} = P_{nm}^*$, a glance at (15.4.12) confirms that $P_{mn} = 0$ whenever $m \neq n$. Consequently,

$$P = \bigoplus_{n=1}^{\infty} P_{nn}.$$

Since P is an orthogonal projection, it is selfadjoint and idempotent. Thus, each P_{nn} is selfadjoint and idempotent, that is, each P_{nn} is an orthogonal projection.

The orthogonal projections P that commute with J are precisely those of the form

$$P = \bigoplus_{n=1}^{\infty} \delta_n I_{nn},$$

in which each $\delta_n \in \{0, 1\}$. This means the reducing subspaces for J are those of the form

$$\bigvee \left(\bigcup_{n \in A} \mathcal{P}_n \right),$$

in which $A \subseteq \mathbb{N} \cup \{0\}$. We assume that above is $\{0\}$ if $A = \varnothing$. Pull this back to $S \otimes S^*$ and conclude the following.

Theorem 15.4.13. *The reducing subspaces for $S \otimes S^*$ are precisely*

$$\bigvee \left(\bigcup_{n \in A} \{ \mathbf{e}_j \otimes \mathbf{e}_k : j + k = n \} \right),$$

in which $A \subseteq \mathbb{N} \cup \{0\}$.

15.5 Notes

The von Neumann–Wold decomposition implies that for a pure isometry T, every invariant subspace $\mathcal{M}$ is generated by its corresponding wandering subspace $\mathcal{M} \cap (T\mathcal{M})^\perp$. The wandering subspace property holds for the Hardy, Dirichlet, and Bergman shifts, despite the fact that the latter two are not isometric. For other non-isometries, this wandering subspace property was explored, with both positive and negative results, in [79, 137, 138].

This chapter covered the operators $S \oplus S^*$ and $S \otimes S^*$. The operator $S \oplus S$ is unitarily equivalent to S^2 (Exercise 15.6.9) while $S \otimes S$ is unitarily equivalent to a shift operator on the infinite direct sum $(H^2)^{(\infty)}$ (Exercise 15.6.21).

Theorem 15.2.2 gave the spectral representation of the selfadjoint Toeplitz operator $T_{\cos \theta}$ and showed that its spectral measure is absolutely continuous with respect to Lebesgue measure on $\mathbb{R}$. For $n \geqslant 1$, the spectral measure for $T_{\cos n\theta}$ is also absolutely continuous [277]. See [312] for the spectral representation of other selfadjoint Toeplitz operators.

More on tensor products of Hilbert spaces and operators is in the classic text [112]. Our analysis for $S \otimes S^*$ comes from [142]. We discussed the reducing subspaces of $S \otimes S^*$. The invariant subspaces are not yet fully described.

This chapter made a connection with $H^2(\mathbb{D}^2)$, the Hardy space of the bidisk. There is a well-developed theory for this space that parallels that of H^2 [318].

15.6 Exercises

Exercise 15.6.1. If $T \in \mathcal{B}(\mathcal{H})$ is an isometry, prove that $T(\mathcal{H} \cap (T\mathcal{H})^\perp) = T\mathcal{H} \cap (T^2\mathcal{H})^\perp$.

Exercise 15.6.2. If q is an inner function, find the von Neumann–Wold decomposition of M_q (multiplication by q) on H^2.

Exercise 15.6.3. Prove that the constant function $f \equiv 1$ is a cyclic vector for $S + S^*$.

Exercise 15.6.4. Prove that any nonzero polynomial is a cyclic vector for $S + S^*$.

Exercise 15.6.5. Use the following steps to describe the commutant of $S+S^*$. First observe that $S + S^*$ is a cyclic selfadjoint operator (Exercise 15.6.3). The continuous functional calculus from Chapter 8 ensures that $\varphi(S + S^*)$ is a well-defined bounded operator on H^2 for any $\varphi \in C[-1, 1]$.

(a) Prove that $\varphi(S + S^*) \in \{S + S^*\}'$.

(b) Prove that $\varphi(S + S^*)$ is unitarily equivalent to M_φ on $L^2(\rho)$, where $\rho(x) = \sqrt{1-x^2}$.

(c) Prove that the commutant of M_x on $L^2(\rho)$ is $\{M_\psi : \psi \in L^\infty[-1,1]\}$.

(d) Argue via unitary equivalence that $\psi(S + S^*)$ is well defined for all $\psi \in L^\infty[-1,1]$ and that $\{S + S^*\}' = \{\psi(S + S^*) : \psi \in L^\infty[-1,1]\}$.

Exercise 15.6.6. This exercise continues Exercise 15.6.5. Prove that the invariant subspaces of $S + S^*$ are $\{\chi_E(S + S^*)f : f \in H^2\}$, where E is a Lebesgue-measurable subset of $[-1,1]$.

Exercise 15.6.7. For $A, B \in \mathcal{B}(\mathcal{H})$, prove the following.

(a) $(A \oplus B)^* = A^* \oplus B^*$.

(b) $\sigma_p(A \oplus B) = \sigma_p(A) \cup \sigma_p(B)$.

Exercise 15.6.8. Prove that that the map C in (15.3.3) is a conjugation on $H^2 \oplus H^2$.

Exercise 15.6.9. Prove that S^2 on H^2 is unitarily equivalent to $S \oplus S$ on $H^2 \oplus H^2$.

Exercise 15.6.10. Describe the commutant of $S \oplus S$.

Exercise 15.6.11. Let $\mathcal{M} \subseteq H^2 \oplus H^2$ be an invariant subspace for $T = S \oplus S$.

(a) Prove there is a subspace $\mathcal{N}$ of $H^2 \oplus H^2$ such that $\mathcal{M} = \mathcal{N} \oplus T\mathcal{N} \oplus T^2\mathcal{N} \oplus \cdots$.

(b) For $z \in \mathbb{D}$, define $F(z) \in \mathcal{B}(\mathcal{N}, \mathbb{C}^2)$ by evaluating each element of $\mathcal{N}$ at z. Prove that F is a $\mathbb{C}^2$-valued analytic function on $\mathbb{D}$.

(c) Prove that $\mathcal{M} = F\left\{ \begin{bmatrix} f \\ g \end{bmatrix} \in \mathcal{N} \right\}$.

Remark: The *Beurling–Lax theorem* yields more about F [202, p. 115].

Exercise 15.6.12. For $n \in \mathbb{N} \cup \{\infty\}$, $T \in \mathcal{B}(\mathcal{H})$ is a *shift of multiplicity n* if there is a subspace $\mathcal{H}_1$ of $\mathcal{H}$ with $\dim \mathcal{H}_1 = n$ and pairwise orthogonal subspaces $\mathcal{H}_j$ with $j \geqslant 1$ such that $\mathcal{H} = \mathcal{H}_1 \oplus \mathcal{H}_2 \oplus \mathcal{H}_3 \oplus \cdots$ and T maps $\mathcal{H}_j$ isometrically onto $\mathcal{H}_{j+1}$.

(a) For each $n \in \mathbb{N} \cup \{\infty\}$, give an example of a shift of multiplicity n.

(b) Prove that two such shifts are unitarily equivalent if and only if their multiplicities are the same.

Exercise 15.6.13. Here is a bilateral version of $S + S^*$ on H^2 from [25, p. 56]. Define $T : L^2(\mathbb{T}) \to L^2(\mathbb{T})$ by $T = M_\xi + M_\xi^*$.

(a) What is the matrix representation of T with respect to the basis $(\xi^n)_{n=-\infty}^\infty$?

(b) Determine $\sigma(T), \sigma_p(T)$, and $\sigma_{ap}(T)$.

(c) Define $W : L^2[-2,2] \to L^2[-2,2]$ by $(Wf)(x) = xf(x)$, and prove that T is not unitarily equivalent to W.

(d) Prove that T is unitarily equivalent to $W \oplus W$.

Exercise 15.6.14. For any $\mathbf{x}$ and $\mathbf{y}$ in a Hilbert space $\mathcal{H}$, prove that $(\mathbf{x} \otimes \mathbf{y})^* = \mathbf{y} \otimes \mathbf{x}$.

Exercise 15.6.15. Suppose $\mathbf{y}_1, \mathbf{y}_2, ..., \mathbf{y}_n$ are linearly independent vectors in a Hilbert space $\mathcal{H}$ and, for some $\mathbf{x}_1, \mathbf{x}_2, ..., \mathbf{x}_n \in \mathcal{H}$, we have $\sum_{j=1}^{n} \mathbf{x}_j \otimes \mathbf{y}_j = \mathbf{0}$. Prove that $\mathbf{x}_j = \mathbf{0}$ for all $1 \leqslant j \leqslant n$.

Exercise 15.6.16. Suppose $A, B \in \mathcal{B}(\mathcal{H})$. Prove that $\|A \otimes B\| = \|A\|\|B\|$.

Exercise 15.6.17. Suppose $A, B, C, D \in \mathcal{B}(\mathcal{H})$. Prove that $(A \otimes B)(C \otimes D) = (AC) \otimes (BD)$.

Exercise 15.6.18. Prove that the $n \times n$ Jordan block

$$
\begin{bmatrix}
\lambda & 1 & 0 & \cdots & 0 \\
0 & \lambda & 1 & \cdots & 0 \\
\vdots & \vdots & \ddots & \ddots & \vdots \\
0 & 0 & 0 & \lambda & 1 \\
0 & 0 & 0 & 0 & \lambda
\end{bmatrix}
$$

is unitarily equivalent to its adjoint.

Exercise 15.6.19. Prove that the direct sum of complex symmetric operators is complex symmetric.

Exercise 15.6.20. This exercise demonstrates that the converse of Exercise 15.6.19 is false.

(a) Prove that the operator $A \in \mathcal{B}(\mathbb{C}^3)$ induced by the matrix

$$
\begin{bmatrix}
0 & 1 & 0 \\
0 & 0 & 2 \\
0 & 0 & 0
\end{bmatrix}
$$

is a not a complex symmetric operator

(b) Let J be a conjugation on $\mathbb{C}^3$ and define

$$
C = \begin{bmatrix} 0 & J \\ J & 0 \end{bmatrix}.
$$

Prove that $T = A \oplus (JA^*J)$ is C-symmetric, but its direct summand A is not.

Exercise 15.6.21. Prove that $S \otimes S$ is unitarily equivalent to the infinite matrix operator defined on $(H^2)^{(\infty)}$ (see (14.1.2)) by

$$
\begin{bmatrix}
0 & 0 & 0 & 0 & 0 & \cdots \\
S & 0 & 0 & 0 & 0 & \cdots \\
0 & S & 0 & 0 & 0 & \cdots \\
0 & 0 & S & 0 & 0 & \cdots \\
0 & 0 & 0 & S & 0 & \cdots \\
\vdots & \vdots & \vdots & \vdots & \vdots & \ddots
\end{bmatrix}.
$$

Exercise 15.6.22. Prove that $S \otimes S$ is an isometry and compute its von Neumann–Wold decomposition.

Exercise 15.6.23. Prove that $S \otimes S^*$ is unitarily equivalent to the infinite matrix operator defined on $(H^2)^{(\infty)}$ by

$$
\begin{bmatrix}
0 & 0 & 0 & 0 & 0 & \cdots \\
S^* & 0 & 0 & 0 & 0 & \cdots \\
0 & S^* & 0 & 0 & 0 & \cdots \\
0 & 0 & S^* & 0 & 0 & \cdots \\
0 & 0 & 0 & S^* & 0 & \cdots \\
\vdots & \vdots & \vdots & \vdots & \vdots & \ddots
\end{bmatrix}.
$$

Exercise 15.6.24. Prove that $S \otimes S^*$ is a partial isometry. Recall Definition 14.9.8.

Exercise 15.6.25. Let

$$
T = \begin{bmatrix}
0 & \frac{1}{2} & 0 & 0 & 0 & \cdots \\
\frac{1}{2} & 0 & \frac{1}{2} & 0 & 0 & \cdots \\
0 & \frac{1}{2} & 0 & \frac{1}{2} & 0 & \cdots \\
0 & 0 & \frac{1}{2} & 0 & \frac{1}{2} & \cdots \\
0 & 0 & 0 & \frac{1}{2} & 0 & \cdots \\
\vdots & \vdots & \vdots & \vdots & \vdots & \ddots
\end{bmatrix}.
$$

If $(u_n)_{n=0}^{\infty}$ is the sequence of Chebyshev polynomials from (15.2.4), prove that $T = U|T|$ is the polar decomposition of T, where

$$
|T| = \left[\frac{2}{\pi} \int_{-\pi}^{\pi} |x| u_n(x) u_m(x) \sqrt{1 - x^2} \, dx \right]_{m,n=0}^{\infty}
$$

and

$$
U = \left[\frac{2}{\pi} \int_{-\pi}^{\pi} e^{i \arg(x)} u_n(x) u_m(x) \sqrt{1 - x^2} \, dx \right]_{m,n=0}^{\infty}.
$$

Exercise 15.6.26. Find the polar decomposition of $S \oplus S^*$.

Exercise 15.6.27. Find the polar decomposition of $S \otimes S^*$.

Exercise 15.6.28. If

$$
(Tf)(w, z) = z\left(\frac{f(w, z) - f(w, 0)}{w}\right)
$$

on $H^2(\mathbb{D}^2)$, prove that

$$
(T^*f)(w, z) = w\left(\frac{f(w, z) - f(0, z)}{z}\right).
$$

Exercise 15.6.29. Prove that $S \otimes S^*$ has an uncountable collection of distinct reducing subspaces $\mathcal{M}_\alpha$ for $\alpha \in \mathbb{R}$ such that $\mathcal{M}_\alpha \subseteq \mathcal{M}_\beta$ if and only if $\alpha \leqslant \beta$.

Exercise 15.6.30. For each $\lambda \in \mathbb{D}$, prove that $\bigvee\{(S \otimes S^*)^n(z \otimes k_\lambda) : n \geqslant 0\}$ is an invariant subspace for $S \otimes S^*$ that is not reducing.

Exercise 15.6.31. Prove that $S + S^*$ has a square root, meaning there is a $B \in \mathcal{B}(H^2)$ such that $B^2 = S + S^*$.

Exercise 15.6.32. Prove that $S \oplus S^*$ does not have a square root, meaning there is no $B \in \mathcal{B}(H^2 \oplus H^2)$ such that $B^2 = S \oplus S^*$.
Remark: See [96] for more on roots and logarithms of operators.

Exercise 15.6.33. Let $A = U|A|$ and $B = V|B|$ be the polar decompositions of $A, B \in \mathcal{B}(\mathcal{H})$.

(a) Prove that $U \otimes V$ is a partial isometry.

(b) Prove that $|A| \otimes |B|$ is positive.

(c) Is $(U \otimes V)(|A| \otimes |B|)$ the polar decomposition of $A \otimes B$?

15.7 Hints for the Exercises

Hint for Ex. 15.6.1: Use $T^*T = I$.
Hint for Ex. 15.6.4: Examine the proof of Theorem 15.2.2.
Hint for Ex. 15.6.9: Consider $W\left(\sum_{n=0}^{\infty} a_n z^n\right) = \left(\sum_{n=0}^{\infty} a_{2n} z^n, \sum_{n=0}^{\infty} a_{2n+1} z^n\right)$.
Hint for Ex. 15.6.10: See Exercise 5.9.28.
Hint for Ex. 15.6.15: Consider the adjoint of the expression.
Hint for Ex. 15.6.23: Consult Theorem 15.4.9.
Hint for Ex. 15.6.29: Let $\varphi : \mathbb{N} \cup \{0\} \to \mathbb{Q}$ be a bijection. For each $\alpha \in \mathbb{R}$, consider $\{x \in \mathbb{N} \cup \{0\} : \varphi(x) < \alpha\}$. Then use Theorem 15.4.13.
Hint for Ex. 15.6.32: Mimic the ideas in Exercise 5.9.33.

16

· · • · ·

Toeplitz Operators

Key Concepts: Toeplitz matrix, Toeplitz operator, Riesz projection, Brown–Halmos characterization of Toeplitz operators, spectral properties, universal Toeplitz operators.

Outline: This chapter surveys *Toeplitz operators* $T_\varphi : H^2 \to H^2$ defined by

$$T_\varphi f = P_+(\varphi f),$$

where $\varphi \in L^\infty(\mathbb{T})$ and P_+ is the orthogonal projection of $L^2(\mathbb{T})$ onto H^2. We examine the matrix representations of these operators, their spectral properties, and a characterization of them related to the unilateral shift.

16.1 Toeplitz Matrices

We discussed diagonal operators on ℓ^2 in Chapter 2 and more general matrix operators on ℓ^2 in Chapter 3. This chapter covers the operators on ℓ^2 induced by matrices of the form

$$T(\mathbf{a}) = \begin{bmatrix} a_0 & a_{-1} & a_{-2} & a_{-3} & a_{-4} & \cdots \\ a_1 & a_0 & a_{-1} & a_{-2} & a_{-3} & \cdots \\ a_2 & a_1 & a_0 & a_{-1} & a_{-2} & \cdots \\ a_3 & a_2 & a_1 & a_0 & a_{-1} & \cdots \\ a_4 & a_3 & a_2 & a_1 & a_0 & \cdots \\ \vdots & \vdots & \vdots & \vdots & \vdots & \ddots \end{bmatrix}, \tag{16.1.1}$$

where $\mathbf{a} = (a_n)_{n=-\infty}^\infty$ is a doubly infinite sequence of complex numbers. These *Toeplitz matrices* are constant on each diagonal. Does a Toeplitz matrix define a bounded operator on ℓ^2? To answer this, recall from Chapter 5 the Hilbert space $L^2(\mathbb{T})$ of Lebesgue-measurable functions on the circle with norm and inner product

$$\|f\| = \left(\int_{\mathbb{T}} |f|^2 dm \right)^{\frac{1}{2}}, \quad \langle f, g \rangle = \int_{\mathbb{T}} f\overline{g}\, dm, \tag{16.1.2}$$

respectively, along with the Hardy space $H^2 = \{f \in L^2(\mathbb{T}) : \widehat{f}(n) = 0$ for all $n < 0\}$. In the above, m is normalized Lebesgue measure on $\mathbb{T}$ and

$$\widehat{f}(n) = \int_{\mathbb{T}} f(\xi)\overline{\xi}^n \, dm(\xi) \quad \text{for } n \in \mathbb{Z}$$

is the nth Fourier coefficient of f.

Theorem 16.1.3 (Hartman–Wintner [182], Toeplitz [364]). *For a Toeplitz matrix $T(\mathbf{a})$, the following are equivalent.*

(a) *$T(\mathbf{a})$ defines a bounded operator on ℓ^2.*

(b) *There is a $\varphi \in L^\infty(\mathbb{T})$ such that $a_n = \widehat{\varphi}(n)$ for all $n \in \mathbb{Z}$.*

Under these circumstances, $\|T(\mathbf{a})\| = \|\varphi\|_\infty$.

Proof We follow [59]. Proposition 8.1.5 says that the multiplication operator $M_\varphi f = \varphi f$ is bounded if and only if $\varphi \in L^\infty(\mathbb{T})$. Furthermore, $\|M_\varphi\| = \|\varphi\|_\infty$. With respect to the orthonormal basis $(\xi^n)_{n=-\infty}^\infty$ for $L^2(\mathbb{T})$, the matrix representation of M_φ is

$$M(\mathbf{a}) = \begin{bmatrix} \ddots & \vdots & \vdots & \vdots & \vdots & \vdots & \vdots & \iddots \\ \cdots & a_0 & a_{-1} & a_{-2} & a_{-3} & a_{-4} & a_{-5} & \cdots \\ \cdots & a_1 & a_0 & a_{-1} & a_{-2} & a_{-3} & a_{-4} & \cdots \\ \cdots & a_2 & a_1 & a_0 & a_{-1} & a_{-2} & a_{-3} & \cdots \\ \cdots & a_3 & a_2 & a_1 & a_0 & a_{-1} & a_{-2} & \cdots \\ \cdots & a_4 & a_3 & a_2 & a_1 & a_0 & a_{-1} & \cdots \\ \cdots & a_5 & a_4 & a_3 & a_2 & a_1 & a_0 & \cdots \\ \iddots & \vdots & \vdots & \vdots & \vdots & \vdots & \vdots & \ddots \end{bmatrix},$$

where $a_n = \widehat{\varphi}(n)$. This comes from the fact that $\langle M_\varphi \xi^n, \xi^m \rangle = \langle \varphi, \xi^{m-n} \rangle = \widehat{\varphi}(m - n)$, and hence the entries are constant along each diagonal. Notice how the Toeplitz matrix $T(\mathbf{a})$ from (16.1.1) is the lower-right corner of $M(\mathbf{a})$.

(b) $\Rightarrow$ (a) From the discussion in the previous paragraph, $M(\mathbf{a})$ is a bounded operator on $\ell^2(\mathbb{Z})$. For each $n \geq 0$, let P_n denote the orthogonal projection of $\ell^2(\mathbb{Z})$ onto $\bigvee\{e_k : k \geq -n\}$, where $(e_k)_{k=-\infty}^\infty$ is the standard orthonormal basis for $\ell^2(\mathbb{Z})$. In other words, for $\mathbf{x} = (x_i)_{i=-\infty}^\infty \in \ell^2(\mathbb{Z})$,

$$P_n(\mathbf{x}) = (\dots, 0, 0, x_{-n}, x_{-(n-1)}, x_{-(n-2)}, \dots).$$

For $\varphi \in L^\infty(\mathbb{T})$,

$$\|T(\mathbf{a})\| = \|P_0 M(\mathbf{a})|_{\ell^2(\mathbb{N}_0)}\| \leq \|M(\mathbf{a})\| = \|\varphi\|_\infty, \tag{16.1.4}$$

and hence $T(\mathbf{a})$ is a bounded operator on ℓ^2.

(a) $\Rightarrow$ (b) If $T(\mathbf{a})$ is bounded, then for each $n \geqslant 0$, we identify $T(\mathbf{a})$ with the lower-right corner of $P_n M(\mathbf{a}) P_n$. Indeed, note that $T(\mathbf{a})$ is naturally defined on $\bigvee\{\mathbf{e}_m : m \geqslant 0\}$ while the lower right corner of $P_n M(\mathbf{a}) P_n$ is naturally defined on

$$\bigvee\{\mathbf{e}_m : m \geqslant -n\}.$$

One can see this with the matrix:

$$M(\mathbf{a}) = \begin{bmatrix} \ddots & \vdots & \vdots & \vdots & \vdots & \vdots & \vdots & \iddots \\ \cdots & a_0 & a_{-1} & a_{-2} & a_{-3} & a_{-4} & a_{-5} & \cdots \\ \cdots & a_1 & a_0 & a_{-1} & a_{-2} & a_{-3} & a_{-4} & \cdots \\ \cdots & a_2 & a_1 & a_0 & a_{-1} & a_{-2} & a_{-3} & \cdots \\ \cdots & a_3 & a_2 & a_1 & a_0 & a_{-1} & a_{-2} & \cdots \\ \cdots & a_4 & a_3 & a_2 & a_1 & a_0 & a_{-1} & \cdots \\ \cdots & a_5 & a_4 & a_3 & a_2 & a_1 & a_0 & \cdots \\ \iddots & \vdots & \vdots & \vdots & \vdots & \vdots & \vdots & \ddots \end{bmatrix}.$$

Since $T(\mathbf{a})$ and $P_n M(\mathbf{a}) P_n$ have the same matrix representation, we can identify them. For each $\mathbf{x} \in \ell^2(\mathbb{Z})$, observe that $P_n \mathbf{x} \to \mathbf{x}$ and hence $P_n M(\mathbf{a}) P_n \mathbf{x} \to M(\mathbf{a})\mathbf{x}$. Thus, for any unit vector $\mathbf{x} \in \ell^2(\mathbb{Z})$,

$$\begin{aligned} \|M(\mathbf{a})\mathbf{x}\| &= \lim_{n \to \infty} \|P_n M(\mathbf{a}) P_n \mathbf{x}\| \\ &= \liminf_{n \to \infty} \|P_n M(\mathbf{a}) P_n \mathbf{x}\| \\ &\leqslant \liminf_{n \to \infty} \|P_n M(\mathbf{a}) P_n\| \|\mathbf{x}\| \\ &= \|T(\mathbf{a})\| \end{aligned}$$

and hence

$$\|M(\mathbf{a})\| = \sup_{\|\mathbf{x}\|=1} \|M(\mathbf{a})\mathbf{x}\| \leqslant \|T(\mathbf{a})\|. \tag{16.1.5}$$

This shows that $M(\mathbf{a})$ is bounded and hence, by the discussion at the beginning of the proof, $\varphi \in L^\infty(\mathbb{T})$. Furthermore, (16.1.4) and (16.1.5) yield $\|T(\mathbf{a})\| = \|\varphi\|_\infty$. $\blacksquare$

The many fascinating properties of $T(\mathbf{a})$ are difficult to deduce in the matrix setting. The next section develops an equivalent function-theoretic viewpoint via Fourier series that permits us to study the deeper properties of Toeplitz matrices.

16.2 The Riesz Projection

Since H^2 is a subspace of $L^2(\mathbb{T})$, there is an orthogonal projection from $L^2(\mathbb{T})$ onto H^2.

Definition 16.2.1. The *Riesz projection* is the operator $P_+ : L^2(\mathbb{T}) \to L^2(\mathbb{T})$ defined by

$$P_+\left(\sum_{n=-\infty}^{\infty} \widehat{f}(n)\xi^n \right) = \sum_{n=0}^{\infty} \widehat{f}(n)\xi^n.$$

Proposition 16.2.2. P_+ *is an orthogonal projection whose range is* H^2.

Proof Notice from Parseval's formula that

$$\left\| P_+\left(\sum_{n=-\infty}^{\infty} \hat{f}(n)\xi^n \right) \right\|^2 = \left\| \sum_{n=0}^{\infty} \hat{f}(n)\xi^n \right\|^2 = \sum_{n=0}^{\infty} |\hat{f}(n)|^2 \leqslant \sum_{n=-\infty}^{\infty} |\hat{f}(n)|^2$$

and thus P_+ is a bounded operator on $L^2(\mathbb{T})$. By definition, ran $P_+ = H^2$. To prove that P_+ is an orthogonal projection, it suffices to prove that it is selfadjoint and idempotent (Theorem 14.5.4). For each $n \in \mathbb{Z}$,

$$P_+(P_+\xi^n)) = \begin{cases} \xi^n & \text{if } n \geqslant 0, \\ 0 & \text{if } n < 0, \end{cases}$$

which equals $P_+(\xi^n)$. Thus, $P_+^2 = P_+$ on $L^2(\mathbb{T})$. A case by case analysis shows that

$$\langle P_+ f, g \rangle = \langle f, P_+ g \rangle \tag{16.2.3}$$

for $f = \xi^n$ and $g = \xi^m$ with $m, n \in \mathbb{Z}$. Linearity of the inner product in the first position and conjugate linearity in the second ensure that (16.2.3) holds for all $f, g \in L^2(\mathbb{T})$. Therefore, $P_+^* = P_+$. ∎

The next proposition, which arises when studying Hankel operators (Chapter 17), is interesting in its own right.

Proposition 16.2.4. *For* $f \in L^2(\mathbb{T})$, *the following are equivalent.*

(a) $f \in H^2$.

(b) $P_+ f = f$.

(c) $\|P_+ f\| = \|f\|$.

Proof (a) $\Rightarrow$ (b) $\Rightarrow$ (c) These follow from the definitions.
(c) $\Rightarrow$ (a) Since $f = P_+ f + (I - P_+)f$ and $P_+ f \perp (I - P_+)f$, it follows from Proposition 1.4.6 (Pythagorean theorem) that

$$\|f\|^2 = \|P_+ f\|^2 + \|(I - P_+)f\|^2 = \|f\|^2 + \|(I - P_+)f\|^2,$$

so $(I - P_+)f = 0$, and hence $f \in H^2$. ∎

With the representation of H^2 as a space of analytic functions on $\mathbb{D}$ (Chapter 5), one can realize the Riesz projection as an integral operator.

Proposition 16.2.5. *If* $f \in L^2(\mathbb{T})$, *then*

$$(P_+ f)(z) = \int_{\mathbb{T}} \frac{f(\xi)}{1 - \overline{\xi} z} \, dm(\xi) \quad \text{for all } z \in \mathbb{D}.$$

Proof Let $\xi \in \mathbb{T}$ and $z \in \mathbb{D}$. Since $|\bar{\xi}z| < 1$, the series

$$\frac{1}{1 - \bar{\xi}z} = \sum_{n=0}^{\infty} (\bar{\xi}z)^n$$

converges absolutely and uniformly in ξ for fixed z. Thus,

$$\int_{\mathbb{T}} \frac{f(\xi)}{1 - \bar{\xi}z} \, dm(\xi) = \int_{\mathbb{T}} f(\xi) \Big(\sum_{n=0}^{\infty} (\bar{\xi}z)^n \Big) dm(\xi)$$

$$= \sum_{n=0}^{\infty} z^n \int_{\mathbb{T}} f(\xi) \bar{\xi}^n \, dm(\xi)$$

$$= \sum_{n=0}^{\infty} \hat{f}(n) z^n,$$

which is the Taylor expansion of the Riesz projection of f (Definition 16.2.1). ∎

16.3 Toeplitz Operators

One gains more traction studying Toeplitz matrices when recasting Theorem 16.1.3 as a result about operators on the Hardy space H^2.

Theorem 16.3.1 (Brown–Halmos [68]). *For $\varphi \in L^\infty$, the operator $T_\varphi : H^2 \to H^2$ defined by $T_\varphi f = P_+(\varphi f)$ is bounded and $\|T_\varphi\| = \|\varphi\|_\infty$.*

Proof For any $f \in H^2$, use the fact that P_+ is an orthogonal projection (and hence a contraction) to see that $\|T_\varphi f\| = \|P_+(\varphi f)\| \leqslant \|\varphi f\|$. Furthermore,

$$\|\varphi f\|^2 = \int_{\mathbb{T}} |\varphi f|^2 dm \leqslant \|\varphi\|_\infty^2 \int_{\mathbb{T}} |f|^2 dm = \|\varphi\|_\infty^2 \|f\|^2.$$

This yields the upper bound

$$\|T_\varphi\| = \sup_{\|f\|=1} \|T_\varphi f\| \leqslant \|\varphi\|_\infty.$$

For the lower bound, recall from Corollary 5.3.15 that for any $\lambda \in \mathbb{D}$, the reproducing kernel $k_\lambda(z) = (1 - \bar{\lambda}z)^{-1}$ for H^2 satisfies

$$\|k_\lambda\| = \frac{1}{\sqrt{1 - |\lambda|^2}}.$$

Define the normalized reproducing kernel $\tilde{k}_\lambda = k_\lambda / \|k_\lambda\|$ and use the Cauchy–Schwarz inequality to obtain

$$|\langle T_\varphi \tilde{k}_\lambda, \tilde{k}_\lambda \rangle| \leqslant \|T_\varphi \tilde{k}_\lambda\| \|\tilde{k}_\lambda\| \leqslant \|T_\varphi\| \|\tilde{k}_\lambda\| \|\tilde{k}_\lambda\| = \|T_\varphi\|.$$

The selfadjointness of P_+ ensures that

$$
\begin{aligned}
\|T_\varphi\| &\geqslant |\langle T_\varphi \widetilde{k}_\lambda, \widetilde{k}_\lambda \rangle| \\
&= |\langle P_+(\varphi \widetilde{k}_\lambda), \widetilde{k}_\lambda \rangle| \\
&= |\langle \varphi \widetilde{k}_\lambda, P_+ \widetilde{k}_\lambda \rangle| \\
&= |\langle \varphi \widetilde{k}_\lambda, \widetilde{k}_\lambda \rangle| \\
&= \left| \int_{\mathbb{T}} \frac{1 - |\lambda|^2}{|\zeta - \lambda|^2} \varphi(\zeta)\, dm(\zeta) \right| \\
&= |\mathscr{P}(\varphi)(\lambda)|,
\end{aligned}
$$

where $\mathscr{P}(\varphi)$ is the Poisson integral of φ defined by (5.5.3). Finally, let $\lambda = r\xi$, where $\xi \in \mathbb{T}$ and $r \in (0,1)$, and use Fatou's theorem (Theorem 5.5.2) to see that $|\varphi(\xi)| \leqslant \|T_\varphi\|$ for almost every $\xi \in \mathbb{T}$. This implies the desired lower bound $\|\varphi\|_\infty \leqslant \|T_\varphi\|$. ∎

Definition 16.3.2. For $\varphi \in L^\infty(\mathbb{T})$, the operator $T_\varphi f = P_+(\varphi f)$ on H^2 is a *Toeplitz operator* with symbol φ.

The previous result implies that the symbol of a Toeplitz operator is unique.

Corollary 16.3.3. $T_\varphi = T_\psi$ *if and only if* $\varphi = \psi$ *almost everywhere on* $\mathbb{T}$.

Proof If $\varphi = \psi$ almost everywhere, then $T_\varphi f = P_+(\varphi f) = P_+(\psi f) = T_\psi f$ for all $f \in H^2$. Thus, $T_\varphi = T_\psi$. Conversely, if $T_\varphi = T_\psi$, then $T_{\varphi-\psi} = T_\varphi - T_\psi = 0$. Theorem 16.3.1 ensures that $\|\varphi - \psi\|_\infty = 0$ and thus $\varphi = \psi$ almost everywhere. ∎

Toeplitz operators T_φ and Toeplitz matrices $T(\mathbf{a})$ from (16.1.1), where $\mathbf{a} = (a_n)_{n=-\infty}^\infty$ is the sequence of Fourier coefficients of φ, are closely related.

Proposition 16.3.4. *For* $\varphi \in L^\infty(\mathbb{T})$, *the matrix representation of* T_φ *with respect to the orthonormal basis* $(\xi^n)_{n=0}^\infty$ *for* H^2 *is* $T(\mathbf{a})$, *where* $\mathbf{a} = (a_n)_{n=-\infty}^\infty$ *and* $a_n = \widehat{\varphi}(n)$ *for all* $n \in \mathbb{Z}$.

Proof With respect to the standard orthonormal basis $(\xi^n)_{n=0}^\infty$ for H^2, the (m, n) entry of the matrix representation of T_φ is

$$
\begin{aligned}
\langle T_\varphi \xi^n, \xi^m \rangle &= \langle P_+(\varphi \xi^n), \xi^m \rangle \\
&= \langle \varphi \xi^n, P_+ \xi^m \rangle && (P_+ \text{ is selfadjoint}) \\
&= \langle \varphi \xi^n, \xi^m \rangle && (P_+ \xi^m = \xi^m \text{ for } m \geqslant 0) \\
&= \langle \varphi, \xi^{m-n} \rangle && (\text{integral inner product - (16.1.2)}) \\
&= \widehat{\varphi}(m - n). && (16.3.5)
\end{aligned}
$$

This shows that the matrix representation of T_φ is $T(\mathbf{a})$. ∎

16.4 Selfadjoint and Compact Toeplitz Operators

The adjoint of a Toeplitz operator is the Toeplitz operator corresponding to the complex conjugate of its symbol. As a consequence of the next proposition, a Toeplitz operator is selfadjoint if and only if its symbol is a real-valued function in $L^\infty(\mathbb{T})$.

Proposition 16.4.1. $T_\varphi^* = T_{\overline{\varphi}}$ *for any* $\varphi \in L^\infty(\mathbb{T})$.

Proof For any $f, g \in H^2$,

$$
\begin{aligned}
\langle T_\varphi f, g \rangle &= \langle P_+(\varphi f), g \rangle \\
&= \langle \varphi f, P_+ g \rangle && (P_+ \text{ is selfadjoint}) \\
&= \langle \varphi f, g \rangle && (P_+ g = g \text{ since } g \in H^2) \\
&= \langle f, \overline{\varphi} g \rangle && (\text{integral inner product - (16.1.2)}) \\
&= \langle P_+ f, \overline{\varphi} g \rangle && (P_+ f = f \text{ since } f \in H^2) \\
&= \langle f, P_+(\overline{\varphi} g) \rangle && (P_+ \text{ is selfadjoint}) \\
&= \langle f, T_{\overline{\varphi}} g \rangle,
\end{aligned}
$$

which proves the result. ∎

Example 16.4.2. From (5.1.3), observe that $T_z = S$ (the forward shift) and $T_{\overline{z}} = S^*$ (the backward shift). Thus,

$$
T_{z+\overline{z}} = \begin{bmatrix}
0 & 1 & 0 & 0 & 0 & \cdots \\
1 & 0 & 1 & 0 & 0 & \cdots \\
0 & 1 & 0 & 1 & 0 & \cdots \\
0 & 0 & 1 & 0 & 1 & \cdots \\
0 & 0 & 0 & 1 & 0 & \cdots \\
\vdots & \vdots & \vdots & \vdots & \vdots & \ddots
\end{bmatrix}
$$

is selfadjoint. It was shown in Example 3.2.7 that this matrix operator on ℓ^2 has norm 2. One can also confirm this by observing that

$$
\sup_{|z|=1} |z + \overline{z}| = 2
$$

and using Theorem 16.3.1. Also recall the spectral decomposition of

$$
T_{\cos\theta} = \frac{1}{2} T_{\overline{z}+z}
$$

(originally due to Hilbert) in Theorem 15.2.2.

There are no interesting compact Toeplitz operators.

Theorem 16.4.3 (Brown–Halmos [68]). *For* $\varphi \in L^\infty(\mathbb{T})$, *the following are equivalent.*

(a) $\varphi = 0$ *almost everywhere on* $\mathbb{T}$.

(b) T_φ *is compact.*

Proof (a) $\Rightarrow$ (b) Observe that $T_0 = 0$ is compact.

(b) $\Rightarrow$ (a) Since $z^n \to 0$ weakly in H^2 (Exercise 4.5.12) and T_φ is compact, it follows that $\|T_\varphi z^n\| \to 0$ (Exercise 3.6.9). On the other hand,

$$\|T_\varphi z^n\|^2 = \left\| P_+\left(\sum_{k=-\infty}^\infty \widehat{\varphi}(k)\zeta^{k+n} \right) \right\|^2 = \left\| \sum_{k=-n}^\infty \widehat{\varphi}(k)\zeta^{k+n} \right\|^2 = \sum_{k=-n}^\infty |\widehat{\varphi}(k)|^2,$$

which tends to $\|\varphi\|^2$ as $n \to \infty$ (Parseval's theorem). Thus, $\|\varphi\| = 0$ and hence $\varphi = 0$ almost everywhere. ∎

Every operator with finite-dimensional range is compact (Exercise 2.8.19). This yields information about the range of a Toeplitz operator.

Corollary 16.4.4. ran T_φ is infinite dimensional for every $\varphi \in L^\infty(\mathbb{T})\backslash\{0\}$.

16.5 The Brown–Halmos Characterization

When is $A \in \mathcal{B}(H^2)$ a Toeplitz operator? The following theorem is an operator-theoretic characterization that involves the unilateral shift S from Chapter 5.

Theorem 16.5.1 (Brown–Halmos [68]). *For $A \in \mathcal{B}(H^2)$, the following are equivalent.*

(a) *A is a Toeplitz operator.*

(b) *$S^*AS = A$.*

Proof (a) $\Rightarrow$ (b) If $A = T_\varphi$ for some $\varphi \in L^\infty(\mathbb{T})$, then

$$
\begin{aligned}
\langle S^* T_\varphi S\xi^n, \xi^m \rangle &= \langle T_\varphi S\xi^n, S\xi^m \rangle \\
&= \langle T_\varphi \xi^{n+1}, \xi^{m+1} \rangle \\
&= \widehat{\varphi}((m+1) - (n+1)) \qquad\qquad \text{(by (16.3.5))} \\
&= \widehat{\varphi}(m-n) \\
&= \langle T_\varphi \xi^n, \xi^m \rangle. \qquad\qquad\qquad \text{(by (16.3.5))}
\end{aligned}
$$

Thus, the operators $S^* T_\varphi S$ and T_φ have the same matrix representations with respect to the orthonormal basis $(\xi^n)_{n=0}^\infty$ for H^2 and are therefore equal (Exercise 3.6.2).

(b) $\Rightarrow$ (a) Suppose that $S^*AS = A$. Induction ensures that $S^{*k}AS^k = A$ for all $k \geqslant 0$. For all $m, n \geqslant 0$, deduce that

$$\langle A\xi^n, \xi^m \rangle = \langle S^{*k}AS^k\xi^n, \xi^m \rangle = \langle AS^k\xi^n, S^k\xi^m \rangle = \langle A\xi^{n+k}, \xi^{m+k} \rangle.$$

Consequently, the matrix representation of A with respect to $(\xi^n)_{n=0}^\infty$ is a Toeplitz matrix since the entries are constant along each diagonal. Toeplitz's theorem (Theorem 16.1.3) yields a $\varphi \in L^\infty(\mathbb{T})$ such that A has the same matrix representation as T_φ. Thus, $A = T_\varphi$ (Exercise 3.6.2). ∎

16.6 Analytic and Co-analytic Symbols

We have seen a special class of Toeplitz operators in Corollary 5.6.2 when we described the commutant of the unilateral shift S.

Definition 16.6.1. A Toeplitz operator T_φ is *analytic* if $\varphi \in H^\infty$ and *co-analytic* if $\overline{\varphi} \in H^\infty$.

Proposition 16.6.2. *If T_φ is an analytic Toeplitz operator with nonconstant symbol φ, the following hold.*

(a) $T_\varphi f = \varphi f$ *for all* $f \in H^2$.

(b) $\sigma_p(T_\varphi) = \varnothing$.

(c) $\sigma(T_\varphi) = \varphi(\mathbb{D})^-$.

(d) *The matrix representation of T_φ with respect to the basis $(\xi^n)_{n=0}^\infty$ is lower triangular.*

Proof (a) Since $\varphi \in H^\infty$, it follows that $\varphi f \in H^2$ for all $f \in H^2$ and hence $T_\varphi f = P_+(\varphi f) = \varphi f$.

(b) If $\lambda \in \mathbb{C}$ and $(T_\varphi - \lambda I)f = 0$, then (a) ensures that $(\varphi - \lambda)f = 0$. Since $\varphi - \lambda$ and f are analytic, and φ is nonconstant, f is the zero function. Thus, $\sigma_p(T_\varphi) = \varnothing$.

(c) From (5.5.5), it follows that $T_{\overline{\varphi}}k_\lambda = \overline{\varphi(\lambda)}k_\lambda$ for all $\lambda \in \mathbb{D}$. Since $T_\varphi^* = T_{\overline{\varphi}}$ and $\sigma(T_\varphi^*) = \overline{\sigma(T_\varphi)}$, one deduces that $\overline{\varphi(\mathbb{D})} \subseteq \sigma_p(T_{\overline{\varphi}}) \subseteq \sigma(T_{\overline{\varphi}}) = \overline{\sigma(T_\varphi)}$. Since the spectrum of T_φ is compact, it follows that $\varphi(\mathbb{D})^- \subseteq \sigma(T_\varphi)$. If $\lambda \notin \varphi(\mathbb{D})^-$, then $(\varphi - \lambda)^{-1} \in H^\infty$. Thus $T_{(\varphi-\lambda)^{-1}}$ is a bounded operator and is equal to $(T_\varphi - \lambda I)^{-1}$. Therefore, $\sigma(T_\varphi) = \varphi(\mathbb{D})^-$.

(d) The entries above the main diagonal of the matrix representation of T_φ with respect to the basis $(\xi^n)_{n=0}^\infty$ are

$$\langle T_\varphi \xi^k, \xi^j \rangle = \int_{\mathbb{T}} \varphi(\xi)\overline{\xi}^{\,j-k}\, dm(\xi) = \widehat{\varphi}(j-k) = 0 \quad \text{for all } j < k,$$

since $\varphi \in H^\infty$. ∎

Observe that Corollary 5.6.2 (the description of the commutant of the shift on H^2) can be stated in the following equivalent form.

Corollary 16.6.3. *For $\varphi \in L^\infty$, the following are equivalent.*

(a) $\varphi \in H^\infty$.

(b) $T_\varphi S = S T_\varphi$.

Proposition 16.6.4. *If $\varphi \in H^\infty \setminus \{0\}$, then* ran $T_{\overline{\varphi}}$ *is dense.*

Proof Proposition 16.6.2 ensures that ker $T_\varphi = \{0\}$. Proposition 3.1.7 implies that

$$(\text{ran } T_{\overline{\varphi}})^- = (\text{ker } T_{\overline{\varphi}}^*)^\perp = (\text{ker } T_\varphi)^\perp = \{0\}^\perp = H^2,$$

which proves the result. ∎

Recall from Theorem 5.4.12 that the proper invariant subspaces for the backward shift on H^2 are precisely those subspaces of the form $(\varphi H^2)^\perp$, in which φ is an inner function.

Proposition 16.6.5. *If φ is inner, then* $\ker T_{\overline{\varphi}} = (\varphi H^2)^\perp$.

Proof For all $h \in H^2$,

$$
\begin{aligned}
f \in (\varphi H^2)^\perp &\iff \langle f, \varphi h \rangle = 0 \\
&\iff \langle \overline{\varphi} f, h \rangle = 0 \\
&\iff \langle \overline{\varphi} f, P_+ h \rangle = 0 \\
&\iff \langle P_+(\overline{\varphi} f), h \rangle = 0 \\
&\iff \langle T_{\overline{\varphi}} f, h \rangle = 0 \\
&\iff T_{\overline{\varphi}} f = 0 \\
&\iff f \in \ker T_{\overline{\varphi}},
\end{aligned}
$$

which proves the result. ∎

16.7 Universal Toeplitz Operators

Recall the definition of a universal operator (Definition 13.3.1) and Caradus' criterion (Theorem 13.3.2) for universality. Here is a universal Toeplitz operator on H^2.

Proposition 16.7.1. *If φ is inner and not a finite Blaschke product, then $T_{\overline{\varphi}}$ is a universal operator on H^2.*

Proof We use Caradus' theorem (Theorem 13.3.2) and verify that $\operatorname{ran} T_{\overline{\varphi}} = H^2$ and $\dim \ker T_{\overline{\varphi}} = \infty$. For any $g \in H^2$, use the fact that $\varphi \overline{\varphi} = 1$ almost everywhere on $\mathbb{T}$ to see that $T_{\overline{\varphi}}(\varphi g) = P_+(\overline{\varphi} \varphi g) = P_+(g) = g$. Hence $\operatorname{ran} T_{\overline{\varphi}} = H^2$.

Proposition 16.6.5 implies that $\ker T_{\overline{\varphi}} = (\varphi H^2)^\perp$. It remains to argue that $(\varphi H^2)^\perp$ is infinite dimensional. One can verify that $T_z^{*n} \varphi \in (\varphi H^2)^\perp$ for all $n \geqslant 1$ (Exercise 16.9.12). We claim that $\{T_z^{*n} \varphi : n \geqslant 1\}$ is a set of linearly independent vectors. Fix n and suppose there are constants $c_1, c_2, \ldots, c_n$ such that

$$
\sum_{j=1}^{n} c_j T_z^{*j} \varphi = 0.
$$

Then $\varphi \in \ker T_{\overline{p}}$, where $p(z) = \overline{c_1} z + \overline{c_2} z^2 + \cdots + \overline{c_n} z^n$. We leave it to the reader (Exercise 16.9.14) to prove that $\ker T_{\overline{p}}$ consists only of certain rational functions in H^2. Since φ is not a rational function, $c_j = 0$ for all $1 \leqslant j \leqslant n$. Therefore, $\{T_z^{*j} \varphi : 1 \leqslant j \leqslant n\}$ is a linearly independent set of vectors. Since this is true for all $n \geqslant 1$, it follows that $\ker T_{\overline{\varphi}} = (\varphi H^2)^\perp$ is infinite dimensional. ∎

16.8 Notes

The literature on Toeplitz operators is vast and well documented in [60]. A recent and detailed historical survey is [254]. Some of the early work on multiplication operators on $L^2(\mathbb{T})$, although in terms of matrix multiplication operators, was done by Toeplitz in 1910-11 [363, 364]. Hartman and Wintner in 1950 [181] were early adopters of the term "Toeplitz matrix" for the infinite Toeplitz matrix. They also discussed some of the first results on the spectrum of these operators. Furthermore, they proved that a nonzero multiplication operator on $L^2(\mathbb{T})$ is not compact and that nonzero Toeplitz operators are not compact. The proof of Theorem 16.1.3 appears in an appendix of a paper of Hartman and Wintner [182] but was proved decades earlier for selfadjoint Toeplitz matrices by Toeplitz [363, 364]. A proof also appears in [68]. The commutant of S in terms of analytic Toeplitz operators (Corollary 16.6.3) appears in [68].

We explored the spectrum of T_φ for $\varphi \in H^\infty$ (Proposition 16.6.2). For general $\varphi \in L^\infty(\mathbb{T})$, the spectrum of T_φ satisfies $\mathcal{R}_\varphi \subseteq \sigma(T_\varphi) \subseteq \mathrm{co}(\mathcal{R}_\varphi)$, where $\mathcal{R}_\varphi$ is the essential range of φ (Definition 8.1.11) and $\mathrm{co}(\mathcal{R}_\varphi)$ is the closed convex hull of $\mathcal{R}_\varphi$. If φ is real valued, then $\sigma(T_\varphi) = [\mathrm{essinf}\,\varphi, \mathrm{esssup}\,\varphi]$. If $\varphi \in C(\mathbb{T})$, then $\sigma(T_\varphi) = \mathcal{R}(\varphi) \cup \{\lambda \in \mathbb{C} : \mathrm{wind}(\varphi, \lambda) \neq 0\}$, where $\mathrm{wind}(\varphi, \lambda)$ is the winding number of the curve $\theta \mapsto \varphi(e^{i\theta})$ with respect to λ. The book [116] surveys all of the results above. The reader is encouraged to work through the exercises for this chapter to learn more about the spectrum of a Toeplitz operator.

The kernels of Toeplitz operators have undergone intense study. Coburn [88] proved that if $\varphi \in L^\infty(\mathbb{T}) \backslash \{0\}$ then either $\ker T_\varphi = \{0\}$ or $\ker T_{\overline{\varphi}} = \{0\}$. If $\varphi \in H^\infty$, then $\ker T_\varphi = \{0\}$ and $\ker T_{\overline{\varphi}} = (uH^2)^\perp$, where u is the inner factor of φ. For general $\varphi \in L^\infty(\mathbb{T})$, these kernels were described by Sarason [328] in his characterization of nearly-invariant subspaces of H^2 [327]. See [139] for a function-theoretic parametrization of the kernel of a Toeplitz operator. The paper [183] is an informative survey of kernels of Toeplitz operators and their connection to many areas of analysis.

The invertibility of a Toeplitz operator has many characterizations [109, 373]. There are also criteria for a Toeplitz operator to be surjective [184].

Brown and Douglas determined when a Toeplitz operator is a partial isometry [67]. A Toeplitz operator T_φ is an isometry if and only if φ is an inner function (Exercise 16.9.7). Furthermore, $T_{\overline{\varphi}}$ is a partial isometry when φ is inner (Exercise 16.9.8). The main result in [67] is that the partially isometric Toeplitz operators are of the form T_φ or $T_{\overline{\varphi}}$ where φ is inner. The primary tool used to prove this result makes a connection to Exercise 5.9.25, which asks when a unimodular function q on $\mathbb{T}$ can be written as the quotient of two inner functions. If $\lambda > 0$ and φ, ψ are inner functions, then $\lambda T_{\overline{\varphi}} T_\psi$ is norm attaining (Exercise 16.9.17). Moreover, any norm-attaining Toeplitz operator is of this form. It follows (Exercise 16.9.17) that q can be written as the quotient of two inner functions if and only if T_q is norm attaining.

Describing the invariant subspaces for Toeplitz operators is a complicated undertaking. We know a full characterization of them for T_z and $T_{\overline{z}}$ since these are the forward and backward shifts on H^2. For other Toeplitz operators, it is not clear that proper nonzero invariant subspaces exist. On the other hand, certain Toeplitz operators are universal (Proposition 16.7.1) and thus yield an incredibly complicated invariant-subspace structure. For T_φ, where $\varphi \in H^\infty$, we know that uH^2 is an invariant subspace for any inner function u. A paper of Peller [265] gives some geometric criteria guaranteeing the existence of invariant subspaces of Toeplitz operators with certain piecewise-continuous symbols.

An important structural result is one of Coburn [89, 90] which starts with the fact that the semicommutator $T_\varphi T_\psi - T_{\varphi\psi}$ is compact for every $\varphi, \psi \in C(\mathbb{T})$. He uses this to show that if $\mathcal{K}$ is the closed ideal of compact operators on H^2, then the mapping $\varphi \mapsto T_\varphi + \mathcal{K}$ is an isometric $*$-isomorphism between $C(\mathbb{T})$ and the commutative subalgebra $\{T_\varphi + \mathcal{K} : \varphi \in C(\mathbb{T})\}$ of $\mathcal{B}(H^2)/\mathcal{K}$.

For $\varphi \in L^\infty(\mathbb{R})$, define a Toeplitz operator on $H^2(\mathbb{R})$ by $T_\varphi f = P(\varphi f)$, where P denotes the orthogonal projection from $L^2(\mathbb{R})$ onto $H^2(\mathbb{R})$. Many analogues of standard results for Toeplitz operators on H^2 hold, but some are slightly different. For example, the Brown–Halmos characterization from Theorem 16.5.1 needs to be replaced by the following: if $A \in \mathcal{B}(H^2(\mathbb{R}))$, then A is a Toeplitz operator if and only if $A = T_{e^{-i\lambda t}} A T_{e^{i\lambda t}}$ for all $\lambda > 0$ (see [251, p. 273] and Exercise 16.9.28).

One can also explore Toeplitz operators on other function spaces such as the Bergman space A^2 (see Chapter 10). Indeed, for $\varphi \in L^\infty(dA)$, define T_φ on A^2 by $T_\varphi f = P_{A^2}(\varphi f)$, where P_{A^2} denotes the orthogonal projection of $L^2(dA)$ onto A^2. A good place to get started in this area is the book [379] (see also Exercises 16.9.31, 16.9.32, and 16.9.33).

16.9 Exercises

Exercise 16.9.1. Find the matrix representation of the Riesz projection P_+ on $L^2(\mathbb{T})$ with respect to the orthonormal basis $(\xi^n)_{n=-\infty}^\infty$.

Exercise 16.9.2. Prove the following variation of Proposition 16.2.4 for $P_- = I - P_+$. For $f \in H^2$, the following conditions are equivalent: (a) $f \in \overline{H_0^2}$; (b) $P_- f = f$; (c) $\|P_- f\| = \|f\|$.

Exercise 16.9.3. Consider the multiplication operator M_φ on $L^2(\mathbb{T})$ whose matrix representation with respect to the orthonormal basis $(\xi^n)_{n=-\infty}^\infty$ is

$$\begin{bmatrix} \ddots & \vdots & \vdots & \vdots & \vdots & \vdots & \vdots & \iddots \\ \cdots & 1 & 2 & 1 & 0 & 0 & 0 & \cdots \\ \cdots & 0 & 1 & 2 & 1 & 0 & 0 & \cdots \\ \cdots & 0 & 0 & 1 & 2 & 1 & 0 & \cdots \\ \cdots & 0 & 0 & 0 & 1 & 2 & 1 & \cdots \\ \cdots & 0 & 0 & 0 & 0 & 1 & 2 & \cdots \\ \cdots & 0 & 0 & 0 & 0 & 0 & 1 & \cdots \\ \iddots & \vdots & \vdots & \vdots & \vdots & \vdots & \vdots & \ddots \end{bmatrix}.$$

(a) Compute φ.

(b) Compute $\|M_\varphi\|$.

(c) Compute $\sigma(M_\varphi)$.

Exercise 16.9.4. For each $\alpha, \beta \in \mathbb{D}$, consider the Toeplitz matrix

$$\begin{bmatrix} 1 & \alpha & \alpha^2 & \alpha^3 & \alpha^4 & \cdots \\ \beta & 1 & \alpha & \alpha^2 & \alpha^3 & \cdots \\ \beta^2 & \beta & 1 & \alpha & \alpha^2 & \cdots \\ \beta^3 & \beta^2 & \beta & 1 & \alpha & \cdots \\ \beta^4 & \beta^3 & \beta^2 & \beta & 1 & \cdots \\ \vdots & \vdots & \vdots & \vdots & \vdots & \ddots \end{bmatrix}.$$

(a) Compute the symbol of the corresponding Toeplitz operator.

(b) Compute the norm of this operator.

Exercise 16.9.5. For $\gamma \in \mathbb{C} \backslash \mathbb{Z}$, consider the *Cauchy–Laurent matrix L* associated with γ as the doubly infinite matrix whose entries are

$$L_{jk} = \frac{1}{j - k + \gamma} \quad \text{for } j, k \in \mathbb{Z}.$$

(a) Prove that L is the matrix (with respect to the standard basis) that corresponds to the multiplication operator M_{φ_γ} on $L^2(\mathbb{T})$, where

$$\varphi_\gamma(e^{i\theta}) = \frac{\pi}{\sin \pi \gamma} e^{i\pi\gamma} e^{-i\gamma\theta} \quad \text{for } -\pi \leqslant \theta < \pi.$$

(b) Compute the norm and spectrum of L.

Exercise 16.9.6. For a Toeplitz operator T_φ on H^2, prove the following.

(a) $T_\varphi \geqslant 0$ if and only if $\varphi \geqslant 0$ almost everywhere.

(b) T_φ is selfadjoint if and only if φ is real valued almost everywhere.

(c) T_φ is an orthogonal projection if and only if $T_\varphi = 0$ or $T_\varphi = I$.

(d) T_φ is normal if and only if $\varphi = a\psi + b$, where $a, b \in \mathbb{C}$ and $\psi \in L^\infty(\mathbb{T})$ is real valued.

Remark: See [68] for more on algebraic properties of Toeplitz operators.

Exercise 16.9.7. Let $\varphi \in L^\infty(\mathbb{T})$.

(a) Prove that T_φ is an isometry if and only if φ is an inner function.

(b) Prove that T_φ is unitary if and only if φ is a constant function of modulus one.

Exercise 16.9.8. If φ is inner, prove that $T_{\overline{\varphi}}$ is a partial isometry.
Remark: Brown and Douglas proved that the only Toeplitz operators (other than the zero operator) that are partial isometries are T_φ and T_φ^*, where φ is an inner function [67].

Exercise 16.9.9. Give an example of a partially isometric Toeplitz operator that is not an isometry.

Exercise 16.9.10. Suppose $T \in \mathcal{B}(H^2)$ satisfies $T(uH^2) \subseteq uH^2$ for every inner function u. Prove that T is an analytic Toeplitz operator.

Exercise 16.9.11. If $\varphi \in H^\infty$ and T_φ is cyclic, prove that φ is injective.

Exercise 16.9.12. For any inner function φ, prove that $T_z^{*n}\varphi \in (\varphi H^2)^\perp$ for all $n \geqslant 1$.

Exercise 16.9.13. Consider the Toeplitz matrix

$$T = \begin{bmatrix} 0 & 1 & \frac{1}{2} & \frac{1}{3} & \frac{1}{4} & \cdots \\ -1 & 0 & 1 & \frac{1}{2} & \frac{1}{3} & \cdots \\ -\frac{1}{2} & -1 & 0 & 1 & \frac{1}{2} & \cdots \\ -\frac{1}{3} & -\frac{1}{2} & -1 & 0 & 1 & \cdots \\ -\frac{1}{4} & -\frac{1}{3} & -\frac{1}{2} & -1 & 0 & \cdots \\ \vdots & \vdots & \vdots & \vdots & \vdots & \ddots \end{bmatrix}.$$

(a) Prove that $\|T\| = \pi$.

(b) Prove that the associated lower-triangular Toeplitz matrix

$$\begin{bmatrix} 0 & 0 & 0 & 0 & 0 & \cdots \\ -1 & 0 & 0 & 0 & 0 & \cdots \\ -\frac{1}{2} & -1 & 0 & 0 & 0 & \cdots \\ -\frac{1}{3} & -\frac{1}{2} & -1 & 0 & 0 & \cdots \\ -\frac{1}{4} & -\frac{1}{3} & -\frac{1}{2} & -1 & 0 & \cdots \\ \vdots & \vdots & \vdots & \vdots & \vdots & \ddots \end{bmatrix}$$

does not define a bounded operator on ℓ^2.

Exercise 16.9.14. This exercise computes $\ker T_p^*$, where p is a polynomial.

(a) Prove that if q is a polynomial with no zeros in $\mathbb{D}$, then $\ker T_q^* = \{0\}$.

(b) Let $\lambda_1, \lambda_2, \ldots, \lambda_n$ be the zeros of p in $\mathbb{D}$ with corresponding multiplicities $m_1, m_2, \ldots, m_n$ and let B be the finite Blaschke product

$$B(z) = \left(\frac{z - \lambda_1}{1 - \overline{\lambda_1}z}\right)^{m_1} \left(\frac{z - \lambda_2}{1 - \overline{\lambda_2}z}\right)^{m_2} \cdots \left(\frac{z - \lambda_n}{1 - \overline{\lambda_n}z}\right)^{m_n}.$$

(c) Prove that $\ker T_B^* = H^2 \cap (BH^2)^\perp$.

(d) Use the Cauchy integral formula to prove that

$$H^2 \cap (BH^2)^\perp = \bigvee\left\{\frac{1}{(1 - \overline{\lambda_j}z)^k} : 1 \leqslant j \leqslant n, \, 1 \leqslant k \leqslant m_j\right\}.$$

(e) Write $T_p = T_B T_{p/B}$ and prove that $\ker T_p^* = \ker T_B^*$.

Exercise 16.9.15. If $\varphi \in \overline{H^\infty}$ or $\psi \in H^\infty$, prove that $T_\varphi T_\psi = T_{\varphi\psi}$.
Remark: A result from [68] says that for $\varphi, \psi \in L^\infty$, $T_\varphi T_\psi$ is a Toeplitz operator if and only if $\varphi \in \overline{H^\infty}$ or $\psi \in H^\infty$. Moreover, if either occurs, $T_\varphi T_\psi = T_{\varphi\psi}$.

Exercise 16.9.16. For an inner function φ and $f \in H^\infty$, prove that $\operatorname{ran} T_{\overline{\varphi f}} = \operatorname{ran} T_{\overline{f}}$.

Exercise 16.9.17. If φ, ψ are inner and $\lambda > 0$, prove that $T = \lambda T_\varphi^* T_\psi$ is a norm-attaining Toeplitz operator. In other words, T is a Toeplitz operator and there is an $f \in H^2$ such that $\|Tf\| = \|T\|$.

Exercise 16.9.18. Suppose that both f and $1/f$ belong to $L^\infty(\mathbb{T})$. Prove that T_f is invertible if and only if $T_{f/|f|}$ is invertible.
Remark: This exercises needs a detail concerning *outer functions* [149, p. 64]: if $g \in L^\infty(\mathbb{T})$ and $\log|g| \in L^\infty(\mathbb{T})$, there is an $G \in H^\infty$ such that $|G(\xi)| = |g(\xi)|$ for almost every $\xi \in \mathbb{T}$.

Exercise 16.9.19. For $\varphi, \psi \in L^\infty(\mathbb{T})$, prove that $T_\varphi T_\psi = 0$ if and only if $\varphi = 0$ or $\psi = 0$ almost everywhere.
Remark: This problem requires the following result of Riesz [118, p. 17] (see also Exercise 5.9.37): if $f \in H^\infty$ is zero on a subset of $\mathbb{T}$ of positive measure, then $f \equiv 0$. Although significantly harder to prove, the zero-product result for two Toeplitz operators can be generalized to the following statement: $T_{\varphi_1} T_{\varphi_2} \cdots T_{\varphi_n} = 0$ if and only if at least one of the φ_j is zero [18].

Exercise 16.9.20. Recall from (6.4.1) that $A \in \mathcal{B}(\mathcal{H})$ is hyponormal if $A^*A - AA^* \geqslant 0$.

(a) Prove that $T_{z+\overline{z}/2}$ is hyponormal.

(b) Prove that if $A \in \mathcal{B}(\mathcal{H})$ is hyponormal, then $A + \lambda A^*$ is hyponormal if and only if $|\lambda| \leqslant 1$.

Exercise 16.9.21. Prove that if $(\varphi_n)_{n=1}^\infty$ is a sequence in $L^\infty(\mathbb{T})$ such that $T_{\varphi_n} \to T$ (WOT) (recall the weak operator topology from Exercise 4.5.24), then T is a Toeplitz operator.

Exercise 16.9.22. Let $\varphi \in H^\infty \backslash \{0\}$.

(a) If $\mathcal{M}$ is a nonzero invariant subspace for T_φ, show that $\dim \mathcal{M} = \infty$.

(b) If $A \in \mathcal{B}(H^2)$ is compact and $AT_\varphi = T_\varphi A$, prove that $\sigma(A) = \{0\}$.

Exercise 16.9.23. Let $\varphi \in L^\infty(\mathbb{T})$.

(a) Prove that $\sigma(M_\varphi) \subseteq \sigma(T_\varphi)$.

(b) Give an example of a φ such that $\sigma(M_\varphi) \subsetneq \sigma(T_\varphi)$.

Exercise 16.9.24. Let $\varphi \in L^\infty(\mathbb{T})$.

(a) Prove that T_φ is quasinilpotent if and only if $\varphi = 0$.

(b) Prove that if $\sigma(T_\varphi)$ is a singleton, then φ is constant.

Exercise 16.9.25. If $\varphi \in L^\infty(\mathbb{T})$ is nonconstant, prove that $\sigma(T_\varphi)$ is not a set consisting of two points. Use the following steps from [68].

(a) Prove that if T_φ is selfadjoint, then $\sigma_p(T_\varphi) = \varnothing$.

(b) Prove that if $\sigma(T_\varphi) \subseteq \mathbb{R}$, then T_φ is selfadjoint.

(c) If $\sigma(T_\psi)$ lies on a line, prove that $\alpha T_\psi + \beta I$ is selfadjoint for some $\alpha, \beta \in \mathbb{C}$.

(d) If $\sigma(T_\psi) = \{a, b\}$, use (a) - (c) to obtain a contradiction.

Remark: The spectrum of a Toeplitz operator is connected [374].

Exercise 16.9.26. Consider the numerical range $W(T_\varphi)$ of a Toeplitz operator.

(a) For $A \in \mathcal{B}(\mathcal{H})$, prove that $\sigma(A) \subseteq W(A)^-$.

(b) Prove that $W(T_\varphi) \subseteq W(M_\varphi)$.

(c) Prove that $\sigma(M_\varphi) \subseteq W(T_\varphi)^-$.

(d) Prove that the convex hull of the essential range $\mathcal{R}_\varphi$ is contained in $W(T_\varphi)^-$.

Remark: One actually has equality in (d) [68].

Exercise 16.9.27. Let

$$
f_n(x) = \begin{cases} \dfrac{1}{\sqrt{\pi}} \dfrac{(x-i)^n}{(x+i)^{n+1}} & \text{if } n \geqslant 0, \\[3mm] \dfrac{1}{\sqrt{\pi}} \dfrac{(x+i)^{-n-1}}{(x-i)^{-n}} & \text{if } n \leqslant -1. \end{cases}
$$

Recall from Exercise 11.10.10 that $(f_n)_{n=-\infty}^{\infty}$ is an orthonormal basis for $L^2(\mathbb{R})$ and that $(f_n)_{n=0}^{\infty}$ is an orthonormal basis for $H^2(\mathbb{R})$.

(a) Find an integral representation for P, the orthogonal projection from $L^2(\mathbb{R})$ onto $H^2(\mathbb{R})$, with respect to the orthonormal basis $(f_n)_{n=-\infty}^{\infty}$.

(b) For the Toeplitz operator $T_\Phi f = P(\Phi f)$ on $H^2(\mathbb{C}_+)$, defined in the endnotes of this chapter, compute the matrix representation of T_Φ with respect to the basis $(f_n)_{n=0}^{\infty}$.

Exercise 16.9.28. If $A \in \mathcal{B}(H^2(\mathbb{C}_+))$, prove that A is a Toeplitz operator on $H^2(\mathbb{C}_+)$ if and only if $A = T_{e^{-i\lambda t}} A T_{e^{i\lambda t}}$ for all $\lambda > 0$.

Exercise 16.9.29. For $\lambda > 0$, define

$$
\varphi_\lambda(z) = \exp\left(-\lambda \frac{1+z}{1-z} \right).
$$

Prove that $A \in \mathcal{B}(H^2)$ is a Toeplitz operator if and only if $A = T_{\varphi_\lambda}^* A T_{\varphi_\lambda}$ for all $\lambda > 0$.
Remark: See [245] for more on this.

Exercise 16.9.30. For a Toeplitz operator T_φ, consider

$$\|T_\varphi\|_e := \inf_{K \in \mathcal{K}} \|T_\varphi + K\|,$$

where $\mathcal{K}$ is the space of compact operators on H^2. The quantity $\|T_\varphi\|_e$ is the *essential norm* of T_φ and it measures the distance from T_φ to the compact operators. The goal of this problem is to prove that $\|T_\varphi\|_e = \|\varphi\|_\infty$.

(a) Prove that $\|T_\varphi\|_e \leqslant \|\varphi\|_\infty$.

(b) For each $n \geqslant 1$, prove that $P_n = S^n S^{*n}$ is the orthogonal projection of H^2 onto $z^n H^2$.

(c) For $A \in \mathcal{B}(H^2)$, prove that $\|S^n A S^{*n}\| = \|A\|$ for all $n \geqslant 0$.

(d) Prove that $\|T_\varphi + K\| \geqslant \|T_\varphi + S^{*n} K S^n\|$ for all $n \geqslant 1$ and $K \in \mathcal{K}$.

(e) Use Exercise 5.9.26 to prove that $\|T_\varphi + K\| \geqslant \|T_\varphi\|$ and hence $\|T_\varphi\|_e = \|\varphi\|_\infty$.

Remark: We encounter the essential norm for Hankel operators in Exercise 17.10.13.

Exercise 16.9.31. Theorem 16.4.3 says there are no nonzero compact Toeplitz operators on H^2. For the Bergman space, the situation is different. These next two problems explore this phenomenon. Let $C(\mathbb{D}^-)$ denote the space of all complex-valued continuous functions φ on $\mathbb{D}^-$, endowed with the norm $\|\varphi\|_\infty = \sup_{z \in \mathbb{D}^-} |\varphi(z)|$. Recall from Exercise 10.7.12 the properties of the Bergman projection P_{A^2}.

(a) If $\varphi \in C(\mathbb{D}^-)$, prove that the Toeplitz operator $T_\varphi f = P_{A^2}(\varphi f)$ on the Bergman space A^2 is bounded and $\|T_\varphi\| \leqslant \|\varphi\|_\infty$.

(b) For $N \geqslant 1$, find the matrix representation of $T_{\bar{z}^N}$ with respect to the orthonormal basis $(\sqrt{n+1}\, z^n)_{n=0}^\infty$ for A^2 (Proposition 10.1.8).

Exercise 16.9.32. This is a continuation of Exercise 16.9.31. If $\varphi \in C(\mathbb{D}^-)$ and $\varphi|_\mathbb{T} = 0$, prove that T_φ is a compact operator on A^2 as follows.

(a) Given $\varepsilon > 0$, prove there exists a $\psi \in C(\mathbb{D}^-)$ whose support K is a compact subset of $\mathbb{D}$ and such that $\sup_{z \in \mathbb{D}^-} |\varphi(z) - \psi(z)| < \varepsilon$.

(b) Prove that if a sequence $(f_n)_{n=1}^\infty$ in A^2 converges weakly to zero, then $f_n(\lambda) \to 0$ for each $\lambda \in \mathbb{D}$.

(c) Use Exercise 10.7.20 and the principle of uniform boundedness (Theorem 2.2.3) to prove that $f_n \to 0$ uniformly on K.

(d) Prove there is a $C > 0$ such that $\|T_\psi f_n\| \leqslant C \sup_{z \in K} |f_n(z)|$ for all $n \geqslant 1$.

(e) Prove that T_ψ is compact.

(f) Prove that T_φ is compact.

Remark: See [30] for a generalization of this result to Bergman spaces of general domains.

Exercise 16.9.33. Recall the Berezin transform $\widetilde{T}(z)$ of a bounded operator T on A^2 from Exercise 10.7.27.

(a) If $\varphi \in C(\mathbb{D}^-)$, prove that $\widetilde{T_\varphi} \in C(\mathbb{D}^-)$ and $\widetilde{T_\varphi}(\xi) = \varphi(\xi)$ for all $\xi \in \mathbb{T}$.

(b) Use (a) to prove a converse to Exercise 16.9.32, namely, if $\varphi \in C(\mathbb{D}^-)$ and T_φ is compact, then $\varphi|_\mathbb{T} = 0$.

Remark: See [31, 91] for more on this.

16.10 Hints for the Exercises

Hint for Ex. 16.9.4: Write the matrix as an infinite linear combination of S^n and S^{*n}.

Hint for Ex. 16.9.8: Consult Proposition 14.9.14.

Hint for Ex. 16.9.10: Theorem 14.3.4 implies that $T_{z^n}TT_{z^n} = TT_{z^n}$ for all $n \geqslant 0$.

Hint for Ex. 16.9.16: Consult Corollary 14.6.6.

Hint for Ex. 16.9.18: Exercise 16.9.15 says that $T_f T_g = T_{fg}$ if either $f \in \overline{H^\infty}$ or $g \in H^\infty$. Let h be the outer function with $|h| = |f|^{-1/2}$ almost everywhere on $\mathbb{T}$. What is $T_{\overline{h}}T_f T_h$?

Hint for Ex. 16.9.19: Consult Exercise 16.9.15.

Hint for Ex. 16.9.21: Consult Theorem 16.5.1.

Hint for Ex. 16.9.22. For (a), observe that a linear transformation on a finite-dimensional vector space has a eigenvalue. For (b), observe that $\dim(A - \lambda I) < \infty$ for all $\lambda \neq 0$. Also consult Theorem 2.6.9.

Hint for Ex. 16.9.23: Consider $\sigma_{ap}(M_\varphi)$ from Proposition 8.1.12.

Hint for Ex. 16.9.24: For (a), consult Exercise 16.9.23.

Hint for Ex. 16.9.26: For (a) show that if $0 \in \sigma(A)$, then $0 \in W(A)^-$. Consult Lemma 2.3.5.

Hint for Ex. 16.9.28: Use the Brown–Halmos characterization (Theorem 16.5.1) along with Exercise 11.10.7.

Hint for Ex. 16.9.29: Use the fact that φ_λ is inner, the identity $T_{\varphi_\lambda}^* T_\varphi T_{\varphi_\lambda} = T_\varphi$ (Exercise 16.9.15), and Exercise 16.9.28.

Hint for Ex. 16.9.30: For (d), start with $\|T_\varphi + K\| \geqslant \|P_n(T_\varphi + K)P_n\|$. Also consult Exercise 16.9.15.

17

$\cdot$ $\cdot$ $\bullet$ $\bullet$ $\cdot$ $\cdot$

Hankel Operators

Key Concepts: Hilbert matrix, Hankel operator, Nehari's theorem, Hilbert's inequality, Carathéodory–Fejér problem, Nevanlinna–Pick problem.

Outline: In this chapter we survey *Hankel operators* $H_\varphi : H^2 \to \overline{H_0^2}$ which are defined by

$$H_\varphi f = (I - P_+)(\varphi f),$$

where H^2 is the Hardy space, $H_0^2 = zH^2$ (the subspace of functions in H^2 vanishing at zero), P_+ is the Riesz projection from L^2 onto H^2, and $\varphi \in L^\infty(\mathbb{T})$. These operators generalize the Hilbert matrix and connect to several problems in function theory such as the Nehari, Carathéodory–Fejér, and Nevanlinna–Pick problems.

17.1 The Hilbert Matrix

In 1894, Hilbert examined the following polynomial approximation problem [194]. Given a closed interval $[\alpha, \beta]$ and an $\varepsilon > 0$, can one find a nonzero polynomial p with integer coefficients such that

$$\int_\alpha^\beta |p(x)|^2 \, dx < \varepsilon?$$

If $p(x) = c_1 x^{n-1} + c_2 x^{n-2} + \cdots + c_n$, then the integral above can be expressed as a positive quadratic form

$$\sum_{j,k=1}^n a_{jk} c_j c_k, \qquad (17.1.1)$$

where

$$a_{jk} = \int_\alpha^\beta x^{2n-j-k} dx.$$

The analysis of Hilbert's problem involves

$$\det \begin{bmatrix} a_{11} & a_{12} & \cdots & a_{1n} \\ a_{21} & a_{22} & \cdots & a_{2n} \\ \vdots & \vdots & \ddots & \vdots \\ a_{n1} & a_{n2} & \cdots & a_{nn} \end{bmatrix},$$

which can be expressed as

$$\left(\frac{\beta-\alpha}{2}\right)^{n^2} 2^{n^2} \det \begin{bmatrix} 1 & \frac{1}{2} & \frac{1}{3} & \frac{1}{4} & \cdots & \frac{1}{n} \\ \frac{1}{2} & \frac{1}{3} & \frac{1}{4} & \frac{1}{5} & \cdots & \frac{1}{n+1} \\ \frac{1}{3} & \frac{1}{4} & \frac{1}{5} & \frac{1}{6} & \cdots & \frac{1}{n+2} \\ \frac{1}{4} & \frac{1}{5} & \frac{1}{6} & \frac{1}{7} & \cdots & \frac{1}{n+3} \\ \vdots & \vdots & \vdots & \vdots & \ddots & \vdots \\ \frac{1}{n} & \frac{1}{n+1} & \frac{1}{n+2} & \frac{1}{n+3} & \cdots & \frac{1}{2n-1} \end{bmatrix}. \tag{17.1.2}$$

Hilbert found that the determinant in (17.1.2) equals

$$\frac{\left(1^{n-1}2^{n-2}\cdots(n-2)^2(n-1)^1\right)^4}{1^{2n-1}2^{2n-2}\cdots(2n-2)^2(2n-1)^1}.$$

If $\beta - \alpha < 4$, then the quantity in (17.1.2) tends to zero as $n \to \infty$. Thus,

$$\int_\alpha^\beta |p(x)|^2\,dx$$

can be as small as desired.

The $n \times n$ *Hilbert matrix* in (17.1.2) has many other fascinating properties [83]. From here, one is inspired to consider the infinite Hilbert matrix

$$H = \begin{bmatrix} 1 & \frac{1}{2} & \frac{1}{3} & \frac{1}{4} & \cdots \\ \frac{1}{2} & \frac{1}{3} & \frac{1}{4} & \frac{1}{5} & \cdots \\ \frac{1}{3} & \frac{1}{4} & \frac{1}{5} & \frac{1}{6} & \cdots \\ \frac{1}{4} & \frac{1}{5} & \frac{1}{6} & \frac{1}{7} & \cdots \\ \vdots & \vdots & \vdots & \vdots & \ddots \end{bmatrix}. \tag{17.1.3}$$

Notice that the entries of H are constant on the reverse diagonals, that is, a_{jk} only depends on $j + k$. The matrix H is an example of a *Hankel matrix*, the focus of this chapter. An inequality of Hilbert [196] (see (17.5.3) below), later improved by Schur [333] (Example 3.3.4), implies that H defines a bounded linear operator on ℓ^2 with norm π. Further work of Magnus [237] shows that $\sigma(H) = [0, \pi]$ and $\sigma_p(H) = \varnothing$.

Although we do not get too far into the details, let us mention some connections the Hilbert matrix H makes with the integral and multiplication operators studied earlier in this book. We survey some results from [310, 311, 354].

Theorem 17.1.4 (Rosenblum). *The Hilbert matrix H on ℓ^2 is unitarily equivalent to the integral operator*

$$(Kf)(x) = \int_0^\infty \frac{f(y)}{x+y} e^{-y} dy$$

on $L^2((0, \infty), e^{-x} dx)$.

Proof We outline the proof from [354]. The sequence of Laguerre polynomials $(L_n(x))_{n=0}^\infty$ form an orthonormal basis for $L^2((0, \infty), e^{-x} dx)$ [160] in the sense that

$$\langle L_m, L_n \rangle = \int_0^\infty L_m(x) L_n(x) e^{-x} dx = \delta_{m,n} \quad \text{for all } m, n \geqslant 0. \tag{17.1.5}$$

These polynomials are created by applying the Gram–Schmidt process to the monomials $1, x, x^2, \ldots$ An explicit formula for $L_n(x)$ is

$$L_n(x) = \sum_{k=0}^n \binom{n}{k} \frac{(-1)^k}{k!} x^k$$

and the Laplace transform of L_n satisfies

$$\int_0^\infty e^{-xt} L_n(t) \, dt = (x-1)^n x^{-n-1} \quad \text{for } x > 0.$$

From here one can see that for any $m, n \geqslant 0$,

$$\langle L_m, KL_n \rangle = \int_0^\infty L_m(y)(KL_n)(y) e^{-y} dy$$

$$= \int_0^\infty \int_0^\infty \frac{L_m(y) L_n(x)}{x+y} e^{-x-y} dx \, dy$$

$$= \int_0^\infty \int_0^\infty \frac{1}{x+y} L_m(y) L_n(x) e^{-x-y} dx \, dy$$

$$= \int_0^\infty \int_0^\infty \left(\int_0^\infty e^{-t(x+y)} dt \right) L_m(y) L_n(x) e^{-x-y} dx \, dy$$

$$= \int_0^\infty \left(\int_0^\infty e^{-(t+1)x} L_n(x) \, dx \right) \left(\int_0^\infty e^{-(t+1)y} L_m(y) \, dy \right) dt$$

$$= \int_0^\infty t^{m+n}(t+1)^{-m-n-2} dt$$

$$= \int_0^1 u^{m+n} du$$

$$= \frac{1}{m+n+1}.$$

This shows that the matrix representation of K with respect to the Laguerre basis for $L^2((0, \infty), e^{-x} dx)$ is the Hilbert matrix H. Thus, K is unitarily equivalent to H. ∎

A deeper analysis from [311], using a certain second-order differential operator, gives the spectral representation of the Hilbert matrix H. We provide a sketch of the proof.

Theorem 17.1.6 (Rosenblum). *The Hilbert matrix H on ℓ^2 is unitarity equivalent to the multiplication operator $M_{\pi/\cosh(\pi x)}$ on $L^2((0, \infty), dx)$.*

Proof As noted in (17.1.5), the sequence of Laguerre polynomials $(L_n)_{n=0}^{\infty}$ forms an orthonormal basis for $L^2((0, \infty), e^{-x}dx)$. Integral substitution shows that the map $(Qf)(x) = e^{-x/2}f(x)$ defines a unitary operator from $L^2((0, \infty), e^{-x}dx)$ onto $L^2((0, \infty), dx)$. Thus,

$$QL_n = e^{-x/2}L_n(x) \quad \text{for } n \geqslant 0,$$

is an orthonormal basis for $L^2((0, \infty), dx)$.

Lebedev [226, 227] proved that if $K_\nu(z)$ is defined by

$$K_\nu(z) = \int_0^{\infty} e^{-z\cosh t} \cosh \nu t \, dt \quad \text{for Re } z > 0$$

(the modified Bessel function of the third kind), then the operator

$$(Uf)(\tau) = \frac{1}{\sqrt{\pi}} \int_0^{\infty} \frac{\sqrt{2\tau \sinh \pi\tau}}{\sqrt{x}} K_{i\tau}\left(\frac{x}{2}\right) f(x) \, dx$$

is a unitary operator from $L^2((0, \infty), dx)$ to itself. Thus,

$$w_n(x) = UQL_n \quad \text{for } n \geqslant 0,$$

is an orthonormal basis for $L^2((0, \infty), dx)$. Rosenblum proved that if

$$h(\tau) = \frac{\pi}{\cosh \pi\tau},$$

then

$$\langle M_h w_m, w_n \rangle_{L^2((0,\infty),dx)} = \frac{1}{n + m + 1} \quad \text{for } m, n \geqslant 0,$$

which are the entries of the Hilbert matrix. In summary, the linear transformation $W : \ell^2 \to L^2((0, \infty), dx)$ defined by

$$W((a_n)_{n \geqslant 0}) = \sum_{n=0}^{\infty} a_n w_n$$

is unitary with $WHW^* = M_h$. ∎

From here (see Figure 17.1.1), note that $\|H\| = \pi$, $\sigma(H) = [0, \pi]$, and $\sigma_p(H) = \emptyset$. We noted this before in our discussion of multiplication operators in Chapter 8. In addition, one also sees that $H \geqslant 0$ and that H is a cyclic operator (Theorem 8.2.8).

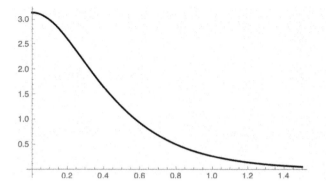

Figure 17.1.1 The graph of $\pi/\cosh \pi x$.

17.2 Doubly Infinite Hankel Matrices

From our survey of Toeplitz operators in Chapter 16, recall that a doubly infinite sequence

$$\mathbf{a} = (..., a_{-3}, a_{-2}, a_{-1}, \boxed{a_0}, a_1, a_2, a_3, ...)$$

of complex numbers (the boxed entry denotes the 0th position) gives rise to the doubly infinite Toeplitz matrix

$$
M(\mathbf{a}) =
\begin{bmatrix}
\ddots & \vdots & \vdots & \vdots & \vdots & \vdots & \vdots & \iddots \\
\cdots & a_0 & a_{-1} & a_{-2} & a_{-3} & a_{-4} & a_{-5} & \cdots \\
\cdots & a_1 & a_0 & a_{-1} & a_{-2} & a_{-3} & a_{-4} & \cdots \\
\cdots & a_2 & a_1 & a_0 & a_{-1} & a_{-2} & a_{-3} & \cdots \\
\cdots & a_3 & a_2 & a_1 & a_0 & a_{-1} & a_{-2} & \cdots \\
\cdots & a_4 & a_3 & a_2 & a_1 & a_0 & a_{-1} & \cdots \\
\cdots & a_5 & a_4 & a_3 & a_2 & a_1 & a_0 & \cdots \\
\iddots & \vdots & \vdots & \vdots & \vdots & \vdots & \vdots & \ddots
\end{bmatrix}.
\qquad (17.2.1)
$$

This defines a bounded operator on $\ell^2(\mathbb{Z})$ if and only if $\mathbf{a} = (\widehat{\varphi}(n))_{n=-\infty}^{\infty}$ for some $\varphi \in L^\infty(\mathbb{T})$ (Theorem 16.1.3). By identifying $\ell^2(\mathbb{Z})$ with $L^2(\mathbb{T})$ via Parseval's theorem, one can view $M(\mathbf{a})$ as the multiplication operator M_φ on $L^2(\mathbb{T})$ with

$$\|M(\mathbf{a})\| = \|M_\varphi\| = \|\varphi\|_\infty.$$

Now consider the "flip" operator $F : L^2(\mathbb{T}) \to L^2(\mathbb{T})$ defined by

$$(Ff)(\xi) = f(\overline{\xi}),$$

and observe that the matrix representation of F with respect to the orthonormal basis $(\xi^n)_{n=-\infty}^{\infty}$ for $L^2(\mathbb{T})$ is

$$F = \begin{bmatrix} \ddots & \vdots & \vdots & \vdots & \vdots & \vdots & \cdot^{\cdot^{\cdot}} \\ \cdots & 0 & 0 & 0 & 0 & 1 & \cdots \\ \cdots & 0 & 0 & 0 & 1 & 0 & \cdots \\ \cdots & 0 & 0 & \boxed{1} & 0 & 0 & \cdots \\ \cdots & 0 & 1 & 0 & 0 & 0 & \cdots \\ \cdots & 1 & 0 & 0 & 0 & 0 & \cdots \\ \cdot_{\cdot_{\cdot}} & \vdots & \vdots & \vdots & \vdots & \vdots & \ddots \end{bmatrix}.$$

The boxed entry denotes the $(0, 0)$ position.

Proposition 17.2.2. *The flip operator F is a selfadjoint unitary operator with*

$$\sigma(F) = \sigma_p(F) = \{-1, 1\}.$$

Proof For any $f \in L^2(\mathbb{T})$, observe that $\widehat{Ff}(n) = \hat{f}(-n)$, and hence Parseval's theorem implies that $\|Ff\| = \|f\|$. One can also see that $F^2 = I$, which implies that F is surjective. Since F is isometric and surjective, it is unitary. For any $f, g \in L^2(\mathbb{T})$,

$$\langle Ff, g \rangle = \int_0^{2\pi} f(e^{-i\theta})\overline{g(e^{i\theta})}\, \frac{d\theta}{2\pi}$$

$$= -\int_0^{-2\pi} f(e^{it})\overline{g(e^{-it})}\, \frac{dt}{2\pi}$$

$$= \int_0^{2\pi} f(e^{it})\overline{g(e^{-it})}\, \frac{dt}{2\pi}$$

$$= \langle f, Fg \rangle,$$

which says that F is selfadjoint.

Since F is selfadjoint, its spectrum is real (Theorem 8.5.1). Since F is unitary, its spectrum is contained in $\mathbb{T}$ (Exercise 8.10.10). Thus, $\sigma(F) \subseteq \{-1, 1\}$. Now observe that

$$F1 = 1 \quad \text{and} \quad F\left(\frac{\xi - \bar{\xi}}{2i}\right) = -\left(\frac{\xi - \bar{\xi}}{2i}\right),$$

hence $\sigma(F) = \sigma_p(F) = \{-1, 1\}$. ∎

Consider the doubly infinite *Hankel matrix*

$$H(\mathbf{a}) = \begin{bmatrix} \ddots & \vdots & \vdots & \vdots & \vdots & \vdots & \iddots \\ \cdots & \alpha_4 & \alpha_3 & \alpha_2 & \alpha_1 & \alpha_0 & \cdots \\ \cdots & \alpha_3 & \alpha_2 & \alpha_1 & \alpha_0 & \alpha_{-1} & \cdots \\ \cdots & \alpha_2 & \alpha_1 & \boxed{\alpha_0} & \alpha_{-1} & \alpha_{-2} & \cdots \\ \cdots & \alpha_1 & \alpha_0 & \alpha_{-1} & \alpha_{-2} & \alpha_{-3} & \cdots \\ \cdots & \alpha_0 & \alpha_{-1} & \alpha_{-2} & \alpha_{-3} & \alpha_{-4} & \cdots \\ \iddots & \vdots & \vdots & \vdots & \vdots & \vdots & \ddots \end{bmatrix}$$

and observe that the entries are constant on the reverse diagonals.

Proposition 17.2.3. *For a doubly infinite Hankel matrix $H(\mathbf{a})$ corresponding to $\mathbf{a} = (a_n)_{n=-\infty}^{\infty}$, the following are equivalent.*

(a) *$H(\mathbf{a})$ defines a bounded operator on $\ell^2(\mathbb{Z})$.*

(b) *There is a $\varphi \in L^{\infty}(\mathbb{T})$ such that $a_n = \hat{\varphi}(n)$ for all $n \in \mathbb{Z}$.*

Furthermore, $\|H(\mathbf{a})\| = \|\varphi\|_{\infty}$.

Proof Use $FH(\mathbf{a}) = M(\mathbf{a})$ and the discussion following (17.2.1). ∎

17.3 Hankel Operators

Recall from Proposition 5.3.12 that the Hardy space H^2 can be realized as $H^2 = \{f \in L^2(\mathbb{T}) : \hat{f}(n) = 0$ for all $n < 0\}$. Form the spaces

$$H_0^2 = \xi H^2 \quad \text{and} \quad \overline{H_0^2} = \{f \in L^2(\mathbb{T}) : \overline{f} \in \xi H^2\}$$

and observe that $\overline{H_0^2} = (H^2)^{\perp}$. Let P_+ denote the orthogonal projection of $L^2(\mathbb{T})$ onto H^2 (Definition 16.2.1) and note that

$$P_- := I - P_+$$

is the orthogonal projection of $L^2(\mathbb{T})$ onto $\overline{H_0^2}$.

Definition 17.3.1. For $\varphi \in L^{\infty}(\mathbb{T})$, the *Hankel operator* with *symbol* φ is

$$H_{\varphi} : H^2 \to \overline{H_0^2}, \quad H_{\varphi} f = P_-(\varphi f).$$

For $f \in H^2$, observe that $\|H_{\varphi} f\| = \|P_-(\varphi f)\| \leq \|\varphi f\| \leq \|\varphi\|_{\infty} \|f\|$ and hence

$$\|H_{\varphi}\| \leq \|\varphi\|_{\infty}. \tag{17.3.2}$$

The norm of a Hankel operator is often smaller than $\|\varphi\|_{\infty}$ and this fact connects Hankel operators to various approximation problems. We discuss this momentarily. We first compute the adjoint of a Hankel operator.

Proposition 17.3.3. *For $\varphi \in L^\infty(\mathbb{T})$, the adjoint $H_\varphi^* : \overline{H_0^2} \to H^2$ is $H_\varphi^* f = P_+(\overline{\varphi} f)$.*

Proof If $g \in H^2$ and $f \in \overline{H_0^2}$, then

$$\langle H_\varphi g, f \rangle = \langle P_-(\varphi g), f \rangle = \langle \varphi g, P_- f \rangle = \langle \varphi g, f \rangle = \langle P_+ g, \overline{\varphi} f \rangle = \langle g, P_+(\overline{\varphi} f) \rangle,$$

which proves the result. ∎

The reader should exercise caution here. With Toeplitz operators, $T_\varphi^* = T_{\overline{\varphi}}$. For Hankel operators, H_φ^* and $H_{\overline{\varphi}}$ are not the same since they have different domains.

To obtain a matrix representation for H_φ, equip H^2 with the orthonormal basis $(\xi^n)_{n=0}^\infty$ and $\overline{H_0^2}$ with the orthonormal basis $(\xi^{-m})_{m=1}^\infty$. Then

$$\begin{aligned}
\langle H_\varphi \xi^n, \xi^{-m} \rangle &= \langle P_-(\varphi \xi^n), \xi^{-m} \rangle \\
&= \langle \varphi \xi^n, P_-(\xi^{-m}) \rangle \\
&= \langle \varphi \xi^n, \xi^{-m} \rangle \\
&= \langle \varphi, \xi^{-m-n} \rangle \\
&= \widehat{\varphi}(-m - n).
\end{aligned} \tag{17.3.4}$$

Thus, the entry in position (m, n) of the matrix representation of H_φ with respect to the two bases above is $\widehat{\varphi}(-m - n)$, where $m \geqslant 1$ and $n \geqslant 0$. In other words, H_φ is represented by the infinite Hankel matrix

$$\begin{bmatrix}
\alpha_{-1} & \alpha_{-2} & \alpha_{-3} & \alpha_{-4} & \alpha_{-5} & \cdots \\
\alpha_{-2} & \alpha_{-3} & \alpha_{-4} & \alpha_{-5} & \alpha_{-6} & \cdots \\
\alpha_{-3} & \alpha_{-4} & \alpha_{-5} & \alpha_{-6} & \alpha_{-7} & \cdots \\
\alpha_{-4} & \alpha_{-5} & \alpha_{-6} & \alpha_{-7} & \alpha_{-8} & \cdots \\
\alpha_{-5} & \alpha_{-6} & \alpha_{-7} & \alpha_{-8} & \alpha_{-9} & \cdots \\
\vdots & \vdots & \vdots & \vdots & \vdots & \ddots
\end{bmatrix}, \tag{17.3.5}$$

where $\alpha_n = \widehat{\varphi}(n)$.

17.4 The Norm of a Hankel Operator

From Proposition 17.2.3, the norm of the doubly infinite Hankel matrix $H(\mathbf{a})$ is $\|\varphi\|_\infty$. As one can see from (17.3.5), "half" of the Fourier coefficients of φ are missing from the matrix representation of the Hankel operator. Thus, computing the norm of H_φ is more difficult than the same problem for Toeplitz operators (Theorem 16.3.1). From (17.3.2), we see that $\|H_\varphi\| \leqslant \|\varphi\|_\infty$. An important feature of Hankel operators is that the symbol φ defining H_φ is not unique. This fact is made clear in the following proposition.

Proposition 17.4.1. *For $\varphi \in L^\infty(\mathbb{T})$, the following are equivalent.*

(a) $H_\varphi = 0$.

(b) $\varphi \in H^\infty$.

Proof (a) $\Rightarrow$ (b) If $H_\varphi f = 0$ for all $f \in H^2$, then the constant function 1 belongs to H^2 and $H_\varphi 1 = P_-(\varphi) = 0$. Thus, $\varphi \in H^\infty$.

(b) $\Rightarrow$ (a) If $\varphi \in H^\infty$, then $f\varphi \in H^2$ for all $f \in H^2$. Thus, $H_\varphi(f) = P_-(f\varphi) = 0$. ∎

The previous proposition shows that

$$H_\varphi = H_{\varphi - \eta} \quad \text{for all } \eta \in H^\infty \text{ and } \varphi \in L^\infty, \tag{17.4.2}$$

which implies that $\|H_\varphi\| = \|H_{\varphi - \eta}\| \leqslant \|\varphi - \eta\|_\infty$ for all $\eta \in H^\infty$. Take the infimum with respect to η and obtain

$$\|H_\varphi\| \leqslant \text{dist}(\varphi, H^\infty) = \inf\{\|\varphi - \eta\|_\infty : \eta \in H^\infty\}. \tag{17.4.3}$$

Nehari proved that (17.4.3) is an equality. The reader might want to review Lemma 3.4.3 and Parrott's theorem (Theorem 14.8.1) before proceeding to the proof below.

Theorem 17.4.4 (Nehari [247]). *Suppose that $(\alpha_n)_{n=1}^\infty$ is a sequence of complex numbers such that the Hankel matrix*

$$A = \begin{bmatrix} \alpha_1 & \alpha_2 & \alpha_3 & \cdots \\ \alpha_2 & \alpha_3 & \alpha_4 & \cdots \\ \alpha_3 & \alpha_4 & \alpha_5 & \cdots \\ \vdots & \vdots & \vdots & \ddots \end{bmatrix} \tag{17.4.5}$$

defines a bounded operator on ℓ^2. Then the following hold.

(a) *There exists a $\varphi \in L^\infty(\mathbb{T})$ such that $\hat{\varphi}(-n) = \alpha_n$ for all $n \geqslant 1$.*

(b) *For any such φ, $\|A\| = \text{dist}(\varphi, H^\infty) \leqslant \|\varphi\|_\infty$.*

(c) *There is a $\psi \in L^\infty$ such that $\hat{\psi}(-n) = \alpha_n$ for all $n \geqslant 1$ and $\|A\| = \text{dist}(\psi, H^\infty) = \|\psi\|_\infty$.*

Proof Without loss of generality, assume that $\|A\| = 1$. Add one column to the left of the matrix A as follows:

$$\begin{bmatrix} \boxed{\alpha_0} & \alpha_1 & \alpha_2 & \alpha_3 & \alpha_4 & \cdots \\ \alpha_1 & \alpha_2 & \alpha_3 & \alpha_4 & \alpha_5 & \cdots \\ \alpha_2 & \alpha_3 & \alpha_4 & \alpha_5 & \alpha_6 & \cdots \\ \alpha_3 & \alpha_4 & \alpha_5 & \alpha_6 & \alpha_7 & \cdots \\ \alpha_4 & \alpha_5 & \alpha_6 & \alpha_7 & \alpha_8 & \cdots \\ \vdots & \vdots & \vdots & \vdots & \vdots & \ddots \end{bmatrix}, \tag{17.4.6}$$

in which the entry α_0 is unspecified. Write this matrix as

$$\begin{bmatrix} \alpha_0 & B \\ C & D \end{bmatrix},$$

where

$$B = [\alpha_1\ \alpha_2\ \alpha_3\ \alpha_4 \cdots], \quad C = \begin{bmatrix} \alpha_1 \\ \alpha_2 \\ \alpha_3 \\ \alpha_4 \\ \vdots \end{bmatrix}, \quad \text{and} \quad D = \begin{bmatrix} \alpha_2 & \alpha_3 & \alpha_4 & \alpha_5 & \cdots \\ \alpha_3 & \alpha_4 & \alpha_5 & \alpha_6 & \cdots \\ \alpha_4 & \alpha_5 & \alpha_6 & \alpha_7 & \cdots \\ \alpha_5 & \alpha_6 & \alpha_7 & \alpha_8 & \cdots \\ \vdots & \vdots & \vdots & \vdots & \ddots \end{bmatrix}.$$

Observe that the conditions of Parrott's theorem (Theorem 14.8.1) are fulfilled since A is assumed to be a contraction and hence $[C\ D]$ and $[B\ D]^T$ are also contractions. Hence there is an $\alpha_0 \in \mathbb{C}$ such that the matrix (17.4.6) defines a contraction on ℓ^2. Repeat the same process with the matrix (17.4.6) and obtain an $\alpha_{-1} \in \mathbb{C}$ such that

$$\begin{bmatrix} \boxed{\alpha_{-1}} & \alpha_0 & \alpha_1 & \alpha_2 & \alpha_3 & \alpha_4 & \cdots \\ \alpha_0 & \alpha_1 & \alpha_2 & \alpha_3 & \alpha_4 & \alpha_5 & \cdots \\ \alpha_1 & \alpha_2 & \alpha_3 & \alpha_4 & \alpha_5 & \alpha_6 & \cdots \\ \alpha_2 & \alpha_3 & \alpha_4 & \alpha_5 & \alpha_6 & \alpha_7 & \cdots \\ \alpha_3 & \alpha_4 & \alpha_5 & \alpha_6 & \alpha_7 & \alpha_8 & \cdots \\ \alpha_4 & \alpha_5 & \alpha_6 & \alpha_7 & \alpha_8 & \alpha_9 & \cdots \\ \vdots & \vdots & \vdots & \vdots & \vdots & \vdots & \ddots \end{bmatrix}$$

is a contraction on ℓ^2. By induction, we obtain a sequence $(\alpha_n)_{n=-\infty}^{\infty}$ such that the Hankel matrix

$$\begin{bmatrix} \boxed{\alpha_{n_0}} & \alpha_{n_0+1} & \alpha_{n_0+2} & \alpha_{n_0+3} & \alpha_{n_0+4} & \alpha_{n_0+5} & \cdots \\ \alpha_{n_0+1} & \alpha_{n_0+2} & \alpha_{n_0+3} & \alpha_{n_0+4} & \alpha_{n_0+5} & \alpha_{n_0+6} & \cdots \\ \alpha_{n_0+2} & \alpha_{n_0+3} & \alpha_{n_0+4} & \alpha_{n_0+5} & \alpha_{n_0+6} & \alpha_{n_0+7} & \cdots \\ \alpha_{n_0+3} & \alpha_{n_0+4} & \alpha_{n_0+5} & \alpha_{n_0+6} & \alpha_{n_0+7} & \alpha_{n_0+8} & \cdots \\ \alpha_{n_0+4} & \alpha_{n_0+5} & \alpha_{n_0+6} & \alpha_{n_0+7} & \alpha_{n_0+8} & \alpha_{n_0+9} & \cdots \\ \alpha_{n_0+5} & \alpha_{n_0+6} & \alpha_{n_0+7} & \alpha_{n_0+8} & \alpha_{n_0+9} & \alpha_{n_0+10} & \cdots \\ \vdots & \vdots & \vdots & \vdots & \vdots & \vdots & \ddots \end{bmatrix}$$

is a contraction on ℓ^2 for each $n_0 \in \mathbb{Z}$. Continue this process and form the doubly infinite Hankel matrix

$$H(\mathbf{a}) = \begin{bmatrix} \ddots & \vdots & \vdots & \vdots & \vdots & \vdots & \iddots \\ \cdots & \alpha_4 & \alpha_3 & \alpha_2 & \alpha_1 & \alpha_0 & \cdots \\ \cdots & \alpha_3 & \alpha_2 & \alpha_1 & \alpha_0 & \alpha_{-1} & \cdots \\ \cdots & \alpha_2 & \alpha_1 & \boxed{\alpha_0} & \alpha_{-1} & \alpha_{-2} & \cdots \\ \cdots & \alpha_1 & \alpha_0 & \alpha_{-1} & \alpha_{-2} & \alpha_{-3} & \cdots \\ \cdots & \alpha_0 & \alpha_{-1} & \alpha_{-2} & \alpha_{-3} & \alpha_{-4} & \cdots \\ \iddots & \vdots & \vdots & \vdots & \vdots & \vdots & \ddots \end{bmatrix}$$

with the boxed α_0 in the $(0, 0)$ position. The discussion above and Lemma 3.4.3 shows that $H(\mathbf{a})$ is a contraction on $\ell^2(\mathbb{Z})$. Proposition 17.2.3 provides a $\varphi \in L^\infty(\mathbb{T})$ such that

$$\hat{\varphi}(n) = \alpha_{-n} \quad \text{for all } n \in \mathbb{Z}. \tag{17.4.7}$$

This proves (a).

Moreover, $H(\mathbf{a})$ is the matrix representation of the multiplication operator M_φ on $L^2(\mathbb{T})$ (via $FH(\mathbf{a}) = M_\varphi$) with respect to the standard basis $(\xi^n)_{n=-\infty}^{\infty}$. Furthermore,

$$\|\varphi\|_\infty = \|M_\varphi\| \leqslant 1 = \|A\|. \tag{17.4.8}$$

Consider the unitary operators $U : \ell^2 \to H^2$ defined by

$$U\left(\sum_{k=0}^{\infty} a_k \mathbf{e}_k\right) = \sum_{k=0}^{\infty} a_k \xi^k,$$

and $V : \ell^2 \to \overline{H_0^2}$ defined by

$$V\left(\sum_{k=0}^{\infty} a_k \mathbf{e}_k\right) = \sum_{k=0}^{\infty} a_k \overline{\xi^{k+1}}.$$

We have shown that there is a $\varphi \in L^\infty(\mathbb{T})$ such that (a) holds. Since only the negatively indexed Fourier coefficients of φ, that is $\hat{\varphi}(-n)$ for $n \geqslant 1$, appear in the matrix representation of H_φ in (17.3.5), the choice of φ is not unique. However, for any such φ, the matrix representations of H_φ and A show that

$$A = V^* H_\varphi U.$$

From (17.4.3), it follows that

$$\|A\| = \|H_\varphi\| \leqslant \operatorname{dist}(\varphi, H^\infty). \tag{17.4.9}$$

This proves (b).

If φ_1 and φ_2 are two such representing symbols for A, then $H_{\varphi_1} = H_{\varphi_2}$, and hence $\varphi_1 - \varphi_2 \in H^\infty$ (Proposition 17.4.1). This implies that

$$\operatorname{dist}(\varphi_1, H^\infty) = \operatorname{dist}(\varphi_2, H^\infty). \tag{17.4.10}$$

Therefore, neither $\|H_\varphi\|$ nor $\operatorname{dist}(\varphi, H^\infty)$ depend on the symbols φ for which $A = H_\varphi$. By (17.4.8) the symbol φ obtained in (17.4.7) therefore satisfies $\|A\| = \operatorname{dist}(\varphi, H^\infty) = \|\varphi\|_\infty$. The relation (17.4.10) shows that $\|A\| = \operatorname{dist}(\varphi, H^\infty)$ holds for all symbols φ that give rise to the same Hankel operator. This proves (c). ∎

Corollary 17.4.11. *For $\varphi \in L^\infty(\mathbb{T})$, $\|H_\varphi\| = \operatorname{dist}(\varphi, H^\infty)$. Moreover, there is an $\eta \in H^\infty$ such that $\|H_\varphi\| = \|\varphi - \eta\|_\infty$.*

Proof From (17.3.2) it follows that H_φ is a bounded operator. The equality $\|H_\varphi\| = \operatorname{dist}(\varphi, H^\infty)$ is from Theorem 17.4.4b. By Theorem 17.4.4c, there is an $\psi \in L^\infty(\mathbb{T})$ such that $H_\varphi = H_\psi$ and $\|H_\psi\| = \operatorname{dist}(\psi, H^\infty) = \|\psi\|_\infty$. Since $H_\varphi = H_\psi$, it follows that $\eta = \varphi - \psi \in H^\infty$ (Proposition 17.4.1). Thus, $\|H_\varphi\| = \|H_\psi\| = \|\psi\|_\infty = \|\varphi - \eta\|_\infty$. ∎

17.5 Hilbert's Inequality

The inequality $\|H_\varphi\| \leqslant \|\varphi\|_\infty$ implies that

$$|\langle H_\varphi f, g\rangle| \leqslant \|\varphi\|_\infty \|f\| \|g\| \quad \text{for all } f \in H^2 \text{ and } g \in \overline{H_0^2}. \tag{17.5.1}$$

Now observe that if $f \in H^2$ and $g \in \overline{H_0^2}$,

$$
\begin{aligned}
\langle H_\varphi f, g\rangle &= \langle P_-(\varphi f), g\rangle \\
&= \langle \varphi f, P_- g\rangle &&(\text{since } P_-^* = P_-) \\
&= \langle \varphi f, g\rangle &&(P_- g = g \text{ since } g \in \overline{H_0^2}).
\end{aligned}
$$

From here it follows that (17.5.1) is equivalent to

$$\left| \sum_{\substack{n=0 \\ m=1}}^{\infty} \hat{\varphi}(-m-n)\hat{f}(n)\overline{\hat{g}(-m)} \right| \leqslant \|\varphi\|_\infty \left(\sum_{n=0}^{\infty} |\hat{f}(n)|^2 \right)^{\frac{1}{2}} \left(\sum_{m=1}^{\infty} |\hat{g}(-m)|^2 \right)^{\frac{1}{2}}.$$

Since the Hilbert spaces H^2, $\overline{H_0^2}$, and ℓ^2 are isometrically isomorphic, the previous inequality is equivalent to

$$\left| \sum_{i,j=1}^{\infty} \hat{\varphi}(-i-j+1)x_i\overline{y_j} \right| \leqslant \|\varphi\|_\infty \left(\sum_{i=1}^{\infty} |x_i|^2 \right)^{\frac{1}{2}} \left(\sum_{j=1}^{\infty} |y_j|^2 \right)^{\frac{1}{2}} \tag{17.5.2}$$

for all $(x_i)_{i=1}^\infty, (y_i)_{i=1}^\infty \in \ell^2(\mathbb{N})$.

In the special case

$$\varphi(e^{it}) = -i(\pi - t) \quad \text{for } 0 < t < 2\pi,$$

one can check that

$$\hat{\varphi}(n) = \begin{cases} -\dfrac{1}{n} & \text{if } n \neq 0, \\[2mm] 0 & \text{if } n = 0. \end{cases}$$

Hence, the matrix corresponding to the Hankel operator H_φ is

$$H = \begin{bmatrix} 1 & \frac{1}{2} & \frac{1}{3} & \frac{1}{4} & \cdots \\ \frac{1}{2} & \frac{1}{3} & \frac{1}{4} & \frac{1}{5} & \cdots \\ \frac{1}{3} & \frac{1}{4} & \frac{1}{5} & \frac{1}{6} & \cdots \\ \vdots & \vdots & \vdots & \vdots & \ddots \end{bmatrix},$$

the Hilbert matrix (17.1.3). Since $\|\varphi\|_\infty = \pi$, the inequality (17.5.2) becomes

$$\left| \sum_{i,j=1}^{\infty} \frac{x_i\overline{y_j}}{i+j-1} \right| \leqslant \pi \left(\sum_{i=1}^{\infty} |x_i|^2 \right)^{\frac{1}{2}} \left(\sum_{j=1}^{\infty} |y_j|^2 \right)^{\frac{1}{2}} \tag{17.5.3}$$

for all $(x_i)_{i=1}^\infty, (y_i)_{i=1}^\infty \in \ell^2(\mathbb{N})$. That is, H defines a bounded operator on $\ell^2(\mathbb{N})$ and its norm is at most π. This is not easy to verify directly.

17.6 The Nehari Problem

The *Nehari Problem* is the following: for a given $\varphi \in L^\infty(\mathbb{T})$, find an $\eta \in H^\infty(\mathbb{T})$ such that $\|\varphi - \eta\|_\infty$ is minimal. In this section we show that a solution exists and is unique. From a theoretical point of view, the answer to Nehari's problem is implicitly contained in Corollary 17.4.11. This result even provides a formula for the distance between φ and H^∞, that is, $\text{dist}(\varphi, H^\infty) = \|H_\varphi\|$. In its proof, we outlined a procedure to construct an $\eta \in H^\infty$ such that $\text{dist}(\varphi, H^\infty) = \|\varphi - \eta\|_\infty$. The construction of η ultimately uses Theorem 17.4.4, which requires us to inductively obtain the parameters α_n when creating the Hankel matrix (17.4.5). Although this construction works in theory, it is not practical. Thus, we seek another method.

Definition 17.6.1. A *maximizing vector* for $T \in \mathcal{B}(\mathcal{H}_1, \mathcal{H}_2)$ is an $\mathbf{x} \in \mathcal{H}_1 \backslash \{0\}$ such that

$$\|T\mathbf{x}\|_{\mathcal{H}_2} = \|T\|_{\mathcal{H}_1 \to \mathcal{H}_2} \|\mathbf{x}\|_{\mathcal{H}_1}.$$

Not every operator has a maximizing vector (Exercise 17.10.12). An operator with a maximizing vector enjoys special properties. Below is such an example. A theorem of Riesz [118, p. 17] (see also Exercise 5.9.37) says that if $f \in H^2 \backslash \{0\}$, then $\{\xi \in \mathbb{T} : f(\xi) = 0\}$ has measure zero. This detail allows us to state the following theorem.

Theorem 17.6.2. *For $\varphi \in L^\infty(\mathbb{T})$, suppose that the Hankel operator H_φ has a maximizing vector $f \in H^2 \backslash \{0\}$. Then*

$$\psi = \frac{H_\varphi f}{f} \tag{17.6.3}$$

is defined almost everywhere on $\mathbb{T}$ and satisfies the following.

(a) *$|\psi|$ is constant almost everywhere on $\mathbb{T}$.*

(b) *$H_\varphi = H_\psi$ and $\text{dist}(\varphi, H^\infty) = \text{dist}(\psi, H^\infty) = \|\psi\|_\infty$.*

(c) *If $\omega \in L^\infty(\mathbb{T})$ is such that $H_\varphi = H_\omega$ and $\text{dist}(\omega, H^\infty) = \|\omega\|_\infty$, then $\omega = \psi$ almost everywhere.*

Proof The existence of a symbol of minimal norm is guaranteed (even without the extra assumption of having a maximizing vector) by Corollary 17.4.11. To establish the uniqueness of this symbol, suppose that a maximizing vector f exists for H_φ and let $\omega \in L^\infty(\mathbb{T})$ be such that $H_\varphi = H_\omega$ and $\|H_\varphi\| = \|H_\omega\| = \text{dist}(\omega, H^\infty) = \|\omega\|_\infty$. Then for any $g \in H^2$,

$$\|H_\varphi g\| = \|H_\omega g\| = \|P_-(\omega g)\| \leqslant \|\omega g\| \leqslant \|\omega\|_\infty \|g\| = \|H_\varphi\| \|g\|.$$

In particular, the specific choice $g = f$ (the maximizing vector for H_φ) yields

$$\|P_-(\omega f)\| = \|\omega f\| = \|\omega\|_\infty \|f\|.$$

Exercise 16.9.2 implies that $\omega f \in \overline{H_0^2}$ and hence $H_\varphi f = P_-(\omega f) = \omega f$. Since $f(\xi) \neq 0$ for almost every $\xi \in \mathbb{T}$, one can divide by f to obtain $\omega = \psi$. The identity $\|\omega f\| = \|\omega\|_\infty \|f\|$ ensures that $|\omega|$ is constant on $\mathbb{T}$. ∎

Example 17.6.4. There are symbols $\varphi \in L^\infty(\mathbb{T})$ for which the Nehari problem has several solutions. Theorem 17.6.2 implies that the corresponding H_φ does not have a maximizing vector. On the other hand, if f_1 and f_2 are maximizing vectors of H_φ, the formula (17.6.3) for ψ shows that

$$\frac{H_\varphi f_1}{f_1} = \frac{H_\varphi f_2}{f_2}$$

almost everywhere on $\mathbb{T}$.

In some elementary cases, we can guess a maximizing vector and solve Nehari's problem. For example, if

$$\varphi(z) = \frac{1 - \overline{\alpha}\, z}{\alpha - z} \quad \text{for } \alpha \in \mathbb{D} \text{ and } z \in \mathbb{T},$$

then

$$\varphi(z) = \overline{\alpha} + \frac{1 - |\alpha|^2}{\alpha - z}.$$

Thus, for each $f \in H^2$,

$$\varphi(z)\,f(z) = \overline{\alpha}\,f(z) - (1 - |\alpha|^2)\frac{f(z) - f(\alpha)}{z - \alpha} - (1 - |\alpha|^2)\frac{f(\alpha)}{z - \alpha}$$

$$= \overline{\alpha}\,f(z) - (1 - |\alpha|^2)\,Q_\alpha f(z) - (1 - |\alpha|^2)\frac{f(\alpha)}{z - \alpha},$$

where

$$(Q_\alpha f)(z) = \frac{f(z) - f(\alpha)}{z - \alpha},$$

which one can check is a bounded operator on H^2 (Exercise 5.9.16). Moreover, the Fourier-series representation

$$g(\xi) = \frac{1}{\xi - \alpha} = \sum_{n=1}^{\infty} \alpha^{n-1}\overline{\xi}^{\,n} \quad \text{for } \xi \in \mathbb{T},$$

shows that $g \in \overline{H_0^2}$ and

$$\|g\| = \frac{1}{\sqrt{1 - |\alpha|^2}}. \tag{17.6.5}$$

We conclude that

$$H_\varphi f = -(1 - |\alpha|^2)\,f(\alpha)g = -(1 - |\alpha|^2)\langle f, k_\alpha \rangle g,$$

where $k_\alpha(z) = (1 - \overline{\alpha}z)^{-1}$. This can be written using tensor notation as

$$H_\varphi = -(1 - |\alpha|^2)\,g \otimes k_\alpha,$$

which yields $\|H_\varphi\| = (1 - |\alpha|^2)\|g\|\|k_\alpha\| = 1$ (Exercise 3.6.3). Set $f = k_\alpha$ in the above to get $H_\varphi k_\alpha = -g$. Hence,

$$\|H_\varphi k_\alpha\| = \|g\| = \|k_\alpha\|.$$

In other words, k_α is a maximizing vector for H_φ. Theorem 17.6.2 says that the best approximation $\psi \in H^\infty$ to φ is

$$\psi = \varphi - \frac{H_\varphi k_\alpha}{k_\alpha} = \varphi + \frac{g}{k_\alpha} = \varphi - \varphi = 0.$$

It follows that $\mathrm{dist}(\varphi, H^\infty) = \|\varphi\|_\infty = 1$ and $\|\varphi - \eta\|_\infty > 1$ for every $\eta \in H^\infty \setminus \{0\}$.

17.7 The Carathéodory–Fejér Problem

In the *Carathéodory–Fejér problem*, a polynomial $p(z) = \alpha_0 + \alpha_1 z + \cdots + \alpha_n z^n$ is given and the goal is to determine complex numbers $\alpha_{n+1}, \alpha_{n+2}, \ldots$ such that

$$\sum_{k=0}^\infty \alpha_k z^k \in H^\infty$$

and

$$\left\| \sum_{k=0}^\infty \alpha_k z^k \right\|_\infty$$

is minimized. One can always satisfy the first condition. The second condition is more troublesome. Write

$$\sum_{k=0}^\infty \alpha_k z^k = p(z) + z^{n+1} f(z),$$

and see that the coefficients α_k for $k \geq n+1$ are the Taylor coefficients of a typical $f \in H^\infty$. Thus, we should compute

$$\inf_{f \in H^\infty} \|p + z^{n+1} f\|_\infty.$$

Since z^{n+1} is unimodular on $\mathbb{T}$, $\|p + z^{n+1} f\|_\infty = \|p z^{-n-1} + f\|_\infty$, and hence

$$\inf_{f \in H^\infty} \|p + z^{n+1} f\|_\infty = \mathrm{dist}(p z^{-n-1}, H^\infty).$$

Corollary 17.4.11 provides a $g \in H^\infty$ such that

$$\|p + z^{n+1} g\|_\infty = \inf_{f \in H^\infty} \|p + z^{n+1} f\|_\infty = \|H_{p z^{-n-1}}\|.$$

But on $\mathbb{T}$, $z^{-n-1} p(z) = \alpha_n z^{-1} + \alpha_{n-1} z^{-2} + \cdots + \alpha_0 z^{-n-1}$ is a trigonometric polynomial, and thus $H_{p z^{-n-1}}$ has finite rank (Exercise 17.10.8). Therefore,

$$\inf_{f \in H^\infty} \|p + z^{n+1} f\|_\infty$$

equals the norm of the infinite Hankel matrix

$$
\left[\begin{array}{ccccccc|ccccccc}
\alpha_n & \alpha_{n-1} & \alpha_{n-2} & \cdots & \alpha_2 & \alpha_1 & \alpha_0 & 0 & 0 & 0 & 0 & 0 & 0 & \cdots \\
\alpha_{n-1} & \alpha_{n-2} & \alpha_{n-3} & \cdots & \alpha_1 & \alpha_0 & 0 & 0 & 0 & 0 & 0 & 0 & 0 & \cdots \\
\alpha_{n-2} & \alpha_{n-3} & \alpha_{n-4} & \cdots & \alpha_0 & 0 & 0 & 0 & 0 & 0 & 0 & 0 & 0 & \cdots \\
\vdots & \vdots & \vdots & \ddots & \vdots & \vdots & \vdots & \vdots & \vdots & \vdots & \vdots & \vdots & \vdots & \ddots \\
\alpha_2 & \alpha_1 & \alpha_0 & \cdots & 0 & 0 & 0 & 0 & 0 & 0 & 0 & 0 & 0 & \cdots \\
\alpha_1 & \alpha_0 & 0 & \cdots & 0 & 0 & 0 & 0 & 0 & 0 & 0 & 0 & 0 & \cdots \\
\alpha_0 & 0 & 0 & \cdots & 0 & 0 & 0 & 0 & 0 & 0 & 0 & 0 & 0 & \cdots \\ \hline
0 & 0 & 0 & \cdots & 0 & 0 & 0 & 0 & 0 & 0 & 0 & 0 & 0 & \cdots \\
0 & 0 & 0 & \cdots & 0 & 0 & 0 & 0 & 0 & 0 & 0 & 0 & 0 & \cdots \\
0 & 0 & 0 & \cdots & 0 & 0 & 0 & 0 & 0 & 0 & 0 & 0 & 0 & \cdots \\
0 & 0 & 0 & \cdots & 0 & 0 & 0 & 0 & 0 & 0 & 0 & 0 & 0 & \cdots \\
0 & 0 & 0 & \cdots & 0 & 0 & 0 & 0 & 0 & 0 & 0 & 0 & 0 & \cdots \\
0 & 0 & 0 & \cdots & 0 & 0 & 0 & 0 & 0 & 0 & 0 & 0 & 0 & \cdots \\
\vdots & \vdots & \vdots & \ddots & \vdots & \vdots & \vdots & \vdots & \vdots & \vdots & \vdots & \vdots & \vdots & \ddots
\end{array}\right],
$$

which has the same norm as the finite Hankel matrix (Exercise 14.11.11)

$$
\left[\begin{array}{ccccccc}
\alpha_n & \alpha_{n-1} & \alpha_{n-2} & \cdots & \alpha_2 & \alpha_1 & \alpha_0 \\
\alpha_{n-1} & \alpha_{n-2} & \alpha_{n-3} & \cdots & \alpha_1 & \alpha_0 & 0 \\
\alpha_{n-2} & \alpha_{n-3} & \alpha_{n-4} & \cdots & \alpha_0 & 0 & 0 \\
\vdots & \vdots & \vdots & \ddots & \vdots & \vdots & \vdots \\
\alpha_2 & \alpha_1 & \alpha_0 & \cdots & 0 & 0 & 0 \\
\alpha_1 & \alpha_0 & 0 & \cdots & 0 & 0 & 0 \\
\alpha_0 & 0 & 0 & \cdots & 0 & 0 & 0
\end{array}\right].
$$

17.8 The Nevanlinna–Pick Problem

Suppose that $z_1, z_2, \ldots, z_n$ are distinct points in $\mathbb{D}$ and $w_1, w_2, \ldots, w_n \in \mathbb{C}$. There are many $f \in H^\infty$ such that

$$f(z_k) = w_k \quad \text{for all } 1 \leqslant k \leqslant n. \tag{17.8.1}$$

For example, the Lagrange interpolation theorem provides a unique polynomial p of degree at most $n - 1$ which satisfies (17.8.1). The *Nevanlinna–Pick problem* asks for an interpolating function f such that $\|f\|_\infty$ is minimal.

On one hand, if f satisfies (17.8.1) and p is the interpolating polynomial described above, then $f - p \in H^\infty$ and, moreover, $(f - p)(z_k) = 0$ for all $1 \leqslant k \leqslant n$. Hence, $f - p = Bh$, where B is the finite Blaschke product whose zeros are the distinct points $z_1, z_2, \ldots, z_n$, and $h \in H^\infty$. On the other hand, $f = p + Bh$ satisfies (17.8.1) for any $h \in H^\infty$. Thus, the solutions to (17.8.1) are parameterized by $f = p + Bh$, where p is the Lagrange interpolating polynomial, B is the finite Blaschke product with zeros $z_1, z_2, \ldots, z_n$, and $h \in H^\infty$. To solve the Nevanlinna–Pick problem, we must compute

$$\inf_{h \in H^\infty} \|p + Bh\|_\infty.$$

Since B is unimodular on $\mathbb{T}$, $\|p + Bh\|_\infty = \|p\bar{B} + h\|_\infty$ and hence

$$\inf_{h \in H^\infty} \|p + Bh\|_\infty = \text{dist}(p\bar{B}, H^\infty).$$

This problem reduces to Nehari's problem from the previous section. In the discussion above, p can be chosen to be any solution of the interpolation problem and the rest of the analysis is the same.

17.9 Notes

The Hilbert matrix H from (17.1.3) continues to be an active area of study. We already know that H acts on ℓ^2 with operator norm equal to π. Hardy showed that for $1 < p < \infty$, the Hilbert matrix defines a bounded operator $\mathbf{a} \mapsto H\mathbf{a}$ on the sequence space ℓ^p (see Example 1.8.1) and

$$\|H\|_{\ell^p \to \ell^p} = \frac{\pi}{\sin(\pi/p)}.$$

We know that $\|H\|_{\ell^2 \to \ell^2} = \pi$ from the discussion at the beginning of this chapter. Working with power series, one can define the Hilbert matrix H as an operator on various spaces of analytic functions as follows. For $f(z) = \sum_{k=0}^\infty a_k z^k$, define

$$(Hf)(z) = \sum_{n=0}^\infty \left(\sum_{k=0}^\infty \frac{a_k}{n+k+1} \right) z^n.$$

In other words, the Taylor coefficients of $(Hf)(z)$ are the entries of the column vector $H\mathbf{a}$. For the Hardy spaces H^p, where $1 < p < \infty$, the operator $f \mapsto Hf$ is bounded and

$$\|H\|_{H^p \to H^p} = \frac{\pi}{\sin(\pi/p)}.$$

For the Bergman spaces A^p, where $1 < p < \infty$, the operator $f \mapsto Hf$ is bounded for $2 < p < \infty$, and

$$\|H\|_{A^p \to A^p} = \frac{\pi}{\sin(2\pi/p)} \quad \text{for } 4 \leqslant p < \infty.$$

When $2 < p < 4$, the exact value of the norm of H on A^p is unknown. The key to proving many of these results is to represent the function Hf above as an integral (see Exercise 17.10.2). The book [205] contains the proofs of the results above and further references. The spectral properties of the Hilbert matrix for these spaces are studied in [13, 64, 345].

Peller's book [267] is a comprehensive and authoritative text on Hankel operators. Two other good sources are Partington [262] and Power [274].

In his 1861 doctoral dissertation [173], Hankel explored finite Hankel matrices. In 1881, Kronecker [225] studied finite-rank Hankel operators (see Exercise 17.10.8). The use of the terms "Toeplitz operator" and "Hankel operator" originated in a paper of Hartman and Wintner [181], where they discussed the boundedness and compactness properties of certain types of Hankel operators. These results were extended in 1958 by Hartman [180], who showed that a Hankel operator is compact if and only if it can be represented (in the sense of Theorem 17.4.4) by a continuous symbol. A Hankel operator can belong to various other classes of operators [267].

One can realize the boundedness and compactness of a Hankel matrix

$$
\begin{bmatrix}
\alpha_1 & \alpha_2 & \alpha_3 & \cdots \\
\alpha_2 & \alpha_3 & \alpha_4 & \cdots \\
\alpha_3 & \alpha_4 & \alpha_5 & \cdots \\
\vdots & \vdots & \vdots & \ddots
\end{bmatrix}
\tag{17.9.1}
$$

in a function theoretic way. A function $\varphi \in L^1(\mathbb{T})$ is of *bounded mean oscillation* if

$$
\sup_{I \subseteq \mathbb{T}} \frac{1}{m(I)} \int_I |f - f_I|\, dm < \infty.
$$

In the above, I is an arc of $\mathbb{T}$ and

$$
f_I = \frac{1}{m(I)} \int_I f\, dm
$$

is the *mean* of f on I. The space of functions with bounded mean oscillation is denoted by BMO. One can show that $L^\infty(\mathbb{T}) \subseteq$ BMO. However, the reverse containment is not true since $\log |p| \in$ BMO whenever p is a trigonometric polynomial with $p \not\equiv 0$. A $\varphi \in L^1(\mathbb{T})$ is of *vanishing mean oscillation* (VMO) if

$$
\lim_{a \to 0^+} \sup_{|I| \leqslant a} \frac{1}{m(I)} \int_I |f - f_I|\, dm = 0.
$$

Note that $C(\mathbb{T}) \subseteq$ VMO. A classical theorem of Riesz says that if $1 < p < \infty$, then the Riesz projection satisfies $P_+ L^p(\mathbb{T}) \subseteq L^p(\mathbb{T})$ (in fact $P_+ L^p(\mathbb{T}) = H^p$, the L^p version of the Hardy space). This no longer holds if $p = \infty$. Here one has $P_+ L^\infty(\mathbb{T}) \subseteq$ BMO and $P_+ C(\mathbb{T}) \subseteq$ VMO. The Hankel matrix from (17.9.1) is bounded on ℓ^2 if and only if $\varphi = \sum_{k=0}^\infty \alpha_k \xi^k$ belongs to BMO and compact if and only if $\varphi \in$ VMO.

One may also consider Hankel operators on the Bergman space [21, 27, 232]. This first requires the orthogonal decomposition

$$
L^2(dA) = A^2 \oplus (A^2)^\perp.
$$

For an appropriately chosen symbol φ on $\mathbb{D}$, define $H_\varphi : A^2 \to (A^2)^\perp$ by

$$
H_\varphi f = (I - P)(\varphi f),
$$

where P is the orthogonal projection of $L^2(dA)$ onto A^2.

17.10 Exercises

Exercise 17.10.1. For each $(x_n)_{n=1}^\infty \in \ell^2(\mathbb{N})$, prove that $\lim\limits_{m \to \infty} \sum\limits_{n=1}^\infty \dfrac{x_n}{m+n} = 0$.

Exercise 17.10.2. Consider the infinite Hilbert matrix H from (17.1.3). It follows from (17.5.3) that H defines a bounded operator on ℓ^2. Since ℓ^2 is unitarily isomorphic to the Hardy space H^2 via $\mathbf{a} = (a_n)_{n=0}^\infty \mapsto a(z) = \sum_{n=0}^\infty a_n z^n$ (Proposition 5.3.1), one can think of H as acting on H^2. Use the following steps to express H as an integral operator.

(a) For a polynomial f, define

$$(\mathfrak{H}f)(z) = \int_0^1 \frac{f(t)}{1 - tz}\,dt \quad \text{for } z \in \mathbb{D}.$$

Prove that H and $\mathfrak{H}$ agree on the polynomials, meaning if

$$\mathbf{a} = (a_0, a_1, ..., a_n, 0, 0, ...) \quad \text{and} \quad a(z) = \sum_{j=0}^n a_j z^j,$$

then $H\mathbf{a}$ equals $\mathfrak{H}a$ in the sense that $H\mathbf{a}$ is the sequence of Taylor coefficients of $\mathfrak{H}a$.

(b) An inequality of Fejér and Riesz [118, p. 46] says that

$$\int_{-1}^1 |f(x)|^2 dx \leqslant \frac{1}{2} \int_{\mathbb{T}} |f(\xi)|^2 dm(\xi) \quad \text{for all } f \in H^2.$$

Use this to prove that for $f \in H^2$, the integral $\mathfrak{H}f$ converges for all $z \in \mathbb{D}$ and defines an analytic function on $\mathbb{D}$. Thus, $\mathfrak{H}$ provides an integral representation of the Hilbert-matrix operator.

Remark: This integral representation allows the Hilbert matrix to be defined on various other spaces of analytic functions [13, 110, 205]. See also Exercise 18.8.14.

Exercise 17.10.3. Consider the Hilbert matrix H as an operator on ℓ^2.

(a) Prove that H is not a finite-rank operator.

(b) Prove that H is not compact.

Exercise 17.10.4.

(a) For $f \in C[0, 1]$, prove that $\sum\limits_{n=0}^\infty \left| \int_0^1 t^n f(t)\,dt \right|^2 \leqslant \pi \int_0^1 |f(t)|^2\,dt.$

(b) Show that the constant π is best possible.

Exercise 17.10.5. If H is the Hilbert matrix from Exercise 17.10.2, C is the Cesàro matrix C from (6.2.5), and $B = [b_{jk}]_{j,k=0}^{\infty}$, where

$$b_{jk} = \frac{k+1}{(j+k+1)(j+k+2)} \quad \text{for } j, k \geqslant 0,$$

prove that $H = BC$.

Remark: See [40] for more on this.

Exercise 17.10.6. Here is another way to prove the boundedness of the Hilbert matrix $H = [h_{jk}]_{j,k=0}^{\infty}$ on ℓ^2 [40] (see also [83]).

(a) Let L be the matrix

$$L = \begin{bmatrix} 1 & \frac{1}{2} & \frac{1}{3} & \cdots \\ \frac{1}{2} & \frac{1}{2} & \frac{1}{3} & \cdots \\ \frac{1}{3} & \frac{1}{3} & \frac{1}{3} & \cdots \\ \vdots & \vdots & \vdots & \ddots \end{bmatrix}.$$

In other words,

$$L_{ij} = \frac{1}{\max\{i, j\} + 1}.$$

Prove that $L = CC^*$, where C is the Cesàro matrix.

(b) Prove that $\|H\| \leqslant \|L\|$.

(c) Prove that H is bounded on ℓ^2 and $\|H\| \leqslant 4$.

Remark: L is a special type of "L-shaped matrix" (see Chapter 6).

Exercise 17.10.7. Let $S \in \mathcal{B}(\ell^2)$ be the unilateral shift and H be the Hilbert matrix. Prove that $S^*HS - H$ is a Hilbert–Schmidt operator.

Exercise 17.10.8. This exercise proves a result of Kronecker from 1881 [225]. Let

$$H(\mathbf{a}) = \begin{bmatrix} a_0 & a_1 & a_2 & a_3 & a_4 & \cdots \\ a_1 & a_2 & a_3 & a_4 & a_5 & \cdots \\ a_2 & a_3 & a_4 & a_5 & a_6 & \cdots \\ a_3 & a_4 & a_5 & a_6 & a_7 & \cdots \\ a_4 & a_5 & a_6 & a_7 & a_8 & \cdots \\ \vdots & \vdots & \vdots & \vdots & \vdots & \ddots \end{bmatrix}$$

denote the Hankel matrix corresponding to $\mathbf{a} = (a_n)_{n=0}^{\infty}$. The goal is to show that if $H(\mathbf{a})$ has finite rank, then $f(z) = \sum_{n=0}^{\infty} a_n z^n$ is a rational function. Use this approach from [266].

(a) For a power series $f(z) = \sum_{n=0}^{\infty} a_n z^n$, let S and B denote the formal forward and backward shift operators $S(a_0 + a_1 z + a_2 z^2 + \cdots) = a_0 z + a_1 z^2 + a_2 z^3 + \cdots$ and $B(a_0 + a_1 z + a_2 z^2 + \cdots) = a_1 + a_2 z + a_3 z^2 + \cdots$, respectively. Prove that

$$S^n B^k f = S^{n-k} f - S^{n-k} \sum_{j=0}^{k-1} a_j z^j \quad \text{for } 0 \leqslant k \leqslant n.$$

(b) If rank $H(\mathbf{a}) = n$, look at the first $n+1$ rows of $H(\mathbf{a})$ and prove that there are constants $c_0, c_1, \ldots, c_n$, not all zero, such that $c_0 f + c_1 B f + c_2 B^2 f + \cdots + c_n B^n f = 0$.

(c) Prove that $\sum_{k=0}^{n} c_k S^{n-k} f = p$, where p is a polynomial of degree at most $n - 1$.

(d) Let $q(z) = \sum_{j=0}^{n} c_{n-j} z^j$ and prove that $qf = p$.

Remark: The converse of this result is true. See [266] for the details.

Exercise 17.10.9. For $\varphi \in L^\infty(\mathbb{T})$, prove that the following conditions are equivalent.

(a) The Hankel operator H_φ has finite rank.

(b) $(I - P_+)\varphi$ is a rational function.

(c) There exists a finite Blaschke product B such that $B\varphi \in H^\infty$.

Exercise 17.10.10. Prove the following version of the Brown–Halmos theorem (Theorem 16.5.1) for Hankel operators. Suppose that $A \in \mathcal{B}(H^2, \overline{H_0^2})$, S is the unilateral shift on H^2, and M_ξ is the bilateral shift on $L^2(\mathbb{T})$. Prove that A is a Hankel operator if and only if $P_- M_\xi A = AS$.

Exercise 17.10.11. Let $\mathbf{a} = (a_n)_{n=0}^{\infty}$ be a sequence of complex numbers. If $\sum_{n=1}^{\infty} n|a_n|^2 < \infty$, prove that the Hankel matrix $H(\mathbf{a})$ in Exercise 17.10.8 is compact on ℓ^2.

Exercise 17.10.12. Prove that the diagonal operator $\operatorname{diag}(\frac{1}{2}, \frac{2}{3}, \frac{3}{4}, \ldots)$ does not have a maximizing vector (Definition 17.6.1).

Exercise 17.10.13. Let $\varphi \in L^\infty(\mathbb{T})$. Nehari's theorem (Theorem 17.4.4) says that $\|H_\varphi\| = \operatorname{dist}(\varphi, H^\infty)$. Use the following argument from [28] to outline a proof that

$$\|H_\varphi\|_e = \operatorname{dist}(\varphi, H^\infty + C(\mathbb{T})),$$

where $H^\infty + C(\mathbb{T}) = \{f + g : f \in H^\infty, g \in C(\mathbb{T})\}$ and $\|H_\varphi\|_e = \inf_{K \in \mathcal{K}} \|H_\varphi - K\|$ is the *essential norm* of H_φ. In the infimum above, $\mathcal{K}$ denotes the set of compact operators $K : H^2 \to \overline{H_0^2}$. Thus, $\|H_\varphi\|_e$ is the distance from H_φ to $\mathcal{K}$. We examined the essential norm of a Toeplitz operator in Exercise 16.9.30.

(a) For each $K \in \mathcal{K}$, prove that $\|H_\varphi S^n\| \leqslant \|(H_\varphi - K)S^n\| + \|KS^n\|$ for all $n \geqslant 1$.

(b) Use Exercise 5.9.26 to prove that $\limsup_{n \to \infty} \|H_\varphi S^n\| \leqslant \|H_\varphi\|_e$.

(c) Using the matrix representation of $H_\varphi S^n$, prove that $\|H_\varphi S^n\| = \|H_\varphi\|$.

(d) Prove that $\|H_\varphi S^n\| \geqslant \|H_\varphi\|_e$ for all $n \geqslant 1$.

(e) Prove that $\lim_{n\to\infty} \|H_\varphi S^n\| = \|H_\varphi\|_e$.

(f) Use Nehari's theorem to prove that

$$\|H_\varphi S^n\| = \|H_{z^n\varphi}\| = \operatorname{dist}(z^n\varphi, H^\infty) = \operatorname{dist}(\varphi, \overline{z}^n H^\infty).$$

(g) Conclude that

$$\|H_\varphi\|_e = \lim_{n\to\infty} \|H_\varphi S^n\| = \lim_{n\to\infty} \operatorname{dist}(\varphi, \overline{z}^n H^\infty) = \operatorname{dist}(\varphi, H^\infty + C(\mathbb{T})).$$

(h) Finally, conclude that H_φ is compact if and only if $\varphi \in H^\infty + C(\mathbb{T})$.

Remark: The original proof of this is from [3].

Exercise 17.10.14. Let $\varphi \in H^\infty$ and let $\psi = \xi^{-n}\varphi$ for $n \geqslant 0$. Prove that the matrix representation of H_ψ has a finite number of nonzero entries.

Exercise 17.10.15. Suppose H is a Hankel operator whose matrix representation has a finite number of nonzero entries. Prove there is a finite Blaschke product and an $n \geqslant 0$ such that $H = H_{\xi^{-n}B}$.

Exercise 17.10.16. Show that a compact Hankel operator is the operator-norm limit of a sequence of finite-rank Hankel operators.

Exercise 17.10.17. The next three exercises discuss an operator on H^2 that is related to a Hankel operator. Define $J : L^2(\mathbb{T}) \to L^2(\mathbb{T})$ by $(Jf)(\xi) = \overline{\xi} f(\overline{\xi})$ and prove the following.

(a) $J^2 = I$.

(b) $J(fg)(\xi) = \xi (Jf)(\xi)(Jg)(\xi)$.

(c) $JP_+J = I - P_+$.

(d) $JH^2 = \overline{H_0^2}$.

Remark: See [254] for more on this.

Exercise 17.10.18. This is a continuation of Exercise 17.10.17. Let $\varphi \in L^\infty(\mathbb{T})$ and define $\Gamma_\varphi : H^2 \to H^2$ by $\Gamma_\varphi = JH_\varphi$. Prove the following.

(a) $\|\Gamma_\varphi\| = \|H_\varphi\|$.

(b) $\Gamma_\varphi = 0$ if and only if $\varphi \in H^\infty$.

Exercise 17.10.19. This is a continuation of Exercise 17.10.18.

(a) For $\varphi, \psi \in L^\infty(\mathbb{T})$, prove that $\Gamma_\varphi T_\psi + T_{\xi J\varphi}\Gamma_\psi = \Gamma_{\varphi\psi}$, where $T_{\xi J\varphi}$ and T_ψ are Toeplitz operators.

(b) Prove that $\Gamma_\varphi \Gamma_\psi = 0$ if and only if $\Gamma_\varphi = 0$ or $\Gamma_\psi = 0$.

(c) For $\varphi \in L^\infty(\mathbb{T})$, prove that $S^* \Gamma_\varphi = \Gamma_\varphi S$.

(d) If $A \in \mathcal{B}(H^2)$ satisfies $S^* A = AS$, prove that $A = \Gamma_\varphi$ for some $\varphi \in L^\infty(\mathbb{T})$.

Exercise 17.10.20. Using the ideas from Exercise 17.10.17, one can define Hankel operators on $H^2(\mathbb{C}_+)$ as follows [273]. Let P denote the orthogonal projection of $L^2(\mathbb{R})$ onto $H^2(\mathbb{R})$ and let F be the unitary operator $(Ff)(x) = f(-x)$ (this plays the role of J from Exercise 17.10.17 for $L^2(\mathbb{R})$). For $\varphi \in L^\infty(\mathbb{R})$, define $\mathcal{H}_\varphi : H^2(\mathbb{R}) \to H^2(\mathbb{R})$ by

$$\mathcal{H}_\varphi f = PFM_\varphi f.$$

(a) Prove that $\mathcal{H}_\varphi$ is bounded on $H^2(\mathbb{R})$.

(b) Prove that $\mathcal{H}_\varphi = 0$ if and only if $\varphi \in H^\infty(\mathbb{R}) = H^2(\mathbb{R}) \cap L^\infty(\mathbb{R})$.

(c) Exercise 11.10.7 shows that

$$(Ug)(x) = \frac{1}{\sqrt{\pi}} \frac{1}{x+i} g\left(\frac{x-i}{x+i}\right)$$

defines a unitary operator from $L^2(\mathbb{T})$ onto $L^2(\mathbb{R})$ and a unitary operator from $H^2(\mathbb{T})$ onto $H^2(\mathbb{R})$. Prove that $U^* FU = J$.

(d) For such $\mathcal{H}_\varphi$, prove there is a $\psi \in L^\infty$ such that $\mathcal{H}_\varphi$ and Γ_ψ are unitarily equivalent.

Exercise 17.10.21. Write the Hankel matrix

$$H_0 = \frac{2}{\pi} \begin{bmatrix} 1 & 0 & \frac{1}{3} & 0 & \frac{1}{5} & \cdots \\ 0 & \frac{1}{3} & 0 & \frac{1}{5} & 0 & \cdots \\ \frac{1}{3} & 0 & \frac{1}{5} & 0 & \frac{1}{7} & \cdots \\ 0 & \frac{1}{5} & 0 & \frac{1}{7} & 0 & \cdots \\ \frac{1}{5} & 0 & \frac{1}{7} & 0 & \frac{1}{9} & \cdots \\ \vdots & \vdots & \vdots & \vdots & \vdots & \ddots \end{bmatrix}$$

as a Γ_φ operator from Exercise 17.10.17.
Remark: See [273] for more on this matrix.

Exercise 17.10.22. The Hankel matrix H_0 from Exercise 17.10.21 (regarded as an operator in H^2) is unitarily equivalent to the integral operator $A : L^2(\mathbb{R}_+) \to L^2(\mathbb{R}_+)$ defined by

$$(Af)(x) = \frac{1}{\pi} \int_0^\infty \frac{f(t)}{t+x} dt,$$

which can be regarded as the continuous version of H_0. Follow these steps from [273] to see why.

(a) Define the operator $A_1 : L^1[-1, 1] \to L^2[-1, 1]$ by

$$(A_1 g)(x) = \frac{1}{\pi} \int_{-1}^{1} \frac{g(y)}{1 - xy} dy$$

and prove that A is unitarily equivalent to A_1 via the unitary operator $U : L^2[-1, 1] \to L^2(\mathbb{R}_+)$ defined by

$$(Ug)(s) = \frac{\sqrt{2}}{s + 1} g\left(\frac{1 - s}{1 + s}\right).$$

(b) Prove that the restriction map $R : H^2 \to L^2[-1, 1]$ defined by $Rh = h|_{[-1,1]}$ is bounded and has dense range.

(c) Prove that $RH_0 = A_1 R$.

(d) Prove that H_0 and A_1 are unitarily equivalent.

(e) Conclude that H_0 and A are unitarily equivalent.

Exercise 17.10.23. This is a continuation of Exercise 17.10.22. It determines the spectrum of H_0.

(a) Prove that A is unitarily equivalent to the operator $A_2 : L^2(\mathbb{R}) \to L^2(\mathbb{R})$ defined by

$$(A_2 f)(x) = \frac{1}{\pi} \int_{-\infty}^{\infty} \frac{f(t)}{\cosh(x - t)} dt.$$

(b) Observe that A_2 is a convolution operator. Use properties of the Fourier transform to prove that A_2 is unitarily equivalent to M_φ on $L^2(\mathbb{R})$, where

$$\varphi(x) = \frac{1}{\cosh\left(\frac{\pi}{2} x\right)}.$$

(c) Prove that $\sigma(M_\varphi) = [0, 1]$.

Exercise 17.10.24. If H is the Hilbert matrix from Exercise 17.10.2, prove that $H^2 = T_{(\arg z - \pi)^2} - T_{\arg z - \pi}^2$, where T_φ is the matrix representation of the corresponding Toeplitz operator with respect to the basis $(z^n)_{n=0}^{\infty}$.
Remark: From here one can fashion another proof that $\sigma(H) = [0, \pi]$ [273].

Exercise 17.10.25.

(a) With respect to the decomposition $L^2(\mathbb{T}) = H^2 \oplus \overline{H_0^2}$, prove that

$$M_\varphi = \begin{bmatrix} T_\varphi & H_{\overline{\varphi}}^* \\ H_\varphi & S_\varphi \end{bmatrix},$$

where $S_\varphi = P_- M_\varphi P_-$.

(b) Prove that $T_{\varphi\psi} = T_\varphi T_\psi + H_{\overline{\varphi}}^* H_\psi$ and $H_{\overline{\varphi\psi}}^* = T_\varphi H_{\overline{\psi}}^* + H_{\overline{\varphi}}^* S_\psi$.

Exercise 17.10.26. If $\varphi \in L^\infty(\mathbb{T})$ and $\psi \in C(\mathbb{T})$, prove that $T_{\varphi\psi} - T_\varphi T_\psi$ is compact.

Exercise 17.10.27. There is a notion of asymptotic Toeplitz operators from [34]. Recall from Theorem 16.5.1 that every Toeplitz operator T_φ satisfies $S^* T_\varphi S = T_\varphi$. Then $T \in \mathcal{B}(H^2)$ is an *asymptotic Toeplitz operator* if $S^{*n} T S^n$ converges in the strong operator topology as $n \to \infty$, that is, there is an operator $T_\infty \in \mathcal{B}(H^2)$ such that

$$\|S^{*n} T S^n f - T_\infty f\| \to 0 \quad \text{for all } f \in H^2.$$

(a) Show that every Toeplitz operator is an asymptotic Toeplitz operator.

(b) Show that every compact operator on H^2 is an asymptotic Toeplitz operator.

(c) Show that Γ_φ from Exercise 17.10.18 is an asymptotic Toeplitz operator for every $\varphi \in L^\infty(\mathbb{T})$.

(d) For all $\varphi, \psi \in L^\infty(\mathbb{T})$, prove that $\Gamma_\varphi \Gamma_\psi$ is an asymptotic Toeplitz operator.

(e) For all $\varphi, \psi \in L^\infty(\mathbb{T})$, prove that $T_\varphi T_\psi$ is an asymptotic Toeplitz operator.

Remark: See [128] for more on this where one considers when $(S^{*n} T S^n)_{n=0}^\infty$ converges in the norm or weak operator topologies.

Exercise 17.10.28. This is a continuation of Exercise 17.10.27. Suppose D_Λ is a diagonal operator on ℓ^2, where $\Lambda = (\lambda_n)_{n=0}^\infty$, and $S \in \mathcal{B}(\ell^2)$ is the unilateral shift. Prove that $S^{*n} D_\Lambda S^n \to 0$ in the strong operator topology if and only if $(\lambda_n)_{n=0}^\infty$ converges.

17.11 Hints for the Exercises

Hint for Ex. 17.10.4: The Gram matrix $[\langle t^j, t^i \rangle]_{i,j=0}^\infty$ is the Hilbert matrix $[1/(i+j+1)]_{i,j=0}^\infty$.

Hint for Ex. 17.10.7: Consult Exercise 3.6.31.

Hint for Ex. 17.10.11: Consult Theorem 3.4.1.

Hint for Ex. 17.10.20: For (d), consider the operator U.

Hint for Ex. 17.10.22: For (b), the boundedness comes from the Fejér–Riesz inequality (see Exercise 17.10.2). For part (d), consult Exercise 8.10.28.

Hint for Ex. 17.10.23: For (a), consider an appropriate substitution. For (c), consult Exercise 8.10.1 and Proposition 8.1.12.

Hint for Ex. 17.10.26: Consult Exercise 17.10.25.

18

· · ● · ·

Composition Operators

Key Concepts: Composition operator on the Hardy space, Littlewood subordination principle, norm of a composition operator, compact composition operator, adjoint and spectrum of a composition operator, universal composition operator.

Outline: This chapter explores composition operators $C_\varphi f = f \circ \varphi$ on the Hardy space H^2, where the symbol φ is an analytic map from the open unit disk $\mathbb{D}$ to itself. We cover representative results concerning the boundedness, compactness, and spectral properties of composition operators.

18.1 A Motivating Example

Before heading into the subject of composition operators on H^2, let us work through some examples that provide a baseline for the types of results to expect. This allows us to connect with the infinite matrices material covered in Chapter 3. If $|\beta| \leqslant 1$ and $f(z) = \sum_{n=0}^{\infty} a_n z^n$ belongs to H^2, that is $\|f\|^2 = \sum_{n=0}^{\infty} |a_n|^2 < \infty$, then

$$\|f(\beta z)\|^2 = \sum_{n=0}^{\infty} |a_n|^2 |\beta|^{2n} \leqslant \|f\|^2.$$

Thus, the composition operator $f(z) \mapsto f(\beta z)$, denoted by $C_{\beta z}$, is bounded on H^2. In fact, it is a contraction. Furthermore, its matrix representation with respect to the standard orthonormal basis $(z^k)_{k=0}^{\infty}$ for H^2 is the diagonal operator

$$
\begin{bmatrix}
1 & 0 & 0 & 0 & \cdots \\
0 & \beta & 0 & 0 & \cdots \\
0 & 0 & \beta^2 & 0 & \cdots \\
0 & 0 & 0 & \beta^3 & \cdots \\
\vdots & \vdots & \vdots & \vdots & \ddots
\end{bmatrix}.
$$

Proposition 2.1.1, Theorem 2.4.7, along with Theorem 2.5.1 yield the following result; see Figure 18.1.1.

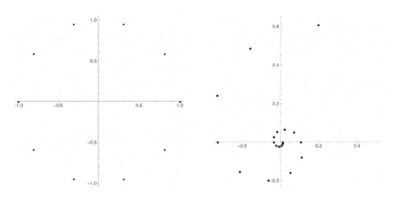

Figure 18.1.1 The spectrum of the composition operator $C_{\beta z}$, where $\beta = e^{-i2\pi/5}$ (left) and $C_{\beta z}$, where $\beta = \frac{4}{5}e^{-i2\pi/5}$ (right).

Proposition 18.1.1. *Let* $|\beta| \leqslant 1$.

(a) *The composition operator* $(C_{\beta z}f)(z) = f(\beta z)$ *is bounded on* H^2 *and* $\|C_{\beta z}\| = 1$.

(b) $\sigma_p(C_{\beta z}) = \{1, \beta, \beta^2, ...\}$.

(c) $\sigma(C_{\beta z}) = \{1, \beta, \beta^2, ...\}^-$.

(d) $C_{\beta z}$ *is compact if and only if* $|\beta| < 1$.

Now consider the composition operator

$$(C_{\alpha+\beta z}f)(z) = f(\alpha + \beta z)$$

on H^2, where $|\alpha| + |\beta| < 1$ (this guarantees that $z \mapsto \alpha + \beta z$ maps $\mathbb{D}$ to another disk whose closure lies in $\mathbb{D}$: see Figure 18.1.2). With respect to the basis $(z^k)_{k=0}^{\infty}$, this operator has the matrix representation

$$\begin{bmatrix} 1 & \alpha & \alpha^2 & \alpha^3 & \cdots \\ 0 & \beta & 2\alpha\beta & 3\alpha^2\beta & \cdots \\ 0 & 0 & \beta^2 & 3\alpha\beta^2 & \cdots \\ 0 & 0 & 0 & \beta^3 & \cdots \\ \vdots & \vdots & \vdots & \vdots & \ddots \end{bmatrix}.$$

The (j, k) entry of this matrix is

$$a_{jk} = \begin{cases} 0 & \text{if } j > k, \\ \binom{k}{j}\alpha^{k-j}\beta^j & \text{if } j \leqslant k, \end{cases} \tag{18.1.2}$$

(see Exercise 18.8.7).

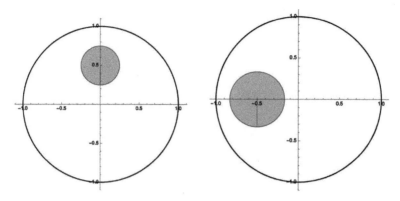

Figure 18.1.2 The images of the maps $z \mapsto \frac{i}{2} + \frac{i}{4}z$ (left) and $z \mapsto -\frac{1}{2} - \frac{i}{3}z$ on $\mathbb{D}$ (right).

Proposition 18.1.3. *Let α, β satisfy $|\alpha| + |\beta| = r < 1$ and $\beta \neq 0$.*

(a) *$C_{\alpha+\beta z}$ on H^2 is compact. In particular, $C_{\alpha+\beta z}$ is bounded.*

(b) *$\sigma_p(C_{\alpha+\beta z}) = \{1, \beta, \beta^2, \ldots\}$.*

(c) *$\sigma(C_{\alpha+\beta z}) = \{1, \beta, \beta^2, \ldots\} \cup \{0\}$.*

Proof (a) Since $|\alpha| + |\beta| = r < 1$, the a_{jk} from (18.1.2) satisfy

$$\sum_{j,k=0}^{\infty} |a_{jk}| = \sum_{k=0}^{\infty} \sum_{j=0}^{k} \binom{k}{j} |\alpha|^{k-j} |\beta|^j = \sum_{k=0}^{\infty} (|\alpha| + |\beta|)^k = \sum_{k=0}^{\infty} r^k = \frac{1}{1-r} < \infty.$$

Apply Theorem 3.4.1 and deduce that $C_{\alpha+\beta z}$ is compact. In particular, $C_{\alpha+\beta z}$ is bounded (Exercise 2.8.18).

(b) If

$$g_n(z) = \left(z - \frac{\alpha}{1-\beta}\right)^n \quad \text{for all } n \geqslant 0,$$

then $C_{\alpha+\beta z} g_n = \beta^n g_n$ and thus $\sigma_p(C_{\alpha+\beta z}) \supseteq \{1, \beta, \beta^2, \ldots\}$. Let us now show equality. A calculation verifies that if $\varphi(z) = \alpha + \beta z$, then

$$z_0 = \frac{\alpha}{1-\beta}$$

belongs to $\mathbb{D}$ and satisfies $\varphi(z_0) = z_0$. If $\lambda \neq 0$ is an eigenvalue of C_φ, then for some $f \in H^2 \backslash \{0\}$, we have $(f \circ \varphi)(z) = \lambda f(z)$ for all $z \in \mathbb{D}$. Evaluating this at $z = z_0$ yields $f(z_0) = \lambda f(z_0)$, which implies that $f(z_0) = 0$. Since $f \not\equiv 0$, there is an analytic function g on $\mathbb{D}$ and a positive integer m such that $g(z_0) \neq 0$ and

$$f(z) = (z - z_0)^m g(z).$$

Use the eigenvalue equation $f \circ \varphi = \lambda f$ to obtain

$$\lambda(z - z_0)^m g(z) = (\varphi(z) - z_0)^m g(\varphi(z)),$$

which can be written as

$$\lambda g(z) = \left(\frac{\varphi(z) - z_0}{z - z_0}\right)^m g(\varphi(z)).$$

Let $z \to z_0$ in the expression above and use the fact that $\varphi(z_0) = z_0$ to obtain $\lambda g(z_0) = \varphi'(z_0)^m g(z_0)$. Since $g(z_0) \neq 0$, it follows that $\lambda = \varphi'(z_0)^m$. Since $\varphi'(z_0) = \beta$, one concludes that $\{1, \beta, \beta^2, ...\} \supseteq \sigma_p(C_{\alpha+\beta z})$. Note that 0 not an eigenvalue since $f \circ \varphi \equiv 0$ implies that $f \equiv 0$.

(c) From part (a), $C_{\alpha+\beta z}$ is compact. Thus, Riesz's theorem (Theorem 2.6.9) implies that $\sigma(C_{\alpha+\beta z}) = \sigma_p(C_{\alpha+\beta z}) \cup \{0\}$. ∎

The discussion of the spectrum of $C_{\alpha+\beta z}$ when $|\alpha| + |\beta| = 1$ is more complicated and is found in a paper of Deddens [108].

18.2 Composition Operators on H^2

If the composition operator $f \mapsto f \circ \varphi$ is bounded on H^2, then $f \circ \varphi$ is analytic on $\mathbb{D}$ for all $f \in H^2$. Apply this to $f(z) = z$ and conclude that the symbol φ is an analytic function from $\mathbb{D}$ to itself. Such maps are *analytic self maps* of $\mathbb{D}$. For a given analytic self map φ, it is not immediately clear that $f \circ \varphi$ belongs to H^2 whenever $f \in H^2$. As a consequence of a more general function-theory result, the Littlewood subordination theorem [229], it does. In the language of composition operators, a proof is given below.

Theorem 18.2.1. *If* $\varphi : \mathbb{D} \to \mathbb{D}$ *is analytic, then the composition operator* $C_\varphi f = f \circ \varphi$ *is bounded on* H^2 *and*

$$1 \leqslant \|C_\varphi\| \leqslant \left(\frac{1 + |\varphi(0)|}{1 - |\varphi(0)|}\right)^{\frac{1}{2}}.$$

Proof This proof is from [143, Ch.6]. We first prove this when $\varphi(0) = 0$. Let $f = \sum_{n=0}^\infty a_n z^n \in H^2$ and let S^* denote the backward shift

$$(S^* f)(z) = \frac{f(z) - f(0)}{z}, \tag{18.2.2}$$

the adjoint of the unilateral shift S (recall Exercise 5.9.14). Then $(S^{*k} f)(z) = \sum_{n=0}^\infty a_{n+k} z^n$ for every $k \geqslant 0$. This implies that

$$(S^{*k} f)(0) = a_k \quad \text{for } k \geqslant 0. \tag{18.2.3}$$

The formula in (18.2.2) can be rewritten as

$$f(z) = f(0) + z(S^* f)(z) \quad \text{for } z \in \mathbb{D}.$$

Since φ maps $\mathbb{D}$ into itself, replace z with $\varphi(z)$ in the preceding identity to obtain

$$f(\varphi(z)) = f(0) + \varphi(z)(S^* f)(\varphi(z)).$$

This identity can be written as

$$C_\varphi f = f(0) + T_\varphi C_\varphi S^* f, \tag{18.2.4}$$

where T_φ is an analytic Toeplitz operator (Chapter 16). One may question the validity of this identity since we do not yet know if C_φ maps H^2 into itself. To take this into account, we initially apply (18.2.4) to polynomials f, in which case the right-hand side of (18.2.4) is well defined.

The assumption that $\varphi(0) = 0$, along with (18.2.4), implies that $(T_\varphi C_\varphi S^* f)(0) = 0$, and thus $T_\varphi C_\varphi S^* f$ is orthogonal to the constant functions. Hence by (18.2.4),

$$\begin{aligned}
\|C_\varphi f\|^2 &= |f(0)|^2 + \|T_\varphi C_\varphi S^* f\|^2 \\
&\leqslant |f(0)|^2 + \|T_\varphi\|^2 \|C_\varphi S^* f\|^2 \\
&\leqslant |f(0)|^2 + \|C_\varphi S^* f\|^2.
\end{aligned}$$

Note the use of the fact that $\|T_\varphi\| \leqslant 1$ since $\|\varphi\|_\infty \leqslant 1$ (Theorem 16.3.1). Replace f with $S^{*k} f$ in the previous estimate and use (18.2.3) to see that

$$\|C_\varphi S^{*k} f\|^2 \leqslant |a_k|^2 + \|C_\varphi S^{*(k+1)} f\|^2 \quad \text{for } k \geqslant 0. \tag{18.2.5}$$

Sum both sides for $0 \leqslant k \leqslant n = \deg f$ and see that

$$\sum_{k=0}^n \left\| C_\varphi S^{*k} f \right\|^2 \leqslant \sum_{k=0}^n |a_k|^2 + \sum_{k=0}^n \left\| C_\varphi S^{*(k+1)} f \right\|^2.$$

A telescoping-series argument, along with the fact that $S^{*(n+1)} f \equiv 0$, yields

$$\|C_\varphi f\|^2 \leqslant \sum_{k=0}^n |a_k|^2 = \|f\|^2. \tag{18.2.6}$$

Hence, $\|C_\varphi f\| \leqslant \|f\|$ when f is a polynomial.

To establish $\|C_\varphi f\| \leqslant \|f\|$ for $f \in H^2$, let $f = \sum_{k=0}^\infty a_k z^k \in H^2$ and define

$$f_n = \sum_{k=0}^n a_k z^k \quad \text{for } n \geqslant 0.$$

The sequence of Taylor polynomials $(f_n)_{n=0}^\infty$ converges uniformly on compact subsets of $\mathbb{D}$ to f (Proposition 5.3.8). For a fixed $r \in [0, 1)$, observe that

$$\int_{\mathbb{T}} |(f_n \circ \varphi)(r\xi)|^2 dm(\xi) \leqslant \|f_n \circ \varphi\|^2 \leqslant \|f_n\|^2 \leqslant \|f\|^2.$$

Note the use of (18.2.6) and Parseval's theorem (Theorem 1.4.9) in the above.

Since φ maps $r\mathbb{T}$ to a compact subset of $\mathbb{D}$, let $n \to \infty$ in the previous inequality and obtain

$$\int_{\mathbb{T}} |(f \circ \varphi)(r\xi)|^2 dm(\xi) \leqslant \|f\|^2.$$

The estimate above holds uniformly in $r \in [0, 1)$. Thus, $f \circ \varphi \in H^2$ (Corollary 5.3.10) and $\|f \circ \varphi\| \leqslant \|f\|$. In other words, $\|C_\varphi\| \leqslant 1$. Since $C_\varphi 1 = 1$, it follows that $\|C_\varphi\| = 1$. Our next step is to estimate the norm of C_{τ_w}, where $w \in \mathbb{D}$ and

$$\tau_w(z) = \frac{w - z}{1 - \bar{w}z} \quad \text{for } z \in \mathbb{D},$$

is an automorphism of $\mathbb{D}$ such that $(\tau_w \circ \tau_w)(z) = z$ for every $z \in \mathbb{D}$. Use $C_{\tau_w} 1 = 1$ and deduce that $\|C_{\tau_w}\| \geqslant 1$. To obtain the upper bound, fix any $f \in H^2$ and use the change of variables

$$e^{is} = \tau_w(e^{it}) \quad \Leftrightarrow \quad e^{it} = \tau_w(e^{is}) \quad \text{and} \quad dt = \frac{1 - |w|^2}{|1 - \bar{w}e^{is}|^2} ds$$

to obtain

$$\|f \circ \tau_w\|^2 = \int_0^{2\pi} |f(\tau_w(e^{it}))|^2 \frac{dt}{2\pi}$$

$$= \int_0^{2\pi} |f(e^{is})|^2 \frac{1 - |w|^2}{|1 - \bar{w}e^{is}|^2} \frac{ds}{2\pi}$$

$$\leqslant \frac{1 - |w|^2}{(1 - |w|)^2} \int_0^{2\pi} |f(e^{is})|^2 \frac{ds}{2\pi}$$

$$= \frac{1 + |w|}{1 - |w|} \|f\|^2.$$

Consequently,

$$1 \leqslant \|C_{\tau_w}\| \leqslant \left(\frac{1 + |w|}{1 - |w|}\right)^{\frac{1}{2}}.$$

So far we have shown that if $\varphi(0) = 0$, then $\|C_\varphi\| = 1$ and $\|C_{\tau_w}\|$ satisfies the estimate above. Now let φ be an analytic self map of the disk. Since $C_\varphi 1 = 1$, we have the lower bound $\|C_\varphi\| \geqslant 1$. To obtain the upper bound, let $w = \varphi(0)$ and $\psi = \tau_w \circ \varphi$. Then ψ is an analytic self map with $\psi(0) = 0$. Thus, $\|C_\psi\| = 1$.

The identity $\psi = \tau_w \circ \varphi$ is equivalent to $\varphi = \tau_w \circ \psi$ and the latter implies that $C_\varphi = C_\psi C_{\tau_w}$. Hence, by our previous estimates

$$\|C_\varphi\| \leqslant \|C_{\tau_w}\| \, \|C_\psi\| \leqslant \left(\frac{1 + |w|}{1 - |w|}\right)^{\frac{1}{2}} = \left(\frac{1 + |\varphi(0)|}{1 - |\varphi(0)|}\right)^{\frac{1}{2}},$$

which completes the proof. ∎

A result from the doctoral thesis of H. J. Schwartz provides an improvement of the lower bound in Theorem 18.2.1.

Proposition 18.2.7 (Schwartz [334]). *For any analytic $\varphi : \mathbb{D} \to \mathbb{D}$,*

$$\left(\frac{1}{1 - |\varphi(0)|^2}\right)^{\frac{1}{2}} \leqslant \|C_\varphi\|.$$

Proof The Cauchy integral formula implies that

$$(C_\varphi f)(0) = \int_{\mathbb{T}} (f \circ \varphi)(\xi)\,dm(\xi)$$

and hence, by the Cauchy–Schwarz inequality,

$$|(C_\varphi f)(0)| \leqslant \left(\int_{\mathbb{T}} |C_\varphi f|^2 dm\right)^{\frac{1}{2}} \left(\int_{\mathbb{T}} 1 \cdot dm\right)^{\frac{1}{2}}.$$

Thus, $|(C_\varphi f)(0)| \leqslant \|C_\varphi f\| \leqslant \|C_\varphi\|\|f\|$. Apply this inequality to the function

$$f(z) = \frac{1}{1 - \overline{\varphi(0)}z}$$

and obtain

$$\frac{1}{1 - |\varphi(0)|^2} = |(C_\varphi f)(0)| \leqslant \|C_\varphi\|\left(\frac{1}{1 - |\varphi(0)|^2}\right)^{\frac{1}{2}}.$$

Note the use of Corollary 5.3.15 above. The desired result now follows. ∎

We encourage the reader to work through Exercise 18.8.28 for an improvement of the proposition above. The next result of Nordgren, from one of the earliest papers on composition operators, computes $\|C_\varphi\|$ when φ is an inner function. Theorem 18.2.1 proves part of the next corollary. However, in order to introduce the reader to a clever integration technique that relies on the fact that φ is an inner function, we include the original proof of Nordgren.

Corollary 18.2.8 (Nordgren [255]). *If φ is an inner function, then*

$$\|C_\varphi\| = \left(\frac{1 + |\varphi(0)|}{1 - |\varphi(0)|}\right)^{\frac{1}{2}}.$$

Proof Since φ is an inner function, it has unimodular radial boundary values almost everywhere on $\mathbb{T}$. Thus, the measure $m \circ \varphi^{-1}$ on $\mathbb{T}$ is well defined. For any $n \geqslant 0$, the Cauchy integral formula yields

$$\int_{\mathbb{T}} \xi^n d(m \circ \varphi^{-1}) = \int_{\mathbb{T}} \varphi(\xi)^n\,dm(\xi) = \varphi(0)^n.$$

Furthermore, if

$$P_\lambda(\xi) = \frac{1 - |\lambda|^2}{|\xi - \lambda|^2} \quad \text{for } \xi \in \mathbb{T} \text{ and } \lambda \in \mathbb{D}$$

is the Poisson kernel, then

$$\int_{\mathbb{T}} \xi^n P_{\varphi(0)}(\xi)\,dm(\xi) = \varphi(0)^n \quad \text{for all } n \geqslant 0.$$

This last equality follows from the fact that z^n is harmonic on $\mathbb{D}$ and the Poisson integral above yields its value at $\varphi(0)$ (Theorem 12.1.6). Thus,

$$\int_{\mathbb{T}} \xi^n d(m \circ \varphi^{-1}) = \int_{\mathbb{T}} \xi^n P_{\varphi(0)}(\xi)\,dm(\xi) \quad \text{for all } n \geqslant 0.$$

Taking complex conjugates of both sides of the previous equation shows that the above holds for all $n \in \mathbb{Z}$. Hence $d(m \circ \varphi^{-1}) = P_{\varphi(0)}\,dm$. Since $P_{\varphi(0)}$ is bounded on $\mathbb{T}$, use the density of the trigonometric polynomials in $L^1(\mathbb{T})$ to conclude that

$$\int_{\mathbb{T}} g(\xi)\,d(m \circ \varphi^{-1}) = \int_{\mathbb{T}} g(\xi)P_{\varphi(0)}(\xi)\,dm(\xi) \quad \text{for all } g \in L^1(\mathbb{T}). \tag{18.2.9}$$

To compute the norm of $\|C_\varphi\|$, let $f \in H^2$ and observe that

$$\|C_\varphi f\|^2 = \int_{\mathbb{T}} |f \circ \varphi|^2 dm$$

$$= \int_{\mathbb{T}} |f|^2 d(m \circ \varphi^{-1})$$

$$= \int_{\mathbb{T}} |f|^2 P_{\varphi(0)}\,dm. \tag{18.2.10}$$

Note the use of (18.2.9) in the last equality.

For fixed $\lambda \in \mathbb{D}$, the Poisson kernel $\xi \mapsto P_\lambda(\xi)$ achieves its maximum when $\xi = \lambda/|\lambda|$. Moreover,

$$P_\lambda(\lambda/|\lambda|) = \frac{1 + |\lambda|}{1 - |\lambda|}.$$

Combine this with (18.2.10) to see that

$$\|C_\varphi\| \leqslant \left(\frac{1 + |\varphi(0)|}{1 - |\varphi(0)|}\right)^{\frac{1}{2}}. \tag{18.2.11}$$

For the lower bound, consider the family of unit vectors

$$f_\lambda(\xi) = \frac{\sqrt{1 - |\lambda|^2}}{1 - \bar{\lambda}\xi} \quad \text{for } \lambda \in \mathbb{D},$$

in H^2. Since $|f_\lambda(\xi)|^2 = P_\lambda(\xi)$, use (18.2.10) to obtain

$$\|C_\varphi\|^2 \geqslant \|C_\varphi f_\lambda\|^2 = \int_{\mathbb{T}} P_\lambda(\xi)P_{\varphi(0)}(\xi)\,dm(\xi) = P_{\varphi(0)}(\lambda).$$

If $\varphi(0) = 0$, then the above shows that $\|C_\varphi\| \geqslant 1$ since $P_0 \equiv 1$. Combine this with (18.2.11) to conclude that $\|C_\varphi\| = 1$. Otherwise, let $\lambda \to \varphi(0)/|\varphi(0)|$ radially to see that $P_{\varphi(0)}(\lambda)$ approaches

$$P_{\varphi(0)}(\varphi(0)/|\varphi(0)|) = \frac{1 + |\varphi(0)|}{1 - |\varphi(0)|}.$$

Thus,

$$\|C_\varphi\| \geqslant \left(\frac{1 + |\varphi(0)|}{1 - |\varphi(0)|}\right)^{\frac{1}{2}}.$$

Combine this with (18.2.11) to obtain the desired identity, ∎

Corollary 18.2.8 says that $\|C_\varphi\| = 1$ whenever $\varphi(0) = 0$. The next corollary says more.

Corollary 18.2.12. *If φ is inner and $\varphi(0) = 0$, then C_φ is an isometry.*

Proof The computation (18.2.10) yields

$$\|C_\varphi f\|^2 = \int_{\mathbb{T}} |f|^2 P_{\varphi(0)}\, dm.$$

If $\varphi(0) = 0$, then $P_{\varphi(0)} = P_0 \equiv 1$, so $\|C_\varphi f\| = \|f\|^2$. ∎

18.3 Compact Composition Operators

The literature on compact composition operators is large and we do not attempt to cover it all here. Instead, we give a few illustrative results which demonstrate the general principle that C_φ is compact when $\varphi(\mathbb{D})$ is small in a certain sense. Recall from Definition 2.5.3 that $T \in \mathcal{B}(\mathcal{H})$ is compact if $(Tx_n)_{n=1}^\infty$ has a convergent subsequence for every bounded sequence $(x_n)_{n=1}^\infty$. From Chapter 2 there are several equivalent characterizations of when $T \in \mathcal{B}(\mathcal{H})$ is compact.

(a) $(Tx_n)_{n=1}^\infty$ has a convergent subsequence for every bounded sequence $(x_n)_{n=1}^\infty$.

(b) T is the norm limit of finite-rank operators.

(c) If $x_n \to x$ weakly, then $Tx_n \to Tx$ in norm.

Here is our first result that relates the "smallness" of $\varphi(\mathbb{D})$ to compactness.

Proposition 18.3.1. *If $\varphi : \mathbb{D} \to \mathbb{D}$ is analytic and*

$$\sup_{z \in \mathbb{D}} |\varphi(z)| < 1,$$

then C_φ is compact.

Proof We follow the proof from [339]. For each $n \geqslant 0$, define $T_n : H^2 \to H^2$ by

$$T_n f = \sum_{j=0}^{n} \widehat{f}(j)\varphi^j.$$

Notice that $T_n = C_\varphi P_n$, where $P_n : H^2 \to \operatorname{span}\{z^j : 0 \leqslant j \leqslant n\}$, defined by

$$(P_n f)(z) = \sum_{j=0}^{n} \widehat{f}(j)z^j,$$

is the orthogonal projection of H^2 onto the finite-dimensional subspace $\operatorname{span}\{z^j : 0 \leqslant j \leqslant n\}$. Since $\operatorname{ran} T_n \subseteq \operatorname{span}\{\varphi^j : 0 \leqslant j \leqslant n\}$, it follows that T_n has finite rank and is therefore bounded (Exercise 2.8.17). Let $\alpha = \sup\{|\varphi(z)| : z \in \mathbb{D}\} < 1$. For any $f \in H^2$,

$$
\begin{aligned}
\|(C_\varphi - T_n)f\| &= \left\| \sum_{j=n+1}^{\infty} \widehat{f}(j)\varphi^j \right\| \\
&\leqslant \sum_{j=n+1}^{\infty} |\widehat{f}(j)| \|\varphi^j\| \\
&\leqslant \sum_{j=n+1}^{\infty} |\widehat{f}(j)| \|\varphi^j\|_\infty \\
&= \sum_{j=n+1}^{\infty} |\widehat{f}(j)| \alpha^j \\
&\leqslant \left(\sum_{j=n+1}^{\infty} |\widehat{f}(j)|^2 \right)^{\frac{1}{2}} \left(\sum_{j=n+1}^{\infty} \alpha^{2j} \right)^{\frac{1}{2}} \\
&\leqslant \|f\| \left(\sum_{j=n+1}^{\infty} \alpha^{2j} \right)^{\frac{1}{2}}.
\end{aligned}
$$

Therefore,

$$\|C_\varphi - T_n\| \leqslant \left(\sum_{j=n+1}^{\infty} \alpha^{2j} \right)^{\frac{1}{2}} = \left(\frac{\alpha^{2n+2}}{1 - |\alpha|^2} \right)^{\frac{1}{2}} \to 0 \quad \text{as } n \to \infty.$$

The above says that C_φ is the norm limit of the finite-rank operators T_n and is therefore compact (Exercise 2.8.22). ∎

Example 18.3.2. Proposition 18.1.3 shows that $C_{\alpha+\beta z}$ is compact if $|\alpha| + |\beta| < 1$. The previous proposition provides another proof of this. Notice how the image of $\alpha + \beta z$ is a disk whose closure is contained in $\mathbb{D}$; see Figure 18.3.1.

Example 18.3.3. One can check that $\varphi(z) = \frac{1}{4}\left(\frac{1}{2} - z\right)^2$ maps $\mathbb{D}$ onto a domain whose closure is contained in $\mathbb{D}$. Thus, C_φ is compact; see Figure 18.3.1.

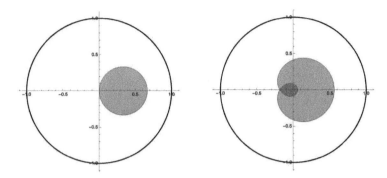

Figure 18.3.1 $\varphi(\mathbb{D})$ where $\varphi(z) = \frac{1}{3}(1-z)$ (left); $\varphi(\mathbb{D})$ where $\varphi(z) = \frac{1}{4}(\frac{1}{2} - z)^2$ (right).

Example 18.3.4. For $|\beta| < 1$, Proposition 18.3.1 ensures that $C_{\beta z^2}$ is compact. This also follows from the matrix representation of $C_{\beta z^2}$ with respect to $(z^n)_{n=0}^{\infty}$:

$$
\begin{bmatrix}
1 & 0 & 0 & 0 & 0 & 0 & 0 & \cdots \\
0 & 0 & 0 & 0 & 0 & 0 & 0 & \cdots \\
0 & \beta & 0 & 0 & 0 & 0 & 0 & \cdots \\
0 & 0 & 0 & 0 & 0 & 0 & 0 & \cdots \\
0 & 0 & \beta^2 & 0 & 0 & 0 & 0 & \cdots \\
0 & 0 & 0 & 0 & 0 & 0 & 0 & \cdots \\
0 & 0 & 0 & \beta^3 & 0 & 0 & 0 & \cdots \\
\vdots & \vdots & \vdots & \vdots & \vdots & \vdots & \vdots & \ddots
\end{bmatrix}.
$$

From here, one can see that the sum of the absolute values of the matrix entries is finite and hence $C_{\beta z^2}$ is compact (Theorem 3.4.1).

Exercise 18.8.6 improves Proposition 18.3.1 to the following.

Corollary 18.3.5. *If $\varphi : \mathbb{D} \to \mathbb{D}$ is analytic and*

$$
\int_{\mathbb{T}} \frac{dm(\xi)}{1 - |\varphi(\xi)|^2} < \infty,
$$

then C_φ is compact.

In his doctoral thesis [334], Schwartz explored a converse to Proposition 18.3.1. Recall that if $\varphi : \mathbb{D} \to \mathbb{D}$ is analytic, then it is bounded and has radial limits almost everywhere on $\mathbb{T}$ (Theorem 5.4.3).

Proposition 18.3.6 (Schwartz [334]). *If C_φ is compact, then $|\varphi(\xi)| < 1$ for almost every $\xi \in \mathbb{T}$.*

Proof Suppose toward a contradiction that there is a measurable set $E \subseteq \mathbb{T}$ such that $m(E) = \delta > 0$ and $|\varphi(\xi)| = 1$ for $\xi \in E$. Since $\langle g, \xi^n \rangle = \hat{g}(n) \to 0$ for every $g \in H^2$,

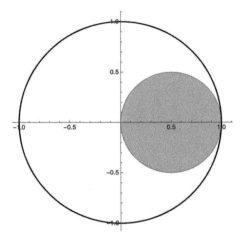

Figure 18.3.2 $\varphi(\mathbb{D})$ (shaded) where $f(z) = \frac{1}{2}(1 + z)$.

it follows that $f_n(z) = z^n$ converges weakly to zero in H^2. If C_φ is compact, then $C_\varphi f_n \to 0$ in the norm of $L^2(\mathbb{T})$ (Exercise 3.6.9). However,

$$\|C_\varphi f_n\|^2 = \|f_n \circ \varphi\|^2 = \|\varphi^n\|^2 = \int_{\mathbb{T}} |\varphi|^{2n} dm \geqslant \int_E |\varphi|^{2n} dm = \int_E dm \geqslant \delta,$$

which does not approach zero as $n \to \infty$. Thus, $|\varphi(\xi)| < 1$ for almost every $\xi \in \mathbb{T}$. ∎

The previous proposition can be used to show that certain composition operators are not compact.

Corollary 18.3.7. *If φ is inner, then C_φ is not compact.*

Proposition 18.3.6 is not a complete characterization of compactness for composition operators on H^2; see Figure 18.3.2 and the next proposition.

Proposition 18.3.8 (Schwartz [334]). *If*

$$\varphi(z) = \frac{1 + z}{2},$$

then φ maps $\mathbb{D}$ into itself and $|\varphi(\xi)| < 1$ for all $\xi \in \mathbb{T}\backslash\{1\}$, but C_φ is not compact.

Proof For $z \in \mathbb{D}$,

$$|\varphi(z)| = \left|\frac{1 + z}{2}\right| \leqslant \frac{1 + |z|}{2} < 1.$$

Thus, φ is an analytic self map of $\mathbb{D}$. One can also see that

$$\varphi(e^{it}) = \frac{1}{2} + \frac{1}{2}e^{it} \quad \text{for } t \in [0, 2\pi],$$

is a parameterization of the circle with center $\frac{1}{2}$ and radius $\frac{1}{2}$, which lies in $\mathbb{D}$ except for $t = 0$. We claim that C_φ is not compact.

Let

$$f_n(z) = \frac{1}{\sqrt{n}} \cdot \frac{1}{1 - z(1 - \frac{1}{n})} = \frac{1}{\sqrt{n}} k_{1 - \frac{1}{n}}(z) \quad \text{for } n \geqslant 2,$$

where k_λ is the reproducing kernel for H^2. For any $g \in H^2$, Exercise 5.9.9 yields

$$|g(\lambda)| = o\left(\frac{1}{\sqrt{1 - |\lambda|}}\right). \tag{18.3.9}$$

Thus,

$$\langle g, f_n \rangle = \frac{1}{\sqrt{n}} g\left(1 - \frac{1}{n}\right) = \sqrt{1 - \left(1 - \frac{1}{n}\right)} \cdot g\left(1 - \frac{1}{n}\right) \to 0 \quad \text{as } n \to \infty.$$

This shows that $f_n \to 0$ weakly in H^2. However,

$$(C_\varphi f_n)(z) = \frac{2\sqrt{n}}{n + 1} \cdot \frac{1}{1 - (\frac{n-1}{n+1})z}.$$

Since

$$\left\| \frac{1}{1 - az} \right\| = \frac{1}{\sqrt{1 - |a|^2}} \quad \text{for all } |a| < 1,$$

a calculation shows that $\|C_\varphi f_n\| = 1$ for all $n \geqslant 1$. Thus, C_φ is not compact since $f_n \to 0$ weakly, but $\|C_\varphi f_n\| = 1$ for all $n \geqslant 1$. ∎

There are other types of compactness results. We mention a few here whose proofs are found in [339].

Theorem 18.3.10. *Each of the following imply that C_φ is compact.*

(a) $\varphi(\mathbb{D})$ *is the interior of a polygon inscribed in* $\mathbb{T}$.

(b) φ *is injective and*

$$\lim_{|z| \to 1^-} \frac{1 - |\varphi(z)|}{1 - |z|} = \infty.$$

(c) φ *is injective and has no radial limits of modulus one.*

Notice how each of the conditions above indicate that C_φ is compact when $\varphi(\mathbb{D})$ is a "small" subset of $\mathbb{D}$. We end this section with a result from [334].

Theorem 18.3.11. *If C_φ is compact, then*

$$\|C_\varphi\| < \left(\frac{1 + |\varphi(0)|}{1 - |\varphi(0)|}\right)^{\frac{1}{2}}.$$

Recall that if φ is inner, then

$$\|C_\varphi\| = \left(\frac{1 + |\varphi(0)|}{1 - |\varphi(0)|}\right)^{\frac{1}{2}}.$$

Proposition 18.3.6 showed that C_φ is not compact when φ is inner. The previous theorem gives another reason for this.

18.4 Spectrum of a Composition Operator

There is no general formula for the spectrum of a composition operator, so any analysis must be done on a case-by-case basis. Since $C_\varphi 1 = 1$ for any composition operator, it follows that

$$1 \in \sigma_p(C_\varphi). \tag{18.4.1}$$

We have seen particular examples of the next result in Proposition 18.1.1 and Proposition 18.1.3.

Theorem 18.4.2 (Schwartz [334]). *Suppose that $\varphi(\mathbb{D}) \subseteq r\mathbb{D}$ for some $0 < r < 1$. Then there is a unique point $z_0 \in \mathbb{D}$ such that*

$$\sigma_p(C_\varphi) = \{\varphi'(z_0)^n : n \geqslant 1\} \cup \{1\} \quad \text{and} \quad \sigma(C_\varphi) = \sigma_p(C_\varphi) \cup \{0\}.$$

We outline the proof of this theorem, which relies on the following result of Koenigs [218] (see also [339, p. 93]).

Theorem 18.4.3 (Koenigs). *Suppose $\varphi : \mathbb{D} \to \mathbb{D}$ is analytic and $\varphi(z_0) = z_0$ for some $z_0 \in \mathbb{D}$. Then the following hold.*

(a) *Suppose $\varphi'(z_0) = 0$. Then $f \circ \varphi = \lambda f$ for some nonzero analytic function f on $\mathbb{D}$ if and only if $\lambda = 1$.*

(b) *Suppose $\varphi'(z_0) \neq 0$. Then $f \circ \varphi = \lambda f$ for some nonzero analytic function f on $\mathbb{D}$ if and only if $\lambda = \varphi'(z_0)^m$ for some $m \geqslant 1$.*

We are now ready for a sketch of the proof of Theorem 18.4.2.

Proof From (18.4.1), it follows that $1 \in \sigma_p(C_\varphi)$. Since $\varphi(\mathbb{D}) \subseteq r\mathbb{D}$ for some $0 < r < 1$, it follows that $\varphi(\mathbb{D})^-$ is a compact subset of $\mathbb{D}$. Proposition 18.3.1 asserts that C_φ is compact. Exercise 2.8.28 ensures that $0 \in \sigma(C_\varphi)$. Since C_φ is compact, the nonzero elements of its spectrum are the eigenvalues of C_φ (Theorem 2.6.9).

We now proceed to compute these nonzero eigenvalues. *Brouwer's fixed point theorem* [320, p. 143] says that if g is a continuous map from a closed disk to itself, then there is a point w_0 in the closed disk such that $g(w_0) = w_0$. Apply this to the continuous function φ, which maps $|z| \leqslant r$ to itself, to produce a $z_0 \in \mathbb{D}$ such that $\varphi(z_0) = z_0$.

Theorem 18.4.3 implies that if $f(\varphi(z)) = \lambda f(z)$ for some $\lambda \neq 1$ and $f \in H^2\backslash\{0\}$, there is a positive integer m such that $\lambda = \varphi'(z_0)^m$ and $\varphi'(z_0) \neq 0$. Thus,

$$\sigma(C_\varphi) \subseteq \{0, 1\} \cup \{\varphi'(z_0)^m : m \geqslant 1\}.$$

On the other hand, if $\lambda = \varphi'(0)^m \neq 0$ for some positive integer m, Theorem 18.4.3 says there is a nonzero analytic function f such that $f(\varphi(z)) = \lambda f(z)$. To show that this f belongs to H^2, use the fact that $\varphi(\mathbb{D}) \subseteq r\mathbb{D}$ and obtain

$$\sup_{z \in \mathbb{D}} |f(z)| = \sup_{z \in \mathbb{D}} \left| \frac{f(\varphi(z))}{\lambda} \right| = \frac{1}{|\lambda|} \sup_{z \in \mathbb{D}} |(f \circ \varphi)(z)| \leqslant \frac{1}{|\lambda|} \sup_{|w| \leqslant r} |f(w)|.$$

Since f is continuous on the compact set $|w| \leqslant r$, the last supremum is finite. We conclude that f is a bounded analytic function on $\mathbb{D}$, and therefore $f \in H^2$. Thus, $\lambda = \varphi'(0)^m \in \sigma(C_\varphi)$ for every $m \geqslant 1$ and hence $\sigma(C_\varphi) = \{0, 1\} \cup \{\varphi'(z_0)^m : m \geqslant 1\}$, which completes the proof. ∎

It is possible that $\sigma(C_\varphi) = \{0, 1\}$ when C_φ is compact. Just take $\varphi(z) = \beta z^2$ (see Example 18.3.4) with $|\beta| < 1$. Then C_φ is compact, $z_0 = 0$ is the unique fixed point in $\mathbb{D}$, and $\varphi'(0) = 0$.

Nordgren [255] computed the spectrum of C_φ when φ is a disk automorphism

$$\varphi(z) = \xi \frac{z - a}{1 - \bar{a}z} \quad \text{for } \xi \in \mathbb{T} \text{ and } a \in \mathbb{D},$$

and showed that $\sigma(C_\varphi)$ is either the closure of $\{\zeta^n : n \in \mathbb{Z}\}$ (where ζ is some unimodular constant), the unit circle, or an annulus. The choice depends on the parameters a and ξ.

18.5 Adjoint of a Composition Operator

In this section, we mention a few facts about the adjoint of a composition operator. We start with the following general result.

Proposition 18.5.1. *If $\varphi : \mathbb{D} \to \mathbb{D}$ is analytic and $k_\lambda(z) = (1 - \bar{\lambda}z)^{-1}$, then $C_\varphi^* k_\lambda = k_{\varphi(\lambda)}$.*

Proof For each $w \in \mathbb{D}$,

$$\begin{aligned}
(C_\varphi^* k_\lambda)(w) &= \langle C_\varphi^* k_\lambda, k_w \rangle \\
&= \langle k_\lambda, C_\varphi k_w \rangle \\
&= \langle k_\lambda, k_w \circ \varphi \rangle \\
&= \overline{\langle k_w \circ \varphi, k_\lambda \rangle} \\
&= \overline{k_w(\varphi(\lambda))} \\
&= k_{\varphi(\lambda)}(w),
\end{aligned}$$

which completes the proof. ∎

This theorem has a converse [77].

Proposition 18.5.2. *Suppose $A \in \mathcal{B}(H^2)$ and A^* maps each kernel function k_λ to another kernel function. Then A is a composition operator.*

Proof Since A^* maps kernel functions to kernel functions, there is a function $\varphi : \mathbb{D} \to \mathbb{D}$ such that $A^* k_\lambda = k_{\varphi(\lambda)}$ for each $\lambda \in \mathbb{D}$. It follows that

$$(Af)(\lambda) = \langle Af, k_\lambda \rangle = \langle f, A^* k_\lambda \rangle = \langle f, k_{\varphi(\lambda)} \rangle = f(\varphi(\lambda)) \quad \text{for all } f \in H^2.$$

Applying this to $f(z) = z$ shows that φ is an analytic self map of $\mathbb{D}$ and $A = C_\varphi$. ■

There is the following integral formula for C_φ^*.

Corollary 18.5.3. *If $\varphi : \mathbb{D} \to \mathbb{D}$ is analytic, then*

$$(C_\varphi^* f)(z) = \int_{\mathbb{T}} \frac{f(\xi)}{1 - \overline{\varphi(\xi)}z} \, dm(\xi) \quad \text{for all } f \in H^2.$$

Proof For $f \in H^2$,

$$(C_\varphi^* f)(z) = \langle C_\varphi^* f, k_z \rangle = \langle f, C_\varphi k_z \rangle = \int_{\mathbb{T}} \frac{f(\xi)}{1 - \overline{\varphi(\xi)}z} \, dm(\xi),$$

which completes the proof. ■

For a linear fractional transformation

$$\varphi(z) = \frac{az + b}{cz + d} \tag{18.5.4}$$

that maps $\mathbb{D}$ into itself, there is a fascinating formula for the adjoint of C_φ that involves Toeplitz operators.

Theorem 18.5.5 (C. Cowen [100]). *Suppose that*

$$\varphi(z) = \frac{az + b}{cz + d}$$

maps $\mathbb{D}$ into itself and is normalized so that $ad - bc = 1$.

(a) *The function*

$$\sigma(z) = \frac{\bar{a}z - \bar{c}}{-\bar{b}z + \bar{d}}$$

maps $\mathbb{D}$ into itself.

(b) *The functions*

$$g(z) = \frac{1}{-\bar{b}z + \bar{d}} \quad \text{and} \quad h(z) = cz + d$$

belong to H^∞.

(c) $C_\varphi^* = T_g C_\sigma T_h^*$.

Proof We follow Cowen's original proof.

(a) Let $\mathbb{D}_e = \{|z| > 1\} \cup \{\infty\}$ denote the open extended exterior disk. Since φ maps $\mathbb{D}$ into itself, $\gamma(z) = \overline{\varphi(\overline{z})}$ also maps $\mathbb{D}$ into itself. Basic facts about linear fractional maps implies that the inverse function γ^{-1} maps $\mathbb{D}_e$ into itself. A calculation shows that

$$\sigma(z) = \frac{1}{\gamma^{-1}(1/z)}$$

and that σ maps $\mathbb{D}$ into itself.

(b) Since h is a linear function it is bounded on $\mathbb{D}$. For suitable constants α, β with $\alpha \neq 0$, $\sigma = \alpha g + \beta$. From (a), σ is bounded. Thus, $g \in H^\infty$.

(c) It suffices to show that

$$C_\varphi^* k_\lambda = T_g C_\sigma T_h^* k_\lambda \quad \text{for all } \lambda \in \mathbb{D}. \tag{18.5.6}$$

To verify this, observe that

$$(C_\sigma k_\lambda)(z) = \frac{1}{1 - \overline{\lambda}\sigma(z)}$$

$$= \frac{1}{1 - \overline{\lambda}\,\frac{\overline{a}z - \overline{c}}{-\overline{b}z + \overline{d}}}$$

$$= \frac{-\overline{b}z + \overline{d}}{-\overline{b}z + \overline{d} - \overline{\lambda}\overline{a}z + \overline{\lambda}\overline{c}}$$

$$= \frac{-\overline{b}z + \overline{d}}{(\overline{d} + \overline{\lambda}\overline{c}) - (\overline{b} + \overline{\lambda}\overline{a})z}$$

$$= \frac{1}{h(\lambda)}\,\frac{1}{g(z)}\,\frac{1}{1 - \overline{\varphi(\lambda)}z}$$

$$= \frac{1}{h(\lambda)}\,\frac{k_{\varphi(\lambda)}(z)}{g(z)}. \tag{18.5.7}$$

Thus,

$$T_g C_\sigma T_h^* k_\lambda = T_g C_\sigma \overline{h(\lambda)} k_\lambda \qquad \text{(by (5.5.5))}$$

$$= \overline{h(\lambda)} T_g\left(\frac{1}{\overline{h(\lambda)}}\,\frac{k_{\varphi(\lambda)}}{g}\right) \qquad \text{(by (18.5.7))}$$

$$= k_{\varphi(\lambda)}$$

$$= C_\varphi^* k_\lambda \qquad \text{(by Proposition 18.5.1)}.$$

This verifies (18.5.6) and completes the proof. ∎

Example 18.5.8. Let $\varphi(z) = (1 + z)/2$. Then φ is of the form (18.5.4) with $a = b = \frac{1}{\sqrt{2}}$, $c = 0$, and $d = \sqrt{2}$. The functions $\sigma, g,$ and h are

$$\sigma(z) = \frac{\frac{1}{\sqrt{2}}z}{\sqrt{2} - \frac{\sqrt{2}}{2}z}, \quad g(z) = \frac{1}{\sqrt{2} - \frac{\sqrt{2}}{2}z}, \quad \text{and} \quad h(z) = \sqrt{2}.$$

See Exercise 18.8.16 for another approach to computing the adjoint of a composition operator.

18.6 Universal Operators and Composition Operators

Recall the definition of a universal operator (Definition 13.3.1). There are several results concerning universal operators obtained from adjoints of composition operators. Here is an example from [80, Ch. 8] (see also [101]).

Proposition 18.6.1. $C_{z^2}^*$ *is universal for* H^2.

Proof To show $C_{z^2}^*$ is universal, we use Theorem 13.3.2 and verify that $\ker C_{z^2}^*$ is infinite dimensional and $\operatorname{ran} C_{z^2}^* = H^2$. Corollary 18.2.12 ensures that C_{z^2} is an isometry, so $C_{z^2}^* C_{z^2} = I$ (Exercise 3.6.13). Thus, $C_{z^2}^*$ is surjective.

For each $n \geqslant 1$, choose n distinct $\lambda_1, \lambda_2, \ldots, \lambda_n$ in $\mathbb{D} \backslash \{0\}$ and let

$$g_j(z) = k_{\lambda_j}(z) - k_{-\lambda_j}(z) \quad \text{for } 1 \leqslant j \leqslant n.$$

We claim that $g_1, g_2, \ldots, g_n$ are linearly independent. Suppose there are constants $c_1, c_2, \ldots, c_n$ such that

$$\sum_{j=1}^{n} c_j(k_{\lambda_j}(z) - k_{-\lambda_j}(z)) = 0 \quad \text{for all } z \in \mathbb{D}.$$

For any polynomial p, use the reproducing property of $k_\lambda(z)$ to see that

$$\left\langle p, \sum_{j=1}^{n} c_j(k_{\lambda_j} - k_{-\lambda_j}) \right\rangle = \sum_{j=1}^{n} \overline{c_j}(p(\lambda_j) - p(-\lambda_j)).$$

For each $1 \leqslant i \leqslant n$, use Lagrange interpolation to select a polynomial p_i such that $p_i(\lambda_i) = 1$ but $p_i(-\lambda_i) = -1$ and $p_i(\pm\lambda_j) = 0$ when $j \neq i$. The identity above yields $c_i = 0$ for all $1 \leqslant i \leqslant n$. Thus, $g_1, g_2, \ldots, g_n$ are linearly independent. Proposition 18.5.1 implies that

$$C_{z^2}^* g_j = \frac{1}{1 - \overline{(\lambda_j)^2}z} - \frac{1}{1 - \overline{(-\lambda_j)^2}z} = 0,$$

and hence $\ker C_{z^2}^*$ is infinite dimensional. ∎

One can see the universality of $C_{z^2}^*$ from its matrix representation

$$\begin{bmatrix} 1 & 0 & 0 & 0 & 0 & 0 & 0 & 0 & 0 & 0 & \cdots \\ 0 & 0 & 1 & 0 & 0 & 0 & 0 & 0 & 0 & 0 & \cdots \\ 0 & 0 & 0 & 0 & 1 & 0 & 0 & 0 & 0 & 0 & \cdots \\ 0 & 0 & 0 & 0 & 0 & 0 & 1 & 0 & 0 & 0 & \cdots \\ 0 & 0 & 0 & 0 & 0 & 0 & 0 & 0 & 1 & 0 & \cdots \\ 0 & 0 & 0 & 0 & 0 & 0 & 0 & 0 & 0 & 0 & \cdots \\ 0 & 0 & 0 & 0 & 0 & 0 & 0 & 0 & 0 & 0 & \cdots \\ \vdots & \vdots & \vdots & \vdots & \vdots & \vdots & \vdots & \vdots & \vdots & \vdots & \ddots \end{bmatrix} \tag{18.6.2}$$

(which one can obtain from the matrix representation of C_{z^2} and then taking adjoints). Notice that the column space of (18.6.2) contains $\mathbf{e}_n$ for every $n \geqslant 0$, and hence $\operatorname{ran} C_z^*$ contains z^n for every $n \geqslant 0$. The columns of the matrix in (18.6.2) also show that $z^{2k+1} \in \ker C_{z^2}^*$ for every $k \geqslant 0$, so $\ker C_{z^2}^*$ is infinite dimensional. Caradus' theorem implies that $C_{z^2}^*$ is universal.

See Exercise 18.8.10 for another proof of the universality of $C_{z^2}^*$. The proposition above can be extended to the following (Exercise 18.8.11).

Theorem 18.6.3. *If φ is inner, but not an automorphism of $\mathbb{D}$, and $\varphi(0) = 0$, then C_φ^* is universal for H^2.*

18.7 Notes

As noted in the beginning of this chapter, we only covered a small fraction of a large literature on this subject. We refer the reader to the surveys [102, 278, 339] for much more.

The boundedness of C_φ on H^2 can also be obtained from a 1925 subordination result of Littlewood [229]. A consequence of the *Littlewood subordination theorem* (see below) says that if f is analytic on $\mathbb{D}$ and φ is an analytic self map on $\mathbb{D}$ such that $\varphi(0) = 0$, then

$$\int_{\mathbb{T}} |f(\varphi(r\xi))|^p \, dm(\xi) \leqslant \int_{\mathbb{T}} |f(r\xi)|^p \, dm(\xi)$$

for all $0 < r < 1$ and $1 \leqslant p < \infty$. If f and F are analytic on $\mathbb{D}$ and $f(z) = F(w(z))$, where w is analytic on $\mathbb{D}$ with $|w(z)| \leqslant |z|$ on $\mathbb{D}$, then f is *subordinate* to F. Littlewood's principle is that

$$\int_{\mathbb{T}} |F(r\xi)|^p \, dm(\xi) \leqslant \int_{\mathbb{T}} |f(r\xi)|^p \, dm(\xi)$$

for all $0 < r < 1$ and $1 \leqslant p < \infty$. Note that the Schwarz lemma [9, p. 135] says that if φ is an analytic self map of $\mathbb{D}$ with $\varphi(0) = 0$, then $|\varphi(z)| \leqslant |z|$ on $\mathbb{D}$.

Ryff in 1966 [321] and Nordgren in 1968 [255] used this result to prove, amongst other things, that composition operators on H^p for $1 \leqslant p < \infty$ are bounded. Ryff obtained the norm estimate

$$\|C_\varphi\|_{H^p \to H^p} \leqslant \left(\frac{1 + |\varphi(0)|}{1 - |\varphi(0)|} \right)^{\frac{1}{p}}.$$

Nordgren's paper seems to be the first to coin the term "composition operator" for C_φ.

We mentioned a few results about compactness of composition operators in this chapter. The issue was basically settled by J. Shapiro in his 1987 paper [338] in terms of the *Nevanlinna counting function*

$$N_\varphi(w) = \sum_{z \in \varphi^{-1}(\{w\})} \log|z|;$$

see Exercise 18.8.27. The result is that C_φ is compact if and only if

$$\limsup_{|w| \to 1^-} \frac{N_\varphi(w)}{-\log|w|} = 0.$$

Although not covered in this chapter, one can also consider when a composition operator belongs to one of the Schatten classes [233, 341].

There is a considerable amount of work on the spectra of various types of composition operators on H^2 in [77, 98, 100, 108, 255, 334]. For compact composition operators, there is the following extension of Theorem 18.4.2 from [77]. If C_φ is compact, there is a unique point $z_0 \in \mathbb{D}$ such that $\varphi(z_0) = z_0$ and $|\varphi'(z_0)| < 1$. If $\varphi'(z_0) = 0$, then $\sigma(C_\varphi) = \{0, 1\}$. If $\varphi'(z_0) \neq 0$, then $\sigma(C_\varphi)$ contains the eigenvalues $\{\varphi'(z_0)^m : m \geqslant 0\}$.

For various classes of symbols, composition operators are defined and bounded on other Hilbert spaces of analytic functions such as the Bergman and Dirichlet spaces (see Exercises 18.8.18, 18.8.19, and 18.8.20). From there one can study concepts such as norm estimates, compactness, and spectral properties, as surveyed here for H^2. We also point out some work on composition operators on spaces of Dirichlet series [35, 157, 278].

18.8 Exercises

Exercise 18.8.1. Prove that $T \in \mathcal{B}(H^2)$ is a composition operator if and only if $T(z^n) = (T(z))^n$ for all $n \geqslant 0$.

Exercise 18.8.2. Prove that $T \in \mathcal{B}(H^2)$ is a composition operator if and only if $T(fg) = (Tf)(Tg)$ for all $f, g \in \mathbb{C}[z]$.

Exercise 18.8.3. Let φ be an analytic self map of $\mathbb{D}$ and define the linear transformation $C_\varphi f = f \circ \varphi$ from the space of analytic functions on $\mathbb{D}$ to itself. Prove that C_φ is invertible if and only if φ is invertible.

Exercise 18.8.4. If φ is an analytic self map of $\mathbb{D}$, $\psi \in H^\infty$, and T_ψ is the analytic Toeplitz operator, prove that $C_\varphi T_\psi = T_\psi C_\varphi$ if and only if $\psi \circ \varphi = \psi$.

Exercise 18.8.5. If φ is an analytic self map of $\mathbb{D}$ such that $\|\varphi\|_\infty < 1$, prove that C_φ is a Hilbert–Schmidt operator (recall Exercise 3.6.31).

Exercise 18.8.6. If $\varphi : \mathbb{D} \to \mathbb{D}$ is analytic and

$$\int_{\mathbb{T}} \frac{1}{1 - |\varphi(\xi)|^2} \, dm(\xi) < \infty,$$

prove that C_φ is compact.

Exercise 18.8.7. Consider composition operators of the type $C_{\alpha+\beta z}$, where $|\alpha| + |\beta| < 1$. Show that the matrix representation of $C_{\alpha+\beta z}$ with respect to the standard basis $(z^n)_{n=0}^{\infty}$ for H^2 is

$$\begin{bmatrix} 1 & \alpha & \alpha^2 & \alpha^3 & \cdots \\ 0 & \beta & 2\alpha\beta & 3\alpha^2\beta & \cdots \\ 0 & 0 & \beta^2 & 3\alpha\beta^2 & \cdots \\ 0 & 0 & 0 & \beta^3 & \cdots \\ \vdots & \vdots & \vdots & \vdots & \ddots \end{bmatrix}.$$

Remark: These types of composition operators were explored by Deddens in [108].

Exercise 18.8.8. If $|a| < 1$ and

$$\varphi(z) = \frac{a - z}{1 - \bar{a}z},$$

prove that C_φ is invertible on H^2.

Exercise 18.8.9. One can discuss composition operators on $L^2(\mu)$, where $(X, \mathcal{A}, \mu)$ is a measure space [346, 347]. Let $\varphi : X \to X$ be a μ-measurable function.

(a) Prove that $C_\varphi f = f \circ \varphi$ is a bounded composition operator on $L^2(\mu)$ if and only if there is a constant M such that $\mu(\varphi^{-1}(A)) \leqslant M\mu(A)$ for every $A \in \mathcal{A}$.

(b) Prove that if φ is measure preserving, that is to say, $\mu(A) = \mu(\varphi^{-1}(A))$ for every $A \in \mathcal{A}$, then C_φ is an isometry.

Exercise 18.8.10.

(a) Prove that C_{z^2} is unitarily equivalent to the shift U on $(H^2)^{(\infty)}$ (recall the definition from (14.1.2)) defined by $U(f_0, f_1, f_2, f_3, \ldots) = (0, f_0, f_1, f_2, f_3, \ldots)$.

(b) Use (a) to prove that $C_{z^2}^*$ is universal for H^2.

Exercise 18.8.11. If φ is inner, but not an automorphism of $\mathbb{D}$, and $\varphi(0) = 0$, prove that C_φ^* is universal for H^2.

Exercise 18.8.12. Suppose φ is an analytic self map of $\mathbb{D}$ such that $\varphi(0) = 0$. Prove that $z^n H^2$ is an invariant subspace for C_φ for each $n \geqslant 0$.

Exercise 18.8.13. For $|\beta| \leqslant 1$, write the polar decomposition of $C_{\beta z}$ in terms of two composition operators.

Exercise 18.8.14. This is a continuation of Exercise 17.10.2. Recall that the Hilbert matrix operator on H^2 can be written as

$$(\mathfrak{H}f)(z) = \int_0^1 \frac{1}{1 - tz} f(t)\, dt.$$

(a) Prove that for each $t \in [0, 1]$, $\varphi_t(z) = \dfrac{t}{(t-1)z + 1}$ is an analytic self map of $\mathbb{D}$.

(b) For a fixed $z \in \mathbb{D}$, describe the curve $t \mapsto \varphi_t(z)$.

(c) For each $t \in (0, 1]$, prove that $w_t(z) = \dfrac{1}{(t-1)z + 1}$ is a bounded analytic function on $\mathbb{D}$.

(d) Prove that the weighted composition operator $T_t = M_{w_t} C_{\varphi_t}$ is bounded on H^2.

(e) For $f \in H^2$, prove that $(\mathfrak{H}f)(z) = \displaystyle\int_0^1 (T_t f)(z)\, dt$ for all $z \in \mathbb{D}$.

Remark: One can estimate the norm of T_t and show that the Hilbert transform is bounded on the H^p spaces when $1 < p < \infty$. See [110] for further details.

Exercise 18.8.15. This is a continuation of Exercise 6.7.11. Recall from Exercise 6.7.11 that for the Cesàro operator C on H^2, the resolvent is

$$((\lambda I - C)^{-1}h)(z) = \frac{h(z)}{\lambda} + \frac{1}{\lambda^2} z^{\frac{1}{\lambda} - 1}(1 - z)^{-\frac{1}{\lambda}} \int_0^z w^{-\frac{1}{\lambda}}(1 - w)^{\frac{1}{\lambda} - 1} h(w)\, dw.$$

(a) Define

$$\varphi(z) = \frac{z}{1 - z} \quad \text{for } z \in \mathbb{D},$$

and consider the family of functions $\varphi_t(z) = \varphi^{-1}(e^{-t}\varphi(z))$ for $t \geq 0$. Prove that each φ_t is an analytic self map of $\mathbb{D}$.

(b) For each $z \in \mathbb{D}$, prove that $\varphi_t(z) \to 0$ as $t \to \infty$.

(c) Let the path of integration be $\gamma(t) = \varphi_t(z)$ in the formula

$$(Cf)(z) = \frac{1}{z} \int_0^z \frac{f(w)}{1 - w}\, dw$$

and prove that

$$(Cf)(z) = \int_0^\infty (S_t f)(z)\, dt,$$

where S_t is the weighted composition operator

$$(S_t f)(z) = \frac{\varphi_t(z)}{z}(C_{\varphi_t} f)(z).$$

(d) Use a similar computation to prove that

$$((\lambda I - C)^{-1}h)(z) = \frac{h(z)}{\lambda} + \frac{1}{\lambda^2} \int_0^\infty e^{\frac{t}{\lambda}}(S_t h)(z)\, dt.$$

Remark: See [268] for more on this in other settings.

Exercise 18.8.16. Another way to compute the adjoint of a composition operator involves *Aleksandrov–Clark measures*. A theorem of Herglotz [118] says that if f is analytic on $\mathbb{D}$ and $\mathrm{Re}\, f > 0$, then there is a positive finite Borel measure μ on $\mathbb{T}$ such that

$$f(z) = \int_{\mathbb{T}} \frac{\xi + z}{\xi - z} d\mu(\xi) + i\, \mathrm{Im}\, f(0).$$

(a) Assuming Herglotz's theorem, prove that if φ is an analytic self map of $\mathbb{D}$ and $\varphi(0) = 0$, then for each $\alpha \in \mathbb{T}$ there is a unique probability measure μ_α on $\mathbb{T}$ such that

$$\int_{\mathbb{T}} \frac{\xi + z}{\xi - z} d\mu_\alpha(\xi) = \frac{\alpha + \varphi(z)}{\alpha - \varphi(z)} \quad \text{for all } z \in \mathbb{D}.$$

The measures $\{\mu_\alpha : \alpha \in \mathbb{T}\}$ form the family of Aleksandrov–Clark measures for φ.

(b) Manipulate the expression in (a) and prove that

$$\int_{\mathbb{T}} \frac{1}{1 - \lambda \bar{\xi}} d\mu_\alpha(\xi) = \frac{1}{1 - \bar{\alpha}\varphi(\lambda)} \quad \text{for all } \lambda \in \mathbb{D}.$$

(c) For functions in H^2 that are continuous on $\mathbb{T}$, define

$$(A_\varphi f)(\alpha) = \int_{\mathbb{T}} f(\xi) d\mu_\alpha(\xi) \quad \text{for all } \alpha \in \mathbb{T}.$$

Prove that $A_\varphi k_\lambda = C_\varphi^* k_\lambda$.

(d) Prove that A_φ extends to a bounded operator on H^2 that is the adjoint of C_φ.

Remark: See [143] for more on this important topic.

Exercise 18.8.17. Let us compute the adjoint for C_{z^2} using Exercise 18.8.16.

(a) Prove that for each $\alpha \in \mathbb{T}$, the Aleksandrov–Clark measure for $\varphi(z) = z^2$ is $\mu_\alpha = \frac{1}{2}\delta_{\sqrt{\alpha}} + \frac{1}{2}\delta_{-\sqrt{\alpha}}$.

(b) Prove that $(C_{z^2}^* k_\lambda)(\alpha) = \frac{1}{2} k_\lambda(\sqrt{\alpha}) + \frac{1}{2} k_\lambda(-\sqrt{\alpha})$.

(c) Let $f \in H^2$ be continuous on $\mathbb{T}$. Prove that $(C_{z^2}^* f)(\alpha) = \frac{1}{2} f(\sqrt{\alpha}) + \frac{1}{2} f(-\sqrt{\alpha})$.

Remark: Aleksandrov proved that

$$(A_\varphi f)(\alpha) = \int_{\mathbb{T}} f(\xi)\, d\mu_\alpha(\xi)$$

is defined for *m*-almost every $\alpha \in \mathbb{T}$ and so

$$(C_{z^2}^* f)(\alpha) = \frac{1}{2} f(\sqrt{\alpha}) + \frac{1}{2} f(-\sqrt{\alpha})$$

makes sense for all $f \in H^2$. This can be used to compute the essential norm of a composition operator [85].

Exercise 18.8.18.

(a) Prove that if φ is an analytic self map of $\mathbb{D}$, then the composition operator C_φ is bounded on the Bergman space A^2.

(b) Compute the matrix representation of C_{z^2} on A^2 with respect to the orthonormal basis $(\sqrt{n+1}\, z^n)_{n=0}^\infty$.

Exercise 18.8.19. Let Ω_1 and Ω_2 be domains in the complex plane and let $\varphi : \Omega_1 \to \Omega_2$ be a conformal mapping.

(a) For each analytic map $f : \Omega_2 \to \mathbb{C}$, prove that

$$\int_{\Omega_1} |(f \circ \varphi)'(z)|^2 \, dA(z) = \int_{\Omega_2} |f'(w)|^2 \, dA(w),$$

where dA is two-dimensional area measure.

(b) If f is analytic on $\mathbb{D}$ and

$$\varphi(z) = e^{i\theta} \frac{\alpha - z}{1 - \bar{\alpha}z}, \quad \text{where } \theta \in \mathbb{R} \text{ and } \alpha \in \mathbb{D},$$

prove that $D(f \circ \varphi) = D(f)$, in which $D(f)$ is the Dirichlet integral from (9.1.1).

Remark: A result in [20] ensures that the Dirichlet space $\mathcal{D}$ is the unique function space on $\mathbb{D}$ for which the composition operators C_φ, where φ is an automorphism of $\mathbb{D}$, act isometrically. However, many function spaces, for example the Hardy space and the Bergman space, are not isometrically invariant under C_φ.

Exercise 18.8.20. For any analytic self map φ of the disk, the composition operator C_φ is bounded on H^2. Exercise 18.8.19 showed that C_φ is bounded on $\mathcal{D}$ when φ is a disk automorphism. Produce a specific example of an analytic self map φ such that $C_\varphi \mathcal{D} \not\subseteq \mathcal{D}$. *Remark:* The papers [123, 235] give various conditions that ensure a self map yields a bounded composition operator on the Dirichlet space.

Exercise 18.8.21. Consider the normalized reproducing kernel $\tilde{k}_\lambda(z) = \dfrac{\sqrt{1 - |\lambda|^2}}{1 - \bar{\lambda}z}$ for H^2.

(a) Prove that $\tilde{k}_\lambda \to 0$ weakly in H^2.

(b) For an analytic self map φ of $\mathbb{D}$, prove that $\|C_\varphi^* \tilde{k}_\lambda\| = \sqrt{\dfrac{1 - |\lambda|^2}{1 - |\varphi(\lambda)|^2}}$.

(c) If $\liminf\limits_{|\lambda| \to 1} \dfrac{1 - |\lambda|}{1 - |\varphi(\lambda)|} > 0$, prove that C_φ is not compact.

Exercise 18.8.22. Show that if φ is an analytic self map of $\mathbb{D}$ that is also a linear fractional transformation, then $\ker C_\varphi^* = \{0\}$.

Exercise 18.8.23. Let φ be an analytic self map of $\mathbb{D}$ and $\psi(z) = \varphi(z^d)$ for $d \geqslant 2$. Prove the following.

(a) $\ker C_\psi^* \supseteq \bigvee\{z^k : k \notin d\mathbb{N}\}$.

(b) If $\ker C_\psi^* = \{0\}$, then $\ker C_\varphi^* = \bigvee\{z^k : k \notin d\mathbb{N}\}$.

Remark: See [244] for other results along these lines.

Exercise 18.8.24. Use matrix representations to prove the formula $C_{\frac{1+z}{2}}^* = T_{\frac{1}{2-z}} C_{\frac{z}{2-z}}$.

Exercise 18.8.25. Prove the adjoint formula

$$C_{\sqrt{\frac{1-z}{2}}}^* = T_{\frac{2}{2-z^2}} C_{\frac{z^2}{z^2-2}} + T_{\frac{2z}{2-z^2}} C_{\frac{z^2}{z^2-2}} T_{\sqrt{\frac{1-z}{2}}}^*$$

by using the following procedure from [162]. Let

$$\ell_1(z) = \frac{1-z}{2}, \quad \ell_2(z) = \frac{z}{z-2}, \quad \text{and} \quad \varphi(z) = \sqrt{\ell_1(z)}.$$

(a) For each $\lambda \in \mathbb{D}$, prove that

$$\frac{1}{1 - \bar{\lambda}\ell_1(z)} = \frac{2}{2 - \bar{\lambda}} \frac{1}{1 - \ell_2(\lambda)z} \quad \text{for all } z \in \mathbb{D}.$$

(b) Use

$$k_\lambda(z) = \frac{1}{1 - \bar{\lambda}^2 z^2} + \frac{\bar{\lambda}z}{1 - \bar{\lambda}^2 z^2} \quad \text{and} \quad (C_\varphi^* f)(\lambda) = \langle f, k_\lambda \circ \varphi \rangle$$

to prove that

$$(C_\varphi^* f)(\lambda) = \left\langle f, \frac{1}{1 - \bar{\lambda}^2 \ell_1} \right\rangle + \lambda \left\langle f, \frac{\varphi}{1 - \bar{\lambda}^2 \ell_1} \right\rangle.$$

(c) Use (a) to prove

$$(C_\varphi^* f)(\lambda) = \frac{2}{2 - \lambda^2} \left\langle f, \frac{1}{1 - \ell_2(\lambda^2)z} \right\rangle + \frac{2\lambda}{2 - \lambda^2} \left\langle f, \frac{\varphi}{1 - \ell_2(\lambda^2)z} \right\rangle.$$

(d) Use the fact that $\langle f, \varphi h \rangle = \langle P_+(\overline{\varphi}f), h \rangle$ for all $f, h \in H^2$ to prove the desired identity.

Exercise 18.8.26. Use Exercise 18.8.25 to derive a formula for $C_{\sqrt[3]{\frac{1-z}{2}}}^*$.

Exercise 18.8.27.

(a) For $f \in H^2$, prove that $\|f\|^2 = |f(0)|^2 + \frac{2}{\pi} \int_{\mathbb{D}} |f'(z)|^2 \log \frac{1}{|z|} \, dA(z)$.

(b) If $\varphi : \mathbb{D} \to \mathbb{D}$ is analytic, define

$$N_\varphi(\lambda) = \sum_{z \in \varphi^{-1}(\{\lambda\})} \log \frac{1}{|z|} \quad \text{for } \lambda \in \varphi(\mathbb{D}) \backslash \{\varphi(0)\}.$$

Prove that $\|C_\varphi f\|^2 = |f(\varphi(0))|^2 + \dfrac{2}{\pi} \displaystyle\int_{\mathbb{D}} |f'(z)|^2 N_\varphi(z) \, dA(z)$ for all $f \in H^2$.

Remark: The formula in (a) is the *Littlewood–Paley formula* and the function N_φ in (b) is the *Nevanlinna counting function*.

Exercise 18.8.28. This exercise provides an improvement of Proposition 18.2.7 for the norm of a composition operator C_φ on H^2.

(a) Use Proposition 18.5.1 to prove that

$$\left(\frac{1}{1 - |\varphi(\lambda)|^2} \right)^{\frac{1}{2}} \leqslant \|C_\varphi\| \left(\frac{1}{1 - |\lambda|^2} \right)^{\frac{1}{2}} \quad \text{for all } \lambda \in \mathbb{D}.$$

(b) Prove that $\|C_\varphi\| \geqslant \displaystyle\sup_{\lambda \in \mathbb{D}} \left(\frac{1 - |\lambda|^2}{1 - |\varphi(\lambda)|^2} \right)^{\frac{1}{2}}$.

Exercise 18.8.29. The reader encountered Brouwer's fixed-point theorem when exploring the eigenvalues of a composition operator. The next three problems explore other types of fixed-point theorems. Let V be a Banach space. Then $T \in \mathcal{B}(V)$ is a *strict contraction* if there is a $c \in [0, 1)$ such that $\|Tv\| \leqslant c\|v\|$ for all $v \in V$. Prove that a strict contraction $T \in \mathcal{B}(V)$ has a *fixed point*; that is, a $v \in V$ such that $Tv = v$.

Exercise 18.8.30. This exercise continues Exercise 18.8.29. Give an example of a strict contraction on a normed linear space that does not have a fixed point (and thus completeness is important).

Exercise 18.8.31. Give an example of an isometry $T \in \mathcal{B}(\mathcal{H})$ that does not have a fixed point (and thus having a strict contraction is important).

Exercise 18.8.32. To explore the eigenvalues of compact composition operators, we used Riesz's theorem (Theorem 2.6.9). The next several problems explore a version of this result. Let $T \in \mathcal{B}(\mathcal{H})$ be compact. Prove that the equation $y + Ty = x$ has a unique solution for every $x \in \mathcal{H}$ or the solution space of $y + Ty = 0$ is nonzero and finite dimensional. *Remark:* This result is known as the *Fredholm alternative*.

Exercise 18.8.33. This exercise continues Exercise 18.8.32. Let $T \in \mathcal{B}(\mathcal{H})$ be compact. Prove that the equations $y + Ty = x$ and $y + T^*y = x'$ have solutions for every $x, x' \in \mathcal{H}$, or the solution spaces of $y + Ty = 0$ y $+ T^*y = 0$ have the same dimension.

Exercise 18.8.34. Let $T \in \mathcal{B}(\mathcal{H})$ be compact. Prove that $y + Ty = x$ has a solution if and only if x is orthogonal to $\{y \in \mathcal{H} : y + T^*y = 0\}$.

Exercise 18.8.35. This exercise is a continuation of Exercise 7.7.14, which considered the *weighted composition operator* T on H^2 defined by $(Tf)(z) = zf(z/2)$. Use the following argument from [113] to characterize the invariant subspaces for T.

(a) For each $k \geqslant 0$ show that $\mathcal{M}_k$, the subspace of H^2 consisting of functions which have a zero of at least k at 0, is an invariant subspace for T.

(b) To prove that $\{\mathcal{M}_k : k \geqslant 1\}$ are all of the T-invariant subspaces, prove that it suffices to show that if $f \in H^2$ and $f(0) \neq 0$, then $\bigvee\{T^n f : n \geqslant 0\} = H^2$.

(c) To show that if $f \in H^2$ with $f(0) \neq 0$, then $\bigvee\{T^n f : n \geqslant 0\} = H^2$, proceed as follows. Let $f \in H^2$ with $f(0) = 1$ and write $f = 1 + g$, where $g \in \mathcal{M}_1$. Define

$$h_n = \sqrt{2}\, 2^{(n^2 - n)} T^n f \quad \text{for } n \geqslant 0$$

and prove that the mapping $W(z^n) = h_n$ extends by linearity to a bounded operator on H^2.

(d) Prove that if W is invertible in $\mathcal{B}(H^2)$, then $\bigvee\{T^n f : n \geqslant 0\} = H^2$.

(e) Prove that $\|(W - I)z^n\| \leqslant \|g\|/2^n$ for all $n \geqslant 0$.

(f) Prove that $W - I$ is compact.

(g) Prove that $\ker W = \{0\}$.

(h) Prove that W is invertible in $\mathcal{B}(H^2)$.

18.9 Hints for the Exercises

Hint for Ex. 18.8.1: Prove that the symbol for the composition operator is $\varphi = Tz$. Now use the fact that $z^n \to 0$ weakly in H^2 to show that φ is a self map of $\mathbb{D}$.
Hint for Ex. 18.8.6: Mimic the proof of Proposition 18.3.1 and use

$$\int_{\mathbb{T}} \frac{1}{1 - |\varphi|^2}\, dm = \sum_{n=0}^{\infty} \int_{\mathbb{T}} |\varphi|^{2n} dm.$$

Hint for Ex. 18.8.10: For (a), consult the von Neumann–Wold decomposition (Theorem 15.1.1). For (b), consider Rota's example from Chapter 13.
Hint for Ex. 18.8.11: Mimic the proof of Proposition 18.6.1.
Hint for Ex. 18.8.21: Consult (18.3.9).
Hint for Ex. 18.8.22: Consult Theorem 18.5.5 and argue that $|h|$ is bounded above and below on $\mathbb{D}$.
Hint for Ex. 18.8.23: Prove that $C_\psi = C_{z^d} C_\varphi$.
Hint for Ex. 18.8.35: For (f), show that $W - I$ is completely continuous. For (g), suppose $f = \sum_{n=0}^{\infty} a_n z^n \in \ker T$. Use a contradiction argument supplied by the fact that $a_0 h_0 + \sum_{n=1}^{\infty} a_n h_n = 0$.

19

· · • · ·

Subnormal Operators

Key Concepts: Subnormal operator, normal extension, hyponormal operator, subnormal weighted shift, representation of a cyclic subnormal operator, Brown's theorem on invariant subspaces.

Outline: A subnormal operator is the restriction of a normal operator to an invariant subspace. This chapter surveys a variety of subnormal operators, a representation theorem that is the analogue of the spectral theorem for normal operators, and a proof of the existence of invariant subspaces for subnormal operators.

19.1 Basics of Subnormal Operators

Since the spectral theorem for normal operators (Theorem 19.2.3) answers many questions about normal operators, it is natural to investigate relatives of the normal operators for which an analogous theory can be developed.

An operator $S \in \mathcal{B}(\mathcal{H})$ is *subnormal* if it satisfies the following three conditions.

(a) $\mathcal{H}$ is a subspace of a Hilbert space $\mathcal{K}$.

(b) There is a normal operator $N \in \mathcal{B}(\mathcal{K})$ such that $N\mathcal{H} \subseteq \mathcal{H}$.

(c) $N|_{\mathcal{H}} = S$.

In short, a subnormal operator is a normal operator restricted to one of its invariant subspaces. The operator N above is a *normal extension* of S. Although a normal extension is never unique (Exercise 19.6.6), there is a minimal normal extension that is unique up to unitary equivalence (Proposition 19.2.7).

Example 19.1.1. The use of the letter S to denote a subnormal operator might appear to invite confusion with the unilateral shift on H^2 from Chapter 5. However, these notational concerns are largely unwarranted since the unilateral shift is a fundamental example of a subnormal operator. Indeed, the unilateral shift S on H^2 is subnormal since the bilateral shift M_ξ on $L^2(\mathbb{T})$ is normal (in fact unitary by Proposition 4.3.1), H^2 is an M_ξ-invariant subspace of $L^2(\mathbb{T})$, and $S = M_\xi|_{H^2}$. Observe that S is not a normal operator (Exercise 5.9.2).

Example 19.1.2. The Bergman shift $(Mf)(z) = zf(z)$ from Chapter 10 is subnormal since M_z on $L^2(dA)$ is normal (Proposition 8.1.14), the Bergman space A^2 is an M_z-invariant subspace of $L^2(dA)$, and $M = M_z|_{A^2}$. Notice that the Bergman shift is not a normal operator (Exercise 10.7.16).

Example 19.1.3. For $\varphi \in H^\infty$, the Toeplitz operator T_φ on H^2 from Chapter 16 is subnormal. Let φ also denote the almost everywhere defined radial boundary function (Theorem 5.4.3) $\varphi(\xi) = \lim_{r \to 1^-} \varphi(r\xi)$ and observe that the multiplication operator M_φ on $L^2(\mathbb{T})$ is normal (Proposition 8.1.14), H^2 is an M_φ-invariant subspace (Proposition 5.5.4), and $T_\varphi = M_\varphi|_{H^2}$ (Proposition 16.6.2). In particular, we obtain the subnormality of the unilateral shift S as the special case $\varphi(z) = z$ (Example 19.1.1).

Example 19.1.4. Any isometry $A \in \mathcal{B}(\mathcal{H})$ is subnormal since the operator

$$N = \begin{bmatrix} A & I - AA^* \\ 0 & -A^* \end{bmatrix}$$

defined on $\mathcal{K} = \mathcal{H} \oplus \mathcal{H}$ is normal (in fact unitary by Exercise 19.6.1) and $N|_{\mathcal{H} \oplus \{0\}} = A$. Observe that

$$N' = \begin{bmatrix} A & I - AA^* \\ 0 & A^* \end{bmatrix}$$

is also normal (Exercise 19.6.2) and $N'|_{\mathcal{H} \oplus \{0\}} = A$. Thus, a normal extension of a subnormal operator is not unique. See Exercise 19.6.6 for other examples.

If $S \in \mathcal{B}(\mathcal{H})$ is subnormal and $N \in \mathcal{B}(\mathcal{K})$ is a normal extension of S on $\mathcal{K} \supseteq \mathcal{H}$, then

$$N = \begin{bmatrix} S & X \\ 0 & T \end{bmatrix}$$

with respect to the orthogonal decomposition $\mathcal{K} = \mathcal{H} \oplus \mathcal{H}^\perp$ (Theorem 14.3.4).

Example 19.1.5. The backward shift operator S^* on H^2 is not subnormal. Suppose, toward a contradiction, that S^* has a normal extension

$$N = \begin{bmatrix} S^* & X \\ 0 & T \end{bmatrix}.$$

Then, with $\star$ denoting an entry whose value is not of interest, we see that

$$\begin{bmatrix} 0 & 0 \\ 0 & 0 \end{bmatrix} = NN^* - N^*N$$

$$= \begin{bmatrix} S^*S - SS^* + XX^* & \star \\ \star & \star \end{bmatrix}$$

$$= \begin{bmatrix} I - SS^* + XX^* & \star \\ \star & \star \end{bmatrix} \qquad \text{(by Exercise 5.9.2)}$$

$$= \begin{bmatrix} 1 \otimes 1 + XX^* & \star \\ \star & \star \end{bmatrix} \qquad \text{(by Exercise 5.9.2).}$$

It follows that $1 \otimes 1 + XX^*$ is the zero operator, which it is not since $1 \otimes 1$ is positive definite and XX^* is positive semidefinite.

Remark 19.1.6. Since the unilateral shift S is subnormal and S^* is not, it follows that subnormality is not preserved under the adjoint operation.

The next proposition shows that subnormality is preserved under unitary equivalence.

Proposition 19.1.7. *An operator that is unitarily equivalent to a subnormal operator is subnormal.*

Proof Let $A \in \mathcal{B}(\mathcal{H}_1)$ be subnormal with normal extension $N_A \in \mathcal{B}(\mathcal{K}_1)$, where $\mathcal{H}_1 \subseteq \mathcal{K}_1$. Suppose $U : \mathcal{H}_1 \to \mathcal{H}_2$ is unitary and $UAU^* = B \in \mathcal{B}(\mathcal{H}_2)$. The Hilbert space $\mathcal{K}_2 = \mathcal{H}_2 \oplus (\mathcal{K}_1 \ominus \mathcal{H}_1)$ contains $\mathcal{H}_2$. Consider the unitary operator

$$V = \begin{bmatrix} U & 0 \\ 0 & I \end{bmatrix} : \mathcal{K}_1 = \mathcal{H}_1 \oplus (\mathcal{K}_1 \ominus \mathcal{H}_1) \to \mathcal{K}_2 = \mathcal{H}_2 \oplus (\mathcal{K}_1 \ominus \mathcal{H}_1).$$

A calculation shows that $VN_A V^*$ is a normal operator on $\mathcal{K}_2$. Furthermore, if $\mathbf{x} \in \mathcal{H}_2$, another calculation reveals that

$$VN_A V^* \begin{bmatrix} \mathbf{x} \\ \mathbf{0} \end{bmatrix} = \begin{bmatrix} B\mathbf{x} \\ \mathbf{0} \end{bmatrix}.$$

Thus, $VN_A V^*|_{\mathcal{H}_2} = B$. Consequently, B has a normal extension and is therefore subnormal. ∎

Example 19.1.8. The Cesàro operator

$$(Cf) = \frac{1}{z} \int_0^z \frac{f(\xi)}{1 - \xi} d\xi$$

on H^2 is subnormal. This is unexpected since there is no obvious normal extension. As shown in Theorem 6.4.7, this operator is unitarily equivalent to $I - M_z$ on a certain $H^2(\mu)$ space, which is subnormal (see below). Now use Proposition 19.1.7.

19.2 Cyclic Subnormal Operators

Suppose μ is a finite positive compactly supported Borel measure on $\mathbb{C}$. The set of polynomials $\mathbb{C}[z]$ is contained in $L^2(\mu)$.

Definition 19.2.1. For μ as above, let $H^2(\mu)$ denote the closure of $\mathbb{C}[z]$ in $L^2(\mu)$.

Example 19.2.2. If m is Lebesgue measure on $\mathbb{T}$, then $H^2(m)$ is the Hardy space H^2. If dA is Lebesgue measure on $\mathbb{D}$, then $H^2(dA)$ is the Bergman space A^2.

Define $N_\mu : L^2(\mu) \to L^2(\mu)$ by

$$(N_\mu f)(z) = zf(z)$$

and recall from Chapter 8 that N_μ is a bounded $*$-cyclic (Definition 8.2.6) normal operator with adjoint $(N_\mu^* f)(z) = \bar{z}f(z)$. Bram's theorem (Theorem 8.2.12) says that N_μ is cyclic, meaning there is some $f \in L^2(\mu)$ such that $\bigvee\{N_\mu^k f : k \geqslant 0\} = L^2(\mu)$.

Theorem 8.7.1 is the spectral theorem for cyclic selfadjoint operators. Here is the corresponding theorem for normal operators [94, Ch. IX]. Readers interested in the proof can consult [94]. The standard proof involves the representation theory of commutative C^*-algebras and would draw us too far afield since our aim here is to present results based on concrete examples.

Theorem 19.2.3 (Spectral theorem for normal operators). *Let $N \in \mathcal{B}(\mathcal{H})$ be a $*$-cyclic normal operator with $*$-cyclic vector $\mathbf{x}$. Then there is a finite positive compactly supported Borel measure μ on $\mathbb{C}$ and a unitary operator $U : \mathcal{H} \to L^2(\mu)$ such that $U\mathbf{x} = 1$ and $N = U^*N_\mu U$.*

Below is a commutative diagram that illustrates the spectral theorem.

$$
\begin{array}{ccc}
\mathcal{H} & \xrightarrow{\ N\ } & \mathcal{H} \\
{\scriptstyle U}\Big\downarrow & & \Big\uparrow{\scriptstyle U^*} \\
L^2(\mu) & \xrightarrow[\ N_\mu\ ]{} & L^2(\mu)
\end{array}
$$

Now define $S_\mu : H^2(\mu) \to H^2(\mu)$ by

$$(S_\mu f)(z) = zf(z) \tag{19.2.4}$$

and observe that

$$N_\mu H^2(\mu) \subseteq H^2(\mu) \quad \text{and} \quad S_\mu = N_\mu|_{H^2(\mu)}.$$

In other words, S_μ is a subnormal operator and N_μ is a normal extension of S_μ. Also notice that S_μ is a cyclic operator with cyclic vector 1. The proof of Proposition 8.1.5 yields the following.

Proposition 19.2.5. $\|S_\mu\| = \sup\{|z| : z \in \operatorname{supp}\mu\}$.

We now discuss a model operator for a cyclic subnormal operator. But first, we need to address another issue: the lack of uniqueness of a normal extension (see Example 19.1.4 and Exercises 19.6.3 and 19.6.6).

Definition 19.2.6. A normal extension $N \in \mathcal{B}(\mathcal{K})$ of a subnormal operator $S \in \mathcal{B}(\mathcal{H})$ is a *minimal normal extension* if $\mathcal{K}$ has no proper subspace that reduces N and contains $\mathcal{H}$.

Proposition 19.2.7. *Let $S \in \mathcal{B}(\mathcal{H})$ be subnormal with a normal extension $N \in \mathcal{B}(\mathcal{K})$, where $\mathcal{H} \subseteq \mathcal{K}$. Then N is a minimal normal extension of S if and only if*

$$\mathcal{K} = \bigvee\{N^{*j}h : j \geqslant 0, h \in \mathcal{H}\}.$$

Proof Let $\mathcal{K}' = \bigvee\{N^{*j}h : j \geqslant 0, h \in \mathcal{H}\}$ and note that $\mathcal{K}' \supseteq \mathcal{H}$ and $\mathcal{K}'$ reduces N. This means that $N|_{\mathcal{K}'}$ defines a normal operator that extends S. Any reducing subspace that contains $\mathcal{H}$ also contains $N^{*j}\mathcal{H}$ for all $j \geqslant 0$, and hence $\mathcal{K}'$ is the smallest reducing subspace of $\mathcal{K}$ that contains $\mathcal{H}$. ∎

We say that $N|_{\mathcal{K}}$ is the *minimal normal extension* of S. A technical detail says that if N and N' are two minimal normal extensions of S, then N is unitarily equivalent to N' [95, p. 39]. Thus, it makes sense to use the term "the" (up to unitary equivalence) minimal normal extension of S.

Here is the main representation theorem for subnormal operators. It mirrors the spectral theorem for cyclic normal operators.

Theorem 19.2.8. *Let $S \in \mathcal{B}(\mathcal{H})$ be cyclic and subnormal. Then there is a finite positive compactly supported Borel measure μ on $\mathbb{C}$ such that S is unitarily equivalent to S_μ on $H^2(\mu)$.*

Proof Let $\mathbf{x}$ be a cyclic vector for S and let $N \in \mathcal{B}(\mathcal{K})$, where $\mathcal{H} \subseteq \mathcal{K}$, denote the minimal normal extension of S. Proposition 19.2.7 and the fact that $\bigvee\{S^n\mathbf{x} : n \geqslant 0\} = \mathcal{H}$ yield

$$\mathcal{K} = \bigvee\{N^{*j}\mathbf{x} : j \geqslant 0, \mathbf{x} \in \mathcal{H}\} = \bigvee\{N^{*j}S^k\mathbf{x} : j, k \geqslant 0\} = \bigvee\{N^{*j}N^k\mathbf{x} : j, k \geqslant 0\}.$$

Thus, $\mathbf{x}$ is a $*$-cyclic vector for N. The spectral theorem for $*$-cyclic normal operators (Theorem 19.2.3) produces a finite positive compactly supported Borel measure ν on $\mathbb{C}$ and a unitary $U : \mathcal{K} \to L^2(\nu)$ such that $UNU^* = N_\nu$.

For any $p \in \mathbb{C}[z]$,

$$\begin{aligned}
\|p(S)\mathbf{x}\|^2 &= \langle p(S)\mathbf{x}, p(S)\mathbf{x}\rangle_{\mathcal{H}} \\
&= \langle p(N)\mathbf{x}, p(N)\mathbf{x}\rangle_{\mathcal{K}} \\
&= \langle Up(N)\mathbf{x}, Up(N)\mathbf{x}\rangle_{L^2(\nu)} \\
&= \langle M_p U\mathbf{x}, M_p U\mathbf{x}\rangle_{L^2(\nu)} \\
&= \int |p|^2 |U\mathbf{x}|^2 d\nu.
\end{aligned}$$

Now define $d\mu = |U\mathbf{x}|^2 d\nu$ and observe that

$$\|p(S)\mathbf{x}\|^2 = \int |p|^2 d\mu.$$

The set $\{p(S)\mathbf{x} : p \in \mathbb{C}[z]\}$ is dense in $\mathcal{H}$ since $\mathbf{x}$ is a cyclic vector. Thus, the linear transformation $V : H^2(\mu) \to \mathcal{H}$, initially defined on $\mathbb{C}[z]$ by

$$Vp = p(S)\mathbf{x}, \tag{19.2.9}$$

is isometric with dense range and thus extends to a unitary operator. For $p \in \mathbb{C}[z]$, we have $VS_\mu p = V(zp) = Sp(S)\mathbf{x} = SVp$. Since $\mathbb{C}[z]$ is dense in $H^2(\mu)$, this extends to all of $H^2(\mu)$. Thus, $S = VS_\mu V^*$. ∎

Remark 19.2.10. From the spectral-radius formula, we also have that $\sigma(S_\mu) \subseteq \|S_\mu\|\mathbb{D}^-$. Proposition 19.2.5 says that the support of μ is contained in $\|S_\mu\|\mathbb{D}^-$.

From (19.2.9), notice that V^* maps the cyclic vector $\mathbf{x}$ to the constant function 1. This becomes important in Theorem 19.3.4. Below is a commutative diagram that illustrates Theorem 19.2.8.

$$
\begin{array}{ccc}
\mathcal{H} & \xrightarrow{\;S\;} & \mathcal{H} \\[4pt]
{\scriptstyle V^*}\big\downarrow & & \big\uparrow{\scriptstyle V} \\[4pt]
H^2(\mu) & \xrightarrow[\;S_\mu\;]{} & H^2(\mu)
\end{array}
$$

19.3 Subnormal Weighted Shifts

Suppose that $\mathcal{H}$ is a Hilbert space with orthonormal basis $(\mathbf{u}_n)_{n=0}^\infty$. If $(\alpha_n)_{n=0}^\infty$ is a sequence of positive numbers, recall from Exercise 3.6.21 the corresponding *weighted shift*

$$A\mathbf{u}_n = \alpha_n \mathbf{u}_{n+1} \quad \text{for } n \geq 0. \tag{19.3.1}$$

This definition extends by linearity to a bounded operator on $\mathcal{H}$ if and only if the *weight sequence* $(\alpha_n)_{n=0}^\infty$ is a bounded sequence. The matrix representation of A with respect to $(\mathbf{u}_n)_{n=0}^\infty$ is

$$
\begin{bmatrix}
0 & 0 & 0 & 0 & \cdots \\
\alpha_0 & 0 & 0 & 0 & \cdots \\
0 & \alpha_1 & 0 & 0 & \cdots \\
0 & 0 & \alpha_2 & 0 & \cdots \\
\vdots & \vdots & \vdots & \vdots & \ddots
\end{bmatrix}
\tag{19.3.2}
$$

and $\|A\| = \sup_{n \geq 0} \alpha_n$.

We have already seen three examples of weighted shifts:

(a) The unilateral shift on H^2 (Chapter 5): $\alpha_n = 1$ and $\mathbf{u}_n = z^n$.

(b) The Dirichlet shift (Chapter 9): $\alpha_n = \sqrt{\dfrac{n+2}{n+1}}$ and $\mathbf{u}_n = z^n/\sqrt{n+1}$.

(c) The Bergman shift (Chapter 10): $\alpha_n = \sqrt{\dfrac{n+1}{n+2}}$ and $\mathbf{u}_n = \sqrt{n+1}\, z^n$.

For $\beta \in \mathbb{R}$, there are also the *Dirichlet-type spaces*

$$\mathcal{D}_\beta = \left\{ f(z) = \sum_{n=0}^\infty a_n z^n : \|f\|_\beta^2 = \sum_{n=0}^\infty |a_n|^2 (n+1)^\beta < \infty \right\}.$$

Note that $\mathcal{D}_0 = H^2$, $\mathcal{D}_1 = \mathcal{D}$, and $\mathcal{D}_{-1} = A^2$. There is a natural orthonormal basis

$$\mathbf{u}_n^{(\beta)}(z) = \frac{z^n}{(n+1)^{\beta/2}} \quad \text{for } n \geqslant 0,$$

for $\mathcal{D}_\beta$, along with a corresponding weighted shift $A_\beta : \mathcal{D}_\beta \to \mathcal{D}_\beta$ defined by

$$A_\beta \mathbf{u}_n^{(\beta)} = \left(\frac{n+2}{n+1}\right)^{\beta/2} \mathbf{u}_{n+1}^{(\beta)}. \tag{19.3.3}$$

For $\beta = 0$ and $\beta = -1$, the corresponding weighted shifts are subnormal because they are M_z on the Hardy space (Example 19.1.1) and Bergman space (Example 19.1.2), respectively. What about the subnormality of other weighted shifts? The following theorem provides the answer. We remind the reader (see (19.2.4)) that S_μ denotes the operator $S_\mu f = zf$ on $H^2(\mu)$, the closure of the polynomials in $L^2(\mu)$.

Theorem 19.3.4 (Gellar–Wallen [152]). *Let A be a weighted shift from (19.3.1) with corresponding positive weight sequence $(\alpha_n)_{n=0}^\infty$. Then the following are equivalent.*

(a) *A is subnormal.*

(b) *There is a finite positive Borel measure μ on $[0, \|A\|]$ whose support contains $\|A\|$ such that*

$$(\alpha_0 \alpha_1 \alpha_2 \cdots \alpha_{n-1})^2 = \int_0^{\|A\|} x^{2n} \, d\mu(x) \quad \text{for all } n \geqslant 1. \tag{19.3.5}$$

(c) *There is a finite positive Borel measure μ on $[0, \|A\|]$ with $\|A\|$ in its support such that A is unitarily equivalent to S_ν on $H^2(\nu)$, where*

$$d\nu = \frac{1}{2\pi} d\theta \, d\mu(r) \tag{19.3.6}$$

is a measure on the disk $\|A\|\mathbb{D}^-$.

Proof We follow the proof from [95, Ch. 2]. Recall that $\|A\| = \sup_{n \geqslant 0} \alpha_n$.

(a) $\Rightarrow$ (b) Since A is cyclic with cyclic vector $\mathbf{u}_0$, Theorem 19.2.8, along with Remark 19.2.10, provides a unitary $U : \mathcal{H} \to H^2(\sigma)$ such that $U\mathbf{u}_0 = 1$ and $UAU^* = S_\sigma$ on $H^2(\sigma)$ for some measure σ on the disk $\|A\|\mathbb{D}^-$. Thus, $UA^n\mathbf{u}_0 = S_\sigma^n U\mathbf{u}_0 = z^n$. Let

$$\gamma_0 = 1 \quad \text{and} \quad \gamma_n = \alpha_0 \alpha_1 \alpha_2 \cdots \alpha_{n-1} \quad \text{for } n \geqslant 1. \tag{19.3.7}$$

Then

$$A\mathbf{u}_0 = \alpha_0 \mathbf{u}_1 = \gamma_1 \mathbf{u}_1,$$
$$A^2 \mathbf{u}_0 = \gamma_1 A\mathbf{u}_1 = \gamma_1 \alpha_1 \mathbf{u}_2 = \gamma_2 \mathbf{u}_2,$$
$$\vdots$$
$$A^n \mathbf{u}_0 = A(\gamma_{n-1} \mathbf{u}_{n-1}) = \gamma_{n-1} \alpha_n \mathbf{u}_n = \gamma_n \mathbf{u}_n.$$

Therefore, $z^n = UA^n \mathbf{u}_0 = \gamma_n U \mathbf{u}_n$ and hence $U \mathbf{u}_n = z^n \gamma_n^{-1}$. Since U is isometric, it follows that $\|U \mathbf{u}_n\| = 1$ and consequently,

$$\frac{1}{\gamma_n^2} \int |z|^{2n} d\sigma = \|U \mathbf{u}_n\|^2 = 1.$$

This yields

$$\gamma_n^2 = \int |z|^{2n} d\sigma.$$

Define a measure μ on $[0, \|A\|]$ by $\mu(B) = \sigma(\{|z| \in B\})$ for any Borel set $B \subseteq [0, \|A\|]$. Then

$$\gamma_n^2 = \int_0^{\|A\|} r^{2n} d\mu(r).$$

Proposition 19.2.5 implies that

$$\|A\| = \|S_\sigma\| = \sup_{z \in \operatorname{supp} \sigma} |z| = \sup_{r \in \operatorname{supp} \mu} r.$$

Thus, $\|A\|$ belongs to the support of μ.

(b) $\Rightarrow$ (c) Let μ and ν be the measures from (19.3.5) and (19.3.6), respectively, and consider the space $H^2(\nu)$. The condition in (19.3.5) implies

$$\|z^n\|^2 = \frac{1}{2\pi} \int_0^{\|A\|} \int_0^{2\pi} |re^{i\theta}|^{2n} d\theta \, d\mu(r) = \int_0^{\|A\|} r^{2n} \, d\mu(r) = \gamma_n^2.$$

Furthermore,

$$\langle z^n, z^m \rangle = \frac{1}{2\pi} \int_0^{\|A\|} \int_0^{2\pi} r^{n+m} e^{i\theta(n-m)} d\theta \, d\mu(r)$$

$$= \delta_{mn} \int_0^{\|A\|} r^{2n} \, d\mu(r)$$

$$= \delta_{mn} \gamma_n^2.$$

Thus, $(z^n \gamma_n^{-1})_{n=0}^\infty$ is an orthonormal basis for $H^2(\nu)$. Now define the unitary operator $V : \mathcal{H} \to H^2(\nu)$ by

$$V \mathbf{u}_n = \frac{z^n}{\gamma_n}, \tag{19.3.8}$$

and note that

$$VAV^* \frac{z^n}{\gamma_n} = VA\mathbf{u}_n \qquad \text{(by (19.3.8))}$$

$$= V\alpha_n \mathbf{u}_{n+1} \qquad \text{(by (19.3.1))}$$

$$= \alpha_n \frac{z^{n+1}}{\gamma_{n+1}} \qquad \text{(by (19.3.8))}$$

$$= z \frac{z^n}{\gamma_n} = S_\nu \frac{z^n}{\gamma_n}.$$

Thus, $A = V^* S_\nu V$.

(c) $\Rightarrow$ (a) Since S_ν is subnormal and A is unitarily equivalent to it, Proposition 19.1.7 ensures that A is subnormal. ∎

Below is a commutative diagram that illustrates Theorem 19.3.4.

$$
\begin{array}{ccc}
\mathcal{H} & \xrightarrow{\;\;A\;\;} & \mathcal{H} \\
\Big\downarrow{\scriptstyle V} & & \Big\uparrow{\scriptstyle V^*} \\
H^2(\nu) & \xrightarrow[\;\;S_\nu\;\;]{} & H^2(\nu)
\end{array}
$$

The next corollary demonstrates a notable difference between the Dirichlet shift, and the Hardy and Bergman shifts.

Corollary 19.3.9. *The Dirichlet shift is not subnormal.*

Proof Suppose toward a contradiction that the Dirichlet shift is subnormal. From our earlier discussion, the Dirichlet shift M_z is a weighted shift with weight sequence

$$
\alpha_n = \sqrt{\frac{n+2}{n+1}}
$$

and $\|M_z\| = \sqrt{2}$. Furthermore, $\gamma_n = \sqrt{n+1}$ as defined in (19.3.7). Thus, there is a positive finite Borel measure μ on $[0, \sqrt{2}]$ with $\sqrt{2} \in \operatorname{supp}\mu$ such that

$$
n + 1 = \int_0^{\sqrt{2}} r^{2n} d\mu(r).
$$

We show this is impossible. For $1 < a < \sqrt{2}$, observe that

$$
n + 1 = \int_0^{\sqrt{2}} r^{2n} d\mu(r) \geqslant \int_a^{\sqrt{2}} r^{2n} d\mu(r) \geqslant a^{2n} \mu([a, \sqrt{2}]).
$$

Thus,

$$
\mu([a, \sqrt{2}]) \leqslant \frac{n+1}{a^{2n}} \quad \text{for all } n \geqslant 1 \text{ and } 1 < a < \sqrt{2}.
$$

Let $n \to \infty$ to conclude that $\mu([a, \sqrt{2}]) = 0$ for all $1 < a < \sqrt{2}$. This contradicts the fact that $\sqrt{2} \in \operatorname{supp}\mu$. ∎

For $\beta > 0$, a similar proof shows that M_z on $\mathcal{D}_\beta$ is not subnormal. One can also give an alternate proof using hyponormality (Exercise 19.6.11). When $\beta < 0$ and

$$
\alpha_n = \left(\frac{n+2}{n+1}\right)^{\beta/2}, \tag{19.3.10}
$$

the corresponding weighted shift A_β from (19.3.3) (equivalently M_z on $\mathcal{D}_\beta$) is hyponormal. A nontrivial result from [39] shows that more is true.

Theorem 19.3.11. M_z on $\mathcal{D}_\beta$ is subnormal for all $\beta < 0$.

A good starting point to learn more about various aspects of weighted shifts, including their corresponding function space realizations, is [342].

19.4 Invariant Subspaces

Subnormal operators have proper nonzero invariant subspaces. Although we do not go into fine detail, we highlight the powerful function theory involved. An important ingredient is the identification of the cyclic subnormal operators with S_μ on $H^2(\mu)$ (Theorem 19.2.8). The original proof of the next theorem is due to Brown [72], but the simplification presented below is due to Thomson [358].

Theorem 19.4.1 (Brown). *If $\mathcal{H}$ is a Hilbert space with* $\dim \mathcal{H} \geqslant 2$ *and* $S \in \mathcal{B}(\mathcal{H})$ *is subnormal, then S has a proper nonzero invariant subspace.*

Proof If S is not cyclic, then $\bigvee \{S^n \mathbf{x} : n \geqslant 0\}$ is a proper nonzero S-invariant subspace for any $\mathbf{x} \in \mathcal{H} \backslash \{\mathbf{0}\}$. So let us assume that S is cyclic. Theorem 19.2.8 says that S is unitarily equivalent to S_μ on $H^2(\mu)$. Thus, it suffices to prove that S_μ has proper nonzero invariant subspaces.

The proof splits into two cases:

$$H^2(\mu) = L^2(\mu) \quad \text{and} \quad H^2(\mu) \subsetneq L^2(\mu).$$

If $H^2(\mu) = L^2(\mu)$, then, since $\dim L^2(\mu) \geqslant 2$, there exists a Borel set A such that $\mu(A) > 0$ and $\mu(\mathbb{C} \backslash A) > 0$. The subspace $\chi_A L^2(\mu)$ is a proper nonzero S_μ-invariant subspace.

The second case is trickier and requires the following technical lemma from [95]. Assuming that $H^2(\mu) \subsetneq L^2(\mu)$, there is a $\lambda \in \mathbb{C}$ with $\mu(\{\lambda\}) = 0$, an $f \in H^2(\mu)$, and a $g \in L^2(\mu)$ such that

$$p(\lambda) = \int p f \overline{g} \, d\mu \quad \text{for all } p \in \mathbb{C}[z]. \tag{19.4.2}$$

Assuming (19.4.2), define

$$\mathcal{M} = \{qf : q \in \mathbb{C}[z], q(\lambda) = 0\}^-.$$

Notice that $\mathcal{M}$ is an S_μ-invariant subspace.

To finish the proof, we need to verify that $\{0\} \subsetneq \mathcal{M} \subsetneq H^2(\mu)$. Apply (19.4.2) to the constant polynomial $p \equiv 1$ to see that $\langle f, g \rangle = 1$. Apply (19.4.2) again for any polynomial p such that $p(\lambda) = 0$ to conclude that $g \perp \mathcal{M}$. Thus, $f \in H^2(\mu) \backslash \mathcal{M}$. Since $f \neq 0$ μ-almost everywhere and $\mu(\{\lambda\}) = 0$, the function $(z - \lambda)f$ belongs to $\mathcal{M} \backslash \{0\}$. Therefore, $\mathcal{M} \neq \{0\}$. ∎

19.5 Notes

Let us mention a few other gems from the theory of subnormal operators. See [95] for all the details along with a wealth of other results that we do not have room to cover here.

Proving that an $S \in \mathcal{B}(\mathcal{H})$ is subnormal by identifying a normal extension can be difficult as was seen by the Cesàro operator. There are a variety of tests one can perform that often prove subnormality. For example, Bram [63] proved that $S \in \mathcal{B}(\mathcal{H})$ is subnormal if and only if

$$\sum_{j,k=0}^{n} \langle S^j \mathbf{x}_k, S^k \mathbf{x}_j \rangle \geq 0 \quad \text{for every } \mathbf{x}_0, \mathbf{x}_1, \ldots, \mathbf{x}_n \in \mathcal{H}.$$

The ambient space $H^2(\mu)$, the closure of the polynomials in $L^2(\mu)$, is worthy of study. A celebrated result of Thomson [359] decomposes the support Δ of a finite positive Borel measure μ into disjoint Borel sets Δ_j for $j \geq 0$ such that

$$H^2(\mu) = L^2(\mu|_{\Delta_0}) \oplus H^2(\mu|_{\Delta_1}) \oplus H^2(\mu|_{\Delta_2}) \oplus \cdots. \tag{19.5.1}$$

This result has important applications to the structure of subnormal operators [95]. More importantly, this description resolves a long-standing conjecture of Brennan [95, Ch. VIII] which says that either $H^2(\mu) = L^2(\mu)$ or there is a $\lambda \in \operatorname{supp}\mu$ and a $C > 0$ such that $|p(\lambda)| \leq C\|p\|_{L^2(\mu)}$ for all $p \in \mathbb{C}[z]$. Permitting a slight abuse of language, such a point λ is called a *bounded point evaluation*.

Sarason [325] described the space $H^\infty(\mu)$, the weak-$*$ closure of the polynomials in $L^\infty(\mu)$. This space is important since it reveals further structure theorems for subnormal operators.

Another gem is a version of von Neumann's inequality for subnormal operators. If $T \in \mathcal{B}(\mathcal{H})$ is a contraction, von Neumann [370] (see also [144]) proved that

$$\|p(T)\| \leq \sup_{z \in \mathbb{D}} |p(z)| \quad \text{for all } p \in \mathbb{C}[z].$$

For any $T \in \mathcal{B}(\mathcal{H})$ and rational function f whose poles are in $\mathbb{C} \backslash \sigma(T)$, one can define $f(T)$ via the Riesz functional calculus. If S is a subnormal operator and K is a compact set containing $\sigma(S)$, then $\|f(S)\| \leq \sup_{z \in K} |f(z)|$. This functional calculus also yields various spectral-mapping theorems for subnormal operators.

There is a version of Corollary 8.3.3 which describes the commutant of S_μ as

$$\{S_\mu\}' = \{M_\varphi : \varphi \in H^2(\mu) \cap L^\infty(\mu)\},$$

where $M_\varphi f = \varphi f$ on $H^2(\mu)$.

Example 19.1.3 shows that any analytic Toeplitz operator T_φ on the Hardy space H^2 is subnormal with normal extension M_φ. Is this the minimal normal extension? As long as φ is nonconstant, the answer is yes. The minimal normal extension for a subnormal operator is also a well-studied topic. One of many results is: if N is the minimal normal extension of a subnormal operator S, then $\varphi(N)$ is the minimal normal extension of $\varphi(S)$ for a rich class of functions φ [95].

19.6 Exercises

Exercise 19.6.1. If $A \in \mathcal{B}(\mathcal{H})$ is an isometry, prove that

$$U = \begin{bmatrix} A & I - AA^* \\ 0 & -A^* \end{bmatrix}$$

is a unitary operator on $\mathcal{H} \oplus \mathcal{H}$.

Exercise 19.6.2. If $A \in \mathcal{B}(\mathcal{H})$ is an isometry, prove that

$$N = \begin{bmatrix} A & I - AA^* \\ 0 & A^* \end{bmatrix}$$

is a normal operator on $\mathcal{H} \oplus \mathcal{H}$.

Exercise 19.6.3.

(a) Let S denote the unilateral shift on ℓ^2. Prove that

$$N(\dots, a_{-2}, a_{-1}, \boxed{a_0}, a_1, a_2, \dots) = (\dots, a_{-3}, a_{-2}, \boxed{a_{-1}}, a_0, a_1, a_2, \dots)$$

and

$$N'(\dots, a_{-2}, a_{-1}, \boxed{a_0}, a_1, a_2, \dots) = (\dots, -a_{-3}, -a_{-2}, \boxed{a_{-1}}, a_0, a_1, a_2, \dots)$$

on $\ell^2(\mathbb{Z})$ are both normal extensions of S. The box denotes the 0th position.

(b) Relate these extensions to those considered in Example 19.1.4.

Exercise 19.6.4. Let $A \in \mathcal{B}(\mathcal{H})$ and suppose that A and A^* are subnormal. Prove that A is normal.

Exercise 19.6.5. If $T \in \mathcal{B}(\mathcal{H})$ is normal and $\mathcal{M}$ is a T-invariant subspace of $\mathcal{H}$ such that $T|_{\mathcal{M}}$ is normal, prove that $\mathcal{M}$ is reducing for $T|_{\mathcal{M}}$.

Exercise 19.6.6.

(a) If $S \in \mathcal{B}(\mathcal{H})$ is subnormal and $N \in \mathcal{B}(\mathcal{H})$ is normal, prove that $S \oplus N \in \mathcal{B}(\mathcal{H} \oplus \mathcal{H})$ is subnormal.

(b) What does this say about the uniqueness of a normal extension of S?

Exercise 19.6.7. Let $d\mu = dA|_{\{\frac{1}{2} < |z| < 1\}}$ denote planar Lebesgue measure on the annulus

$$\{z : \frac{1}{2} < |z| < 1\}.$$

(a) Prove that the restriction of a function in the Bergman space A^2 (of the disk) to the annulus belongs to $H^2(\mu)$.

(b) Prove that each function in $H^2(\mu)$ has an extension to $\mathbb{D}$ that belongs to A^2.

(c) Let $\|f\|_{A^2}$ denote the Bergman-space norm. Prove that there exists $c_1, c_2 > 0$ such that

$$c_1 \|f\|_{A^2}^2 \leqslant \int_{\{\frac{1}{2} < |z| < 1\}} |f|^2 dA \leqslant c_2 \|f\|_{A^2}^2 \quad \text{for all } f \in A^2.$$

Exercise 19.6.8. Let $d\mu = dA|_{\{\frac{1}{2} < |z| < 1\}}$.

(a) Prove that $R^2(\mu)$, the closure of the set of rational functions whose poles lie in $\mathbb{C}\backslash\{\frac{1}{2} \leqslant |z| \leqslant 1\}$, is the space of analytic functions f on $\{\frac{1}{2} < |z| < 1\}$ such that

$$\int_{\{\frac{1}{2} < |z| < 1\}} |f|^2 dA < \infty.$$

(b) Prove that $H^2(\mu) \subsetneq R^2(\mu) \subsetneq L^2(\mu)$.

(c) Prove that M_z on $R^2(\mu)$ is not cyclic.

Exercise 19.6.9. Recall from (6.4.1) that $A \in \mathcal{B}(\mathcal{H})$ is hyponormal if $A^*A - AA^* \geqslant 0$. Prove that any subnormal operator is hyponormal.

Exercise 19.6.10. Let S denote the unilateral shift on H^2. Prove that $T = (S + 2S^*)^2$ is not subnormal.

Exercise 19.6.11. Let $(\alpha_n)_{n=0}^\infty$ be a bounded sequence of positive numbers and W be the weighted shift on ℓ^2 defined by $We_n = \alpha_n e_{n+1}$ (see Exercise 3.6.21).

(a) Prove that W is hyponormal if and only if $\alpha_n \leqslant \alpha_{n+1}$ for all $n \geqslant 0$.

(b) Use this to prove that the Dirichlet shift is not subnormal.

Exercise 19.6.12. For a finite-dimensional Hilbert space $\mathcal{H}$, prove that every hyponormal operator is normal using the following steps.

(a) Suppose that $\mathcal{H} = \mathbb{C}^n$, $A \in M_n$, and $A^*A - AA^* \geqslant 0$. Prove that $\text{tr}(A^*A - AA^*) = 0$.

(b) Use the spectral theorem and the fact that $A^*A - AA^* \geqslant 0$ to show that $A^*A - AA^* = 0$.

Exercise 19.6.13. Prove that if $S \in \mathcal{B}(\mathcal{H})$ is subnormal and compact, then S is normal. Follow this argument from [372].

(a) For a positive $A \in \mathcal{B}(\mathcal{H})$, prove that $|\langle Ax, y \rangle|^2 \leqslant \langle Ax, x \rangle \langle Ay, y \rangle$ for all $x, y \in \mathcal{H}$.

(b) If S is subnormal, then $A = S^*S - SS^*$ is positive (Exercise 19.6.9). Prove that $\ker(S^*S - SS^*) = \{x : \|Sx\| = \|S^*x\|\}$.

(c) Prove that $S = A + iB$, where A and B are compact and selfadjoint.

(d) Prove that $\|(A + i\lambda I)x\| = \|(A - i\lambda I)x\|$ for all $x \in \mathcal{H}$ and $\lambda \in \mathbb{R}$.

(e) Let $\mathbf{x}$ be an eigenvector of B. Prove that $\|S\mathbf{x}\| = \|S^*\mathbf{x}\|$.

(f) Prove that $\ker(S^*S - SS^*)$ contains each eigenspace of B.

(g) Prove that $\ker(S^*S - SS^*) = \mathcal{H}$.

Remark: This exercise shows that every subnormal operator on a finite-dimensional Hilbert space is normal and thus subnormal operators only differ from normal operators in the infinite-dimensional setting.

Exercise 19.6.14. Let $S \in \mathcal{B}(\mathcal{H})$ be subnormal and define $A = S^*S - SS^*$. Prove that $\ker A$ is an invariant subspace for S.
Remark: See [349] for more on this.

Exercise 19.6.15. Exercise 6.7.10 showed that the Cesàro operator is not compact. Use Exercise 19.6.13 and Theorem 6.4.2 to supply another proof.

Exercise 19.6.16. Let $f, g \in H^\infty$. Prove that the Toeplitz operator $T_{f+\bar{g}}$ is hyponormal if and only if $T_f^* T_f - T_f T_f^* \geqslant T_g^* T_g - T_g T_g^*$.

Exercise 19.6.17. Suppose that T is a hyponormal contraction and $\mathcal{M} = \ker(I - TT^*)$. Prove the following.

(a) $\mathcal{M}$ is an invariant subspace for T^*T.

(b) $\mathcal{M}$ is an invariant subspace for T.

(c) $T|_{\mathcal{M}}$ is isometric.

Remark: See [43] for more on this.

Exercise 19.6.18. An operator $S \in \mathcal{B}(\mathcal{H})$ is *quasinormal* if $S(S^*S) = (S^*S)S$. Let $S = U|S|$ be a polar decomposition of S (Theorem 14.9.15). Prove that S is quasinormal if and only if $U|S| = |S|U$.
Remark: See [66] for more on quasinormal operators.

Exercise 19.6.19. Let φ be an analytic self map of $\mathbb{D}$. Prove that if the composition operator C_φ on H^2 (see Chapter 18) is quasinormal, then $\varphi(0) = 0$.
Remark: See [87] for more on quasinormal composition operators.

Exercise 19.6.20. This is a continuation of Exercise 19.6.18. Suppose that $S = U|S|$ is quasinormal and $\ker S = \{0\}$.

(a) Prove that U is an isometry.

(b) Prove that $E = UU^*$ is an orthogonal projection. What is $\operatorname{ran} E$?

(c) Let

$$N = \begin{bmatrix} U & I - E \\ 0 & U^* \end{bmatrix} \begin{bmatrix} |S| & 0 \\ 0 & |S| \end{bmatrix}.$$

Prove that N is a normal extension of S, and thus S is subnormal.

Exercise 19.6.21. The next several problems explore the dual of a subnormal operator [93]. Let $S \in B(\mathcal{H})$ be subnormal with minimal normal extension $N \in B(\mathcal{K})$ and $\mathcal{H} \subseteq \mathcal{K}$. With respect to the decomposition $\mathcal{K} = \mathcal{H} \oplus \mathcal{H}^\perp$, write N as

$$N = \begin{bmatrix} S & X \\ 0 & T^* \end{bmatrix}.$$

Then $T \in B(\mathcal{H}^\perp)$ is the *dual* of S. Prove that T is subnormal by identifying a normal extension.

Exercise 19.6.22. Let S denote the unilateral shift on H^2. Prove that its dual T, as defined in Exercise 19.6.21, is given by $(Tf)(\xi) = \bar{\xi}f(\xi)$ on H_0^2.

Exercise 19.6.23. Let S and T be as in Exercise 19.6.22. Use the following steps to prove that S is unitarily equivalent to T. Such subnormal operators are *self dual*.

(a) Define $W \in B(L^2(\mathbb{T}))$ by $(Wf)(\xi) = \bar{\xi}f(\bar{\xi})$ and prove that W is selfadjoint, $W^2 = I$, and $WM_\xi W = M_{\bar{\xi}}$ on $L^2(\mathbb{T})$.

(b) Prove that $WH^2 = \overline{H_0^2}$.

(c) If $U = W|_{H^2}$, prove that U is unitary and $USU^{-1} = T$.

Exercise 19.6.24. Let $\varphi \in H^\infty$ and let $T_\varphi f = \varphi f$ be the corresponding analytic Toeplitz operator on H^2. Identify the dual of T_φ, as defined in Exercise 19.6.21.

Exercise 19.6.25. Consider $S = M_z$ on the Bergman space A^2.

(a) Prove that if $g \in C_0^\infty(\mathbb{D})$ (the smooth functions with compact support in $\mathbb{D}$) and $\partial_z = \frac{1}{2}(\partial_x - i\partial_y)$, then $\int_{\mathbb{D}} \overline{f} \partial_z g \, dA = 0$ for all $f \in A^2$.

(b) Prove that $\partial_z C_0^\infty(\mathbb{D}) \subseteq (A^2)^\perp$. A technical detail called *Weyl's lemma* [95, p. 172] says that $(A^2)^\perp$ is the $L^2(dA)$ closure of $\partial_z C_0^\infty(\mathbb{D})$.

(c) Prove that the dual (as defined in Exercise 19.6.21) of the Bergman shift is multiplication by $\bar{z}$ on $(A^2)^\perp$.

Exercise 19.6.26. $A \in B(\mathcal{H})$ is *posinormal* if $AA^* = A^*PA$ for some positive operator P called an *interrupter*.

(a) Prove that the unilateral shift S on ℓ^2 is posinormal but S^* is not.

(b) Prove that every unilateral weighted shift W is posinormal.

(c) If A is normal, prove that $\|Ax\| = \|A^*x\|$ for all $x \in \mathcal{H}$.

(d) If A is posinormal with interrupter P, prove that $\|A^*x\| \leqslant \|P^{\frac{1}{2}}\| \|Ax\|$ for all $x \in \mathcal{H}$.

(e) Prove that $\|P^{\frac{1}{2}}A\| = \|A\|$.

Remark: Exercise 6.7.19 examined these operators in relation to the Cesàro operator [290].

Exercise 19.6.27. This is a continuation of Exercise 19.6.26. Use Corollary 14.6.6 to prove that the following are equivalent.

(a) A is posinormal.

(b) $\operatorname{ran} A \subseteq \operatorname{ran} A^*$.

(c) $AA^* \leqslant \lambda^2 A^* A$ for some $\lambda > 0$.

(d) There is a $T \in \mathcal{B}(\mathcal{H})$ such that $A = A^* T$.

Exercise 19.6.28. This is a continuation of Exercise 19.6.26. Use Exercise 19.6.27 to prove the following.

(a) Every hyponormal operator is posinormal.

(b) A is a posinormal operator if and only if A^* belongs to the left ideal in $\mathcal{B}(\mathcal{H})$ generated by A.

(c) Every invertible operator is posinormal.

Exercise 19.6.29. For $\varphi \in H^\infty$, the analytic Toeplitz operator T_φ on H^2 is subnormal (and hence hyponormal). By Exercise 19.6.28, T_φ is posinormal, and hence there is a $T \in \mathcal{B}(H^2)$ such that $T_\varphi = T_\varphi^* T$. Prove that one can choose T to be a Toeplitz operator.

19.7 Hints for the Exercises

Hint for Ex. 19.6.7: For (c), use the closed graph theorem.
Hint for Ex. 19.6.9: Write a normal extension of S as

$$N = \begin{bmatrix} S & X \\ 0 & T \end{bmatrix}$$

and use the identity $N^*N - NN^* = 0$.
Hint for Ex. 19.6.13: For (e), observe that $Sx = (A + iB)x = (A + i\lambda I)x$. For (g), consult Theorem 2.6.7.
Hint for Ex. 19.6.14: Consult Exercise 19.6.13 and use the fact that if N is a normal extension of S, then $\|Sx\| = \|S^*x\|$ if and only if $\|N^*x\| = \|S^*x\|$.
Hint for Ex. 19.6.19: Apply the condition for quasinormality to $k_0 \equiv 1$ and consult Proposition 18.5.1.
Hint for Ex. 19.6.21: Write the matrix of N^* with respect to the decomposition $\mathcal{H}^\perp \oplus \mathcal{H}$.
Hint for Ex. 19.6.25: For (a), prove and use the following version of Green's theorem: for h sufficiently smooth on $\mathbb{D}^-$,

$$-2i \int_{\mathbb{D}} \partial_z h \, dA = \oint_{\mathbb{T}} h \, d\bar{z}.$$

Hint for Ex. 19.6.26: For (a), examine the upper-left corner of S^*S and SPS^* for any positive operator P. For (b), create the interrupter P with a diagonal matrix.
Hint for Ex. 19.6.29: Consult Exercise 16.9.15.

20

. . • . .

The Compressed Shift

Key Concepts: Model space, compressed shift, conjugation, complex symmetric operator, the Volterra operator, matrix representation of a compressed shift, commutant of a compressed shift.

Outline: This chapter examines compressions of the unilateral shift S on H^2 to model spaces $H^2 \cap (uH^2)^\perp$, where u is an inner function. These compressed shift operators provide concrete, function-theoretic models for certain Hilbert-space contractions. As a guiding example, this chapter focuses on the compressed shift on the model space corresponding to an atomic inner function.

20.1 Model Spaces

Recall (Chapter 5) that an inner function u is a bounded analytic function on $\mathbb{D}$ such that

$$u(\xi) = \lim_{r \to 1^-} u(r\xi)$$

satisfies $|u(\xi)| = 1$ for almost every $\xi \in \mathbb{T}$. The limit above exists for almost every $\xi \in \mathbb{T}$ by Fatou's theorem (Theorem 5.4.3). By Theorem 5.4.11, every inner function takes the form $B(z)S_\mu(z)$ with

$$B(z) = \gamma z^N \prod_{i=1}^{\infty} \frac{\overline{a_i}}{|a_i|} \frac{a_i - z}{1 - \overline{a_i}z}$$

and

$$S_\mu(z) = \exp\left(-\int_{\mathbb{T}} \frac{\xi + z}{\xi - z} d\mu(\xi)\right),$$

where $\gamma \in \mathbb{T}$, $N \geqslant 0$, $a_i \in \mathbb{D}\setminus\{0\}$ with $\sum_{i=1}^{\infty}(1 - |a_i|) < \infty$, and μ is a finite positive Borel measure on $\mathbb{T}$ that is singular with respect to Lebesgue measure m. Beurling's theorem (Theorem 5.4.12) says that all nonzero S-invariant subspaces of H^2 are of the form uH^2 where u is inner.

Definition 20.1.1. The *model space* $\mathcal{K}_u$ corresponding to a nonconstant inner function u is the subspace of H^2 defined by

$$\mathcal{K}_u := (uH^2)^\perp = \{f \in H^2 : \langle f, uh \rangle = 0 \text{ for all } h \in H^2\}.$$

Since $S(uH^2) \subseteq uH^2$, it follows that $S^*\mathcal{K}_u \subseteq \mathcal{K}_u$ (Exercise 4.5.1). We invite the reader to work out examples of $\mathcal{K}_u$ for various inner functions u (see Exercises 20.8.18 and 20.8.19) such as finite Blaschke products. The following proposition provides a useful way of describing $\mathcal{K}_u$ in terms of boundary functions on $\mathbb{T}$.

Proposition 20.1.2. *For an inner function u and $f \in H^2$, the following are equivalent.*

(a) $f \in \mathcal{K}_u$.

(b) *There exists a $g \in H^2$ such that $f(\xi) = \overline{g(\xi)\xi}u(\xi)$ for almost every $\xi \in \mathbb{T}$.*

In other words, $\mathcal{K}_u = H^2 \cap u\overline{zH^2}$, where the right side is regarded, via radial boundary values, as a set of functions on $\mathbb{T}$.

Proof By means of radial boundary values, the inner product on H^2 can be written as an $L^2(\mathbb{T})$ inner product

$$\langle f, g \rangle = \int_\mathbb{T} f(\xi)\overline{g(\xi)}\, dm(\xi) \quad \text{for } f, g \in H^2.$$

For each $f \in H^2$,

$$f \in \mathcal{K}_u \iff \langle f, uh \rangle = 0 \text{ for all } h \in H^2$$
$$\iff \langle \overline{u}f, h \rangle = 0 \text{ for all } h \in H^2$$
$$\iff \overline{u}f \in (H^2)^\perp = \overline{zH^2}. \qquad \text{(by (4.2.4))}$$

Since $u\overline{u} = 1$ almost everywhere on $\mathbb{T}$, it follows that $f \in H^2$ belongs to $(uH^2)^\perp$ if and only if $f \in u\overline{zH^2}$ (that is, $f = \overline{gz}u$ for some g in H^2). ∎

Proposition 20.1.6 below asserts that the function g in the statement of the previous proposition also belongs to $\mathcal{K}_u$.

Model spaces are reproducing kernel Hilbert spaces. This stems from the fact that any subspace of a reproducing kernel Hilbert space is a reproducing kernel Hilbert space (Exercise 20.8.3). For a model space, there is a precise formula for the reproducing kernel. For an inner function u, define

$$k_\lambda(z) = \frac{1 - \overline{u(\lambda)}u(z)}{1 - \overline{\lambda}z} \quad \text{for } \lambda, z \in \mathbb{D}. \qquad (20.1.3)$$

Proposition 20.1.4. *For $\lambda \in \mathbb{D}$ and $f \in \mathcal{K}_u$, the following hold.*

(a) $k_\lambda \in \mathcal{K}_u$.

(b) $\langle f, k_\lambda \rangle = f(\lambda)$.

(c) $\|k_\lambda\|^2 = \dfrac{1 - |u(\lambda)|^2}{1 - |\lambda|^2}$.

Proof For the proof below, let

$$c_\lambda(z) = \frac{1}{1 - \bar\lambda z} \qquad (20.1.5)$$

denote the reproducing kernel for H^2.

(a) Observe that

$$k_\lambda = (1 - \overline{u(\lambda)}u)c_\lambda$$

and thus $k_\lambda \in H^\infty \subseteq H^2$. Moreover, for any $h \in H^2$,

$$
\begin{aligned}
\langle uh, (1 - \overline{u(\lambda)}u)c_\lambda \rangle &= u(\lambda)h(\lambda) - u(\lambda)\langle uh, uc_\lambda \rangle \\
&= u(\lambda)h(\lambda) - u(\lambda)\langle h, c_\lambda \rangle \qquad \text{(since } u\bar u = 1 \text{ a.e. on } \mathbb{T}) \\
&= u(\lambda)h(\lambda) - u(\lambda)h(\lambda) \\
&= 0.
\end{aligned}
$$

Therefore, $k_\lambda \in (uH^2)^\perp = \mathcal{K}_u$.

(b) If $f \in \mathcal{K}_u$ and $\lambda \in \mathbb{D}$, it follows that

$$
\begin{aligned}
f(\lambda) &= \langle f, c_\lambda \rangle \\
&= \langle f, c_\lambda \rangle - u(\lambda)\langle f, uc_\lambda \rangle \qquad \text{(since } \langle f, uc_\lambda \rangle = 0) \\
&= \langle f, (1 - \overline{u(\lambda)}u)c_\lambda \rangle \\
&= \langle f, k_\lambda \rangle.
\end{aligned}
$$

(c) For $\lambda \in \mathbb{D}$,

$$
\begin{aligned}
\|k_\lambda\|^2 &= \langle k_\lambda, k_\lambda \rangle \\
&= k_\lambda(\lambda) \qquad \text{(by (b))} \\
&= \frac{1 - |u(\lambda)|^2}{1 - |\lambda|^2},
\end{aligned}
$$

which completes the proof. ∎

There is a natural conjugation on the model space that serves as an important tool for studying the compressed shift. In what follows, we regard the model space $\mathcal{K}_u$ as a subspace of $L^2(\mathbb{T})$.

Proposition 20.1.6. *For an inner function u, the mapping $C : \mathcal{K}_u \to \mathcal{K}_u$ defined by $Cf = u\bar z \bar f$, is well defined, conjugate linear, isometric, and involutive.*

Proof Since $u\bar u = 1$ almost everywhere on $\mathbb{T}$, it follows that C is conjugate linear, isometric, and involutive on $L^2(\mathbb{T})$ (Exercise 20.8.4). It suffices to prove that C maps $\mathcal{K}_u$ onto $\mathcal{K}_u$. For $f \in K_u$, Proposition 20.1.2 produces a $g \in H^2$ such that $f = u\bar z \bar g$ on $\mathbb{T}$. Then

$$Cf = u\bar z \bar f = u\bar z \overline{u\bar z \bar g} = g \in H^2.$$

By definition, $Cf \in u\bar z\overline{H^2}$ and thus $Cf \in H^2 \cap u\bar z\overline{H^2} = \mathcal{K}_u$. ∎

20.2 From a Model Space to $L^2[0, 1]$

Let

$$\Theta(z) = \exp\left(-\frac{1+z}{1-z}\right). \tag{20.2.1}$$

Observe that Θ is analytic on $\mathbb{C}\backslash\{1\}$ and

$$|\Theta(z)| = \exp\left(-\text{Re}\left(\frac{1+z}{1-z}\right)\right) = \exp\left(-\frac{1-|z|^2}{|1-z|^2}\right) < 1 \quad \text{for all } z \in \mathbb{D}.$$

Thus, $\Theta \in H^\infty$. Since $|\Theta(e^{i\theta})| = 1$ for all $\theta \in (0, 2\pi)$, it follows that Θ is an inner function. When $\theta = 0$, the expression for $\Theta(1)$ is undefined. However,

$$\lim_{r \to 1^-} |\Theta(r)| = \lim_{r \to 1^-} \exp\left(-\frac{1+r}{1-r}\right) = 0.$$

There is an essential singularity at $z = 1$ so the Casorati–Weierstrass theorem [9] suggests that the limit along other paths in $\mathbb{D}$ approaching $z = 1$ need not be zero (Exercise 20.8.1).

An important part of our analysis of the model space $\mathcal{K}_\Theta$, where Θ is the function (20.2.1), is the following unitary operator of Sarason. For $t \in [0, 1]$,

$$\Theta(z)^t = \exp\left(-t\frac{1+z}{1-z}\right)$$

is an inner function. To understand the proof of Sarason's result, the reader might want to review the Hardy space of the upper half-plane (Chapter 11).

Theorem 20.2.2 (Sarason [323]). *For each $g \in L^2[0, 1]$, the function*

$$(Wg)(z) = \frac{\sqrt{2}i}{z-1}\int_0^1 g(t)\Theta(z)^t dt, \tag{20.2.3}$$

defined for all $z \in \mathbb{D}$, belongs to $\mathcal{K}_\Theta$. Furthermore, $g \mapsto Wg$ is a unitary operator from $L^2[0, 1]$ onto $\mathcal{K}_\Theta$.

Proof Exercise 11.10.7 ensures that

$$(Af)(\xi) = \frac{2\sqrt{\pi}i}{1-\xi}f\left(i\frac{1+\xi}{1-\xi}\right) \tag{20.2.4}$$

is a unitary operator from $L^2(\mathbb{R})$ onto $L^2(\mathbb{T})$. Furthermore, $AH^2(\mathbb{C}_+) = H^2(\mathbb{D})$. Regard $L^2(0, \infty)$ as a subspace of $L^2(\mathbb{R})$ via the identification $L^2(0, \infty) = \chi_{(0,\infty)}L^2(\mathbb{R})$. For the Fourier transform $\mathscr{F}$ and its adjoint $\mathscr{F}^*$ (Chapter 11), define

$$B = \mathscr{F}^*|_{L^2(0,\infty)} \tag{20.2.5}$$

and observe that B is an isometry from $L^2(0, \infty)$ onto $H^2(\mathbb{C}_+)$ (recall the Paley–Wiener theorem (11.8.2)). Thus, the composition $U = AB$ is a unitary operator from $L^2(0, \infty)$ onto $H^2(\mathbb{D})$.

The next step is to argue that U maps $L^2[0,1] = \chi_{[0,1]}L^2(0,\infty)$ onto $\mathcal{K}_\Theta$. Suppose that f belongs to $\chi_{[1,\infty)}L^2(0,\infty)$. By the Paley–Wiener theorem again, $\mathscr{F}^*f \in H^2(\mathbb{C}_+)$ and

$$(\mathscr{F}^*f)(z) = \frac{1}{\sqrt{2\pi}}\int_1^\infty f(t)e^{itz}\,dt$$

$$= \frac{1}{\sqrt{2\pi}}\int_0^\infty f(t+1)e^{i(t+1)z}\,dt$$

$$= e^{iz}\int_0^\infty f(t+1)e^{itz}\,dt.$$

Thus, $\mathscr{F}^*\chi_{[1,\infty)}L^2(0,\infty) = e^{iz}H^2(\mathbb{C}_+)$, and hence

$$U\chi_{[1,\infty)}L^2(0,\infty) = AB\chi_{[1,\infty)}L^2(0,\infty)$$

$$= A(e^{iz}H^2(\mathbb{C}_+))$$

$$= \exp\left(-\frac{1+z}{1-z}\right)H^2(\mathbb{D})$$

$$= \Theta H^2(\mathbb{D}).$$

From here, observe that

$$UL^2[0,1] = U\big(L^2(0,\infty) \ominus \chi_{[1,\infty)}L^2(0,\infty)\big)$$

$$= UL^2(0,\infty) \ominus U\big(\chi_{[1,\infty)}L^2(0,\infty)\big)$$

$$= H^2 \ominus \Theta H^2$$

$$= \mathcal{K}_\Theta.$$

Finally, for $g \in L^2[0,1]$,

$$(Ug)(z) = (ABg)(z)$$

$$= A\left(\frac{1}{\sqrt{2\pi}}\int_0^1 e^{izt}g(t)\,dt\right)$$

$$= \frac{1}{\sqrt{2\pi}}\,2\sqrt{\pi}i\,\frac{1}{1-z}\int_0^1 \exp\left(iti\frac{1+z}{1-z}\right)g(t)\,dt$$

$$= \frac{\sqrt{2}i}{1-z}\int_0^1 \Theta(z)^t g(t)\,dt,$$

which is the formula (20.2.3) for Wg. ∎

20.3 The Compressed Shift

For an inner function u, the model space $\mathcal{K}_u$ can be regarded as a subspace of $L^2(\mathbb{T})$. Let P_u denote the orthogonal projection from $L^2(\mathbb{T})$ onto $\mathcal{K}_u$ and recall the Riesz projection P_+ from $L^2(\mathbb{T})$ onto H^2 (Definition 16.2.1). The next result furnishes two formulas for P_u.

Proposition 20.3.1. *Let u be an inner function and let k_λ be the reproducing kernel for $\mathcal{K}_u$.*

(a) $(P_u f)(\lambda) = \langle f, k_\lambda \rangle$ *for every $\lambda \in \mathbb{D}$ and $f \in L^2(\mathbb{T})$.*

(b) $P_u f = f - u P_+(\overline{u} f)$ *for every $f \in H^2$.*

Proof (a) Since $k_\lambda \in \mathcal{K}_u$, it follows that $P_u k_\lambda = k_\lambda$, and hence for any $f \in L^2(\mathbb{T})$,

$$
\begin{aligned}
\langle f, k_\lambda \rangle &= \langle f, P_u k_\lambda \rangle \\
&= \langle P_u f, k_\lambda \rangle && (P_u \text{ is selfadjoint}) \\
&= (P_u f)(\lambda) && (\text{by Proposition 20.1.4b}).
\end{aligned}
$$

(b) From (20.1.5), recall that $c_\lambda(z) = (1 - \overline{\lambda} z)^{-1}$. For $f \in H^2$ and $\lambda \in \mathbb{D}$, it follows from (a) that

$$
\begin{aligned}
(P_u f)(\lambda) &= \langle f, k_\lambda \rangle \\
&= \langle f, (1 - \overline{u(\lambda)} u) c_\lambda \rangle \\
&= \langle f, c_\lambda \rangle - u(\lambda) \langle f, u c_\lambda \rangle \\
&= \langle f, c_\lambda \rangle - u(\lambda) \langle \overline{u} f, c_\lambda \rangle \\
&= \langle f, c_\lambda \rangle - u(\lambda) \langle \overline{u} f, P_+ c_\lambda \rangle && (P_+ c_\lambda = c_\lambda) \\
&= \langle f, c_\lambda \rangle - u(\lambda) \langle P_+(\overline{u} f), c_\lambda \rangle && (P_+ \text{ is selfadjoint}) \\
&= f(\lambda) - u(\lambda) P_+(\overline{u} f)(\lambda),
\end{aligned}
$$

which completes the proof. ∎

Recall the unilateral shift operator S from Chapter 5.

Definition 20.3.2. For an inner function u, the corresponding *compressed shift operator* is $S_u : \mathcal{K}_u \to \mathcal{K}_u$, where $S_u = P_u S|_{\mathcal{K}_u}$.

The following proposition shows that the adjoint of the compressed shift is the backward shift S^* restricted to $\mathcal{K}_u$.

Proposition 20.3.3. $S_u^* f = S^* f$ *for all $f \in \mathcal{K}_u$.*

Proof For $f, g \in \mathcal{K}_u$,

$$
\langle S_u^* f, g \rangle = \langle f, S_u g \rangle = \langle f, P_u S g \rangle = \langle P_u f, S g \rangle = \langle f, S g \rangle = \langle S^* f, g \rangle.
$$

Thus, $S_u^* f = S^* f$ for all $f \in \mathcal{K}_u$. ∎

Proposition 20.3.4. *Let C denote the conjugation from Proposition 20.1.6. Then $S_u = C S_u^* C$.*

Proof If $f, g \in \mathcal{K}_u$ then,

$$
\langle C S_u^* C f, g \rangle = \langle C g, S_u^* C f \rangle
$$

$$= \langle Cg, S^*Cf \rangle \qquad \text{(by Proposition 20.3.3)}$$

$$= \langle SCg, Cf \rangle$$

$$= \left\langle \xi u \overline{\xi} g, u \overline{\xi} f \right\rangle$$

$$= \left\langle \xi \overline{g}, \overline{f} \right\rangle$$

$$= \langle \xi f, g \rangle$$

$$= \langle Sf, g \rangle$$

$$= \langle Sf, P_u g \rangle \qquad (P_u g = g \text{ since } g \in \mathcal{K}_u)$$

$$= \langle P_u Sf, g \rangle$$

$$= \langle S_u f, g \rangle,$$

which proves the desired identity. ∎

20.4 A Connection to the Volterra Operator

Recall the Volterra operator

$$(Vg)(x) = \int_0^x g(t)\, dt$$

on $L^2[0,1]$ from Chapter 7. Since $\sigma(V) = \{0\}$ (Proposition 7.2.5), it follows that $I + V$ is invertible and hence $(I - V)(I + V)^{-1}$ is a well-defined bounded operator on $L^2[0,1]$. A fascinating result of Sarason relates this with the compressed shift S_Θ via the unitary operator $W : L^2[0,1] \to \mathcal{K}_\Theta$ from Theorem 20.2.2.

Theorem 20.4.1 (Sarason [323]). $W^* S_\Theta W = (I - V)(I + V)^{-1}$.

Proof We follow a presentation from [280, Ch. 4]. From (20.2.4)

$$(Af)(\xi) = \frac{2\sqrt{\pi i}}{1 - \xi} f\left(i \frac{1 + \xi}{1 - \xi} \right)$$

is a unitary operator from $L^2(\mathbb{R})$ onto $L^2(\mathbb{T})$ and maps $H^2(\mathbb{C}_+)$ onto $H^2(\mathbb{D})$. Invert the formula for A to obtain

$$(A^*g)(x) = \frac{1}{\sqrt{\pi}(x + i)} g\left(\frac{x - i}{x + i} \right) \quad \text{for all } g \in L^2(\mathbb{T}).$$

From (20.2.5), the operator $B = \mathscr{F}^*|_{L^2(0,\infty)}$ is a unitary operator from $L^2(0, \infty)$ onto $H^2(\mathbb{C}_+)$. Thus, $U = AB$ is a unitary operator from $L^2(0, \infty)$ onto $H^2(\mathbb{D})$.

Let

$$\psi(x) = \begin{cases} 0 & \text{if } x < 0, \\ e^{-x} & \text{if } x \geqslant 0, \end{cases}$$

and define the convolution operator (Chapter 11) $X : L^2(0, \infty) \to L^2(0, \infty)$ by

$$(Xg)(x) = \int_0^\infty g(t)\psi(x - t)\,dt.$$

In other words,

$$Xg = g * \psi. \tag{20.4.2}$$

For each $w \in \mathbb{C}_+$, observe that

$$(B\psi)(w) = \frac{1}{\sqrt{2\pi}} \int_0^\infty e^{iwt} e^{-t}\,dt = \frac{1}{\sqrt{2\pi}} \cdot \frac{1}{1 - iw}. \tag{20.4.3}$$

Then for any $f \in H^2(\mathbb{D})$,

$$\begin{aligned}
UXU^*f &= ABXB^*A^*f \\
&= AB\big((B^*A^*f) * \psi\big) &\text{(by (20.4.2))} \\
&= \sqrt{2\pi}A\big((BB^*A^*f)B\psi\big) &\text{(by Proposition 11.3.1)} \\
&= A\Big((A^*f)\frac{1}{1 - iw}\Big) &\text{(by (20.4.3))}.
\end{aligned}$$

Now observe that for each $z \in \mathbb{D}$,

$$\begin{aligned}
\Big(A\big(A^*f\frac{1}{1 - iw}\big)\Big)(z) &= \frac{2\sqrt{\pi}i}{1 - z}(A^*f)\Big(i\frac{1 + z}{1 - z}\Big)\frac{1}{1 - i^2\frac{1+z}{1-z}} \\
&= \sqrt{\pi}i(A^*f)\Big(i\frac{1 + z}{1 - z}\Big) \\
&= \sqrt{\pi}i \cdot \frac{1}{\sqrt{\pi}} \cdot \frac{1}{i\frac{1+z}{1-z} + i}f(z) \\
&= \frac{1 - z}{2}f(z).
\end{aligned}$$

Thus, $UXU^* = \frac{1}{2}(I - S)$. Notice that $W = U|_{L^2[0,1]} = P_\Theta U|_{L^2[0,1]}$ and $U^*|_{\mathcal{K}_\Theta} = W^*$. From here it follows that

$$WXW^* = P_\Theta UXU^*|_{\mathcal{K}_\Theta} = \frac{1}{2}(I - S_\Theta). \tag{20.4.4}$$

Proposition 7.2.9 yields

$$((I + V)^{-1}f)(x) = f(x) - \int_0^x e^{y-x}f(y)\,dy \quad \text{for } f \in L^2[0, 1].$$

This can be written as $(I + V)^{-1}f = (I - X)f$ for $f \in L^2[0, 1]$. Combining this with (20.4.4) yields

$$(I + V)^{-1} = W^*\Big(\frac{1}{2}(I + S_\Theta)\Big)W.$$

Finally,

$$
\begin{aligned}
(I - V)(I + V)^{-1} &= (2I - (I + V))(I + V)^{-1} \\
&= 2(I + V)^{-1} - I \\
&= 2W^* \frac{1}{2}(I + S_\Theta)W - W^* I W \\
&= W^*(I + S_\Theta - I)W \\
&= W^* S_\Theta W,
\end{aligned}
$$

which completes the proof. ∎

Corollary 20.4.5. $\sigma(S_\Theta) = \{1\}$.

Proof Proposition 7.2.5 says that $\sigma(V) = \{0\}$. Since

$$
w(z) = \frac{1 - z}{1 + z}
$$

is analytic in a neighborhood of 0, the spectral mapping theorem says that $w(V)$ is a bounded operator with $\sigma(w(V)) = w(\sigma(V)) = \{1\}$. ∎

There is a connection between the invariant subspaces for S_Θ and those for V. Indeed, one can check that $W\chi_{[a,1]}L^2[0,1] = \Theta^a H^2 \cap \mathcal{K}_\Theta$ for each $a \in [0,1]$. Agmon's theorem (Theorem 7.4.1) describes all of the invariant subspaces of V as $W\chi_{[a,1]}L^2[0,1]$. This connects with a description of the invariant subspaces of S_Θ as $\{\Theta^a H^2 \cap \mathcal{K}_\Theta : 0 \leqslant a \leqslant 1\}$.

20.5 A Basis for the Model Space

We follow a presentation from [140] to produce a basis for $\mathcal{K}_\Theta$ which respects the natural conjugation C. This instructional computation makes use of the Volterra operator. Proposition 7.1.5 says that

$$
(V^* g)(x) = \int_x^1 g(t)\, dt
$$

and Exercise 7.7.15 says that the Volterra operator is complex symmetric with respect to the conjugation $(Jg)(x) = \overline{g(1 - x)}$ on $L^2[0,1]$, that is, $V = JV^* J$.

Lemma 20.5.1. *The conjugation $(Jg)(x) = \overline{g(1 - x)}$ on $L^2[0,1]$ and the conjugation $Cf = \overline{fz\Theta}$ on $\mathcal{K}_\Theta$ are related by the unitary operator W from (20.2.3) via the identity $C = WJW^*$.*

Proof We prove the lemma by establishing that $WJ = CW$. For any g in $L^2[0,1]$, the integrands of the following integrals are all dominated by $\max\{|g(t)|, |g(1-t)|\}$ and it follows that

$$
(WJg)(z) = \frac{\sqrt{2}i}{z - 1} \int_0^1 (Jg)(t)[\Theta(z)]^t\, dt
$$

$$= \frac{\sqrt{2}i}{z-1} \int_0^1 \overline{g(1-t)[\Theta(z)]^t} \, dt$$

$$= \frac{\sqrt{2}i}{z-1} \int_0^1 \overline{g(s)[\Theta(z)]^{1-s}} \, ds$$

$$= \overline{z}\Theta(z) \frac{\sqrt{2}i}{1-\overline{z}} \overline{\int_0^1 g(s)[\Theta(z)]^s \, ds}$$

$$= \overline{(Wg)(z)} z\Theta(z)$$

$$= (CWg)(z)$$

for almost every $z \in \mathbb{T}$. ∎

Suppose that $\mathcal{H}$ is a Hilbert space with a conjugation C. An orthonormal basis $(\mathbf{x}_n)_{n=1}^\infty$ for $\mathcal{H}$ is C-real if $C\mathbf{x}_n = \mathbf{x}_n$ for all $n \geq 1$. It is known that every Hilbert space with a conjugation has an orthonormal basis that is C-real (Exercise 20.8.26). The following is a J-real basis for $L^2[0,1]$.

Lemma 20.5.2. *The vectors*

$$e_n(x) = e^{2\pi i n(x-\frac{1}{2})} \quad \text{for } n \in \mathbb{Z}, \tag{20.5.3}$$

form a J-real orthonormal basis of $L^2[0,1]$.

Proof The following computation shows that each e_n is fixed by J:

$$(Je_n)(x) = \overline{e^{2\pi i n((1-x)-\frac{1}{2})}} = e^{-i2\pi n(\frac{1}{2}-x)} = e^{2\pi i n(x-\frac{1}{2})} = e_n(x).$$

To see that $(e_n)_{n=-\infty}^\infty$ is an orthonormal basis for $L^2[0,1]$, observe the identity

$$e^{2\pi i n(x-\frac{1}{2})} = e^{i(2\pi nx - \pi n)} = (-1)^n e^{2\pi i nx}.$$

Now use the fact that $(e^{2\pi i nx})_{n=-\infty}^\infty$ is an orthonormal basis for $L^2[0,1]$ (Theorem 1.3.9) to complete the proof. ∎

Next we compute $We_n \in \mathcal{K}_\Theta$. Write each basis vector e_n in the form

$$e_n(x) = e^{-2\pi i n/2} e^{2\pi i nx} \tag{20.5.4}$$

to see that it suffices to compute $e^{-i\gamma/2} W e^{i\gamma t}$ for real γ. Once this is done, substitute $\gamma = 2\pi n$ to find the corresponding basis for $\mathcal{K}_\Theta$. This somewhat long computation is presented through a series of lemmas.

The following lemma shows that the image of the basis $(e_n)_{n=-\infty}^\infty$ in $\mathcal{K}_\Theta$ is a family of reproducing kernels corresponding to a sequence of points on the unit circle. Since Θ

is analytically continuable across any arc of the unit circle $\mathbb{T}$ not containing $z = 1$, the reproducing kernel

$$k_\zeta(z) = \frac{1 - \overline{\Theta(\zeta)}\Theta(z)}{1 - \bar\zeta z}$$

corresponding to any boundary point $\zeta \neq 1$ in $\mathbb{T}$ is well defined and belongs to $\mathcal{K}_\Theta$.

Lemma 20.5.5. *If γ is real, then*

$$(We^{i\gamma x})(z) = \frac{(1 - \zeta)}{\sqrt{2i\zeta}}k_\zeta(z),$$

where $k_\zeta(z)$ is the reproducing kernel for $\mathcal{K}_\Theta$ corresponding to the point ζ on $\mathbb{T}$ defined by

$$\zeta = \frac{\gamma + i}{\gamma - i}.$$

In particular, $\Theta(\zeta) = e^{-i\gamma}$.

Proof Fix γ in $\mathbb{R}$, use the definition of W, then observe that

$$
\begin{aligned}
(We^{i\gamma x})(z) &= \frac{\sqrt{2i}}{z - 1}\int_0^1 e^{i\gamma t}\exp\left(t\frac{z+1}{z-1}\right)dt \\
&= \frac{\sqrt{2i}}{z - 1}\int_0^1 \exp\left[t\left(i\gamma + \frac{z+1}{z-1}\right)\right]dt \\
&= \frac{\sqrt{2i}}{z - 1}\frac{1}{i\gamma + \frac{z+1}{z-1}}\left(\exp\left[\left(i\gamma + \frac{z+1}{z-1}\right)\right] - 1\right) \\
&= \frac{\sqrt{2i}}{i\gamma(z - 1) + (z + 1)}\left(e^{i\gamma}\Theta(z) - 1\right) \\
&= \frac{-\sqrt{2i}}{(1 - i\gamma) + (1 + i\gamma)z}\left(1 - \overline{e^{-i\gamma}\Theta(z)}\right) \\
&= \frac{-\sqrt{2i}}{1 - i\gamma}\frac{1}{1 + \frac{1+i\gamma}{1-i\gamma}z}\left(1 - \overline{e^{-i\gamma}\Theta(z)}\right). \tag{20.5.6}
\end{aligned}
$$

Seeking to write (20.5.6) as a scalar multiple of a reproducing kernel, we wish to write $e^{-i\gamma} = \Theta(\zeta)$ for some ζ. Define

$$\zeta = \frac{\gamma + i}{\gamma - i}. \tag{20.5.7}$$

If follows that $\zeta \in \mathbb{T}$, $\Theta(\zeta) = e^{-i\gamma}$, and

$$-\bar\zeta = \frac{1 + i\gamma}{1 - i\gamma}.$$

Substitute these values into (20.5.6) and obtain

$$(We^{i\gamma x})(z) = \frac{-\sqrt{2}i}{1-i\gamma} \cdot \frac{1 - \overline{\Theta(\zeta)}\Theta(z)}{1 - \bar{\zeta}z}$$

$$= \frac{-\sqrt{2}i}{1 + \frac{\zeta+1}{\zeta-1}}k_\zeta(z)$$

$$= \frac{-\sqrt{2}i(\zeta - 1)}{(\zeta - 1) + (\zeta + 1)}k_\zeta(z)$$

$$= \frac{-\sqrt{2}i(\zeta - 1)}{2\zeta}k_\zeta(z)$$

$$= \frac{(1 - \zeta)}{\sqrt{2}i\zeta}k_\zeta(z), \tag{20.5.8}$$

where $k_\zeta(z)$ denotes the boundary kernel evaluated at the boundary point ζ. This proves the desired formula. ∎

Since W is a unitary operator, the image of $We^{i\gamma x}$ in $\mathcal{K}_\Theta$ is a unit vector. This suggests that we could simplify the expression (20.5.8) further to examine the constant appearing in front of $k_\zeta(z)$.

Lemma 20.5.9. *If γ is real, then*

$$(We^{i\gamma x})(z) = \frac{(1 - \zeta)}{|1 - \zeta|\zeta i}\widetilde{k}_\zeta(z),$$

where $\widetilde{k}_\zeta(z)$ is the normalized reproducing kernel for $\mathcal{K}_\Theta$ corresponding to ζ.

Proof Since $\gamma \neq \infty$, we have $\zeta \neq 1$ and $\|k_\zeta\|$ can be explicitly computed in terms of Θ. In fact, if z approaches ζ radially we have

$$\|k_\zeta\|^2 = \lim_{z \to \zeta} \langle k_z, k_z \rangle$$

$$= \lim_{z \to \zeta} \frac{1 - |\Theta(z)|^2}{1 - |z|^2}$$

$$= \lim_{z \to \zeta} \frac{1 - |\Theta(z)|}{1 - |z|} \cdot \frac{1 + |\Theta(z)|}{1 + |z|}$$

$$= \lim_{z \to \zeta} \frac{1 - |\Theta(z)|}{1 - |z|}$$

$$= |\Theta'(\zeta)|.$$

A computation using the definition of Θ shows that

$$\|k_\zeta\| = \frac{\sqrt{2}}{|1 - \zeta|}.$$

With this in mind, continue from (20.5.8) and find that

$$(We^{i\gamma x})(z) = \frac{(1-\zeta)}{|1-\zeta|\zeta i} \cdot \frac{|1-\zeta|}{\sqrt{2}} k_\zeta(z)$$

$$= \frac{(1-\zeta)}{|1-\zeta|\zeta i} \widetilde{k}_\zeta(z).$$

This proves the desired formula. ∎

Although it is evident from the preceding lemma that $We^{i\gamma x}$ is a unit vector in $\mathcal{K}_\Theta$, this was not our primary objective. Indeed, recall that we wanted to show that for any real γ, the function $e^{-i\gamma/2}We^{i\gamma x}$ is fixed by the conjugation C. The apparently unwieldy constant

$$\frac{(1-\zeta)}{|1-\zeta|\zeta i} \tag{20.5.10}$$

in Lemma 20.5.9 addresses this issue.

The square of (20.5.10) equals

$$\frac{-(1-\zeta)^2}{(1-\zeta)(1-\overline{\zeta})\zeta^2} = \overline{\zeta},$$

and hence the constant (20.5.10) is one of the square roots of $\overline{\zeta}$.

We have shown that for any real γ,

$$(We^{i\gamma x})(z) = \overline{\zeta}^{\frac{1}{2}} \widetilde{k}_\zeta(z),$$

where ζ satisfies $\Theta(\zeta) = e^{-i\gamma}$. To complete the evaluation of $e^{-i\gamma/2}We^{i\gamma t}$, we use (20.5.7) to describe the value of the constant $e^{-i\gamma/2}$ in terms of Θ:

$$e^{-i\gamma/2} = e^{\frac{1}{2}\frac{\zeta+1}{\zeta-1}} = [\Theta(\zeta)]^{\frac{1}{2}}.$$

The choice of square root is unimportant. The next lemma summarizes our findings.

Lemma 20.5.11. *If γ is real, then*

$$e^{-i\gamma/2}(We^{i\gamma x})(z) = [\overline{\zeta}\Theta(\zeta)]^{\frac{1}{2}} \widetilde{k}_\zeta(z), \tag{20.5.12}$$

where $\widetilde{k}_\zeta(z)$ is the normalized reproducing kernel for $\mathcal{K}_\Theta$ corresponding to the point $\zeta \in \mathbb{T}$. Each function (20.5.12) is fixed by the conjugation $Cf = \overline{fz\Theta}$ on $\mathcal{K}_\Theta$.

Proof The first portion of the lemma is simply a summary of the previous computations. If $z \in \mathbb{T}\backslash\{1\}$, then

$$\frac{\overline{\widetilde{k}_\zeta(z)}}{\widetilde{k}_\zeta(z)} = \frac{1-\overline{\Theta(\zeta)}\Theta(z)}{1-\overline{\zeta}z} \cdot \frac{1-\zeta\overline{z}}{1-\Theta(\zeta)\overline{\Theta(z)}}.$$

$$= \frac{1 - \overline{\Theta(\zeta)}\Theta(z)}{1 - \Theta(\zeta)\overline{\Theta(z)}} \cdot \frac{1 - \overline{\zeta}z}{1 - \overline{\zeta}z}$$

$$= \overline{\Theta(\zeta)}\Theta(z) \cdot \left(\frac{\Theta(\zeta)\overline{\Theta(z)} - 1}{1 - \Theta(\zeta)\overline{\Theta(z)}}\right) \cdot \overline{\zeta}z \left(\frac{\overline{\zeta}z - 1}{1 - \overline{\zeta}z}\right)$$

$$= \zeta\overline{\Theta(\zeta)}\overline{z}\Theta(z),$$

since $\Theta(z)$ is unimodular. This identity, along with a short calculation, shows that $[\overline{\zeta}\Theta(\zeta)]^{\frac{1}{2}} \widetilde{k}_\zeta(z)$ is fixed by C. ∎

Recall that the orthonormal basis $(e_n)_{n=-\infty}^{\infty}$ of $L^2[0,1]$ defined by (20.5.3) is fixed by the conjugation $(Jf)(x) = \overline{f(1-x)}$ on $L^2[0,1]$. In light of (20.5.12), the image of this basis in $\mathcal{K}_\Theta$ under the Sarason transform

$$(Wg)(z) = \frac{\sqrt{2}i}{z-1} \int_0^1 g(t)[\Theta(z)]^t \, dt$$

is

$$(We_n)(z) = [\overline{\zeta_n}\Theta(\zeta_n)]^{\frac{1}{2}} \widetilde{k}_{\zeta_n}(z) = (-1)^n \overline{\zeta_n}^{-\frac{1}{2}} \widetilde{k}_{\zeta_n}(z),$$

where

$$\zeta_n = \frac{2\pi n + i}{2\pi n - i} \in \mathbb{T}.$$

The functions We_n are fixed by the conjugation $Cf = \overline{f}z\Theta$ on $\mathcal{K}_\Theta$, and hence the matrix representation of the compressed shift on $\mathcal{K}_\Theta$ with respect to the basis $(We_n)_{n=-\infty}^{\infty}$ is complex symmetric. Finally, the points ζ_n are characterized by $\Theta(\zeta_n) = 1$.

20.6 A Matrix Representation

In this section we compute the matrix representation of S_Θ with respect to the orthonormal basis $(v_n)_{n=-\infty}^{\infty}$, where

$$v_n = We_n = (-1)^n \overline{\zeta_n}^{-\frac{1}{2}} \widetilde{k}_{\zeta_n}(z),$$

from the previous section. Indeed, since $W^* S_\Theta W = (I - V)(I + V)^{-1}$ and $We_n = v_n$ for all $n \in \mathbb{Z}$, it follows that

$$\langle S_\Theta v_n, v_m \rangle = \langle W(I - V)(I + V)^{-1} W^* v_n, v_m \rangle$$
$$= \langle (I - V)(I + V)^{-1} W^* v_n, W^* v_m \rangle$$
$$= \langle (I - V)(I + V)^{-1} e_n, e_m \rangle$$

for all $m, n \in \mathbb{Z}$. Thus, the matrix representation of S_Θ with respect to the basis $(v_n)_{n=-\infty}^{\infty}$ is the same as the representation of $(I - V)(I + V)^{-1}$ with respect to the basis $(e_n)_{n=-\infty}^{\infty}$.

Theorem 20.6.1. *The matrix representation of* $(I - V)(I + V)^{-1}$ *with respect to the orthonormal basis*

$$e_n(x) = e^{2\pi i n(x-\frac{1}{2})} \quad for\ n \in \mathbb{Z},$$

is

$$\frac{2(e-1)}{e}\left[\frac{(-1)^{m+n}}{(2\pi m - i)(2\pi n - i)}\right]_{m,n=-\infty}^{\infty}.$$

Proof It follows from Proposition 7.2.9 that

$$((I + V)^{-1}f)(x) = f(x) - \int_0^x e^{y-x}f(y)dy \quad for\ f \in L^2[0, 1].$$

Thus,

$$((I - V)(I + V)^{-1}f)(x) = f(x) - \int_0^x e^{y-x}f(y)\,dy$$

$$- V\left(f(x) - \int_0^x e^{y-x}f(y)\,dy\right)$$

$$= f(x) - \int_0^x e^{y-x}f(y)\,dy$$

$$- \int_0^x f(y)dy + \int_0^x \left(\int_0^t e^{y-t}f(y)\,dy\right)dt$$

$$= f(x) - \int_0^x (e^{y-x} + 1)f(y)\,dy$$

$$+ \int_0^x \left(\int_0^t e^{y-t}f(y)dy\right)dt.$$

An integral computation shows that

$$((I - V)(I + V)^{-1}e_m)(x) = \frac{e^{-x-i\pi m}\left((2\pi m + i)e^{x+2i\pi mx} - 2i\right)}{2\pi m - i}.$$

Now observe that

$$((I - V)(I + V)^{-1}e_m)(x)\overline{e_n(x)}$$

equals

$$\frac{(-1)^{m+n}\left((2\pi m + i)e^{x+2i\pi mx} - 2i\right)e^{(-1-2i\pi n)x}}{2\pi m - i}.$$

Therefore,

$$\langle (I - V)(I + V)^{-1}e_m, e_n\rangle$$

$$= \int_0^1 ((I - V)(I + V)^{-1}e_m)(x)\overline{e_n(x)}\,dx$$

$$= \int_0^1 \frac{(-1)^{m+n}\left((2\pi m + i)e^{x+2i\pi mx} - 2i\right)e^{(-1-2i\pi n)x}}{2\pi m - i}\,dx$$

$$= \frac{2(e-1)}{e}\frac{(-1)^{m+n}}{(2\pi m - i)(2\pi n - i)},$$

which completes the proof. ∎

Observe that the matrix in Theorem 20.6.1 is symmetric in m and n, in other words, it is self transpose. This is the case with the matrix representation of any complex symmetric operator with respect to a C-real orthonormal basis (Exercise 20.8.27).

20.7 Notes

This chapter covered a small sliver of a large field of operator theory [134, 135, 143, 250]. Let us mention a few other important results from the literature.

For an inner function u, the spectrum of the compressed shift S_u is the set

$$\Sigma(u) = \left\{\lambda \in \mathbb{D}^- : \liminf_{z \to \lambda} |u(z)| = 0\right\}$$

and the point spectrum is $\Sigma(u) \cap \mathbb{D} = \{\lambda \in \mathbb{D} : u(\lambda) = 0\}$. The set $\Sigma(u)$ is the *spectrum* of u and has various roles in understanding functions in the model space $\mathcal{K}_u$. For example, if γ is an arc of $\mathbb{T}$ which does not intersect $\Sigma(u)$, then every function in $\mathcal{K}_u$ has an analytic continuation to an open neighborhood of γ. Observe that $\Sigma(\Theta) = \{1\}$ for the inner function Θ from (20.2.1).

The invariant subspaces of S_u are $vH^2 \cap \mathcal{K}_u$, where v is an inner function such that u/v is also inner. Moreover, every invariant subspace is cyclic. Any compressed shift is irreducible in the sense that if $\mathcal{M}$ is a subspace of $\mathcal{K}_u$ with $S_u\mathcal{M} \subseteq \mathcal{M}$ and $S_u^*\mathcal{M} \subseteq \mathcal{M}$, then $\mathcal{M} = \{0\}$ or $\mathcal{M} = \mathcal{K}_u$.

The description of the commutant $\{S_u\}'$ of S_u is one of the gems of operator theory and is a consequence of the commutant lifting theorem [324]. For any $\varphi \in H^\infty$, the operator $\varphi(S_u)$ is well defined and equals $P_u T_\varphi|_{\mathcal{K}_u}$, where T_φ is the analytic Toeplitz operator on H^2 with symbol φ. Furthermore, $\{S_u\}' = \{\varphi(S_u) : \varphi \in H^\infty\}$. See Exercise 20.8.23 for a proof using Hankel operators.

One can show that S_u satisfies the following:

(a) $\|S_u\| \leqslant 1$.

(b) $\operatorname{rank}(I - S_u S_u^*) = \operatorname{rank}(I - S_u^* S_u) = 1$.

(c) $\lim_{n \to \infty} \|S_u^n f\| = 0$ for all $f \in \mathcal{K}_u$.

What makes compressed shifts important is that any Hilbert space operator satisfying (a) - (c) is unitarily equivalent to S_u for some inner function u, hence the use of the term "model space" for $\mathcal{K}_u$ [143].

There is a representation theorem for other types of contractions [134, 135] involving the compression of the shift to a reproducing kernel Hilbert space with kernel

$$k_\lambda^u(z) = \frac{1 - \overline{u(\lambda)}u(z)}{1 - \overline{\lambda}z}.$$

Here u is a general analytic self map of $\mathbb{D}$ and the resulting space is a *de Branges–Rovnyak space*.

20.8 Exercises

Exercise 20.8.1. Consider the circle $\gamma = \{z : |z - \frac{1}{2}| = \frac{1}{2}\}$ and the inner function Θ from (20.2.1). Show that

$$\lim_{\substack{z \to 1 \\ z \in \gamma}} |\Theta(z)| = \frac{1}{e}.$$

Remark: This shows that although $\lim_{r \to 1^-} \Theta(r) = 0$, limits along other paths in $\mathbb{D}$ terminating at $\xi = 1$ can be nonzero.

Exercise 20.8.2. If u is inner, prove that the reproducing kernel $k_\lambda(z)$ for $\mathcal{K}_u$ (see (20.1.3)) has the following property: for distinct $\lambda_1, \lambda_2, ..., \lambda_n \in \mathbb{D}$ and any $c_1, c_2, ..., c_n \in \mathbb{C}$,

$$\sum_{1 \leqslant i, j \leqslant n} c_i \overline{c_j} k_{\lambda_i}(\lambda_j) \geqslant 0.$$

Remark: This is equivalent to the positive semidefiniteness of the matrix $[k_{\lambda_i}(\lambda_j)]_{i,j=1}^n$.

Exercise 20.8.3. Prove that any subspace of a reproducing kernel Hilbert space is a reproducing kernel Hilbert space.

Exercise 20.8.4. Let u be inner and let C be the function $(Cf)(\xi) = u(\xi)\overline{\xi f(\xi)}$ on $L^2(\mathbb{T})$.

(a) Prove that $C^2 = I$.

(b) Prove that $\|Cf\| = \|f\|$ for all $f \in L^2(\mathbb{T})$.

(c) Prove that $C(af + bg) = \overline{a}Cf + \overline{b}Cg$ for all $a, b \in \mathbb{C}$ and $f, g \in L^2(\mathbb{T})$.

Exercise 20.8.5. If u is inner and C is the conjugation on $L^2(\mathbb{T})$ from Exercise 20.8.4, prove that C maps $L^2(\mathbb{T}) \ominus \mathcal{K}_u$ onto itself.
Remark: Proposition 20.1.6 shows that $C\mathcal{K}_u = \mathcal{K}_u$.

Exercise 20.8.6. Prove that the compressed shift S_u satisfies $S_u^n = P_u S^n|_{\mathcal{K}_u}$ for all $n \geqslant 0$.

Exercise 20.8.7. *von Neumann's inequality* [144, p. 213] says that if $p \in \mathbb{C}[z]$ and $T \in \mathcal{B}(\mathcal{H})$ is a contraction, then $p(T)$ satisfies

$$\|p(T)\| \leqslant \sup_{|z| \leqslant 1} |p(z)|.$$

Prove von Neumann's inequality for $T = S_u$.

Exercise 20.8.8. If u is inner, prove the following.

(a) $\bigvee\{S^{*n}u : n \geqslant 1\} = \mathcal{K}_u$.

(b) $\bigvee\{S^{*n}u : n \geqslant 0\} = \mathcal{K}_{zu}$.

Exercise 20.8.9. If u is an inner function, prove that $k_0 = 1 - \overline{u(0)}u$ is a cyclic vector for S_u, that is, $\bigvee\{S_u^n k_0 : n \geqslant 0\} = \mathcal{K}_u$.

Exercise 20.8.10. If u is an inner function and $u(0) = 0$, prove the following.

(a) $1 \in \mathcal{K}_u$.

(b) $S_u S_u^* = I - 1 \otimes 1$.

(c) $S_u^* S_u = I - \dfrac{u}{z} \otimes \dfrac{u}{z}$.

Exercise 20.8.11. Let u be an inner function such that $u(0) = 0$.

(a) Prove that S_u is a partial isometry (Definition 14.9.8).

(b) Prove that $\ker S_u = \text{span}\left\{\dfrac{u}{z}\right\}$ and $(\text{ran } S_u)^\perp = \mathbb{C}$.

Exercise 20.8.12. If u is an inner function and $u(0) = 0$, Exercise 20.8.11 ensures that S_u is a partial isometry. Prove that $U = S_u + 1 \otimes \dfrac{u}{z}$ is unitary.
Remark: U is an example of a *Clark unitary operator* [143, Ch. 11].

Exercise 20.8.13. Let u be an inner function and $u(0) \neq 0$.

(a) Prove that $\ker S_u^* = \{0\}$.

(b) Prove that $\ker S_u = \{0\}$.

Exercise 20.8.14. Recall the Hankel operator H_φ from Chapter 17. If u is inner, prove that $H_{\overline{u}}^* H_{\overline{u}}$ is the orthogonal projection of H^2 onto $\mathcal{K}_u$.

Exercise 20.8.15. For each $n \geqslant 1$ prove that the $n \times n$ matrix

$$T_n = \begin{bmatrix} 0 & 0 & 0 & 0 & 0 & \cdots & 0 \\ 1 & 0 & 0 & 0 & 0 & \cdots & 0 \\ 0 & 1 & 0 & 0 & 0 & \cdots & 0 \\ 0 & 0 & 1 & 0 & 0 & \cdots & 0 \\ 0 & 0 & 0 & 1 & 0 & \cdots & 0 \\ \vdots & \vdots & \vdots & \vdots & \ddots & \ddots & 0 \\ 0 & 0 & 0 & 0 & 0 & 1 & 0 \end{bmatrix}$$

satisfies the following.

(a) $\|T_n\| \leqslant 1$.

(b) $\text{rank}(I - T_n T_n^*) = \text{rank}(I - T_n^* T_n) = 1$.

(c) $\|T_n^k \mathbf{x}\| \to 0$ as $k \to \infty$ for all $\mathbf{x} \in \mathbb{C}^n$.

Remark: If $T \in \mathcal{B}(\mathcal{H})$ is a contraction that satisfies (a) - (c), then T is unitarily equivalent to S_u on some model space $\mathcal{K}_u$ [143, p. 195].

Exercise 20.8.16. For the matrix T_n in Exercise 20.8.15, find an inner function u and an orthonormal basis for $\mathcal{K}_u$ such that a matrix representation of S_u is T_n.

Exercise 20.8.17. Find all the hyperinvariant subspaces for the compressed shift on the model space $\mathcal{K}_{z^n}$.

Exercise 20.8.18. Let u be a finite Blaschke product

$$u(z) = \prod_{j=1}^{n} \frac{z - \lambda_j}{1 - \overline{\lambda_j} z},$$

whose zeros $\lambda_1, \lambda_2, ..., \lambda_n$ are distinct points in $\mathbb{D}$. Prove that

$$\mathcal{K}_u = \bigvee \left\{ \frac{1}{1 - \overline{\lambda_j} z} : 1 \leqslant j \leqslant n \right\}.$$

Exercise 20.8.19. Let u be a finite Blaschke product whose zeros are $\lambda_1, \lambda_2, ..., \lambda_n$, repeated according to multiplicity. Prove that

$$\mathcal{K}_u = \left\{ \frac{p(z)}{(1 - \overline{\lambda_1} z)(1 - \overline{\lambda_2} z) \cdots (1 - \overline{\lambda_n} z)} : p \in \mathscr{P}_{n-1} \right\},$$

where $\mathscr{P}_{n-1}$ denotes the set of polynomials of degree at most $n - 1$.

Exercise 20.8.20. Let u be a finite Blaschke product whose zeros are $\lambda_1, \lambda_2, ..., \lambda_n$, repeated according to multiplicity. Let

$$b_\lambda(z) = \frac{z - \lambda}{1 - \overline{\lambda} z}.$$

Define

$$v_1(z) = \frac{\sqrt{1 - |\lambda_1|^2}}{1 - \overline{\lambda_1} z} \quad \text{and} \quad v_\ell(z) = \left(\prod_{1 \leqslant i \leqslant \ell-1} b_{\lambda_i} \right) \frac{\sqrt{1 - |\lambda_\ell|^2}}{1 - \overline{\lambda_\ell} z} \quad \text{for } 2 \leqslant \ell \leqslant n.$$

(a) Prove that $v_1, v_2, ..., v_n$ is an orthonormal basis for $\mathcal{K}_u$.

(b) Prove that the matrix representation of S_u with respect to the basis $v_1, v_2, ..., v_n$ is lower triangular with $\lambda_1, \lambda_2, ..., \lambda_n$ along the main diagonal (in that order).

Remark: This basis is known as the Takenaka–Malmquist–Walsh basis [143, p. 120].

Exercise 20.8.21. If u is inner and $u = S_\mu B$, where B is an infinite Blaschke product and S_μ is a singular inner function, prove that $\mathcal{K}_u$ is infinite dimensional.
Remark: A more delicate argument shows that $\mathcal{K}_u$ is finite dimensional if and only if u is a finite Blaschke product [143, p. 117].

Exercise 20.8.22. Let u be inner and consider the compression of the analytic Toeplitz operator T_φ on H^2 to $\mathcal{K}_u$. This is the operator A_φ on $\mathcal{K}_u$ defined by $A_\varphi f = P_u T_\varphi f$.

(a) Prove that $\{A_\varphi : \varphi \in H^\infty\} \subseteq \{S_u\}'$.

(b) For $u(z) = z^n$, prove that $\mathcal{K}_u = \mathcal{P}_{n-1}$ (the set of polynomials of degree at most $n-1$).

(c) Find the matrix representation of A_φ with respect to the orthonormal basis $1, z, z^2, \ldots, z^{n-1}$.

(d) Use this matrix representation to prove that $\{A_\varphi : \varphi \in H^\infty\} = \{S_u\}'$.

Remark: The commutant lifting theorem [324] (see also [143]) says that $\{A_\varphi : \varphi \in H^\infty\} = \{S_u\}'$ for any inner function u. We prove this in Exercise 20.8.23.

Exercise 20.8.23. This is a continuation of Exercise 20.8.22 and follows a presentation from [266]. Suppose that u is inner and $T \in \mathcal{B}(\mathcal{K}_u)$ commutes with S_u. Use the following steps to produce a $\varphi \in H^\infty$ such that $T = A_\varphi$ and $\|T\| = \|\varphi\|_\infty$.

(a) For $T \in \mathcal{B}(\mathcal{K}_u)$ let $\widetilde{T} : H^2 \to H_0^2$ be defined by $\widetilde{T} f = \bar{u} T P_u f$. Prove that $T S_u = S_u T$ if and only if $\widetilde{T}$ is a Hankel operator.

(b) Suppose that T commutes with S_u. Prove that $\widetilde{T} = H_\psi$ for some $\psi \in L^\infty$ with $\|H_\psi\| = \|\psi\|_\infty$.

(c) Prove that $H_{\psi u} = 0$ and deduce that $\varphi = \psi u \in H^\infty$.

(d) Prove that $\bar{u} T f = P_-(\bar{u} \varphi f)$ for all $f \in \mathcal{K}_u$.

(e) Prove that $T = A_\varphi$ and $\|T\| = \|\varphi\|_\infty$.

Exercise 20.8.24. For an inner function u and $\varphi \in H^\infty$, deduce from Exercise 20.8.23 that $A_\varphi = M_u H_{\bar{u}\varphi}|\mathcal{K}_u$.
Remark: See [250] for more on this.

Exercise 20.8.25. Let u be an inner function and let $(v_j)_{j \geqslant 1}$ be an orthonormal basis for the model space $\mathcal{K}_u$ (which may be finite or infinite dimensional). Prove that $(u^n v_j)_{n \geqslant 0, j \geqslant 1}$ is an orthonormal basis for H^2.

Exercise 20.8.26. Suppose that C is a conjugation on a complex Hilbert space $\mathcal{H}$. Use the following steps to prove that $\mathcal{H}$ has an orthonormal basis $(\mathbf{u}_n)_{n \geqslant 1}$ such that $C\mathbf{u}_n = \mathbf{u}_n$ for all n.

(a) Prove that $\langle \mathbf{x}, \mathbf{y} \rangle = \langle C\mathbf{y}, C\mathbf{x} \rangle$ for all $\mathbf{x}, \mathbf{y} \in \mathcal{H}$.

(b) Prove that $\mathcal{K} = (I - C)\mathcal{H}$ is a real Hilbert space in the sense that it satisfies the axioms of a Hilbert space except that the field of scalars is $\mathbb{R}$ and not $\mathbb{C}$.

(c) Verify the identity $2\mathbf{x} = (\mathbf{x} + C\mathbf{x}) - i(i\mathbf{x} + C(i\mathbf{x}))$ for all $\mathbf{x} \in \mathcal{H}$.

(d) If $(\mathbf{u}_n)_{n \geqslant 1}$ is an orthonormal basis for $\mathcal{K}$, prove that $C\mathbf{u}_n = \mathbf{u}_n$ for all n.

Remark: Such a basis is a *C-real basis.*

Exercise 20.8.27. Let $T \in \mathcal{B}(\mathcal{H})$ be a complex symmetric operator. That is, there exists a conjugation C on $\mathcal{H}$ such that $T = CT^*C$. Prove that the matrix representation of T with respect to any C-real orthonormal basis of $\mathcal{H}$ is symmetric.

Exercise 20.8.28. Let u be an infinite Blaschke product and consider the analytic Toeplitz operator T_u on H^2.

(a) Use the von Neumann–Wold decomposition (Theorem 15.1.1) to prove that T_u is unitarily equivalent to a block operator on $\bigoplus_{j \geqslant 1} T_u^j \mathcal{K}_u$.

(b) Prove that this block operator is

$$\begin{bmatrix} 0 & 0 & 0 & 0 & 0 & \cdots \\ I & 0 & 0 & 0 & 0 & \cdots \\ 0 & I & 0 & 0 & 0 & \cdots \\ 0 & 0 & I & 0 & 0 & \cdots \\ 0 & 0 & 0 & I & 0 & \cdots \\ \vdots & \vdots & \vdots & \vdots & \ddots & \ddots \end{bmatrix},$$

where 0 is the zero operator on $\mathcal{K}_u$ and I is the identity operator on $\mathcal{K}_u$.

Exercise 20.8.29. Continuing with Exercise 20.8.28, prove that every operator in the commutant of the operator above is of the form

$$\begin{bmatrix} A_0 & 0 & 0 & 0 & 0 & \cdots \\ A_1 & A_0 & 0 & 0 & 0 & \cdots \\ A_2 & A_1 & A_0 & 0 & 0 & \cdots \\ A_3 & A_2 & A_1 & A_0 & 0 & \cdots \\ A_4 & A_3 & A_2 & A_1 & A_0 & \cdots \\ \vdots & \vdots & \vdots & \vdots & \vdots & \ddots \end{bmatrix},$$

where $A_j \in \mathcal{B}(\mathcal{K}_u)$ for all $j \geqslant 0$.

Exercise 20.8.30. Follow, these steps to prove the Nevanlinna–Pick theorem: If $\lambda_1, \lambda_2, ..., \lambda_n$ are distinct points in $\mathbb{D}$ and $w_1, w_2, ..., w_n$ are arbitrary points in $\mathbb{D}$, then there is an analytic self map f of $\mathbb{D}$ such that $f(\lambda_j) = w_j$ for all $1 \leqslant i \leqslant n$ if and only if the matrix

$$Q = \left[\frac{1 - \overline{w_i} w_j}{1 - \overline{\lambda_i} \lambda_j} \right]_{i,j=1}^n$$

is positive semidefinite.

(a) Let

$$u(z) = \prod_{j=1}^n \frac{\lambda_j - z}{1 - \overline{\lambda_j} z}$$

be the finite Blaschke product whose zeros are $\lambda_1, \lambda_2, ..., \lambda_n$ and recall from Exercise 20.8.18 that $\mathcal{K}_u = \text{span}\{k_{\lambda_j} : 1 \leqslant j \leqslant n\}$. Define the operator $R : \mathcal{K}_u \to \mathcal{K}_u$ on the basis elements of $\mathcal{K}_u$ by $Rk_{\lambda_j} = \overline{w_j} k_{\lambda_j}$ for all $1 \leqslant j \leqslant n$. Prove that $R \in \{S_u^*\}'$.

(b) By Exercise 20.8.22 there is a $\varphi \in H^\infty$ such that $R = T_{\overline{\varphi}}|_{\mathcal{K}_u}$ and $\|R\| = \|\varphi\|_\infty$. Use this to prove that $\overline{w_j}k_{\lambda_j} = \overline{\varphi(\lambda_j)}k_{\lambda_j}$ for all $1 \leqslant j \leqslant n$ and thus $\varphi(\lambda_j) = w_j$.

(c) Prove that Q is positive semidefinite if and only if $I - R^*R \geqslant 0$.

(d) Prove that $I - R^*R \geqslant 0$ if and only if $\|\varphi\|_\infty \leqslant 1$.

20.9 Hints for the Exercises

Hint for Ex. 20.8.2: Write $\sum_{1 \leqslant i,j \leqslant n} c_i \overline{c_j} k_{\lambda_i}(\lambda_j)$ as the square of the norm of a function.

Hint for Ex. 20.8.3: If $K(z, w)$ is the reproducing kernel for $\mathcal{H}$ and $P_{\mathcal{M}}$ is the orthogonal projection of $\mathcal{H}$ onto $\mathcal{M}$, examine $P_{\mathcal{M}}K(z, w)$.

Hint for Ex. 20.8.6: Work with adjoints and use the S^*-invariance of $\mathcal{K}_u$.

Hint for Ex. 20.8.7: Since $p(T) = P_u p(S)|_{\mathcal{K}_u}$ and, without loss of generality,

$$\sup_{|z| \leqslant 1} |p(z)| = 1,$$

it suffices to show that $\|p(S)f\| \leqslant \|f\|$ for all $f \in \mathcal{K}_u$.

Hint for Ex. 20.8.9: Let C be the conjugation from Proposition 20.1.6 and prove that $Ck_0 = S^*u$. Then use Exercise 20.8.8.

Hint for Ex. 20.8.10: Prove (b) and then use the conjugation C to prove (c).

Hint for Ex. 20.8.13: Consult Proposition 20.3.3 for (a). Consult Proposition 20.3.4 for (b).

Hint for Ex. 20.8.23: For (a), recall that $\widetilde{T}$ is a Hankel operator if and only if $P_- z\widetilde{T}f = \widetilde{T}Sf$ for all $f \in H^2$ (Exercise 17.10.10) and that $P_u f = uP_-\overline{u}f$ for all $f \in \mathcal{K}_u$ (Proposition 20.3.1).

Hint for Ex. 20.8.25: Consider the Toeplitz operator T_u on H^2 and the von Neumann–Wold decomposition (Theorem 15.1.1).

REFERENCES

[1] M. B. Abrahamse and T. L. Kriete, *The spectral multiplicity of a multiplication operator*, Indiana Univ. Math. J. **22** (1972/73), 845–857.

[2] N. I. Achieser, *Theory of approximation*, Translated by C. J. Hyman, Frederick Ungar Publishing Co., New York, 1956.

[3] V. M. Adamjan, D. Z. Arov, and M. G. Kreĭn, *Analytic properties of the Schmidt pairs of a Hankel operator and the generalized Schur-Takagi problem*, Mat. Sb. (N.S.) **86(128)** (1971), 34–75.

[4] J. Agler, *A disconjugacy theorem for Toeplitz operators*, Amer. J. Math. **112** (1990), no. 1, 1–14.

[5] J. Agler and M. Stankus, *m-isometric transformations of Hilbert space. I*, Integral Equations Operator Theory **21** (1995), no. 4, 383–429.

[6] _____, *m-isometric transformations of Hilbert space. II*, Integral Equations Operator Theory **23** (1995), no. 1, 1–48.

[7] _____, *m-isometric transformations of Hilbert space. III*, Integral Equations Operator Theory **24** (1996), no. 4, 379–421.

[8] S. Agmon, *Sur un problème de translations*, C. R. Acad. Sci. Paris **229** (1949), 540–542.

[9] L. Ahlfors, *Complex analysis*, third ed., McGraw-Hill Book Co., New York, 1978, International Series in Pure and Applied Mathematics.

[10] A. Aleman and J. A. Cima, *An integral operator on H^p and Hardy's inequality*, J. Anal. Math. **85** (2001), 157–176.

[11] A. Aleman, N. S. Feldman, and W. T. Ross, *The Hardy space of a slit domain*, Frontiers in Mathematics, Birkhäuser Verlag, Basel, 2009.

[12] A. Aleman and B. Korenblum, *Volterra invariant subspaces of H^p*, Bull. Sci. Math. **132** (2008), no. 6, 510–528.

[13] A. Aleman, A. Montes-Rodríguez, and A. Sarafoleanu, *The eigenfunctions of the Hilbert matrix*, Constr. Approx. **36** (2012), no. 3, 353–374.

[14] A. Aleman and R. Olin, *Hardy spaces of crescent domains*, unpublished.

[15] A. Aleman and S. Richter, *Simply invariant subspaces of H^2 of some multiply connected regions*, Integral Equations Operator Theory **24** (1996), no. 2, 127–155.

[16] A. Aleman, S. Richter, and W. T. Ross, *Pseudocontinuations and the backward shift*, Indiana Univ. Math. J. **47** (1998), no. 1, 223–276.

[17] A. Aleman, S. Richter, and C. Sundberg, *Beurling's theorem for the Bergman space*, Acta Math. **177** (1996), no. 2, 275–310.

[18] A. Aleman and D. Vukotić, *Zero products of Toeplitz operators*, Duke Math. J. **148** (2009), no. 3, 373–403.

[19] C. Apostol, H. Bercovici, C. Foiaş, and C. Pearcy, *Invariant subspaces, dilation theory, and the structure of the predual of a dual algebra. I*, J. Funct. Anal. **63** (1985), no. 3, 369–404.

[20] J. Arazy and S. D. Fisher, *The uniqueness of the Dirichlet space among Möbius-invariant Hilbert spaces*, Illinois J. Math. **29** (1985), no. 3, 449–462.

[21] J. Arazy, S. D. Fisher, and J. Peetre, *Hankel operators on weighted Bergman spaces*, Amer. J. Math. **110** (1988), no. 6, 989–1053.

[22] N. Arcozzi, R. Rochberg, E. T. Sawyer, and B. D. Wick, *The Dirichlet space and related function spaces*, Mathematical Surveys and Monographs, vol. 239, American Mathematical Society, Providence, RI, 2019.

[23] N. Aronszajn and K. T. Smith, *Invariant subspaces of completely continuous operators*, Ann. of Math. (2) **60** (1954), 345–350.

[24] W. Arveson, *An invitation to C^*-algebras*, Springer-Verlag, New York-Heidelberg, 1976, Graduate Texts in Mathematics, No. 39.

[25] _____, *A short course on spectral theory*, Graduate Texts in Mathematics, vol. 209, Springer-Verlag, New York, 2002.

[26] S. V. Astashkin, *The Rademacher system in function spaces*, Birkhäuser/Springer, Cham, [2020] ©2020.

[27] S. Axler, *The Bergman space, the Bloch space, and commutators of multiplication operators*, Duke Math. J. **53** (1986), no. 2, 315–332.

[28] S. Axler, I. Berg, N. Jewell, and A. Shields, *Approximation by compact operators and the space $H^\infty + C$*, Ann. of Math. (2) **109** (1979), no. 3, 601–612.

[29] S. Axler and P. Bourdon, *Finite-codimensional invariant subspaces of Bergman spaces*, Trans. Amer. Math. Soc. **306** (1988), no. 2, 805–817.

[30] S. Axler, J. B. Conway, and G. McDonald, *Toeplitz operators on Bergman spaces*, Canadian J. Math. **34** (1982), no. 2, 466–483.

[31] S. Axler and D. Zheng, *Compact operators via the Berezin transform*, Indiana Univ. Math. J. **47** (1998), no. 2, 387–400.

[32] S. Banach, *Sur les fonctionnelles linéaires*, Studia Math. **1** (1929), 211–216; 223–239.

[33] _____, *Theory of linear operations*, North-Holland Mathematical Library, vol. 38, North-Holland Publishing Co., Amsterdam, 1987, Translated from the French by F. Jellett, With comments by A. Pełczyński and Cz. Bessaga.

[34] J. Barría and P. R. Halmos, *Asymptotic Toeplitz operators*, Trans. Amer. Math. Soc. **273** (1982), no. 2, 621–630.

[35] F. Bayart, *Compact composition operators on a Hilbert space of Dirichlet series*, Illinois J. Math. **47** (2003), no. 3, 725–743.

[36] E. Bedrosian, *A product theorem for Hilbert transforms*, Proc. IEEE **51** (1963), 868–869.

[37] C. Bénéteau, A. Condori, C. Liaw, D. Seco, and A. Sola, *Cyclicity in Dirichlet-type spaces and extremal polynomials*, J. Anal. Math. **126** (2015), 259–286.

[38] C. Bénéteau, D. Khavinson, C. Liaw, D. Seco, and A. Sola, *Orthogonal polynomials, reproducing kernels, and zeros of optimal approximants*, J. Lond. Math. Soc. (2) **94** (2016), no. 3, 726–746.

[39] C. Benhida, R. Curto, and G. Exner, *Moment infinitely divisible weighted shifts*, Complex Anal. Oper. Theory **13** (2019), no. 1, 241–255.

[40] G. Bennett, *Lower bounds for matrices*, Linear Algebra Appl. **82** (1986), 81–98.

[41] _____, *Factorizing the classical inequalities*, Mem. Amer. Math. Soc. **120** (1996), no. 576, viii+130.

[42] S. K. Berberian, *Introduction to Hilbert space*, University Texts in the Mathematical Sciences, Oxford University Press, New York, 1961.

[43] _____, *A note on hyponormal operators*, Pacific J. Math. **12** (1962), 1171–1175.

[44] F. A. Berezin, *Covariant and contravariant symbols of operators*, Izv. Akad. Nauk SSSR Ser. Mat. **36** (1972), 1134–1167.

[45] S. Bergman, *Über die Entwicklung der harmonischen Funktionen der Ebene und des Raumes nach Orthogonalfunktionen*, Math. Ann. **86** (1922), no. 3-4, 238–271.

[46] _____, *The kernel function and conformal mapping*, Mathematical Surveys, No. 5, American Mathematical Society, New York, N. Y., 1950.

[47] A. Bernstein and A. Robinson, *Solution of an invariant subspace problem of K. T. Smith and P. R. Halmos*, Pacific J. Math. **16** (1966), 421–431.

[48] A. S. Besicovitch, *Almost periodic functions*, Dover Publications, Inc., New York, 1955.

[49] F. Bessel, *Über die Bestimmung des Gesetzes einer periodischen Erscheinung*, Astron. Nachr. **6** (1828), 333–348.

[50] A. Beurling, *Études sur un problème de majoration*, 1933, Thesis (Ph.D.)–Uppsala University.

[51] _____, *Sur les intégrales de Fourier absolument convergentes et leur application à une transformation fonctionelle*, Ninth Scandinavian Math. Congress, Helsingfors (1938), 345–366.

[52] _____, *Ensembles exceptionnels*, Acta Math. **72** (1940), 1–13.

[53] _____, *On two problems concerning linear transformations in Hilbert space*, Acta Math. **81** (1948), 17.

[54] D. P. Blecher and A. M. Davie, *Invariant subspaces for an operator on $L^2(\Pi)$ composed of a multiplication and a translation*, J. Operator Theory **23** (1990), no. 1, 115–123.

[55] S. Bochner, *Über orthogonale Systeme analytischer Funktionen*, Math. Z. **14** (1922), no. 1, 180–207.

[56] H. Bohr, *Almost periodic functions*, Chelsea Publishing Company, New York, N.Y., 1947.

[57] É. Borel, *Leçons sur les fonctions monogènes uniformes d'une variable complexe*, Gauthier-Villars, Paris, 1917.

[58] _____, *Leçons sur la théorie des fonctions*, third ed., Gauthier-Villars, Paris, 1928.

[59] A. Böttcher and S. M. Grudsky, *Toeplitz matrices, asymptotic linear algebra, and functional analysis*, Birkhäuser Verlag, Basel, 2000.

[60] A. Böttcher and B. Silbermann, *Analysis of Toeplitz operators*, second ed., Springer Monographs in Mathematics, Springer-Verlag, Berlin, 2006, Prepared jointly with Alexei Karlovich.

[61] L. Bouthat and J. Mashreghi, *L-matrices with lacunary coefficients*, Oper. Matrices **15** (2021), no. 3, 1045–1053.

[62] _____, *The norm of an infinite L-matrix*, Oper. Matrices **15** (2021), no. 1, 47–58.

[63] J. Bram, *Subnormal operators*, Duke Math. J. **22** (1955), 75–94.

[64] F. Brevig, K.-M. Perfekt, K. Seip, A. Siskakis, and D. Vukotić, *The multiplicative Hilbert matrix*, Adv. Math. **302** (2016), 410–432.

[65] M. S. Brodskiĭ, *On a problem of I. M. Gelfand*, Uspehi Mat. Nauk (N.S.) **12** (1957), no. 2(74), 129–132.

[66] A. Brown, *On a class of operators*, Proc. Amer. Math. Soc. **4** (1953), 723–728.

[67] A. Brown and R. G. Douglas, *Partially isometric Toeplitz operators*, Proc. Amer. Math. Soc. **16** (1965), 681–682.

[68] A. Brown and P. R. Halmos, *Algebraic properties of Toeplitz operators*, J. Reine Angew. Math. **213** (1963/1964), 89–102.

[69] A. Brown, P. R. Halmos, and A. L. Shields, *Cesàro operators*, Acta Sci. Math. (Szeged) **26** (1965), 125–137.

[70] L. Brown and A. L. Shields, *Cyclic vectors in the Dirichlet space*, Trans. Amer. Math. Soc. **285** (1984), no. 1, 269–303.

[71] L. Brown, A. L. Shields, and K. Zeller, *On absolutely convergent exponential sums*, Trans. Amer. Math. Soc. **96** (1960), 162–183.

[72] S. W. Brown, *Some invariant subspaces for subnormal operators*, Integral Equations Operator Theory **1** (1978), no. 3, 310–333.

[73] V. Buniakowsky, *Sur quelques inégalités concernant les intégrales ordinaires et les intégrales aux différences finies*, Mémoires Acad. St. Pétersbourg **1** (1859).

[74] M. C. Câmara and W. T. Ross, *The dual of the compressed shift*, Canad. Math. Bull. **64** (2021), no. 1, 98–111.

[75] S. R. Caradus, *Universal operators and invariant subspaces*, Proc. Amer. Math. Soc. **23** (1969), 526–527.

[76] L. Carleson, *On convergence and growth of partial sums of Fourier series*, Acta Math. **116** (1966), 135–157.

[77] J. G. Caughran and H. J. Schwartz, *Spectra of compact composition operators*, Proc. Amer. Math. Soc. **51** (1975), 127–130.

[78] E. Cesàro, *Sur la multiplication de séries*, Bull. Sci. Math. **14** (1890), 114–120.

[79] I. Chalendar, E. A. Gallardo-Gutiérrez, and J. R. Partington, *Weighted composition operators on the Dirichlet space: boundedness and spectral properties*, Math. Ann. **363** (2015), no. 3-4, 1265–1279.

[80] I. Chalendar and J. R. Partington, *Modern approaches to the invariant-subspace problem*, Cambridge Tracts in Mathematics, vol. 188, Cambridge University Press, Cambridge, 2011.

[81] F. Chamizo, E. A. Gallardo-Gutiérrez, M. Monsalve-López, and Adrián U., *Invariant subspaces for Bishop operators and beyond*, Advances in Mathematics **375** (2020), 107365.

[82] R. Cheng, J. Mashreghi, and W. T. Ross, *Function theory and ℓ^p spaces*, University Lecture Series, vol. 75, American Mathematical Society, Providence, RI, 2020.

[83] M. D. Choi, *Tricks or treats with the Hilbert matrix*, Amer. Math. Monthly **90** (1983), no. 5, 301–312.

[84] C. K. Chui, *An introduction to wavelets*, Wavelet Analysis and its Applications, vol. 1, Academic Press, Inc., Boston, MA, 1992.

[85] J. A. Cima and A. L. Matheson, *Essential norms of composition operators and Aleksandrov measures*, Pacific J. Math. **179** (1997), no. 1, 59–64.

[86] K. Clancey, *Seminormal operators*, Lecture Notes in Mathematics, vol. 742, Springer, Berlin, 1979.

[87] B. Cload, *Composition operators: hyperinvariant subspaces, quasi-normals and isometries*, Proc. Amer. Math. Soc. **127** (1999), no. 6, 1697–1703.

[88] L. A. Coburn, *Weyl's theorem for nonnormal operators*, Michigan Math. J. **13** (1966), 285–288.

[89] ———, *The C^*-algebra generated by an isometry*, Bull. Amer. Math. Soc. **73** (1967), 722–726.

[90] ———, *The C^*-algebra generated by an isometry. II*, Trans. Amer. Math. Soc. **137** (1969), 211–217.

[91] ———, *Singular integral operators and Toeplitz operators on odd spheres*, Indiana Univ. Math. J. **23** (1973/74), 433–439.

[92] J. B. Conway, *Functions of one complex variable*, second ed., Graduate Texts in Mathematics, vol. 11, Springer-Verlag, New York-Berlin, 1978.

[93] ———, *The dual of a subnormal operator*, J. Operator Theory **5** (1981), no. 2, 195–211.

[94] ———, *A course in functional analysis*, Graduate Texts in Mathematics, vol. 96, Springer-Verlag, New York, 1985.

[95] ———, *The theory of subnormal operators*, Mathematical Surveys and Monographs, vol. 36, American Mathematical Society, Providence, RI, 1991.

[96] J. B. Conway and B. B. Morrel, *Roots and logarithms of bounded operators on Hilbert space*, J. Funct. Anal. **70** (1987), no. 1, 171–193.

[97] J. B. Conway and R. F. Olin, *A functional calculus for subnormal operators*, Bull. Amer. Math. Soc. **82** (1976), no. 2, 259–261.

[98] C. C. Cowen, *Composition operators on H^2*, J. Operator Theory **9** (1983), no. 1, 77–106.

[99] _____, *Subnormality of the Cesàro operator and a semigroup of composition operators*, Indiana Univ. Math. J. **33** (1984), no. 2, 305–318.

[100] _____, *Linear fractional composition operators on H^2*, Integral Equations Operator Theory **11** (1988), no. 2, 151–160.

[101] C. C. Cowen and E. A. Gallardo-Gutiérrez, *An introduction to Rota's universal operators: properties, old and new examples and future issues*, Concr. Oper. **3** (2016), no. 1, 43–51.

[102] C. C. Cowen and B. D. MacCluer, *Composition operators on spaces of analytic functions*, Studies in Advanced Mathematics, CRC Press, Boca Raton, FL, 1995.

[103] Ž. Čučković and B. Paudyal, *Invariant subspaces of the shift plus complex Volterra operator*, J. Math. Anal. Appl. **426** (2015), no. 2, 1174–1181.

[104] D. P. Dalzell, *On the completeness of a series of normal orthogonal functions*, J. London Math. Soc. **20** (1945), 87–93.

[105] K. Davidson, *C*-algebras by example*, Fields Institute monographs, American Mathematical Society, Providence, RI, 1996.

[106] A. M. Davie, *Invariant subspaces for Bishop's operators*, Bull. London Math. Soc. **6** (1974), 343–348.

[107] E. Davies, *Linear operators and their spectra*, Cambridge Studies in Advanced Mathematics, vol. 106, Cambridge University Press, Cambridge, 2007.

[108] J. A. Deddens, *Analytic Toeplitz and composition operators*, Canadian J. Math. **24** (1972), 859–865.

[109] A. Devinatz, *Toeplitz operators on H^2 spaces*, Trans. Amer. Math. Soc. **112** (1964), 304–317.

[110] E. Diamantopoulos and A. G. Siskakis, *Composition operators and the Hilbert matrix*, Studia Math. **140** (2000), no. 2, 191–198.

[111] B. W. Dickinson and K. Steiglitz, *Eigenvectors and functions of the discrete Fourier transform*, IEEE Trans. Acoust. Speech Signal Process. **30** (1982), no. 1, 25–31.

[112] J. Dixmier, *Les algèbres d'opérateurs dans l'espace Hilbertien (Algèbres de von Neumann)*, Cahiers scientifiques, Fascicule XXV, Gauthier-Villars, Paris, 1957.

[113] W. F. Donoghue, Jr., *The lattice of invariant subspaces of a completely continuous quasi-nilpotent transformation*, Pacific J. Math. **7** (1957), 1031–1035.

[114] J. Douglas, *Solution of the problem of Plateau*, Trans. Amer. Math. Soc. **33** (1931), no. 1, 263–321.

[115] R. G. Douglas, *On majorization, factorization, and range inclusion of operators on Hilbert space*, Proc. Amer. Math. Soc. **17** (1966), 413–415.

[116] _____, *Banach algebra techniques in operator theory*, Pure and Applied Mathematics, Vol. 49, Academic Press, New York-London, 1972.

[117] P. du Bois-Reymond, *Über die Fourierschen reihen*, Nachr. Kön. Ges. Wiss. Göttingen **21** (1873), 571–582.

[118] P. L. Duren, *Theory of H^p Spaces*, Pure and Applied Mathematics, Vol. 38, Academic Press, New York-London, 1970.

[119] P. L. Duren, E. A. Gallardo-Gutiérrez, and A. Montes-Rodríguez, *A Paley-Wiener theorem for Bergman spaces with application to invariant subspaces*, Bull. Lond. Math. Soc. **39** (2007), no. 3, 459–466.

[120] P. L. Duren and A. Schuster, *Bergman spaces*, Mathematical Surveys and Monographs, vol. 100, American Mathematical Society, Providence, RI, 2004.

[121] O. El-Fallah, K. Kellay, J. Mashreghi, and T. Ransford, *A primer on the Dirichlet space*, Cambridge Tracts in Mathematics, vol. 203, Cambridge University Press, Cambridge, 2014.

[122] O. El-Fallah, K. Kellay, and T. Ransford, *On the Brown-Shields conjecture for cyclicity in the Dirichlet space*, Adv. Math. **222** (2009), no. 6, 2196–2214.

[123] O. El-Fallah, K. Kellay, M. Shabankhah, and H. Youssfi, *Level sets and composition operators on the Dirichlet space*, J. Funct. Anal. **260** (2011), no. 6, 1721–1733.

[124] P. Enflo, *On the invariant subspace problem for Banach spaces*, Acta Math. **158** (1987), no. 3-4, 213–313.

[125] A. Erdélyi, W. Magnus, F. Oberhettinger, and F. Tricomi, *Higher transcendental functions. Vols. I, II*, McGraw-Hill Book Company, Inc., New York-Toronto-London, 1953, Based, in part, on notes left by Harry Bateman.

[126] J. A. Erdos, *The commutant of the Volterra operator*, Integral Equations Operator Theory **5** (1982), no. 1, 127–130.

[127] P. Fatou, *Séries trigonométriques et séries de Taylor*, Acta Math. **30** (1906), 335 – 400.

[128] A. Feintuch, *On asymptotic Toeplitz and Hankel operators*, The Gohberg anniversary collection, Vol. II (Calgary, AB, 1988), Oper. Theory Adv. Appl., vol. 41, Birkhäuser, Basel, 1989, pp. 241–254.

[129] N. S. Feldman, *Pure subnormal operators have cyclic adjoints*, J. Funct. Anal. **162** (1999), no. 2, 379–399.

[130] E. Fischer, *Sur la convergence en moyenne*, C.R. Acad. Sci. Paris **144** (1907), 1022–1024.

[131] J. Fourier, *Théorie analytique de la chaleur*, Cambridge Library Collection - Mathematics, Cambridge University Press, 2009.

[132] M. Fréchet, *Sur les opérations linéaires. III*, Trans. Amer. Math. Soc. **8** (1907), no. 4, 433–446.

[133] _____ , *Sur la définition axiomatique d'une classe d'espaces vectoriels distanciés applicables vectoriellement sur l'espace de Hilbert*, Ann. of Math. (2) **36** (1935), no. 3, 705–718.

[134] E. Fricain and J. Mashreghi, *The theory of H(b) spaces, volume 1*, New Mathematical Monographs, vol. 20, Cambridge University Press, Cambridge, 2014.

[135] _____ , *The theory of H(b) spaces, volume 2*, New Mathematical Monographs, vol. 21, Cambridge University Press, Cambridge, 2015.

[136] B. Fuglede, *A commutativity theorem for normal operators*, Proc. Nat. Acad. Sci. U.S.A. **36** (1950), 35–40.

[137] E. A. Gallardo-Gutiérrez and J. R. Partington, *Multiplication by a finite Blaschke product on weighted Bergman spaces*, preprint.

[138] E. A. Gallardo-Gutiérrez, J. R. Partington, and D. Seco, *On the wandering property in Dirichlet spaces*, Integral Equations Operator Theory **92** (2020), no. 2, Paper No. 16, 11.

[139] S. R. Garcia, *Conjugation, the backward shift, and Toeplitz kernels*, J. Operator Theory **54** (2005), no. 2, 239–250.

[140] _____ , *Conjugation and Clark operators*, Recent advances in operator-related function theory, Contemp. Math., vol. 393, Amer. Math. Soc., Providence, RI, 2006, pp. 67–111.

[141] S. R. Garcia and R. A. Horn, *A second course in linear algebra*, Cambridge University Press, Cambridge, 2017.

[142] S. R. Garcia, B. Lutz, and D. Timotin, *Two remarks about nilpotent operators of order two*, Proc. Amer. Math. Soc. **142** (2014), no. 5, 1749–1756.

[143] S. R. Garcia, J. Mashreghi, and W. T. Ross, *Introduction to model spaces and their operators*, Cambridge Studies in Advanced Mathematics, vol. 148, Cambridge University Press, Cambridge, 2016.

[144] _____ , *Finite Blaschke products and their connections*, Springer, Cham, 2018.

[145] S. R Garcia, M. Patterson, and W. Ross, *Partially isometric matrices: a brief and selective survey*, Proceedings of the 27th International Conference on Operator Theory.

[146] S. R. Garcia, E. Prodan, and M. Putinar, *Mathematical and physical aspects of complex symmetric operators*, J. Phys. A **47** (2014), no. 35, 353001, 54.

[147] S. R. Garcia and M. Putinar, *Complex symmetric operators and applications*, Trans. Amer. Math. Soc. **358** (2006), no. 3, 1285–1315.

[148] ———, *Complex symmetric operators and applications. II*, Trans. Amer. Math. Soc. **359** (2007), no. 8, 3913–3931.

[149] J. Garnett, *Bounded analytic functions*, first ed., Graduate Texts in Mathematics, vol. 236, Springer, New York, 2007.

[150] I. M. Gelfand, *A problem*, Uspekhi Mat. Nauk **5** (1938), 233.

[151] ———, *Normierte Ringe*, Rec. Math. [Mat. Sbornik] N. S. **9 (51)** (1941), 3–24.

[152] R. Gellar and L. J. Wallen, *Subnormal weighted shifts and the Halmos-Bram criterion*, Proc. Japan Acad. **46** (1970), 375–378.

[153] O. Giselsson and A. Olofsson, *On some Bergman shift operators*, Complex Anal. Oper. Theory **6** (2012), no. 4, 829–842.

[154] I. Gohberg and M. G. Kreĭn, *Theory and applications of Volterra operators in Hilbert space*, Translated from the Russian by A. Feinstein. Translations of Mathematical Monographs, Vol. 24, American Mathematical Society, Providence, R.I., 1970.

[155] I. Gohberg, P. Lancaster, and L. Rodman, *Invariant subspaces of matrices with applications*, Classics in Applied Mathematics, vol. 51, Society for Industrial and Applied Mathematics (SIAM), Philadelphia, PA, 2006, Reprint of the 1986 original.

[156] I. J. Good, *Analogues of Poisson's summation formula*, Amer. Math. Monthly **69** (1962), 259–266.

[157] J. Gordon and H. Hedenmalm, *The composition operators on the space of Dirichlet series with square summable coefficients*, Michigan Math. J. **46** (1999), no. 2, 313–329.

[158] P. Gorkin and J. H. Smith, *Dirichlet: his life, his principle, and his problem*, Math. Mag. **78** (2005), no. 4, 283–296.

[159] E. Goursat, *Sur un cas élémentaire de l'équation de Fredholm*, Bull. Soc. Math. France **35** (1907), 163–173.

[160] I. S. Gradshteyn and I. M. Ryzhik, *Table of integrals, series, and products*, seventh ed., Elsevier/Academic Press, Amsterdam, 2007, Translated from the Russian, Translation edited and with a preface by A. Jeffrey and D. Zwillinger, With one CD-ROM (Windows, Macintosh and UNIX).

[161] L. Grafakos, *Classical Fourier analysis*, third ed., Graduate Texts in Mathematics, vol. 249, Springer, New York, 2014.

[162] C. Gu, E. Rizzie, and J. Shapiro, *Adjoints of composition operators with irrational symbol*, Proc. Amer. Math. Soc. **148** (2020), no. 1, 145–155.

[163] D. Guillot, *Fine boundary behavior and invariant subspaces of harmonically weighted Dirichlet spaces*, Complex Anal. Oper. Theory **6** (2012), no. 6, 1211–1230.

[164] K. Guo and L. Zhao, *On unitary equivalence of invariant subspaces of the Dirichlet space*, Studia Math. **196** (2010), no. 2, 143–150.

[165] A. Haar, *Zur Theorie der orthogonalen Funktionensysteme*, Math. Ann. **69** (1910), no. 3, 331–371.

[166] P. R. Halmos, *Normal dilations and extensions of operators*, Summa Brasil. Math. **2** (1950), 125–134.

[167] ———, *Invariant subspaces of polynomially compact operators*, Pacific J. Math. **16** (1966), 433–437.

[168] ———, *Ten problems in Hilbert space*, Bull. Amer. Math. Soc. **76** (1970), 887–933.

[169] ———, *A Hilbert space problem book*, second ed., Graduate Texts in Mathematics, vol. 19, Springer-Verlag, New York-Berlin, 1982, Encyclopedia of Mathematics and its Applications, 17.

[170] P. R. Halmos, G. Lumer, and J. J. Schäffer, *Square roots of operators*, Proc. Amer. Math. Soc. **4** (1953), 142–149.

[171] P. R. Halmos and J. E. McLaughlin, *Partial isometries*, Pacific J. Math. **13** (1963), 585–596.

[172] P. R. Halmos and V. S. Sunder, *Bounded integral operators on L^2 spaces*, Ergebnisse der Mathematik und ihrer Grenzgebiete [Results in Mathematics and Related Areas], vol. 96, Springer-Verlag, Berlin-New York, 1978.

[173] H. Hankel, *Ueber eine besondere Classe von symmetrischen Determinanten*, Ph.D. thesis, Univ. Leipzig, 1861.

[174] G. H. Hardy, *The mean value of the modulus of an analytic function*, Proc. London Math. Soc. **14** (1915), 269–277.

[175] ———, *Note on a theorem of Hilbert*, Math. Z. **6** (1920), no. 3-4, 314–317.

[176] ———, *Notes on some points in the integral calculus. LIX. on Hilbert transforms (continued)*, Mess. Math. **54** (1924), 81–88.

[177] ———, *Notes on some points in the integral calculus. LVIII. on Hilbert transforms*, Mess. Math. **54** (1924), 20–27.

[178] ———, *Divergent series*, Oxford, at the Clarendon Press, 1949.

[179] G. H. Hardy, J. E. Littlewood, and G. Pólya, *Inequalities*, Cambridge Mathematical Library, Cambridge University Press, Cambridge, 1988, Reprint of the 1952 edition.

[180] P. Hartman, *On completely continuous Hankel matrices*, Proc. Amer. Math. Soc. **9** (1958), 862–866.

[181] P. Hartman and A. Wintner, *On the spectra of Toeplitz's matrices*, Amer. J. Math. **72** (1950), 359–366.

[182] ———, *The spectra of Toeplitz's matrices*, Amer. J. Math. **76** (1954), 867–882.

[183] A. Hartmann and M. Mitkovski, *Kernels of Toeplitz operators*, Recent progress on operator theory and approximation in spaces of analytic functions, Contemp. Math., vol. 679, Amer. Math. Soc., Providence, RI, 2016, pp. 147–177.

[184] A. Hartmann, D. Sarason, and K. Seip, *Surjective Toeplitz operators*, Acta Sci. Math. (Szeged) **70** (2004), no. 3-4, 609–621.

[185] F. Hausdorff, *Summationsmethoden und Momentfolgen. I*, Math. Z. **9** (1921), no. 1-2, 74–109.

[186] H. Hedenmalm, *An invariant subspace of the Bergman space having the codimension two property*, J. Reine Angew. Math. **443** (1993), 1–9.

[187] H. Hedenmalm, B. Korenblum, and K. Zhu, *Theory of Bergman spaces*, Graduate Texts in Mathematics, vol. 199, Springer-Verlag, New York, 2000.

[188] H. Hedenmalm, S. Richter, and K. Seip, *Interpolating sequences and invariant subspaces of given index in the Bergman spaces*, J. Reine Angew. Math. **477** (1996), 13–30.

[189] H. Hedenmalm and A. Shields, *Invariant subspaces in Banach spaces of analytic functions*, Michigan Math. J. **37** (1990), no. 1, 91–104.

[190] E. Hellinger, *Neue Begründung der Theorie quadratischer Formen von unendlichvielen Veränderlichen*, J. Reine Angew. Math. **136** (1909), 210–271.

[191] E. Hellinger and O. Toeplitz, *Grundlagen für eine Theorie der unendlichen Matrizen*, Math. Ann. **69** (1910), no. 3, 289–330.

[192] H. Helson, *Lectures on invariant subspaces*, Academic Press, New York-London, 1964.

[193] H. Helton, *Operators with a representation as multiplication by $\times$ on a Sobolev space*, Hilbert space operators and operator algebras (Proc. Internat. Conf., Tihany, 1970), 1972, pp. 279–287. (loose errata) Colloq. Math. Soc. János Bolyai, 5.

[194] D. Hilbert, *Ein Beitrag zur Theorie des Legendre'schen Polynoms*, Acta Math. **18** (1894), no. 1, 155–159.

[195] _____, *Grundzüge einer allgemeinen theorie der linearen integralgle- ichungen*, Nach. Akad. Wissensch. Gottingen. Math.-phys. Klasse **3** (1904), 213–259.

[196] _____, *Grundzüge einer allgemeinen Theorie der linearen Integralgleichungen (Vierte Mitteilung)*, Nachr. Wiss. Gesell. Gött., Math.-Phys. Kl. (1906), 157–227.

[197] _____, *Grundzüge einer allgemeinen Theorie der linearen Integralgleichungen*, Chelsea Publishing Company, New York, N.Y., 1953.

[198] T. H. Hildebrandt, *Linear functional transformations in general spaces*, Bull. Amer. Math. Soc. **37** (1931), no. 4, 185–212.

[199] C. K. Hill, *The Hilbert bound of a certain doubly-infinite matrix*, J. London Math. Soc. **32** (1957), 7–17.

[200] E. Hille and R. S. Phillips, *Functional analysis and semi-groups*, American Mathematical Society Colloquium Publications, vol. 31, American Mathematical Society, Providence, R. I., 1957, rev. ed.

[201] D. Hitt, *Invariant subspaces of H^2 of an annulus*, Pacific J. Math. **134** (1988), no. 1, 101–120.

[202] K. Hoffman, *Banach spaces of analytic functions*, Dover Publications Inc., New York, 1988, Reprint of the 1962 original.

[203] L. Hupert and A. Leggett, *On the square roots of infinite matrices*, Amer. Math. Monthly **96** (1989), no. 1, 34–38.

[204] A. E. Ingham, *A Note on Hilbert's Inequality*, J. London Math. Soc. **11** (1936), no. 3, 237–240.

[205] M. Jevtić, D. Vukotić, and M. Arsenović, *Taylor coefficients and coefficient multipliers of Hardy and Bergman-type spaces*, RSME Springer Series, vol. 2, Springer, Cham, 2016.

[206] P. Jordan and J. von Neumann, *On inner products in linear, metric spaces*, Ann. of Math. (2) **36** (1935), no. 3, 719–723.

[207] G. Julia, *Sur les projections des systèmes orthonormaux de l'espace hilbertien*, C. R. Acad. Sci. Paris **218** (1944), 892–895.

[208] M. Kadec', *The exact value of the Paley-Wiener constant*, Dokl. Akad. Nauk SSSR **155** (1964), 1253–1254.

[209] J.-P. Kahane and Y. Katznelson, *Sur les séries de Fourier uniformément convergentes*, C. R. Acad. Sci. Paris **261** (1965), 3025–3028.

[210] G. K. Kalisch, *A functional anaysis proof of Titchmarsh's theorem on convolution*, J. Math. Anal. Appl. **5** (1962), 176–183.

[211] T. Kato, *Perturbation theory for linear operators*, second ed., Springer-Verlag, Berlin-New York, 1976, Grundlehren der Mathematischen Wissenschaften, Band 132.

[212] E. Kay, H. Soul, and D. Trutt, *Some subnormal operators and hypergeometric kernel functions*, J. Math. Anal. Appl. **53** (1976), no. 2, 237–242.

[213] D. Kershaw, *Operator norms of powers of the Volterra operator*, J. Integral Equations Appl. **11** (1999), no. 3, 351–362.

[214] L. Khadkhuu and D. Tsedenbayar, *A note about Volterra operator*, Math. Slovaca **68** (2018), no. 5, 1117–1120.

[215] _____, *On the numerical range and numerical radius of the Volterra operator*, Izv. Irkutsk. Gos. Univ. Ser. Mat. **24** (2018), 102–108.

[216] F. W. King, *Hilbert transforms*, Encyclopedia of Mathematics and its Applications, vol. 1, Cambridge University Press, 2009.

[217] _____, *Hilbert transforms*, Encyclopedia of Mathematics and its Applications, vol. 2, Cambridge University Press, 2009.

[218] G. Koenigs, *Recherches sur les intégrales de certaines équations fonctionnelles*, Ann. Sci. École Norm. Sup. (3) **1** (1884), 3–41.

[219] A. Kolmogorov, *Une série de Fourier-Lebesgue divergente presque partout*, Fund. Math **4** (1923), 324–328.

[220] P. Koosis, *Introduction to H_p Spaces*, second ed., Cambridge Tracts in Mathematics, vol. 115, Cambridge University Press, Cambridge, 1998, With two appendices by V. P. Havin [Viktor Petrovich Khavin].

[221] B. I. Korenbljum, *Invariant subspaces of the shift operator in a weighted Hilbert space*, Mat. Sb. (N.S.) **89(131)** (1972), 110–137, 166.

[222] T. L. Kriete, *An elementary approach to the multiplicity theory of multiplication operators*, Rocky Mountain J. Math. **16** (1986), no. 1, 23–32.

[223] T. L. Kriete and D. Trutt, *The Cesàro operator in l^2 is subnormal*, Amer. J. Math. **93** (1971), 215–225.

[224] ———, *On the Cesàro operator*, Indiana Univ. Math. J. **24** (1974/75), 197–214.

[225] L. Kronecker, *Zur Theorie der Elimination einer Variablen aus zwei algebraischen Gleichungen*, Monatsber. Königl. Preussischen Akad. Wiss. (Berlin) (1881), 535–600.

[226] N. N. Lebedev, *The analogue of Parseval's theorem for a certain integral transform*, Doklady Akad. Nauk SSSR (N.S.) **68** (1949), 653–656.

[227] ———, *Some singular integral equations connected with integral representations of mathematical physics*, Doklady Akad. Nauk SSSR (N.S.) **65** (1949), 621–624.

[228] G. Leibowitz, *The Cesàro operators and their generalizations: examples in infinite-dimensional linear analysis*, Amer. Math. Monthly **80** (1973), 654–661.

[229] J. E. Littlewood, *On Inequalities in the Theory of Functions*, Proc. London Math. Soc. (2) **23** (1925), no. 7, 481–519.

[230] V. I. Lomonosov, *Invariant subspaces of the family of operators that commute with a completely continuous operator*, Funkcional. Anal. i Priložen. **7** (1973), no. 3, 55–56.

[231] H. Löwig, *Komplexe euklidische Räume von beliebiger endlicher oder transfiniter Dimensionszahl*, Acta Sci. Math. (Szeged) **7** (1934), 1–33.

[232] D. H. Luecking, *Characterizations of certain classes of Hankel operators on the Bergman spaces of the unit disk*, J. Funct. Anal. **110** (1992), no. 2, 247–271.

[233] D. H. Luecking and K. Zhu, *Composition operators belonging to the Schatten ideals*, Amer. J. Math. **114** (1992), no. 5, 1127–1145.

[234] R. Lyons, *A lower bound on the Cesàro operator*, Proc. Amer. Math. Soc. **86** (1982), no. 4, 694.

[235] B. MacCluer and J. Shapiro, *Angular derivatives and compact composition operators on the Hardy and Bergman spaces*, Canad. J. Math. **38** (1986), no. 4, 878–906.

[236] G. W. MacDonald, *Invariant subspaces for Bishop-type operators*, J. Funct. Anal. **91** (1990), no. 2, 287–311.

[237] W. Magnus, *On the spectrum of Hilbert's matrix*, Amer. J. Math. **72** (1950), 699–704.

[238] M. Martin and M. Putinar, *Lectures on hyponormal operators*, Operator Theory: Advances and Applications, vol. 39, Birkhäuser Verlag, Basel, 1989.

[239] J. Mashreghi, *Representation theorems in Hardy spaces*, London Mathematical Society Student Texts, vol. 74, Cambridge University Press, Cambridge, 2009.

[240] J. H. McClellan and T. W. Parks, *Eigenvalue and eigenvector decomposition of the discrete Fourier transform*, IEEE Trans. Audio Electroacoust. **AU-20** (1972), no. 1, 66–74.

[241] J. D. McGovern, *The Hilbert transform*, McMaster University, Hamilton, Ontario, 1980, Thesis (M.S.)–McMaster University.

[242] F. G. Mehler, *Ueber die Entwicklung einer Function von beliebig vielen Variablen nach Laplaceschen Functionen höherer Ordnung*, J. Reine Angew. Math. **66** (1866), 161–176.

[243] A. J. Michaels, *Hilden's simple proof of Lomonosov's invariant subspace theorem*, Adv. Math. **25** (1977), no. 1, 56–58.

[244] B. R. Miller, *Kernels of adjoints of composition operators on Hilbert spaces of analytic functions*, ProQuest LLC, Ann Arbor, MI, 2016, Thesis (Ph.D.)–Purdue University.

[245] W. Mlocek and M. Ptak, *On the reflexivity of subspaces of Toeplitz operators on the Hardy space on the upper half-plane*, Czechoslovak Math. J. **63(138)** (2013), no. 2, 421–434.

[246] A. Mukherjea and K. Pothoven, *Real and functional analysis. Part B*, second ed., Mathematical Concepts and Methods in Science and Engineering, vol. 28, Plenum Press, New York, 1986, Functional analysis.

[247] Z. Nehari, *On bounded bilinear forms*, Ann. of Math. (2) **65** (1957), 153–162.

[248] C. Neumann, *Untersuchungen über das logarithmische und Newton'sche potential*, Teubner, Leipzig, 1877.

[249] M. M. Neumann, *Spectral properties of Cesàro-like operators*, Advanced courses of mathematical analysis. II, World Sci. Publ., Hackensack, NJ, 2007, pp. 123–140.

[250] N. Nikolski, *Treatise on the shift operator*, Springer-Verlag, Berlin, 1986.

[251] _____, *Operators, functions, and systems: an easy reading. Vol. 1*, Mathematical Surveys and Monographs, vol. 92, American Mathematical Society, Providence, RI, 2002, Hardy, Hankel, and Toeplitz, Translated from the French by A. Hartmann.

[252] _____, *Operators, functions, and systems: an easy reading. Vol. 2*, Mathematical Surveys and Monographs, vol. 93, American Mathematical Society, Providence, RI, 2002, Model operators and systems, Translated from the French by A. Hartmann and revised by the author.

[253] _____, *Hardy spaces*, French ed., Cambridge Studies in Advanced Mathematics, vol. 179, Cambridge University Press, Cambridge, 2019.

[254] _____, *Toeplitz matrices and operators*, French ed., Cambridge Studies in Advanced Mathematics, vol. 182, Cambridge University Press, Cambridge, 2020.

[255] E. A. Nordgren, *Composition operators*, Canadian J. Math. **20** (1968), 442–449.

[256] B.-H. Ong, *Invariant subspace lattices for a class of operators*, ProQuest LLC, Ann Arbor, MI, 1978, Thesis (Ph.D.)–University of California, Berkeley.

[257] _____, *Invariant subspace lattices for a class of operators*, Pacific J. Math. **94** (1981), no. 2, 385–405.

[258] R. Paley and N. Wiener, *Fourier transforms in the complex domain*, first ed., American Mathematical Society, Coll. Publ. XIX, New York, 1934.

[259] J. N. Pandey, *The Hilbert transform of Schwartz distributions and applications*, Pure and Applied Mathematics (New York), John Wiley & Sons, Inc., New York, 1996, A Wiley-Interscience Publication.

[260] S. Parrott, *Weighted translation operators*, ProQuest LLC, Ann Arbor, MI, 1965, Thesis (Ph.D.)–University of Michigan.

[261] _____, *On a quotient norm and the Sz.-Nagy - Foiaş lifting theorem*, J. Functional Analysis **30** (1978), no. 3, 311–328.

[262] J. R. Partington, *An introduction to Hankel operators*, London Mathematical Society Student Texts, vol. 13, Cambridge University Press, Cambridge, 1988.

[263] K. Patarroyo, *A digression on Hermite polynomials*, preprint.

[264] V. I. Paulsen and M. Raghupathi, *An introduction to the theory of reproducing kernel Hilbert spaces*, Cambridge Studies in Advanced Mathematics, vol. 152, Cambridge University Press, Cambridge, 2016.

[265] V. V. Peller, *Invariant subspaces of Toeplitz operators with piecewise continuous symbols*, Proc. Amer. Math. Soc. **119** (1993), no. 1, 171–178.

[266] _____, *An excursion into the theory of Hankel operators*, Holomorphic spaces (Berkeley, CA, 1995), Math. Sci. Res. Inst. Publ., vol. 33, Cambridge Univ. Press, Cambridge, 1998, pp. 65–

120.

[267] _____, *Hankel operators and their applications*, Springer Monographs in Mathematics, Springer-Verlag, New York, 2003.

[268] A.-M. Persson, *On the spectrum of the Cesàro operator on spaces of analytic functions*, J. Math. Anal. Appl. **340** (2008), no. 2, 1180–1203.

[269] S. K. Pichorides, *On the best values of the constants in the theorems of M. Riesz, Zygmund and Kolmogorov*, Studia Math. **44** (1972), 165–179. (errata insert).

[270] A. Pietsch, *History of Banach spaces and linear operators*, Birkhäuser Boston, Inc., Boston, MA, 2007.

[271] M. Plancherel, *Contribution à l'étude de la représentation d'une fonction arbitraire par les intégrales définies*, Rendiconti del Circolo Matematico di Palermo **30** (1910), 289–335.

[272] Ch. Pommerenke, *Schlichte Funktionen und analytische Funktionen von beschränkter mittlerer Oszillation*, Comment. Math. Helv. **52** (1977), no. 4, 591–602.

[273] S. C. Power, *Hankel operators on Hilbert space*, Bull. London Math. Soc. **12** (1980), no. 6, 422–442.

[274] _____, *Hankel operators on Hilbert space*, Research Notes in Mathematics, vol. 64, Pitman (Advanced Publishing Program), Boston, Mass.-London, 1982.

[275] I. Privalov, *Sur les fonctions conjuguées*, Bull. de la S. M. F. **44** (1916), 100–103.

[276] C. R. Putnam, *On normal operators in Hilbert space*, Amer. J. Math. **73** (1951), 357–362.

[277] _____, *Commutators and absolutely continuous operators*, Trans. Amer. Math. Soc. **87** (1958), 513–525.

[278] H. Queffélec, *Composition operators in the Dirichlet series setting*, Perspectives in operator theory, Banach Center Publ., vol. 75, Polish Acad. Sci. Inst. Math., Warsaw, 2007, pp. 261–287.

[279] H. Rademacher, *Einige Sätze über Reihen von allgemeinen Orthogonalfunktionen*, Math. Ann. **87** (1922), no. 1-2, 112–138.

[280] H. Radjavi and P. Rosenthal, *Invariant subspaces*, second ed., Dover Publications, Inc., Mineola, NY, 2003.

[281] C. Read, *A solution to the invariant subspace problem*, Bull. London Math. Soc. **16** (1984), no. 4, 337–401.

[282] _____, *A solution to the invariant subspace problem on the space l_1*, Bull. London Math. Soc. **17** (1985), no. 4, 305–317.

[283] M. Reed and B. Simon, *Methods of modern mathematical physics. II. Fourier analysis, self-adjointness*, Academic Press [Harcourt Brace Jovanovich, Publishers], New York-London, 1975.

[284] _____, *Methods of modern mathematical physics. I*, second ed., Academic Press, Inc. [Harcourt Brace Jovanovich, Publishers], New York, 1980, Functional analysis.

[285] H. C. Rhaly, Jr., *Discrete generalized Cesàro operators*, Proc. Amer. Math. Soc. **86** (1982), no. 3, 405–409.

[286] _____, *An averaging operator on the Dirichlet space*, J. Math. Anal. Appl. **98** (1984), no. 2, 555–561.

[287] _____, *Generalized Cesàro matrices*, Canad. Math. Bull. **27** (1984), no. 4, 417–422.

[288] _____, *p-Cesàro matrices*, Houston J. Math. **15** (1989), no. 1, 137–146.

[289] _____, *Terraced matrices*, Bull. London Math. Soc. **21** (1989), no. 4, 399–406.

[290] _____, *Posinormal operators*, J. Math. Soc. Japan **46** (1994), no. 4, 587–605.

[291] B. E. Rhoades and D. Trutt, *Cesàro-like operators*, Sarajevo J. Math. **15(28)** (2019), no. 2, 283–289.

[292] S. Richter, *Unitary equivalence of invariant subspaces of Bergman and Dirichlet spaces*, Pacific J. Math. **133** (1988), no. 1, 151–156.

[293] ———, *A representation theorem for cyclic analytic two-isometries*, Trans. Amer. Math. Soc. **328** (1991), no. 1, 325–349.

[294] S. Richter, W. T. Ross, and C. Sundberg, *Hyperinvariant subspaces of the harmonic Dirichlet space*, J. Reine Angew. Math. **448** (1994), 1–26.

[295] S. Richter and C. Sundberg, *A formula for the local Dirichlet integral*, Michigan Math. J. **38** (1991), no. 3, 355–379.

[296] ———, *Multipliers and invariant subspaces in the Dirichlet space*, J. Operator Theory **28** (1992), no. 1, 167–186.

[297] ———, *Invariant subspaces of the Dirichlet shift and pseudocontinuations*, Trans. Amer. Math. Soc. **341** (1994), no. 2, 863–879.

[298] F. Riesz, *Sur les systèmes orthogonaux de fonctions*, C.R. Acad. Sci. Paris **144** (1907), 615–619.

[299] ———, *Sur une espèce de géométrie analytique des systèmes de fonctions sommables*, C.R. Acad. Sci. Paris **144** (1907), 1409–1411.

[300] ———, *Untersuchungen über Systeme integrierbarer Funktionen*, Math. Ann. **69** (1909), 449–497.

[301] ———, *Über quadratische Formen von unendlich vielen Veränderlichen*, Nachr. Wiss. Gesell. Gött., Math.-Phys. Kl. (1910), 190–195.

[302] ———, *Sur certains systèmes singuliers d'équations intégrales*, Ann. Sci. École Norm. Sup. (3) **28** (1911), 33–62.

[303] ———, *Les systèmes d'équations linéaires à une infinité d'inconnues*, Gauthier–Villars, Paris, 1913.

[304] ———, *Über lineare Funktionalgleichungen*, Acta Math. **41** (1918), 71–98.

[305] ———, *Über die Randwerte einer analytischen Funktion*, Math. Z. **18** (1923), no. 1, 87–95.

[306] ———, *Zur Theorie des Hilbertschen Raumes*, Acta Sci. Math. (Szeged) **7** (1934), 34–38.

[307] F. Riesz and M. Riesz, *Über die Randwerte einer analytischen Funktion*, Comptes Rendus du Quatrième Congrès des Mathématiciens Scandinaves (1916), 27–44.

[308] M. Riesz, *Sur les fonctions conjuguées*, Math. Z. **27** (1928), no. 1, 218–244.

[309] M. Rosenblum, *On a theorem of Fuglede and Putnam*, J. London Math. Soc. **33** (1958), 376–377.

[310] ———, *On the Hilbert matrix. I*, Proc. Amer. Math. Soc. **9** (1958), 137–140.

[311] ———, *On the Hilbert matrix. II*, Proc. Amer. Math. Soc. **9** (1958), 581–585.

[312] ———, *The absolute continuity of Toeplitz's matrices*, Pacific J. Math. **10** (1960), 987–996.

[313] M. Rosenblum and J. Rovnyak, *Hardy classes and operator theory*, Oxford Mathematical Monographs, The Clarendon Press, Oxford University Press, New York, 1985, Oxford Science Publications.

[314] W. T. Ross, *Invariant subspaces of the harmonic Dirichlet space with large co-dimension*, Proc. Amer. Math. Soc. **124** (1996), no. 6, 1841–1846.

[315] ———, *The classical Dirichlet space*, Recent advances in operator-related function theory, Contemp. Math., vol. 393, Amer. Math. Soc., Providence, RI, 2006, pp. 171–197.

[316] G.-C. Rota, *On models for linear operators*, Comm. Pure Appl. Math. **13** (1960), 469–472.

[317] H. L. Royden, *Real analysis*, third ed., Macmillan Publishing Company, New York, 1988.

[318] W. Rudin, *Function theory in polydiscs*, W. A. Benjamin, Inc., New York-Amsterdam, 1969.

[319] ———, *Real and complex analysis*, second ed., McGraw-Hill Series in Higher Mathematics, McGraw-Hill Book Co., New York-Düsseldorf-Johannesburg, 1974.

[320] ———, *Functional analysis*, second ed., International Series in Pure and Applied Mathematics, McGraw-Hill, Inc., New York, 1991.

[321] J. V. Ryff, *Subordinate H^p functions*, Duke Math. J. **33** (1966), 347–354.

[322] L. A. Sakhnovich, *The spectral analysis of Volterra operators and some inverse problems*, Dokl. Akad. Nauk SSSR (N.S.) **115** (1957), 666–669.

[323] D. Sarason, *A remark on the Volterra operator*, J. Math. Anal. Appl. **12** (1965), 244–246.

[324] ———, *Generalized interpolation in H^∞*, Trans. Amer. Math. Soc. **127** (1967), 179–203.

[325] ———, *Weak-star density of polynomials*, J. Reine Angew. Math. **252** (1972), 1–15.

[326] ———, *Invariant subspaces*, Topics in operator theory, 1974, pp. 1–47. Math. Surveys, No. 13.

[327] ———, *Nearly invariant subspaces of the backward shift*, Contributions to operator theory and its applications (Mesa, AZ, 1987), Oper. Theory Adv. Appl., vol. 35, Birkhäuser, Basel, 1988, pp. 481–493.

[328] ———, *Kernels of Toeplitz operators*, Toeplitz operators and related topics (Santa Cruz, CA, 1992), Oper. Theory Adv. Appl., vol. 71, Birkhäuser, Basel, 1994, pp. 153–164.

[329] H. H. Schaefer, *Eine Bemerkung zur Existenz invarianter Teilräume linearer Abbildungen*, Math. Z. **82** (1963), 90.

[330] J. Schauder, *Über lineare, vollstetige Funktionaloperationen*, Studia Math. **2** (1930), 183–196.

[331] E. Schmidt, *Zur Theorie der linearen und nicht linearen Integralgleichungen Zweite Abhandlung*, Math. Ann. **64** (1907), no. 2, 161–174.

[332] ———, *Über die Auflösung linearer Gleichungen mit unendlich vielen Unbekannten*, Rend. Circ. Mat. Palermo **25** (1908), 53–77.

[333] I. Schur, *Bemerkungen zur Theorie der beschränkten Bilinearformen mit unendlich vielen Veränderlichen*, J. Reine Angew. Math. **40** (1911), 1–28.

[334] H. J. Schwartz, *Composition operators on H^p*, ProQuest LLC, Ann Arbor, MI, 1969, Thesis (Ph.D.)–The University of Toledo.

[335] H. Schwarz, *Über ein die Flächen kleinsten Flächeninhalts betreffendes Problem der Variationsrechnung*, Acta Soc. Sci. Fennicea **15** (1885), 315–362.

[336] D. Seco, *Some problems on optimal approximates*, Recent progress on operator theory and approximation in spaces of analytic functions, Contemp. Math., vol. 679, Amer. Math. Soc., Providence, RI, 2016, pp. 193–205.

[337] H. A. Seid, *Cyclic multiplication operators on L_p-spaces*, Pacific J. Math. **51** (1974), 549–562.

[338] J. H. Shapiro, *The essential norm of a composition operator*, Ann. of Math. (2) **125** (1987), no. 2, 375–404.

[339] ———, *Composition operators and classical function theory*, Universitext: Tracts in Mathematics, Springer-Verlag, New York, 1993.

[340] ———, *Volterra adventures*, Student Mathematical Library, vol. 85, American Mathematical Society, Providence, RI, 2018.

[341] J. H. Shapiro and P. D. Taylor, *Compact, nuclear, and Hilbert-Schmidt composition operators on H^2*, Indiana Univ. Math. J. **23** (1973/74), 471–496.

[342] A. L. Shields, *Weighted shift operators and analytic function theory*, Topics in operator theory, 1974, pp. 49–128. Math. Surveys, No. 13.

[343] A. L. Shields and L. J. Wallen, *The commutants of certain Hilbert space operators*, Indiana Univ. Math. J. **20** (1970/71), 777–788.

[344] S. Shimorin, *Wold-type decompositions and wandering subspaces for operators close to isometries*, J. Reine Angew. Math. **531** (2001), 147–189.

[345] B. Silbermann, *On the spectrum of Hilbert matrix operator*, Integral Equations Operator Theory **93** (2021), no. 3, Paper No. 21, 35.

[346] R. K. Singh, *Composition operators induced by rational functions*, Proc. Amer. Math. Soc. **59** (1976), no. 2, 329–333.

[347] R. K. Singh and A. Kumar, *Compact composition operators*, J. Austral. Math. Soc. Ser. A **28** (1979), no. 3, 309–314.

[348] G. Sinnamon, *Norm of the discrete Cesàro operator minus identity*, Math. Inequal. Appl. **25** (2022), no. 1, 41–48.

[349] J. G. Stampfli, *Hyponormal operators and spectral density*, Trans. Amer. Math. Soc. **117** (1965), 469–476.

[350] E. M. Stein, *Singular integrals and differentiability properties of functions*, Princeton Mathematical Series, No. 30, Princeton University Press, Princeton, N.J., 1970.

[351] E. M. Stein and R. Shakarchi, *Complex analysis*, Princeton Lectures in Analysis, vol. 2, Princeton University Press, Princeton, NJ, 2003.

[352] ———, *Fourier analysis*, Princeton Lectures in Analysis, vol. 1, Princeton University Press, Princeton, NJ, 2003.

[353] M. H. Stone, *Linear transformations in Hilbert space*, American Mathematical Society Colloquium Publications, vol. 15, American Mathematical Society, Providence, RI, 1990, Reprint of the 1932 original.

[354] P. Šťovíček, *Spectral analysis of the Hilbert matrix*, 2014.

[355] B. Sz.-Nagy, *Sur les contractions de l'espace de Hilbert*, Acta Sci. Math. (Szeged) **15** (1953), 87–92.

[356] B. Sz.-Nagy, C. Foiaş, H. Bercovici, and L. Kérchy, *Harmonic analysis of operators on Hilbert space*, second ed., Universitext, Springer, New York, 2010.

[357] G. Szegő, *Orthogonal polynomials*, fourth ed., American Mathematical Society, Providence, R.I., 1975, American Mathematical Society, Colloquium Publications, Vol. XXIII.

[358] J. E. Thomson, *Invariant subspaces for algebras of subnormal operators*, Proc. Amer. Math. Soc. **96** (1986), no. 3, 462–464.

[359] ———, *Approximation in the mean by polynomials*, Ann. of Math. (2) **133** (1991), no. 3, 477–507.

[360] D. Timotin, *The invariant subspaces of $S \oplus S^*$*, Concr. Oper. **7** (2020), no. 1, 116–123.

[361] E. C. Titchmarsh, *A Contribution to the Theory of Fourier Transforms*, Proc. London Math. Soc. (2) **23** (1924), no. 4, 279–289.

[362] ———, *Introduction to the theory of Fourier integrals*, third ed., Chelsea Publishing Co., New York, 1986.

[363] O. Toeplitz, *Zur theorie der quadratischen Formen von unendlichvielen Veränderlichen*, Nachr. Kön. Ges. Wiss. Göttingen (1910), 489–506.

[364] ———, *Zur Theorie der quadratischen und bilinearen Formen von unendlichvielen Veränderlichen*, Math. Ann. **70** (1911), no. 3, 351–376.

[365] C. Vallée Poussin, *Extension de la méthode du balayage de Poincaré et problème de Dirichlet*, Annales de l'institut Henri Poincaré **2** (1932), no. 3, 169–232 (fr).

[366] G. Vitali, *Sulla condizione di chiusura di un sistema di funzioni ortogonali*, Rend. Mat. Acc. Lincei **30** (1921), 498–501.

[367] V. Volterra, *Sulla inversione degli integrali definiti*, R. C. Lincei (Series 5) **5** (1896), 177–185.

[368] ———, *Theory of functionals and of integral and integro-differential equations*, With a preface by G. C. Evans, a biography of Vito Volterra and a bibliography of his published works by E. Whittaker, Dover Publications, Inc., New York, 1959.

[369] J. von Neumann, *Allgemeine Eigenwerttheorie Hermitescher Funktionaloperatoren*, Math. Ann. **102** (1930), no. 1, 49–131.

[370] ———, *Eine Spektraltheorie für allgemeine Operatoren eines unitären Raumes*, Math. Nachr. **4** (1951), 258–281.

[371] J. Wermer, *On invariant subspaces of normal operators*, Proc. Amer. Math. Soc. **3** (1952), 270–277.

[372] R. Whitley, *A note on hyponormal operators*, Proc. Amer. Math. Soc. **49** (1975), 399–400.

[373] H. Widom, *Inversion of Toeplitz matrices. II*, Illinois J. Math. **4** (1960), 88–99.

[374] ———, *On the spectrum of a Toeplitz operator*, Pacific J. Math. **14** (1964), 365–375.

[375] N. Wiener, *Generalized harmonic analysis*, Acta Math. **55** (1930), no. 1, 117–258.

[376] ———, *The Fourier integral and certain of its applications*, Dover Publications, Inc., New York, 1959.

[377] W. H. Young, *On the Nature of the Successions formed by the Coefficients of a Fourier Series*, Proc. London Math. Soc. (2) **10** (1912), 344–352.

[378] S. Zaigler and D. P. L. Castrigiano, *Cyclicity for unbounded multiplication operators in L^p- and C_0-spaces*, Ann. Funct. Anal. **6** (2015), no. 2, 33–48.

[379] K. Zhu, *Operator theory in function spaces*, second ed., Mathematical Surveys and Monographs, vol. 138, American Mathematical Society, Providence, RI, 2007.

[380] A. Zygmund, *Trigonometric series. Vol. I, II*, third ed., Cambridge Mathematical Library, Cambridge University Press, Cambridge, 2002, With a foreword by R. A. Fefferman.

AUTHOR INDEX

SUBJECT INDEX